The AutoCAD® 2004 Tutor for Engineering Graphics with AutoCAD® 2005 UPDATE

Project Manual

The AutoCAD® 2004 Tutor for Engineering Graphics with AutoCAD® 2005 UPDATE

Project Manual

ALAN J. KALAMEJA

THOMSON
DELMAR LEARNING

Australia • Canada • Mexico • Singapore • Spain • United Kingdom • United States

The AutoCAD® 2004 Tutor for Engineering Graphics with AutoCAD® 2005 UPDATE

Project Manual

Alan J. Kalameja

Vice President, Technology and Trades SBU:
Alar Elken

Editorial Director:
Sandy Clark

Senior Acquisitions Editor:
James DeVoe

Senior Development Editor:
John Fisher

Marketing Director:
Dave Garza

Channel Manager:
Dennis Williams

Marketing Coordinator:
Casey Bruno

Production Director:
Mary Ellen Black

Production Manager:
Andrew Crouth

Production Editor:
Stacy Masucci

Art and Design Specialist:
Mary Beth Vought

Technology Project Manager:
Kevin Smith

Technology Project Specialist:
Linda Verde

Cover Image:
Getty Images

Printed in Canada
1 2 3 4 5 XX 06 05 04

For more information contact
Thomson Delmar Learning
Executive Woods
5 Maxwell Drive, PO Box 8007,
Clifton Park, NY 12065-8007

Or find us on the World Wide Web at
www.delmarlearning.com

Library of Congress Cataloging-in-Publication Data
Kalameja, Alan J.
The AutoCAD 2004 tutor for engineering graphics with AutoCAD 2005 UPDATE Project Manual / Alan J. Kalameja.

ISBN 1-4018-8391-5

Notice To The Reader

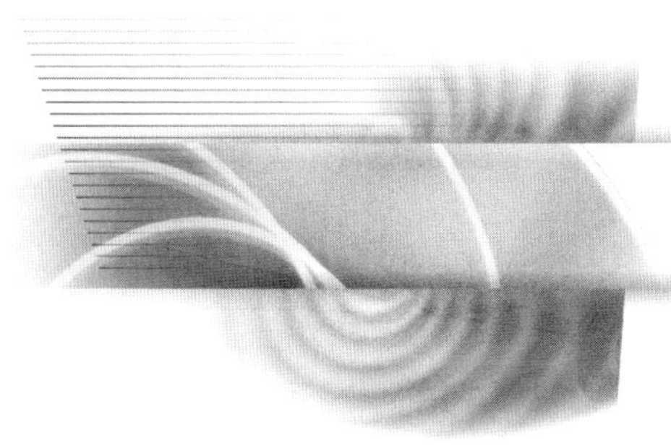

CONTENTS

INTRODUCTION

PROJECT MANUAL FOR THE AUTOCAD 2004 TUTOR FOR ENGINEERING GRAPHICS WITH AUTOCAD 2005 UPDATE

This *Project Manual* is the companion book intended to be used with *The AutoCAD 2004 Tutor for Engineering Graphics with AutoCAD 2005 Update.* Each chapter has a tutorial and exercises designed to allow you to practice and test your skills in the commands and concepts covered in the corresponding chapter of the main text. The combination of these books was written to be used both in the classroom as a textbook as well as in the industry by the professional CADD designer/drafter as a reference and learning tool. Whether you're new to AutoCAD or a seasoned user upgrading your skills, *The AutoCAD 2004 Tutor for Engineering Graphics with AutoCAD 2005 Update* and this *Project Manual* will help you rein in the power of AutoCAD to improve your professional skills and increase your productivity.

Drawing files for use in some of the tutorials and exercises are provided on the CD in back of this book.

If you would like to contribute drawing exercises for possible inclusion in the next version of this project manual, please contact the CADD/Drafting team at Autodesk Press by visiting our website: *www.autodeskpress.com.*

Getting Started with AutoCAD 2004

The exercises in this chapter introduce:

- The AutoCAD interface
- How to begin and save drawings
- Object Snap settings
- Line and Polyline creation
- Straight line (linear based) drafting using various input methods.
- Help File access and Active Assistance
- Erasing in an AutoCAD drawing

TUTORIAL EXERCISE: 01_TEMPLATE.DWG

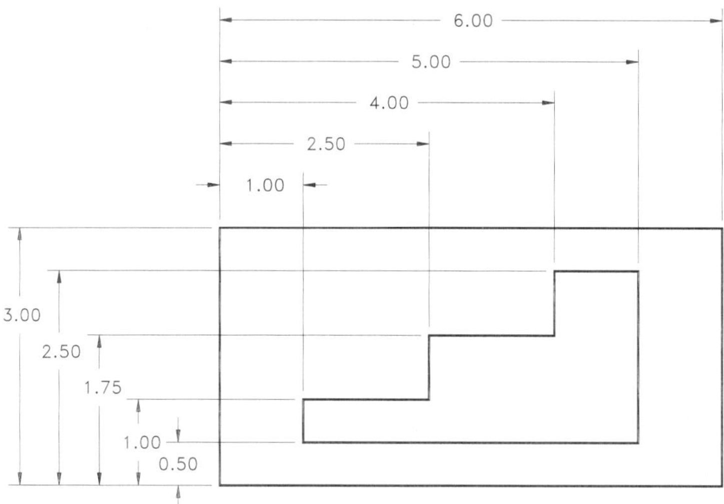

Figure 1–1

Purpose

This tutorial is designed to allow you to construct a one-view drawing of the Template using the Absolute, Relative, and Polar Coordinate modes. See Figure 1–1. The Direct Distance mode can also be used to perform this exercise.

System Settings

Use the current default setting for (0,0). It is located in the lower-left hand corner of the screen.

Use the GRID command and change the grid spacing from 1.0000 to 0.25 units. The grid will be used only as a guide for constructing this object. Do not turn the Snap or Ortho modes on.

Layers

The following layer has already been created:

Name	Color	Linetype
Object	Green	Continuous

Suggested Commands

The LINE command will be used entirely for this tutorial, in addition to a combination of coordinate systems. The ERASE command could be used (however, using this command will force the user to exit out of the LINE command), although a more elaborate method of correcting mistakes while using the LINE command is to execute the undo option. This option allows the user to delete (or undo) previously drawn lines without having to exit the LINE command. The Object Snap From mode will also be used to construct lines from a point of reference. The coordinate mode of entry and the Direct Distance mode will be used throughout this tutorial exercise.

Whenever possible, substitute the appropriate command alias in place of the full AutoCAD command in each tutorial step. For example, use "CP" for the COPY command, "L" for the LINE command, and so on. The complete listing of all command aliases is located in Table 1–2.

STEP 1

Open the drawing file *01_Template.dwg.*

Then use the LINE command to draw the outer perimeter of the box. One method of constructing the box is to use an absolute coordinate point followed by polar coordinates. Use the following command sequence and Figure 1–2 as guides.

@6<180
@3<270
@3<90
@6<0
2.00,2.00

Figure 1–2

Command: **L** *(For LINE)* [Draw>Line]
Specify first point: **2,2**
Specify next point or [Undo]: **@6<0**
Specify next point or [Undo]: **@3<90**
Specify next point or [Close/Undo]:
@6<180
Specify next point or [Close/Undo]: **C**
(To close the shape)

An alternate method is to use the Direct Distance mode. Because the box consists of horizontal and vertical lines, Ortho mode is first turned on; this will force all movements to be in the horizontal or vertical direction. To construct a line segment, move the cursor in the direction in which the line is to be drawn and enter the exact value of the line. The line is drawn at the designated distance in the current direction of the cursor. Repeat this procedure for the other lines that make up the box, as shown in Figure 1–3.

ORTHO Command: **ORTHO** [Button is found on the bottom of the AutoCAD screen, pressed in is ON]
ON/OFF <OFF>: **ON**

Command: **L** *(For LINE)* [Draw>Line]
Specify first point: **2,2**
Specify next point or [Undo]: *(Move the cursor to the right and enter a value of* **6.00** *units)*
Specify next point or [Undo]: *(Move the cursor up and enter a value of* **3.00** *units)*
Specify next point or [Close/Undo]: *(Move the cursor to the left and enter a value of* **6.00** *units)*
Specify next point or [Close/Undo]: **C** *(To close the shape)*

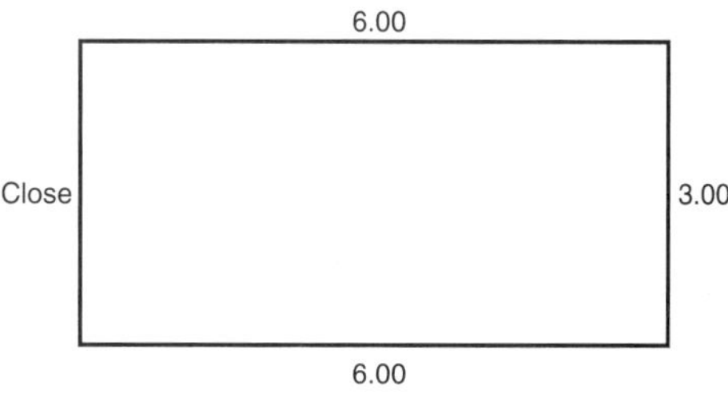

Figure 1–3

STEP 2

The next step is to draw the stair step outline of the template using the LINE command again. However, we first need to identify the starting point of the template.

Absolute coordinates could be calculated, but in more complex objects, this would be difficult. A more efficient method is to use the Object Snap From mode along with the Object Snap to Intersection mode to start the line relative to another point. Both Object Snap selections are found on the Object Snap toolbar. Use the following command sequence and Figure 1–4 as guides for performing this operation.

 Command: **L** *(For LINE)* [Draw>Line]

 Specify first point: **From**

 Base point: **Int**
of *(Pick the intersection at "A")*
<Offset>: **@1.00,0.50**

The relative coordinate offset value begins a new line a distance of 1.00 units in the X direction and 0.50 units in the Y direction. Continue with the LINE command to construct the stair step outline shown in Figure 1–5.

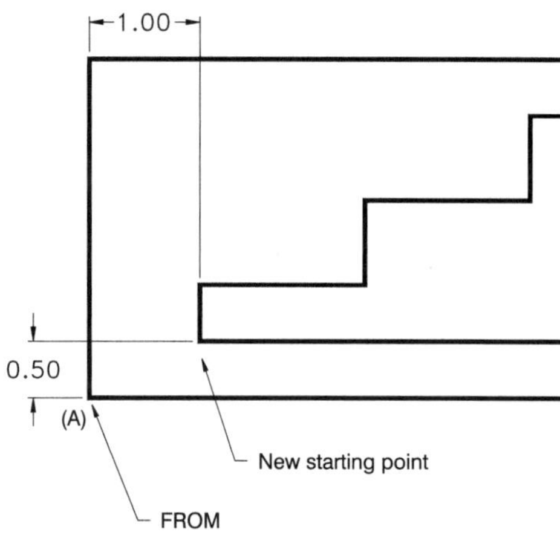

Figure 1–4

Specify next point or [Undo]: **@4.00<0**
Specify next point or [Undo]: **@2.00<90**
Specify next point or [Close/Undo]: **@1.00<180**
Specify next point or [Close/Undo]: **@0.75<270**
Specify next point or [Close/Undo]: **@1.50<180**
Specify next point or [Close/Undo]: **@0.75<270**
Specify next point or [Close/Undo]: **@1.50<180**
Specify next point or [Close/Undo]: **C** *(To close the shape)*

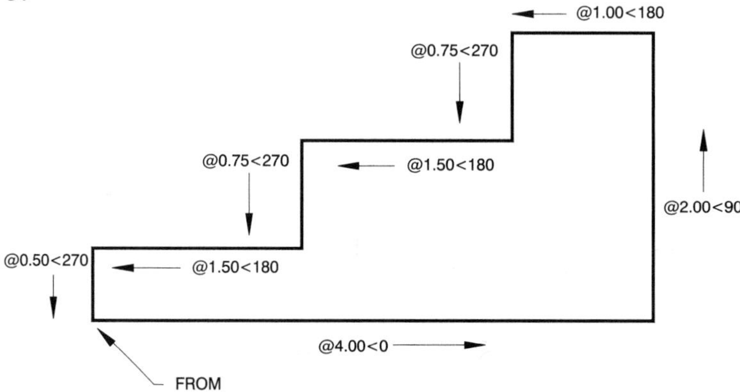

Figure 1–5

An alternate step would be to use the OSNAP-From and OSNAP-Intersect modes in combination with the Direct Distance mode to construct the inner stair step outline (see Figure 1–6). Again, Direct Distance mode is a good choice to use on this object, especially because all lines are either horizontal or vertical. Use the following command sequence to construct the object with this alternate method.

 Command: **L** *(For LINE)* [Draw>Line]

 Specify first point: **From**

 Base point: **Int**

of *(Pick the intersection at "A" in Figure 1–4)*
<Offset>: **@1.00,0.50**
Specify next point or [Undo]: *(Move the cursor to the right and enter a value of 4.00 units)*
Specify next point or [Undo]: *(Move the cursor up and enter a value of* **2.00** *units)*
Specify next point or [Close/Undo]: *(Move the cursor to the left and enter a value of* **1.00** *units)*
Specify next point or [Close/Undo]: *(Move the cursor down and enter a value of* **0.75** *units)*
Specify next point or [Close/Undo]: *(Move the cursor to the left and enter a value of* **1.50** *units)*
Specify next point or [Close/Undo]: *(Move the cursor down and enter a value of* **0.75** *units)*
Specify next point or [Close/Undo]: *(Move the cursor to the left and enter a value of* **1.50** *units)*
Specify next point or [Close/Undo]: **C** *(To close the shape)*

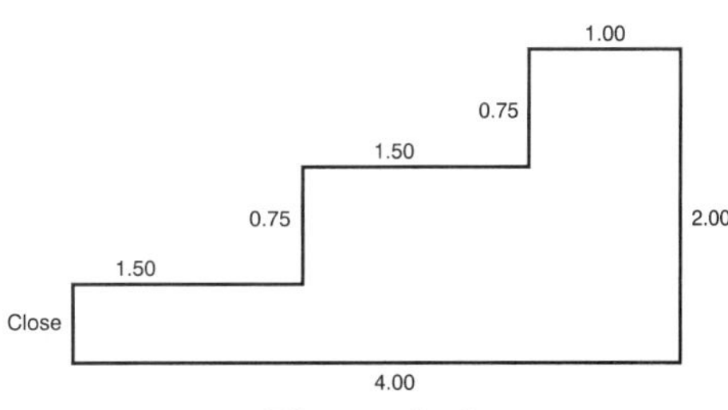

Figure 1–6

STEP 3

Dimensions may be added at a later time upon the request of your instructor. The completed problem is shown in Figure 1–7.

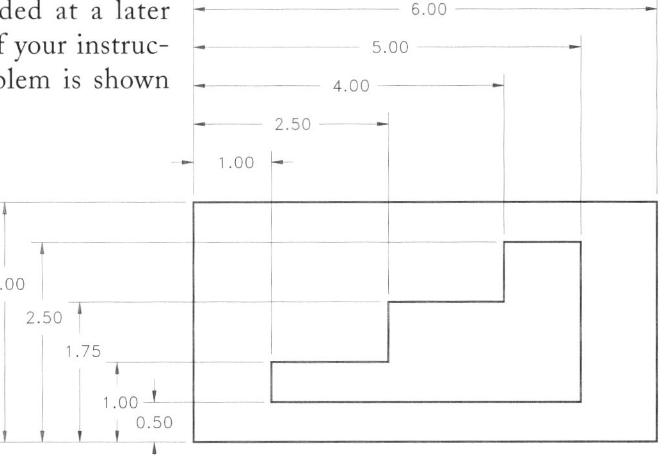

Figure 1–7

Beginning Level Drawings
Directions for Problems 1–1 through 1–6:

Supply the appropriate absolute, relative, and/or polar coordinates for these figures in the matrix that follows each object.

PROBLEM 1–1

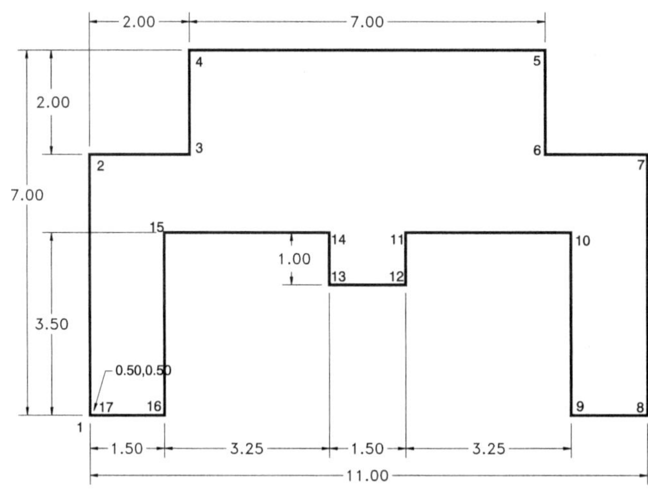

	Absolute	Relative	Polar
From Pt (1)	0.50,0.50	0.50,0.50	0.50,0.50
To Pt (2)			
To Pt (3)			
To Pt (4)			
To Pt (5)			
To Pt (6)			
To Pt (7)			
To Pt (8)			
To Pt (9)			
To Pt (10)			
To Pt (11)			
To Pt (12)			
To Pt (13)			
To Pt (14)			
To Pt (15)			
To Pt (16)			
To Pt (17)			
To Pt	Enter	Enter	Enter

PROBLEM 1–2

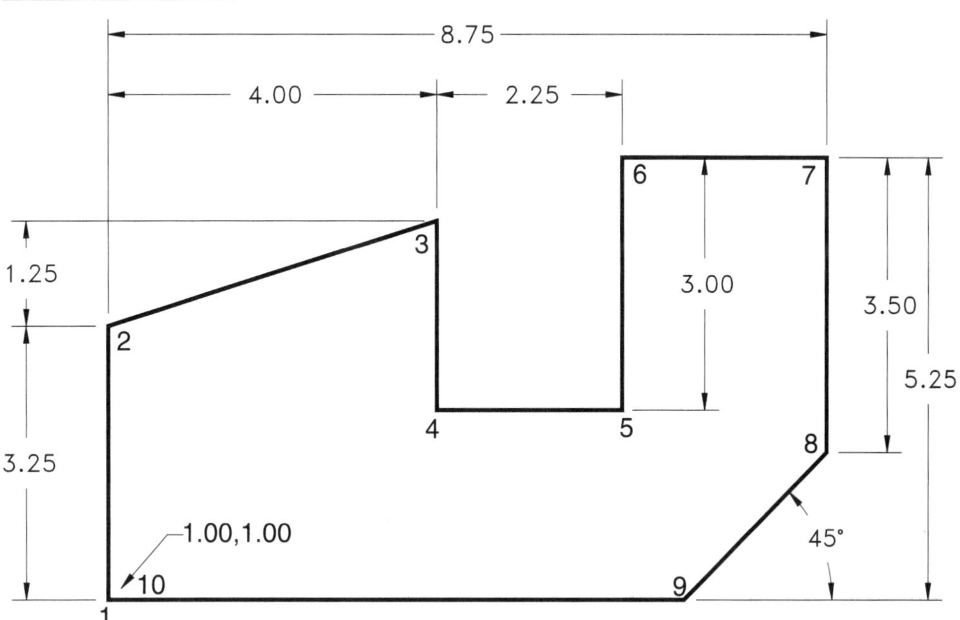

	Absolute	**Relative**
From Pt (1)	1.00,1.00	1.00,1.00
To Pt (2)		
To Pt (3)		
To Pt (4)		
To Pt (5)		
To Pt (6)		
To Pt (7)		
To Pt (8)		
To Pt (9)		
To Pt (10)		
To Pt	Enter	Enter

PROBLEM 1–3

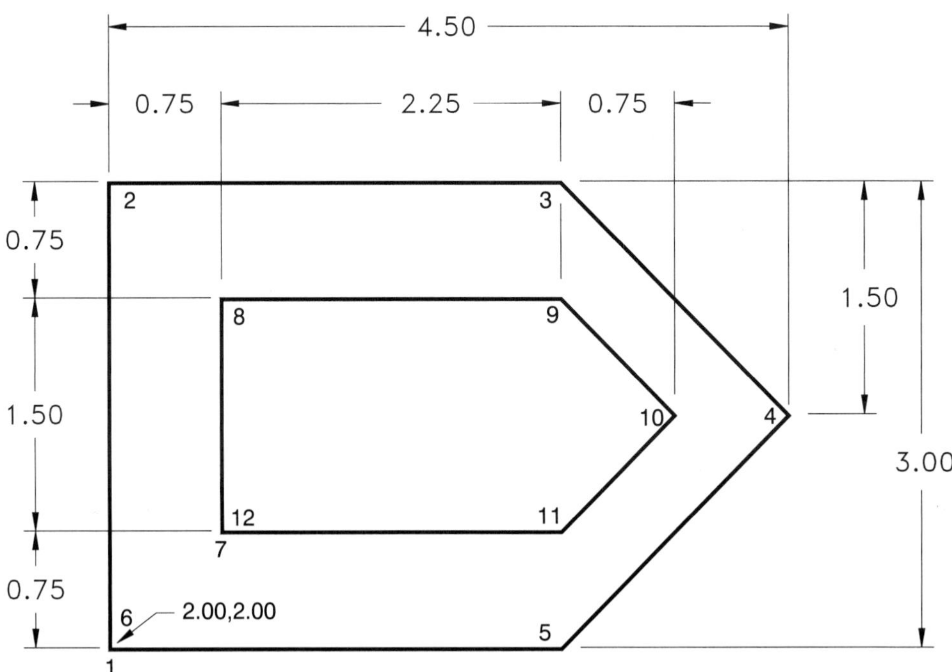

	Absolute	Relative
From Pt (1)	2.00,2.00	2.00,2.00
To Pt (2)		
To Pt (3)		
To Pt (4)		
To Pt (5)		
To Pt (6)		
To Pt	Enter	Enter
From Pt (7)		
To Pt (8)		
To Pt (9)		
To Pt (10)		
To Pt (11)		
To Pt (12)		
To Pt	Enter	Enter

PROBLEM 1–4

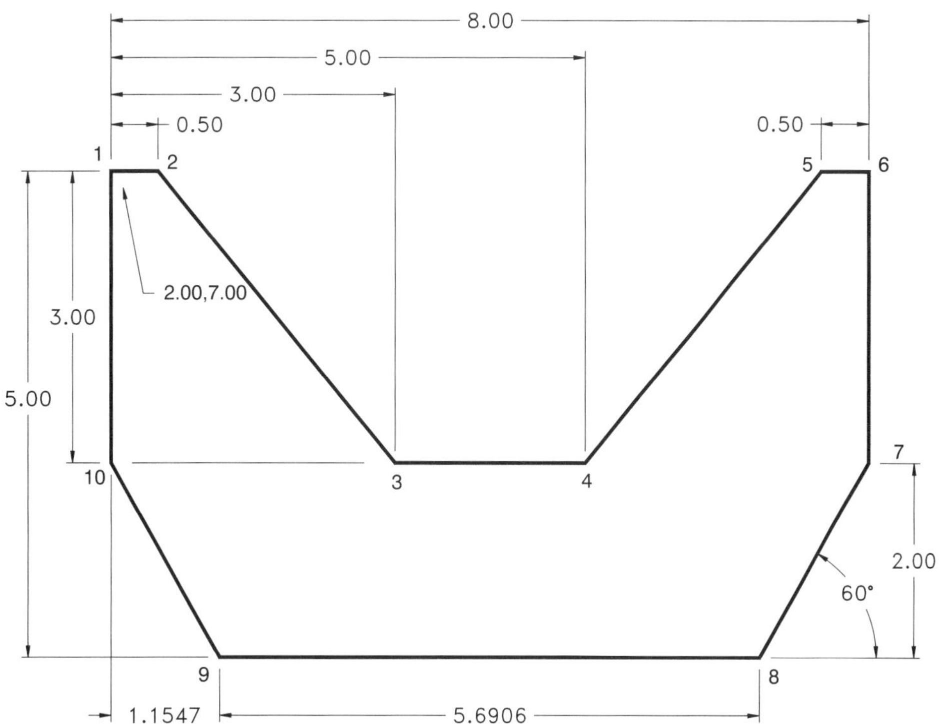

	Absolute
From Pt (1)	2.00,7.00
To Pt (2)	
To Pt (3)	
To Pt (4)	
To Pt (5)	
To Pt (6)	
To Pt (7)	
To Pt (8)	
To Pt (9)	
To Pt (10)	
To Pt	Enter

PROBLEM 1–5

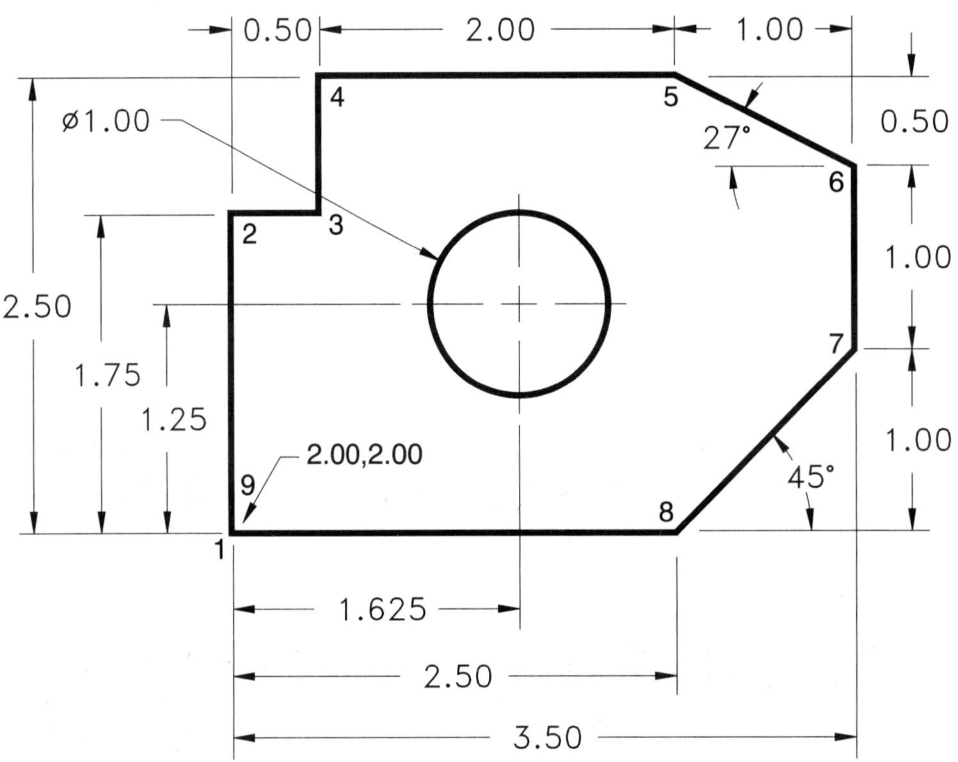

	Absolute	Relative
From Pt (1)	2.00,2.00	2.00,2.00
To Pt (2)		
To Pt (3)		
To Pt (4)		
To Pt (5)		
To Pt (6)		
To Pt (7)		
To Pt (8)		
To Pt (9)		
To Pt	Enter	Enter
Center Pt (10)		

PROBLEM 1–6

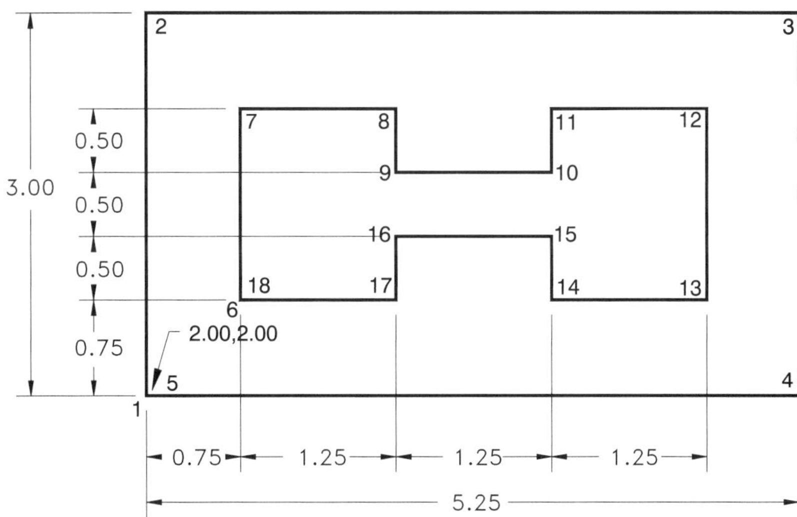

	Absolute	**Relative**	**Polar**
From Pt (1)	2.00,2.00	2.00,2.00	2.00,2.00
To Pt (2)			
To Pt (3)			
To Pt (4)			
To Pt (5)			
To Pt	Enter	Enter	Enter
From Pt (6)			
To Pt (7)			
To Pt (8)			
To Pt (9)			
To Pt (10)			
To Pt (11)			
To Pt (12)			
To Pt (13)			
To Pt (14)			
To Pt (15)			
To Pt (16)			
To Pt (17)			
To Pt (18)			
To Pt	Enter	Enter	Enter

Directions for Problems 1–7 through 1–22:

Construct one-view drawings of the following figures using the LINE command along with coordinate or direct distance modes.

PROBLEM 1–7

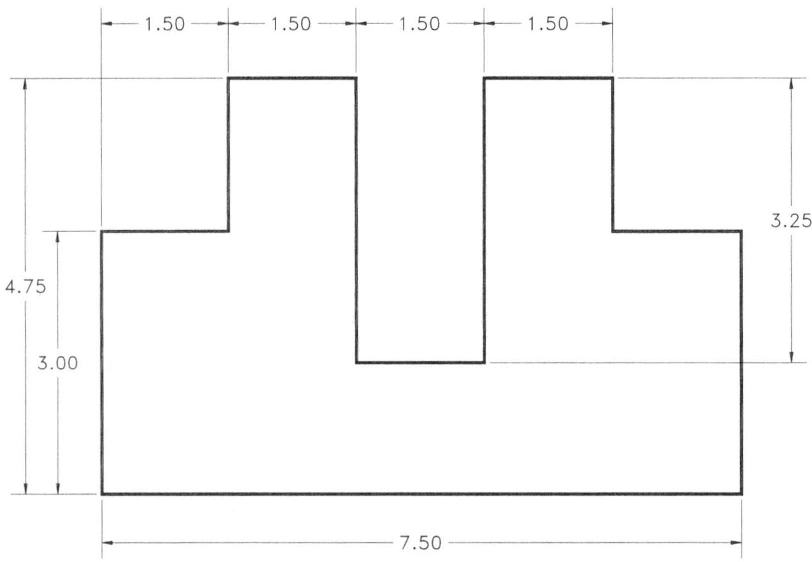

PROBLEM 1–8

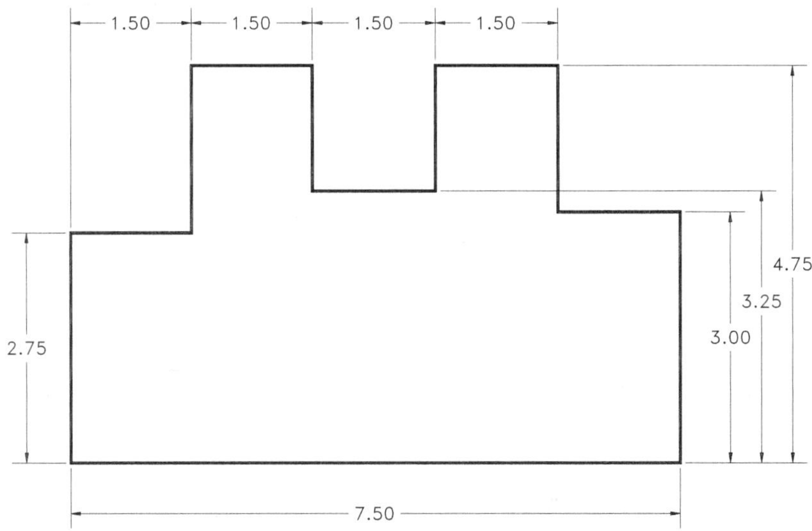

PROBLEM 1–9

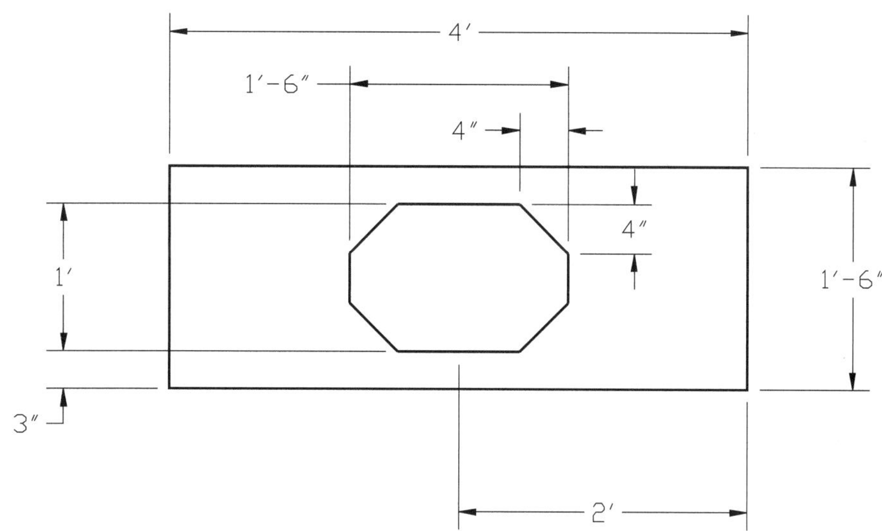

Intermediate Level Drawings

PROBLEM 1–10

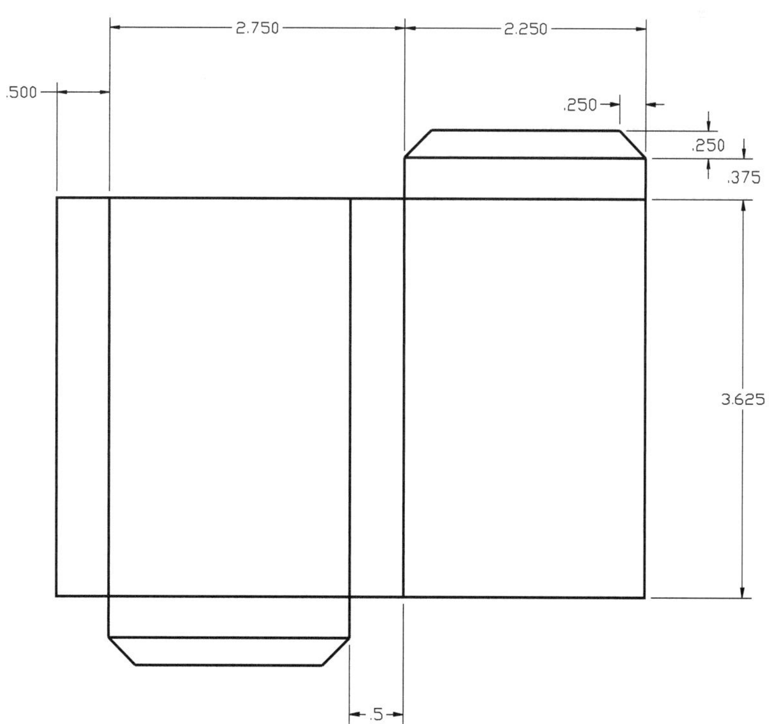

PROBLEM 1–11

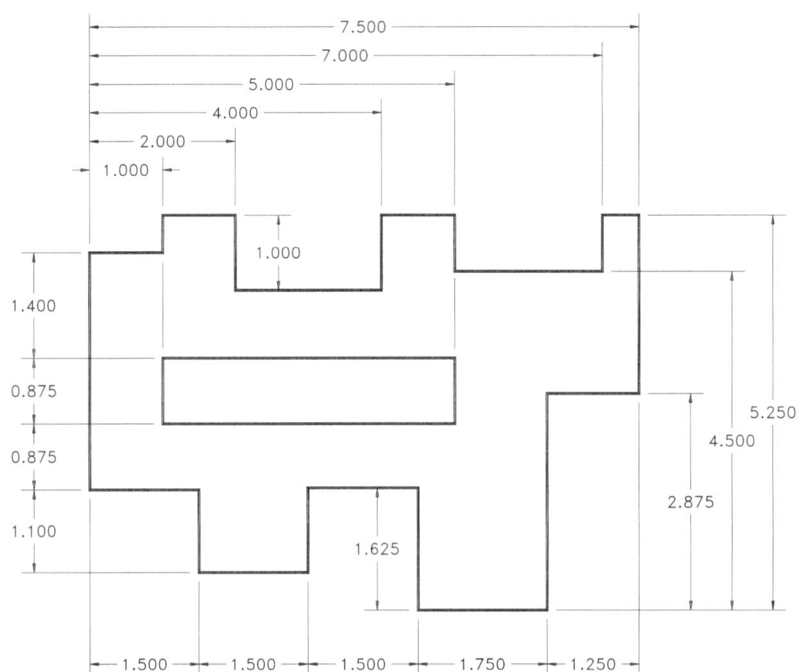

PROBLEM 1–12

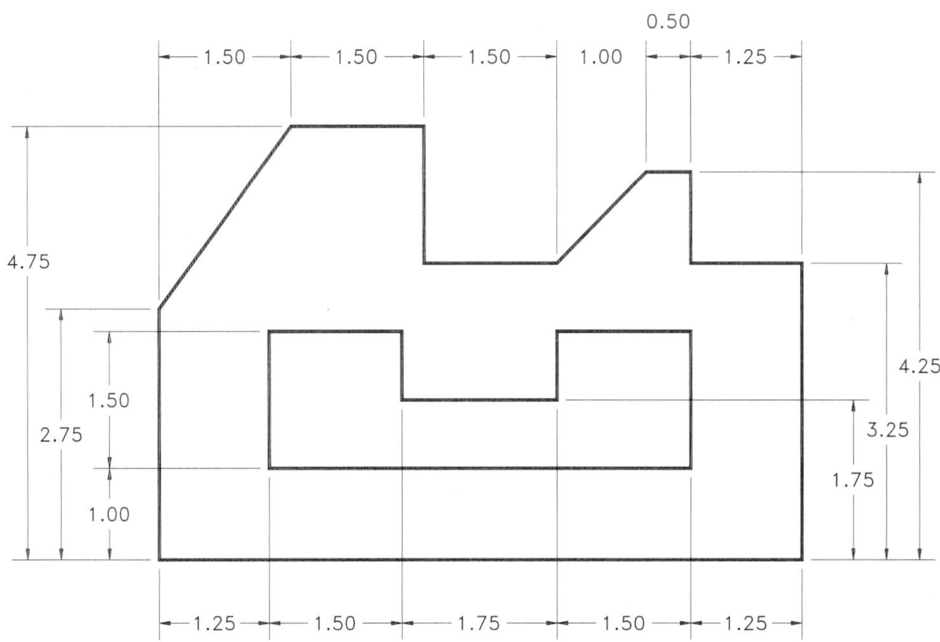

PROBLEM 1–13

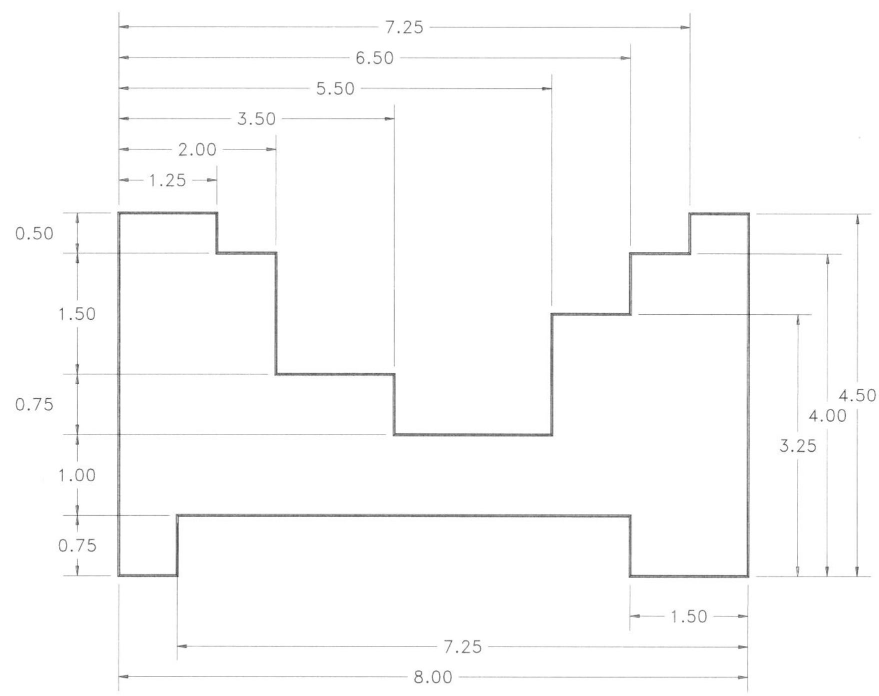

PROBLEM 1–14

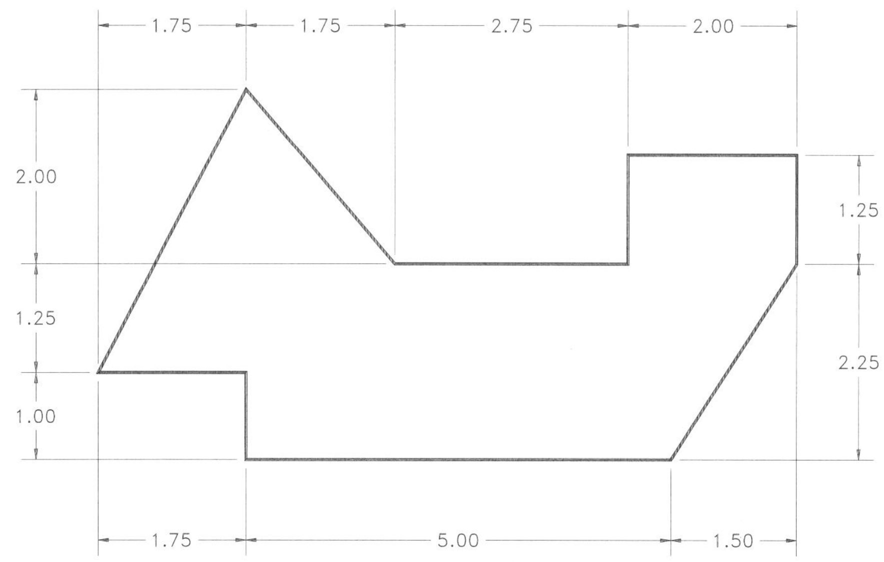

PROBLEM 1–15

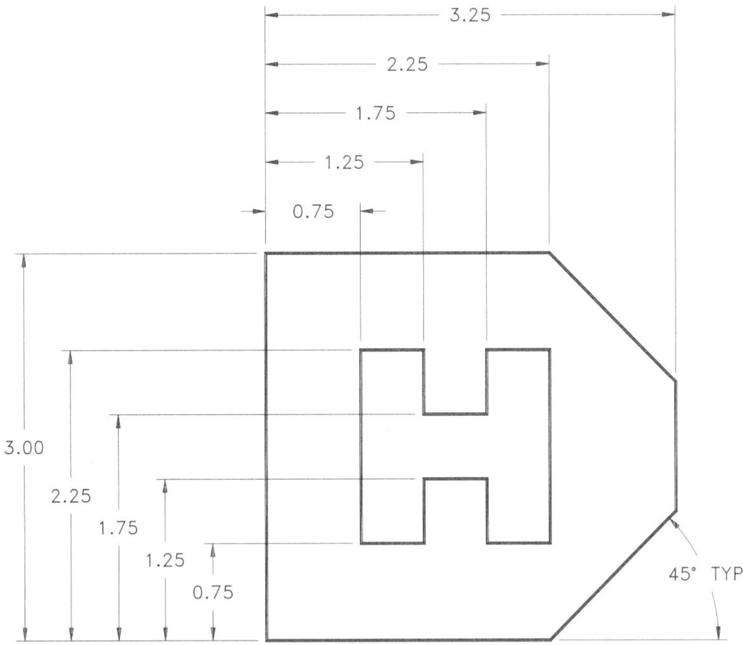

PROBLEM 1–16

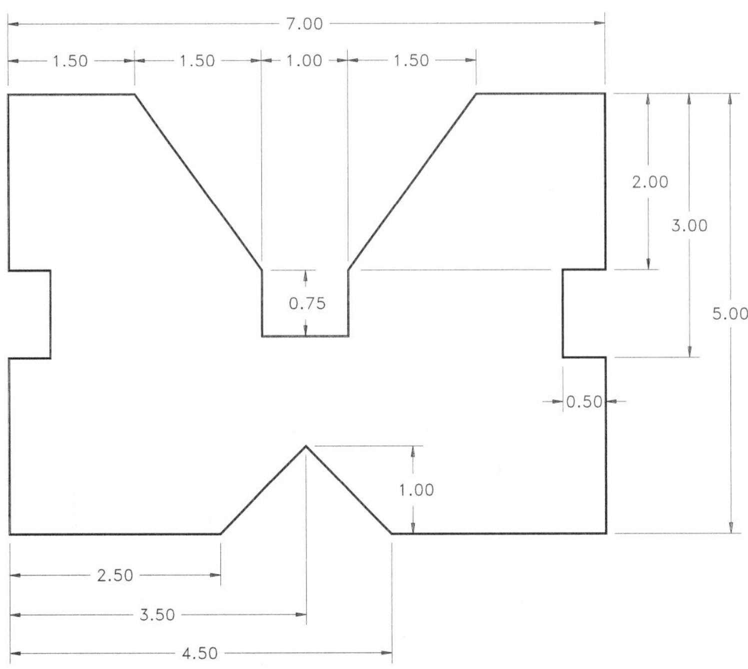

PROBLEM 1–17

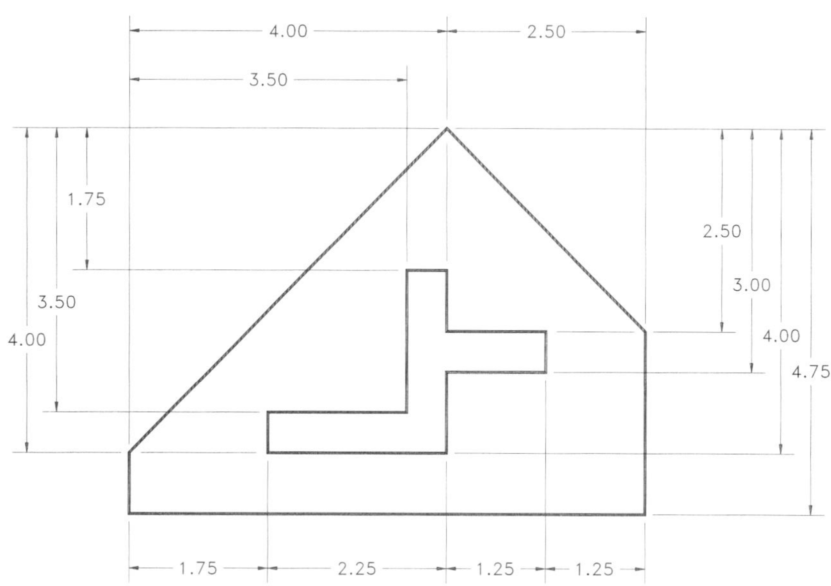

PROBLEM 1–18

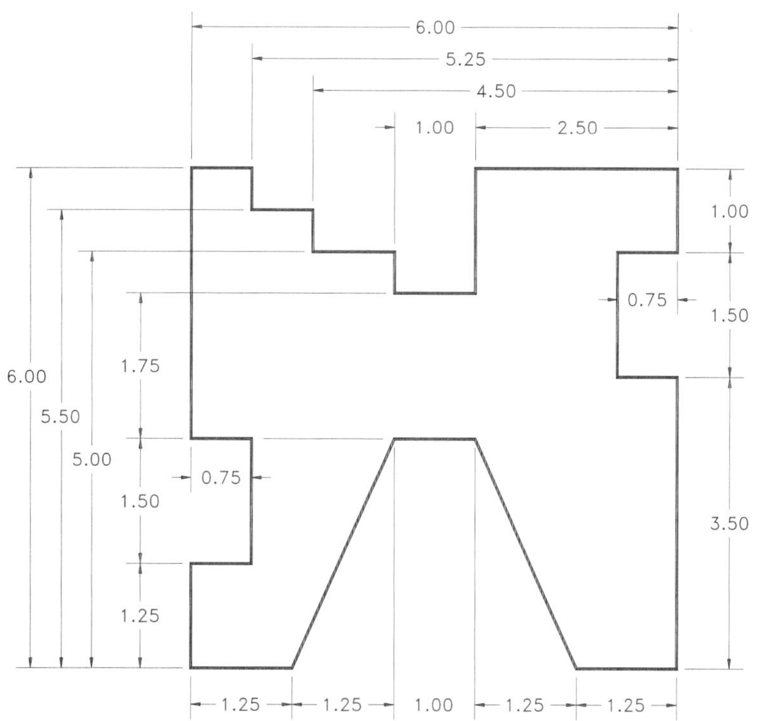

PROBLEM 1–19

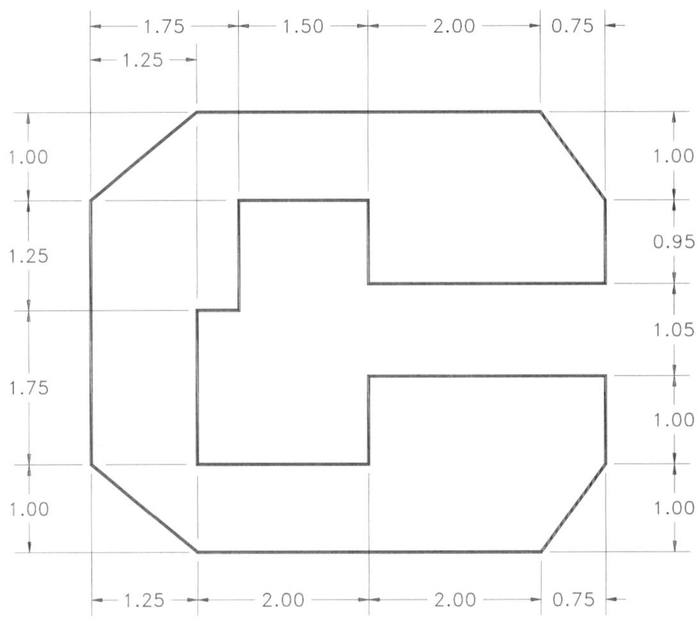

PROBLEM 1–20

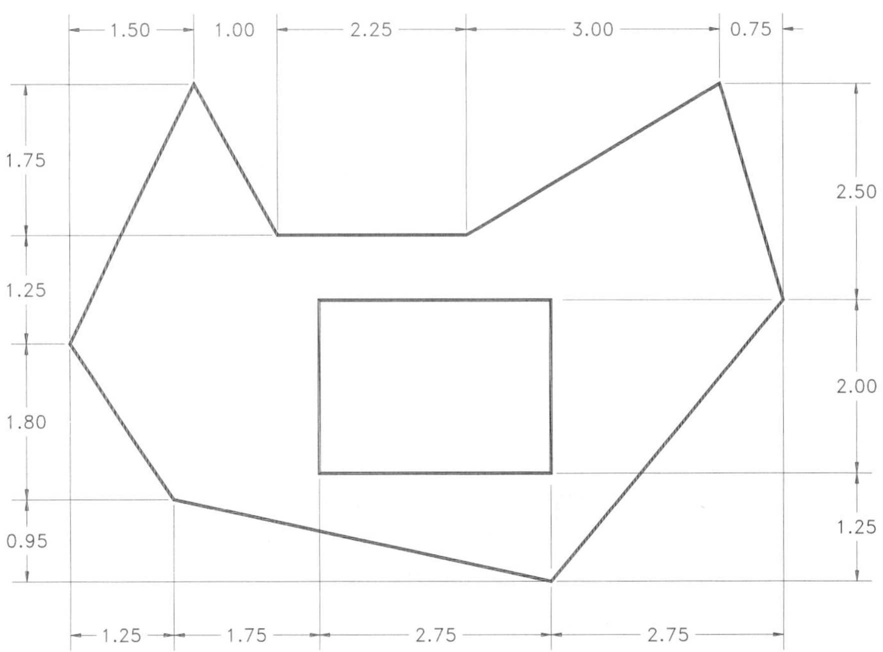

PROBLEM 1–21

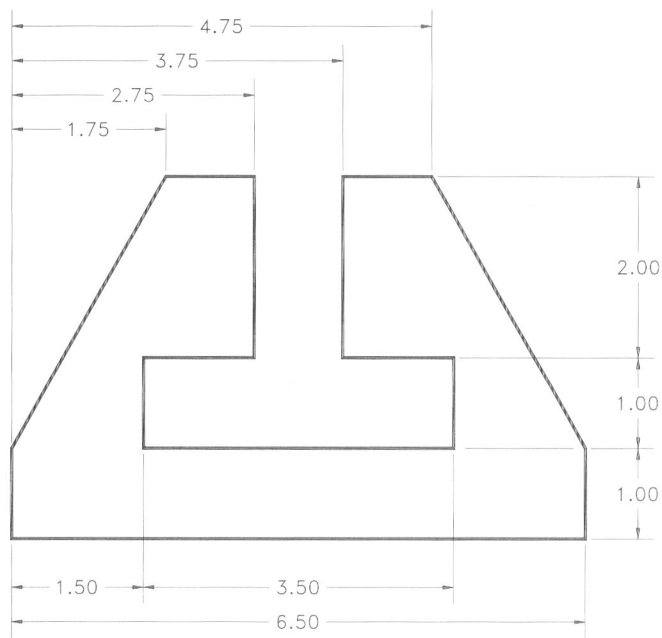

PROBLEM 1–22

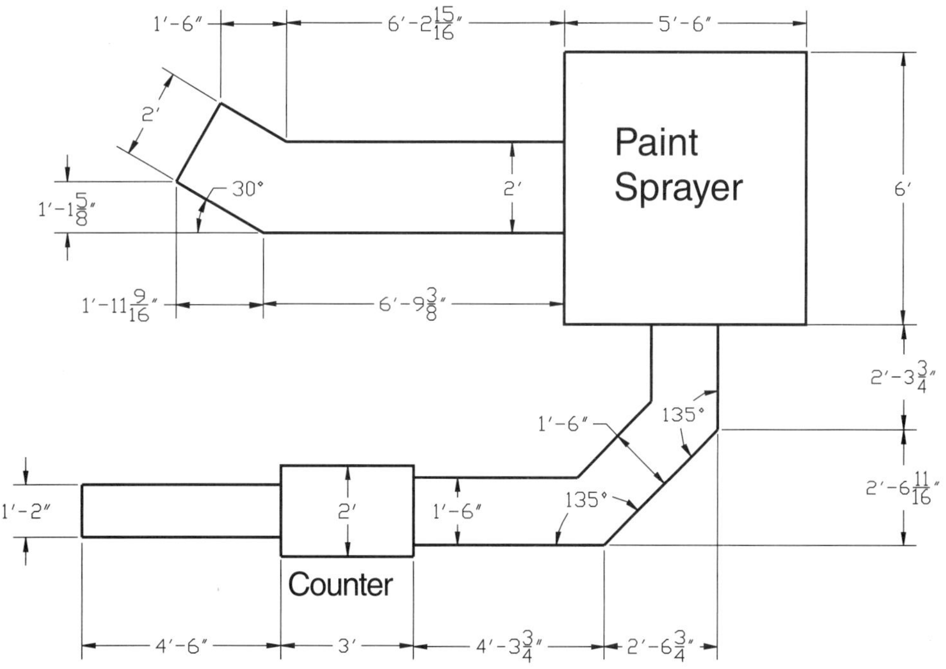

Drawing Setup and Organization

The exercises in this chapter introduce:

- Drawing units
- Layering setup and options
- Layer Standards
- Layer filtering
- Layer Translating

TUTORIAL EXERCISE: 02_LAYER FILTERS.DWG

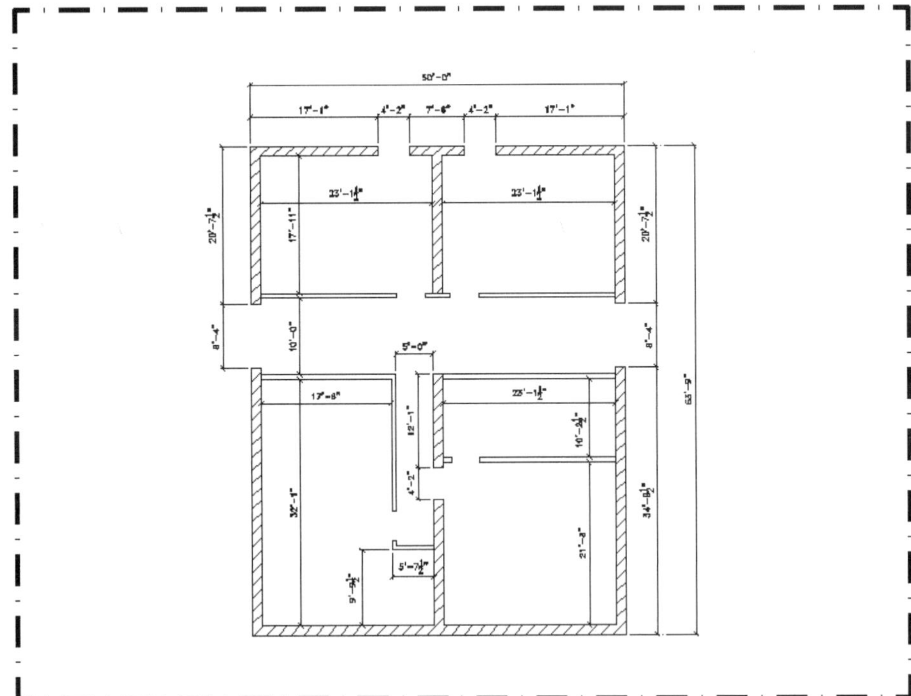

Figure 2–1

Purpose

The purpose of this tutorial is to filter out a number of layers already created in the drawing shown in Figure 2–1.

System Settings

Keep all current system settings.

Layers

Layers have already been created for this tutorial exercise.

Suggested Commands

Activate the Layer properties Manager dialog box (by typing LAYER) and click the button under Named layer filters to display the Named Layer Filters dialog box. Add the desired search requirements for filtering out all other layers, create a filter name, and add this name to the other names already present in the Named layer filters list. Clicking on the layer filter name lists only those layers that match the search criteria.

Whenever possible, substitute the appropriate command alias in place of the full AutoCAD command in each tutorial step. For example, use "CP" for the COPY command, "L" for the LINE command, and so on. The complete listing of all command aliases is located in Table 1–2.

STEP 1

Open the drawing file *02_Layer Filters* shown in Figure 2–1. You need to list all of the layers that are assigned the color CYAN, are currently turned Off, and have a lineweight value of 0.30 mm. Open the Layer Properties Manager dialog box and select the ellipse (a button containing three dots) button, next to the Named layer filters combo box (see Figure 2–2).

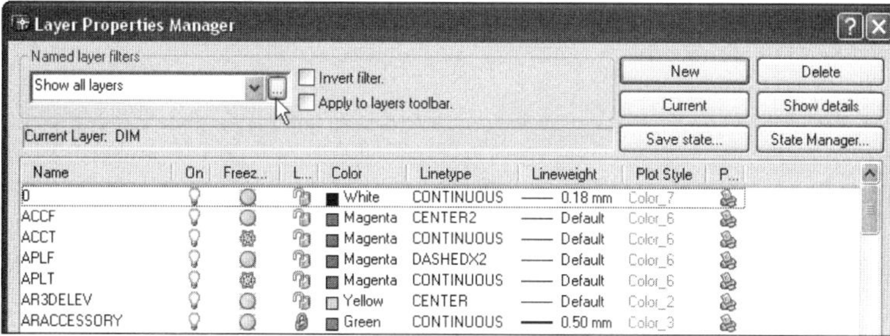

Figure 2–2

STEP 2

This activates the Named Layer Filters dialog box. Make changes to the following fields (see Figure 2–3):

1. Click in the On/Off edit box and set the property to Off.
2. Click in the edit box next to Colors: and add the Cyan color to this area.
3. Click in the edit box next to Lineweight: and type **0.30 mm**.
4. Enter **CYAN Off .3** in the Filter Name: edit box.
5. Pick the Add button.
6. Click the Close button.

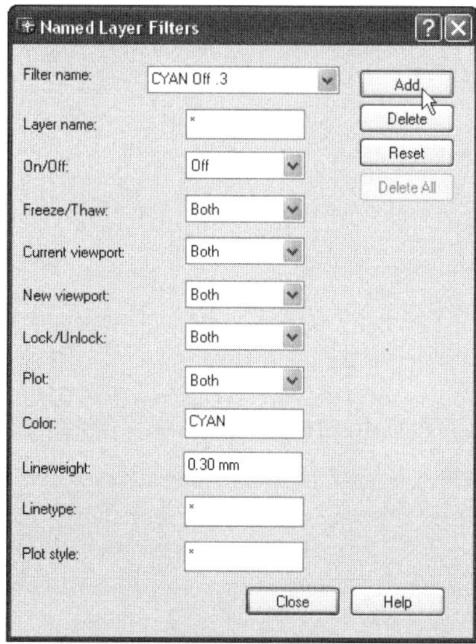

Figure 2–3

STEP 3

Once you return back to the Layer Properties Manager dialog box, click in the edit box under "Named layer filters" and select the layer filter you just created; namely CYAN Off .3 (see Figure 2–4).

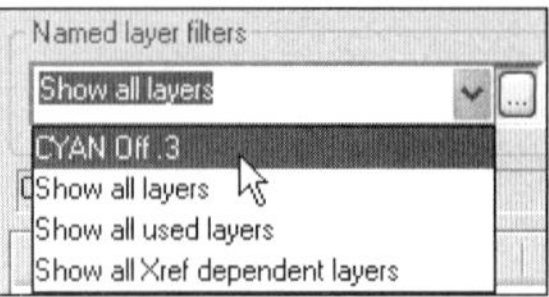

Figure 2–4

STEP 4

The results are displayed in Figure 2–5, showing the Layer Properties Manager dialog box. Only the layers assigned the color Cyan, are turned Off, and have a lineweight of 0.30 mm appear. The following steps review the creation of a layer filter set:

1. Set the On/Off: edit box to **Off**
2. Change the Color: edit box to **CYAN**
3. Change the Lineweight: edit box to **0.30 mm**
4. Set the Filter name: edit box to **CYAN Off .3**
5. Click the Add button to save this name.
6. Click the Close button to return to the Layer Properties Manager dialog box.
7. Select CYAN Off .3 from the Named layer filters list.

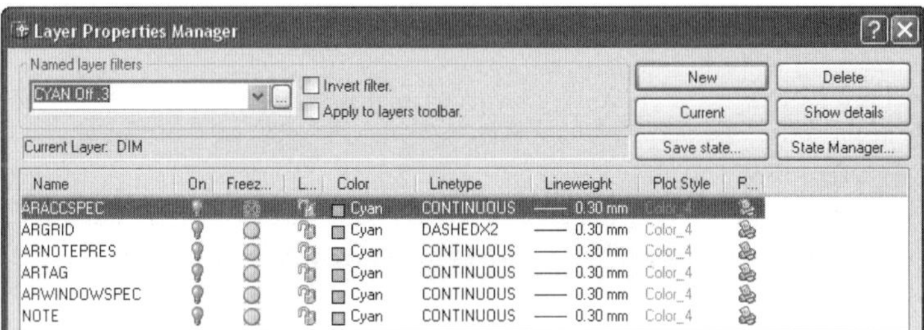

Figure 2–5

STEP 5

To redisplay all layers in the current drawing file, click in the Named layer filters edit box and change the property to "Show all layers", as shown in Figure 2–6. Another technique for listing all layers after performing a filter search, would be to click on the Reset button, as shown in Figure 2–2. This would also list all layers in the drawing.

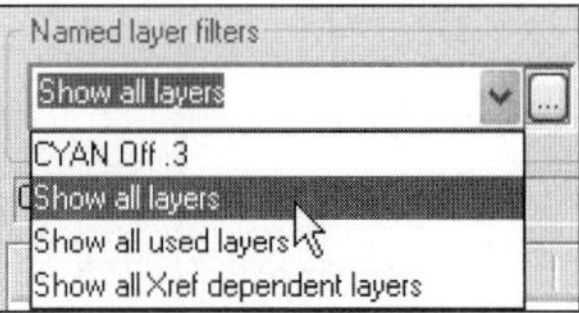

Figure 2–6

Tip: When setting the Lineweight in the Named Layer Filters dialog box, you must enter the exact value along with the units of value. Entering 0.30 mm is a valid unit. A value of .3 or .3 mm is considered invalid.

TUTORIAL EXERCISE: 02_LAYER STATE.DWG

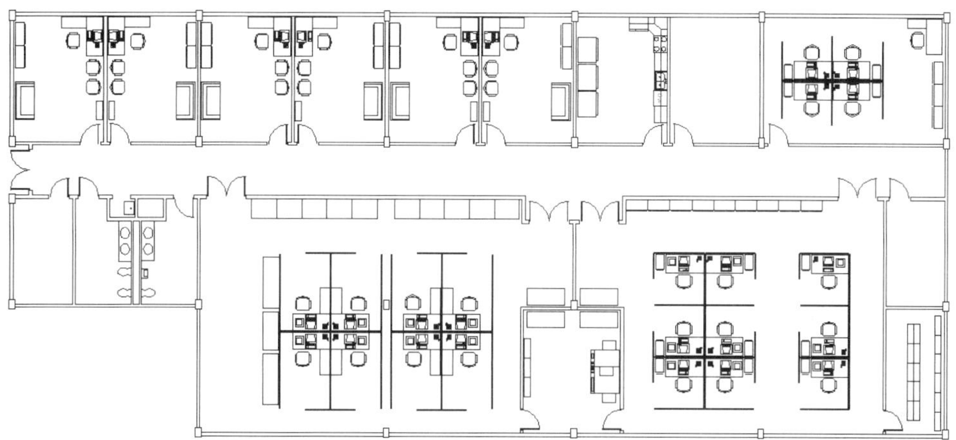

Figure 2–7

Purpose

The purpose of this tutorial is to create a number of layer states and then restore these on the drawing shown in Figure 2–7 and observe the results.

System Settings

Keep all current system settings.

Layers

Layers have already been created for this tutorial exercise.

Suggested Commands

Activate the Layer Properties Manager dialog box and save a number of layer states by clicking on the Save state... button. With a number of layer states created, click on the State Manager... button to display and restore the states. Observe how layer states affect the display of objects in the current drawing.

Whenever possible, substitute the appropriate command alias in place of the full AutoCAD command in each tutorial step. For example, use "CP" for the COPY command, "L" for the LINE command, and so on. The complete listing of all command aliases is located in Table 1–2.

STEP 1

Open the drawing file *02_Layer State*. Your display of the office facility layout should be similar to Figure 2–7. Your task is to create the following three layer states:

1. "Display All Layers" (to display all layers in the drawing)
2. "Display Only Walls" (to isolate only the floor plan walls)
3. "Display Walls and Furniture (to display the walls and the furniture)

STEP 2

Open the Layer Properties Manager dialog box shown in Figure 2–8 and observe the layers and their names. Click the Save state… button to display the Save Layer States dialog box shown in Figure 2–9. Be sure that the On/Off and Frozen/Thawed states are checked. Then type **Display All Layers** in the New layer name state edit box. When finished, click the OK button to return to the Layer Properties Manager dialog box.

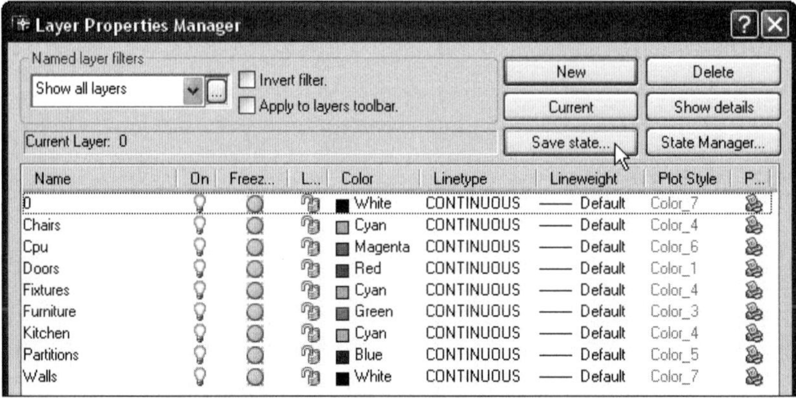

Figure 2–8

Figure 2–9

STEP 3

In the Layer Properties Manager dialog box, turn off all layers except 0 and Walls in Figure 2–10. Then click the Save state... button and enter **Display Only Walls** in the New layer name state edit box, shown in Figure 2–11. Click the OK button to return to the Layer Properties Manager dialog box.

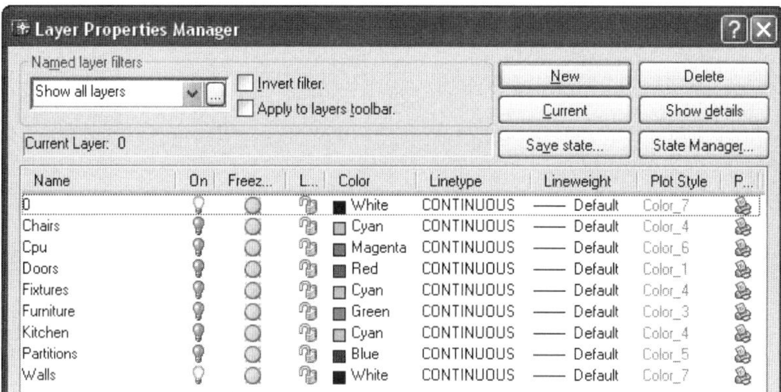

Figure 2–10

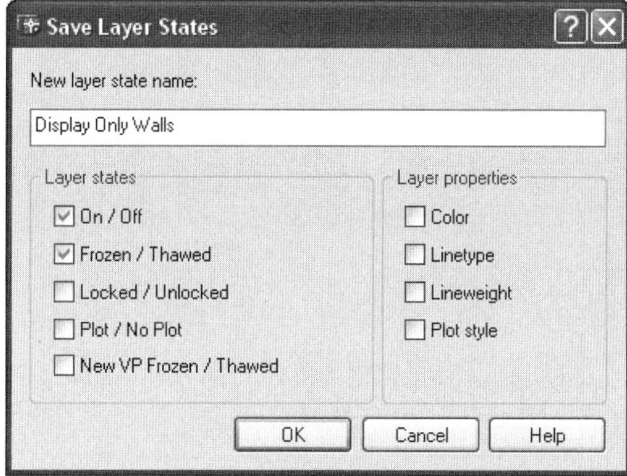

Figure 2–11

STEP 4

In the Layer Properties Manager dialog box, turn off all layers except 0, Furniture, and Walls, as shown in Figure 2–12. Then click the Save state… button and type **Display Walls and Furniture** in the New layer name state edit box, shown in Figure 2–13. Click the OK button. You have now finished creating all three layer states, click the OK button to dismiss the Layer Properties Manager dialog box.

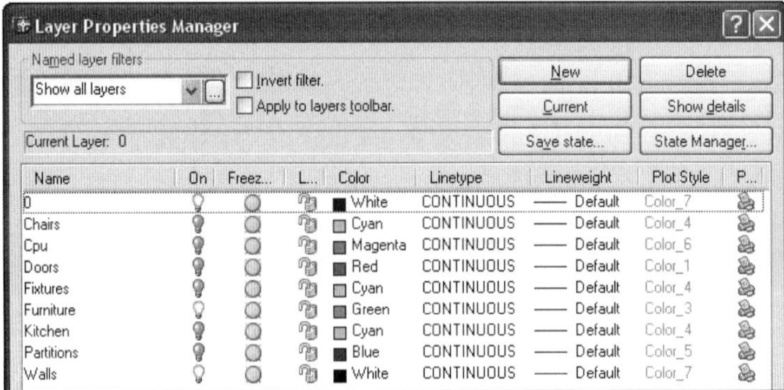

Figure 2–12

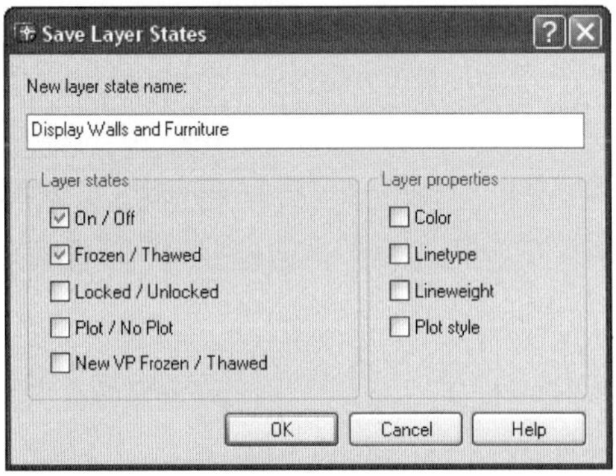

Save Layer States
New layer state name:
Display Walls and Furniture
Layer states
On / Off
Frozen / Thawed
Locked / Unlocked
Plot / No Plot
New VP Frozen / Thawed
Layer properties
Color
Linetype
Lineweight
Plot style
OK
Cancel
Help

Figure 2–13

STEP 5

To test the newly created layer states on this drawing, reopen the Layer Properties Manager dialog box and click on the State Manager... button. Clicking on the layer state "Display Only Walls" and then clicking the Restore button, shown in Figure 2–14, will return you back to the Layer Properties Manager dialog box. Notice that all layers have been turned off except for 0 and Walls. Clicking the OK button will display only the walls in this drawing, as shown in Figure 2–15. Experiment with the other layer states to see how useful this feature can be.

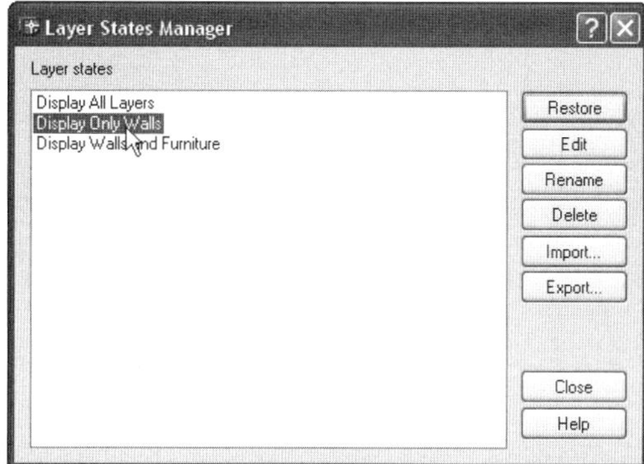

Figure 2–14

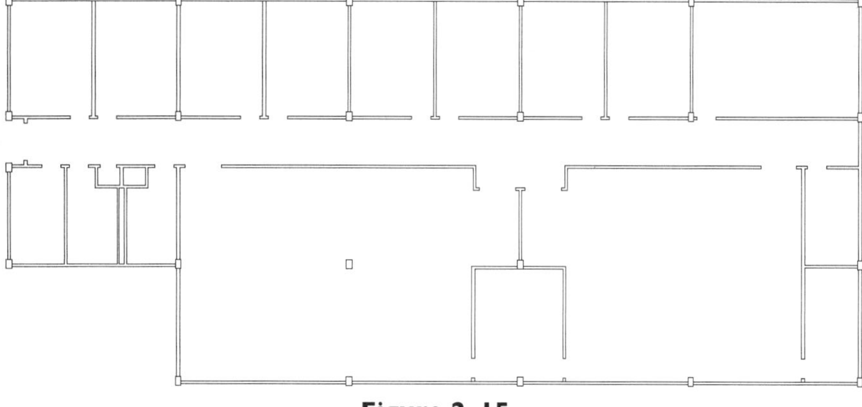

Figure 2–15

STEP 6

To better utilize the Layer State feature, the Layer previous button, shown in Figure 2–16 can be used to display the previous layer state.

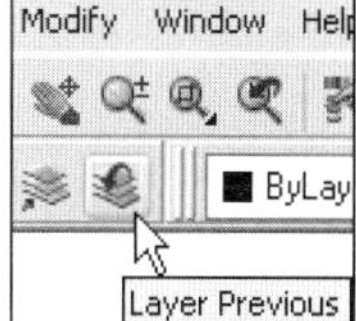

Figure 2–16

TUTORIAL EXERCISE: 02_LAYER TRANSLATE.DWG

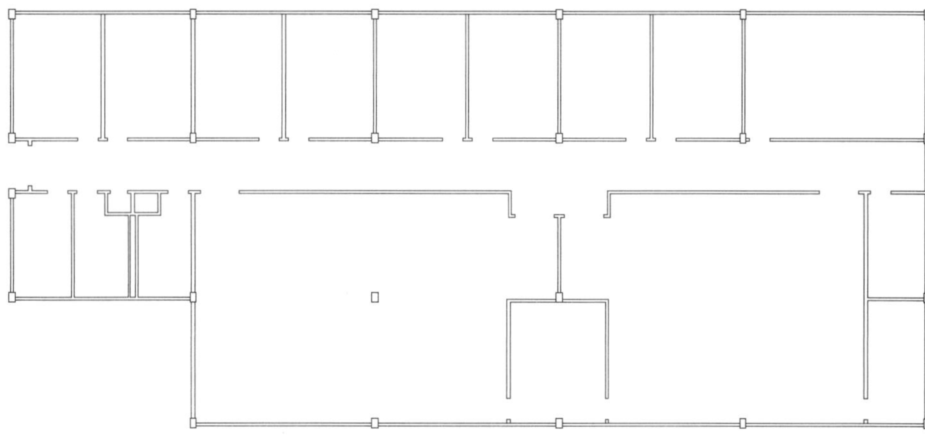

Figure 2–17

Purpose

The purpose of this tutorial is to experiment with the object selection modes on the drawing shown in Figure 2–17.

System Settings

Keep the current limits settings of (0,0) for the lower-left corner and (16,10) for the upper-right corner. Keep all remaining system settings.

Layers

Layers have already been created for this tutorial exercise.

Suggested Commands

This tutorial utilizes the LAYTRANS command. The Layer Translator is used to convert drawing #1 layer names into the existing layer names used in drawing #2. This tool is typically used when external company drawings need to be changed to meet internal company standards.

To view the encoded Translation log file that is generated by the Layer Translator, use Wordpad or Notepad and change the file type to "all documents".

Whenever possible, substitute the appropriate command alias in place of the full AutoCAD command in each tutorial step. For example, use "CP" for the COPY command, "L" for the LINE command, and so on. The complete listing of all command aliases is located in Table 1–2.

STEP 1

Try It! – Open the *02_Layer Translate.dwg* file. After the floor plan displays, examine the layers found in the Layer Properties Manager dialog box. Notice in Figure 2–18 that all layers are listed as a series of numbers. However, it is difficult to determine the function or purpose of these numbered layers. The layer list found in an existing drawing will be used to convert the numbered layers into layers with meaning.

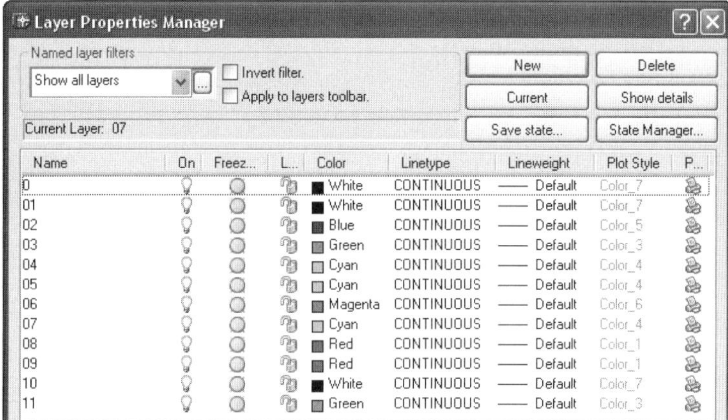

Figure 2–18

STEP 2

With the *02_Layer Translate* drawing still displayed, open the second drawing file: *02_Translation Template*. Examine the list of layers in this drawing (see Figure 2–19). You can immediately identify the purpose of the layers just by reading their names such as "Chairs", "Kitchen" and "Walls". Close the *02_Translation Template* file when finished examining the layers.

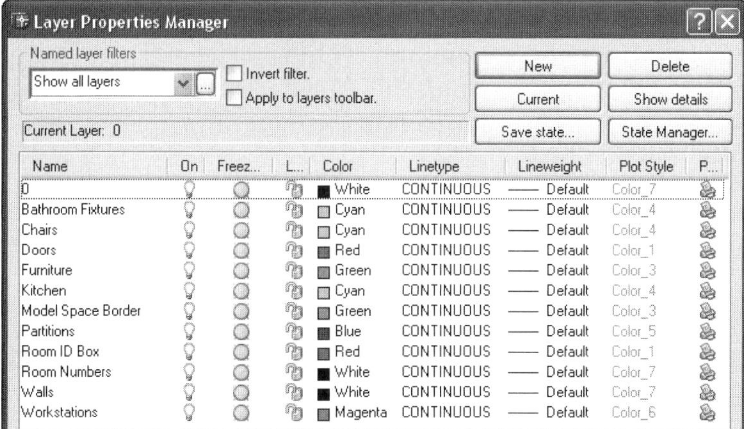

Figure 2–19

STEP 3

Use the table below to convert the layers found in the *02_Layer Translate* drawing ("01", "02", etc) to the layer names found in the *02_ Translation Template* file ("Walls", "Partition" etc.)

Table 2–1 *Layer Translations*

02_Layer Translate		*02_Translation Template*
Layer "01"	To	"Walls"
Layer "02"	To	"Partition"
Layer "03"	To	"Furniture"
Layer "04"	To	"Kitchen"
Layer "05"	To	"Bathroom Fixtures"
Layer "06"	To	"Workstations"
Layer "07"	To	"Chairs"
Layer "08"	To	"Doors"
Layer "09"	To	"Room ID Box"
Layer "10"	To	"Room Numbers"
Layer "11"	To	"Model Space Border"

STEP 4

In the *02_Layer Translate* drawing, activate the Layer Translator dialog box (LAYTRANS), shown in Figure 2–20. Notice that all the layers found in this drawing are listed in the Translate From list box. Click the Load button to bring in the layers from an existing drawing file. This activates the Select Drawing File dialog box, shown in Figure 2–21. There are three types of files that can be loaded: *.dwg* (Drawing File), *.dws* (Standards), and *.dwt* (Drawing Template). Locate the *02_ Translation Template* file and click the Open button.

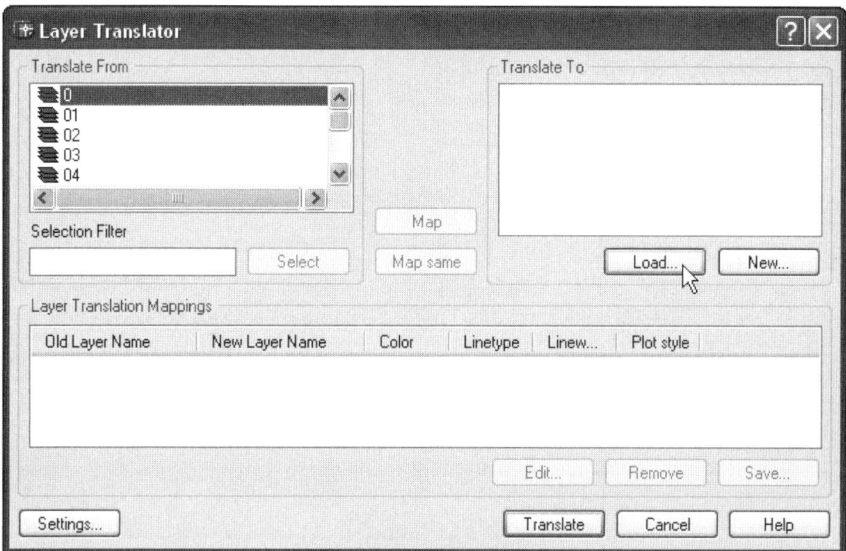

Figure 2–20

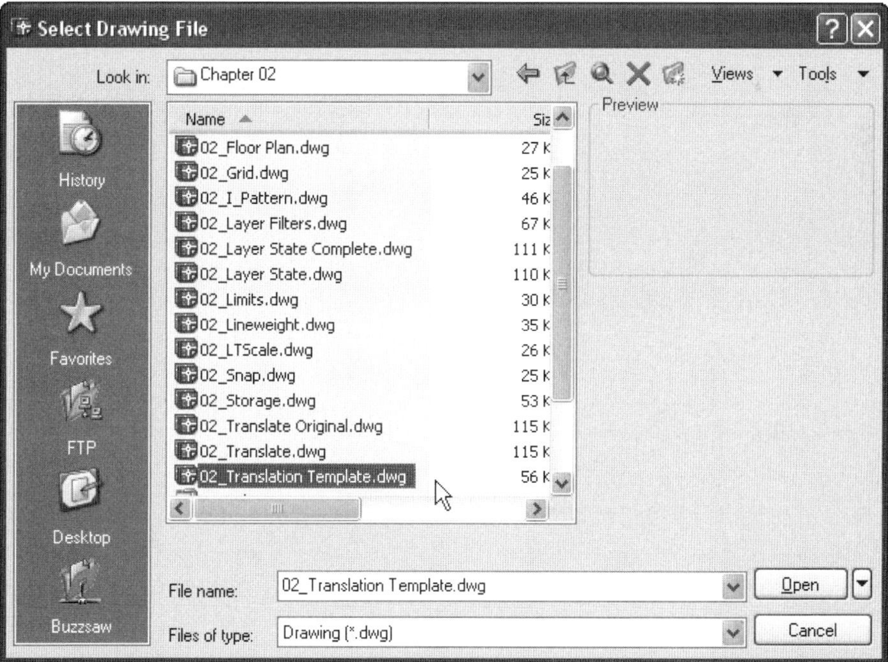

Figure 2–21

STEP 5

Notice in Figure 2–22 that both sets of layers are now loaded in the Layer Translator dialog box. Click on layer "01" in the Translate From list box, then click on "Walls" in the Translate To list box. With both layers highlighted, click the Map button. Both layers are now listed in the Layer Translation Mappings list box, as shown in Figure 2–23. Also, the figure shows that layer "01" has been removed from the Translate From list box.

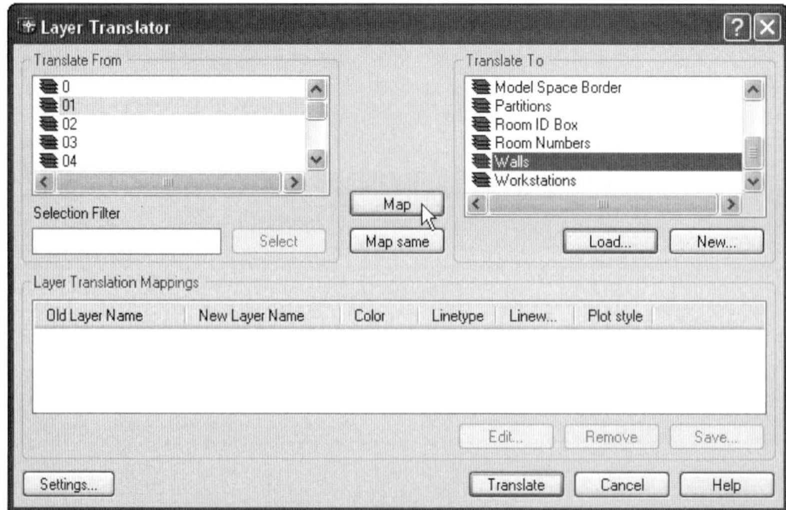

Figure 2–22

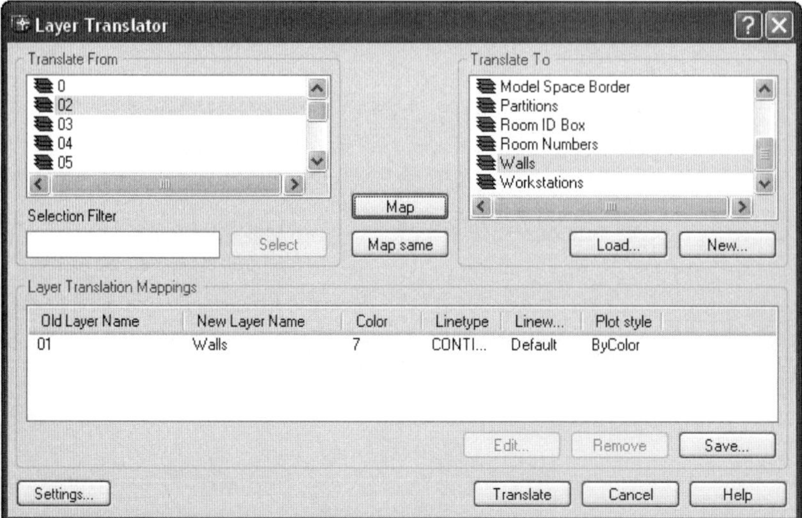

Figure 2–23

STEP 6

Use Table 2–1 as a guide for mapping the remaining layers. After all layers have been mapped, the Layer Translator dialog box should appear similar to Figure 2–24. One other step is necessary to make these layer changes permanent. Click on the Translate button. The Layer Translator Alert dialog box, shown in 2–25, appears alerting you to first save the mapping information to a file. Click the Yes button to do so.

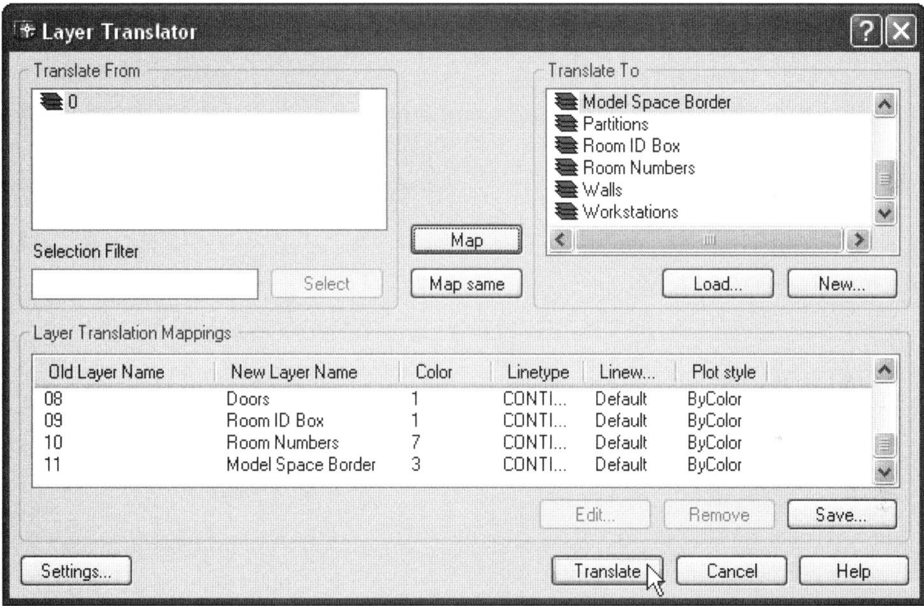

Figure 2–24

Figure 2–25

STEP 7

It is considered good practice to save the mapping information under a file name with a *.dws* (Drawing Standard) extension. This will enable you to call up this file for later use, if other similar layers need to be mapped. Clicking the Yes button in the previous dialog causes the Save Layer Mappings dialog box, shown in Figure 2–26 to appear. The default file name matches the name of the current drawing file. Although you could change the file name to something else, simply lick the Save button to save the file with the name *02_Layer Translate.dws*. This action will not only save the mapping settings, but will permanently translate the layers as well.

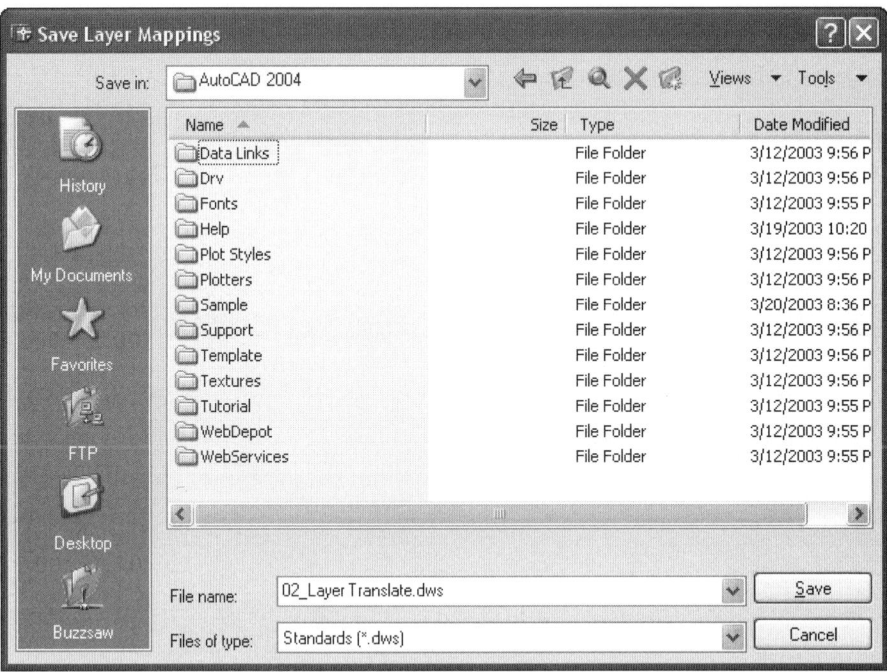

Figure 2–26

STEP 8

In addition to the drawing standard file being created (*.dws*), an encoded log file representing information about the layer translations performed, is also generated. The log file (with the *.log* extension) is saved to the same folder as the drawing file being translated. A sample log file is shown below:

```
============================================================
05157140
Friday, August 01, 2003  11:15:48 AM
051571500514ED80
05157150Administrator
============================================================
05157150
-----------------
05148DE005157160051 51C90
0514802005157160051 51BF0
0514CDB005157160051 51B40
0514CCB005157160051 51AA0
0514CBD005157160051 519F0
0514CAF005157160051 51940
0514CA1005157160051 518A0
0514C93005157160051 51800
0514C85005157160051 51750
0514C77005157160051 516A0
0514C60005157160051 5156005157150
---------------------
0514CFA0
```

PROBLEM 2–1

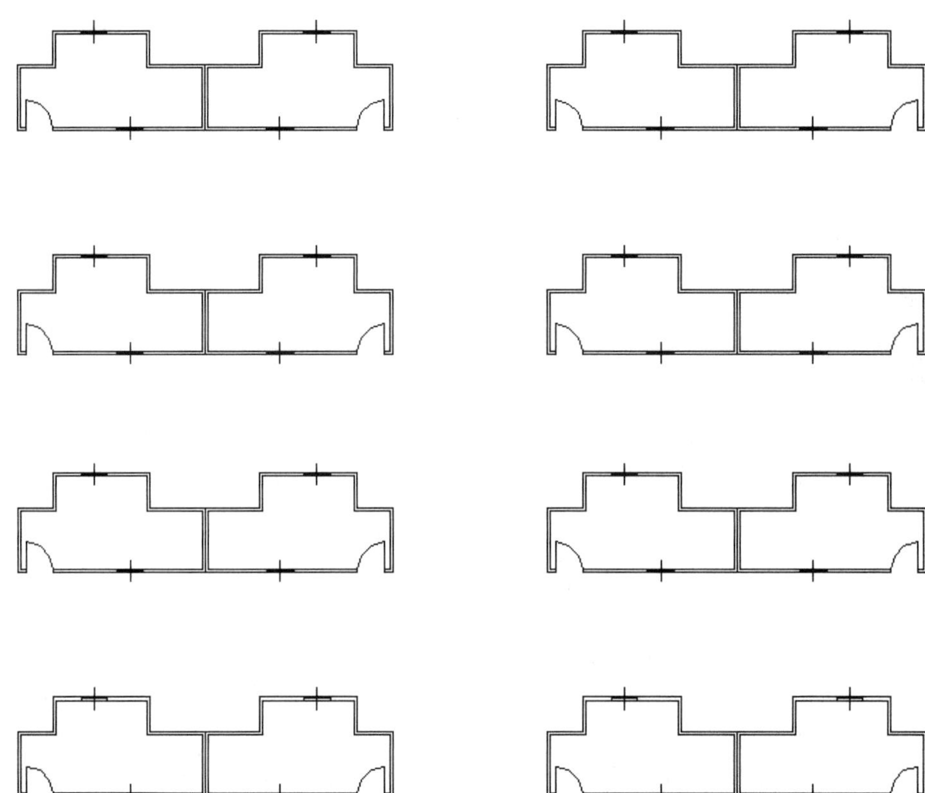

Open the existing drawing file named *02_Storage.dwg*. Use the Layer Properties Manager, along with the Named Layer Filters dialog box, to answer the following layer-related questions:

1. The total number of layers frozen in the database of this drawing is ________.
2. The total number of layers locked in the database of this drawing is ________.
3. The total number of layers turned off in the database of this drawing is ________.
4. The total number of layers assigned the color red and frozen in the database of this drawing is ________.
5. The total number of layers that begin with the letter "L", assigned the color yellow, and turned off in the database of this drawing is ________.

PROBLEM 2–2

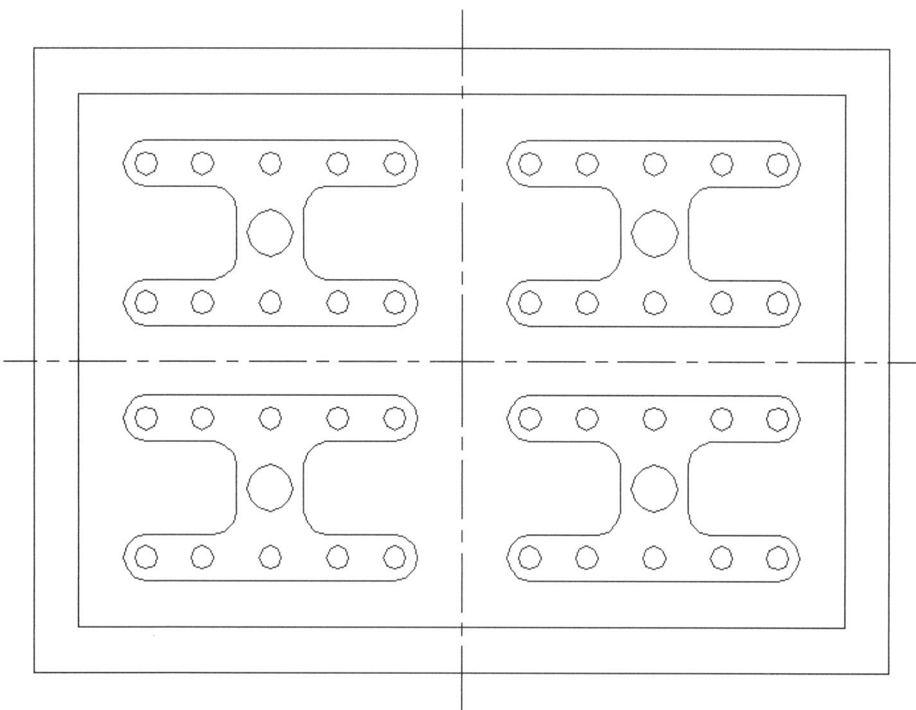

Open the existing drawing named *02_I_Pattern.dwg.* Use the Layer Properties Manager along with the Named Layer Filters dialog box to answer the following layer-related questions:

1. The total number of layers assigned the color red and turned off in the database of this drawing is ________.
2. The total number of layers assigned the color red and locked in the database of this drawing is ________.
3. The total number of layers assigned the color white and frozen in the database of this drawing is ________.
4. The total number of layers assigned the color yellow and assigned the hidden linetype in the database of this drawing is ________.
5. The total number of layers assigned the color red and assigned the phantom linetype in the database of this drawing is ________.
6. The total number of layers assigned the center linetype in the database of this drawing is ________.

PROBLEM 2–3

This project applies the concepts introduced in the earlier tutorials. Use the following projects as a guide for future projects. The following three layer examples will provide guidance for Mechanical, Architectural and Civil based drawings. Provided below are screenshots of the Layer Properties Manager dialog box for each drawing style.

The user or client determines the color and lineweight choices for each layer; the ones shown below are only a representation.

Create a blank drawing with each layer configuration to be used as a starting point for projects presented later in the book. Once the drawings are created and saved the drawings can be converted to a template. Information on creating a template drawing is found in this chapter.

As shown below, Mechanical drawings are based on linetype styles.

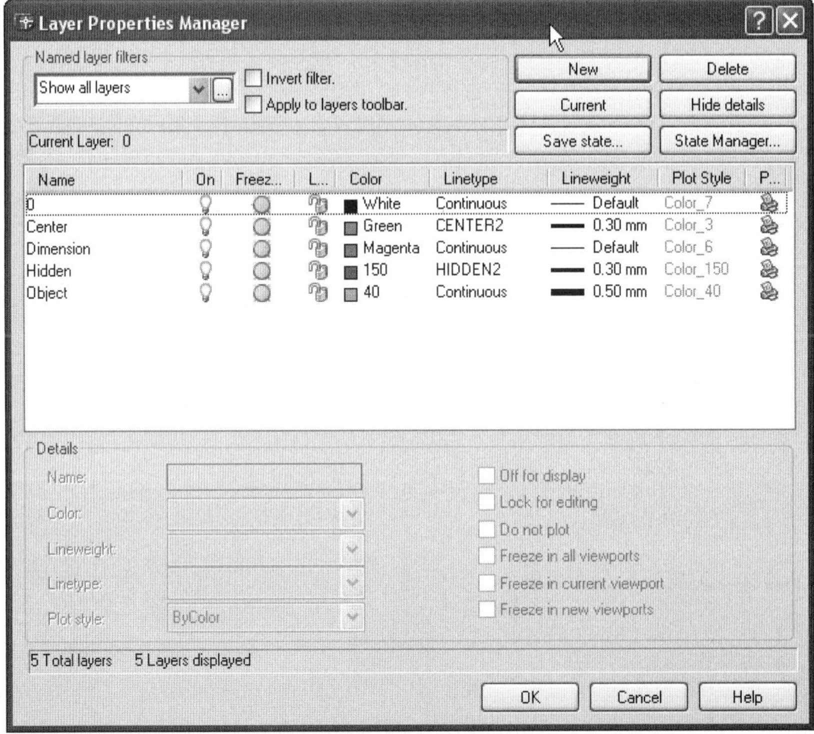

Architectural drawing layers are based on the actual component groups that are being drawn. This Layer Properties Manager shown in the next figure is based on Project 2–4. For complete details on the setup and application of layers in Architectural drawings, the American Institute of Architects (AIA) has developed a complete set of guidelines that are used in the application of Layers in architectural drawings.

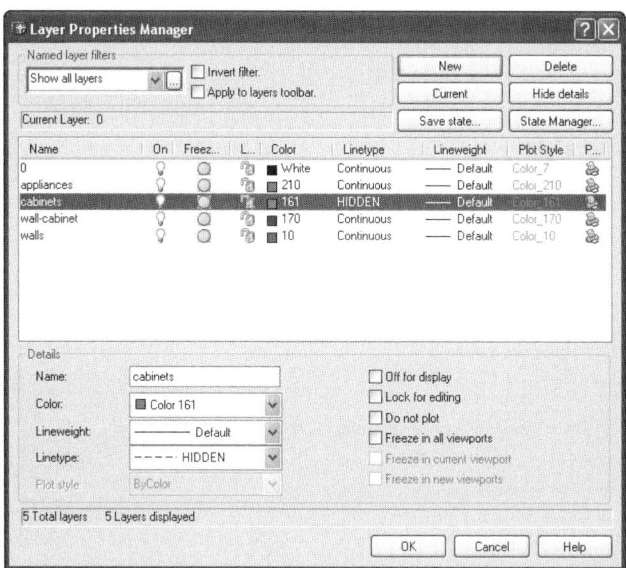

The following figure demonstrates how Civil based drawings use component level layers (similar to Architectural drawings). However, Civil drawings, fall into different categories; such as land (roads, site development and contouring) and civil components (bridges, wells, pumps, piping, wiring, and dams).

The Layer Properties Manager shown below is a typical setup for a road layout drawing.

Notice that the color groupings are based on objects that are easily identifiable on the drawing sheet.

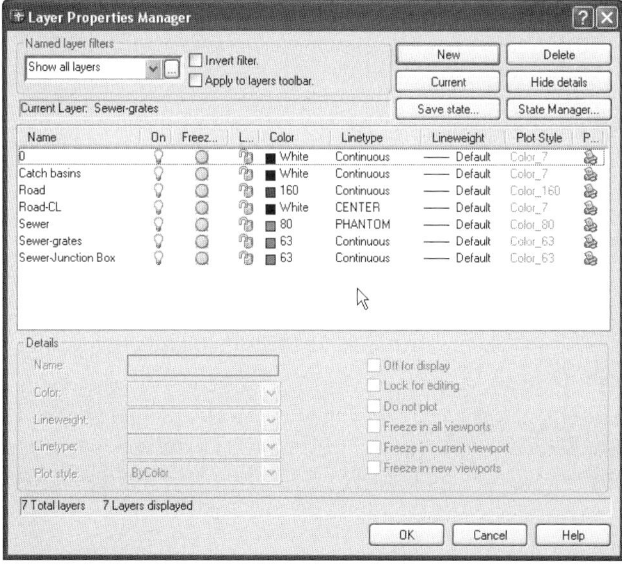

PROBLEM 2–4

Create the following kitchen layout using the Architectural layer guide created in Problem 2–3.

Name	On	Freez...	L...	Color	Linetype	Lineweight	Plot Style	P...
0				White	Continuous	—— Default	Color_7	
appliances				210	Continuous	—— Default	Color_210	
cabinets				161	HIDDEN	—— Default	Color_161	
wall-cabinet				170	Continuous	—— Default	Color_170	
walls				10	Continuous	—— Default	Color_10	

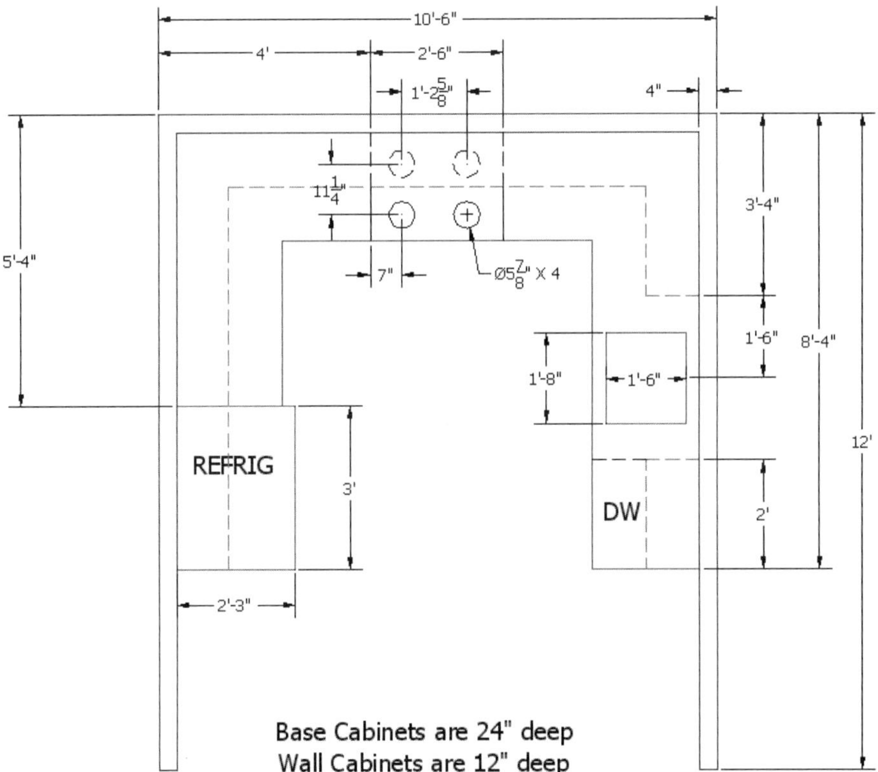

If additional drawing difficulty is desired, 12" x 12" floor tile can be placed on a separate layer in the drawing.

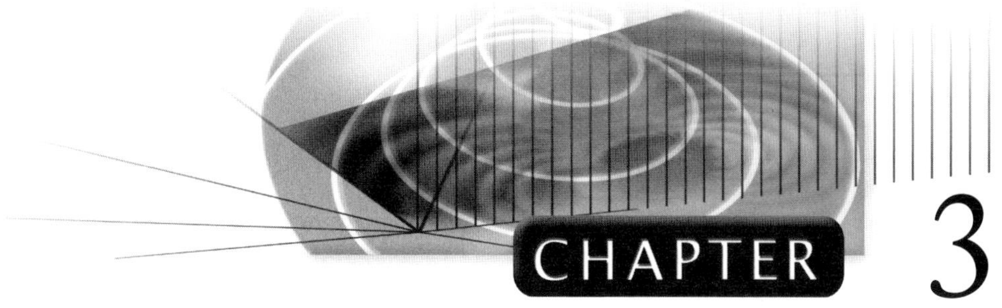

AutoCAD Display and Basic Selection Operations

The exercises in this chapter introduce:

- Zoom
- Pan
- Aerial view
- Mouse Zooming and Panning
- Creating and using Views
- Object Selection Sets

TUTORIAL EXERCISE: 03_HOUSE PLAN.DWG

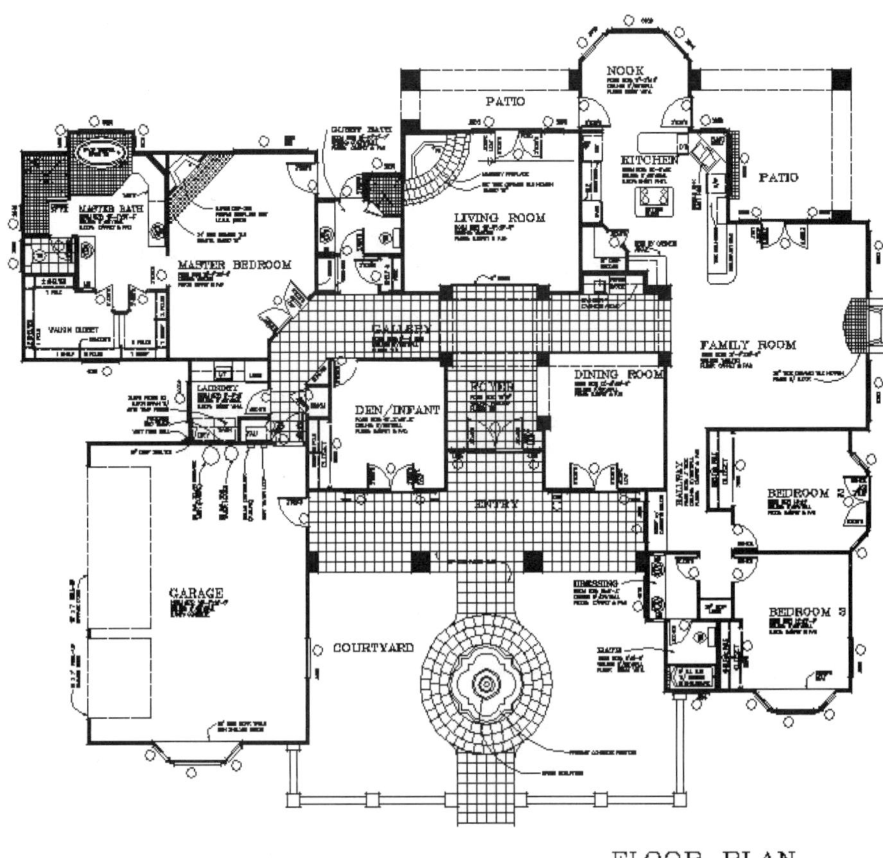

Figure 3–1

Purpose

The purpose of this tutorial is to create a number of named views based on the drawing of the house plan shown in Figure 3–1.

System Settings

Keep all default setting for this tutorial exercise.

Layers

Layers have already been created for this tutorial exercise.

Suggested Commands

This tutorial utilizes the View dialog box as a means of saving pre-defined zoomed portions of a drawing under a name that you define. These pre-defined zooms can then be restored, enabling you to maneuver around your drawing in a quick and efficient means.

Whenever possible, substitute the appropriate command alias in place of the full AutoCAD command in each tutorial step. For example, use "CP" for the COPY command, "L" for the LINE command, and so on. The complete listing of all command aliases is located in Table 1–2.

STEP 1

Open the drawing *03_House Plan.dwg* shown in Figure 3–1 and observe the many detailed rooms of the house plan. Instead of using the ZOOM command to magnify certain portions of the plan and then zoom back to the full plan, the View dialog box will be used to save pre-defined zoomed positions under a name that you create. Activate the View dialog box by clicking on the Named Views button located in the View toolbar, as shown in Figure 3–2. The View dialog box as shown in Figure 3–3 will apear. You could also activate this dialog box from the following pull-down menu sequence: View>Named Views.... A final method of activating the View dialog box is to type **V** from the keyboard, which is the short cut key to the VIEW command.

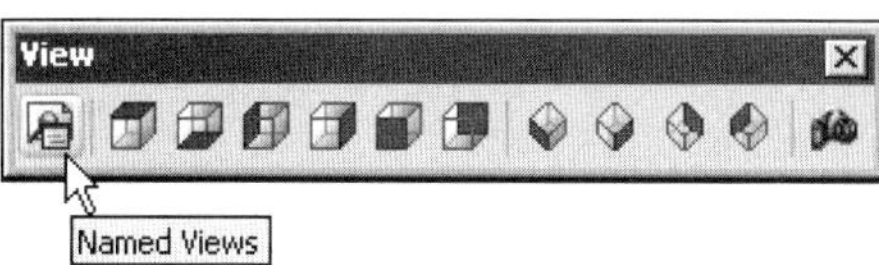

Figure 3–2

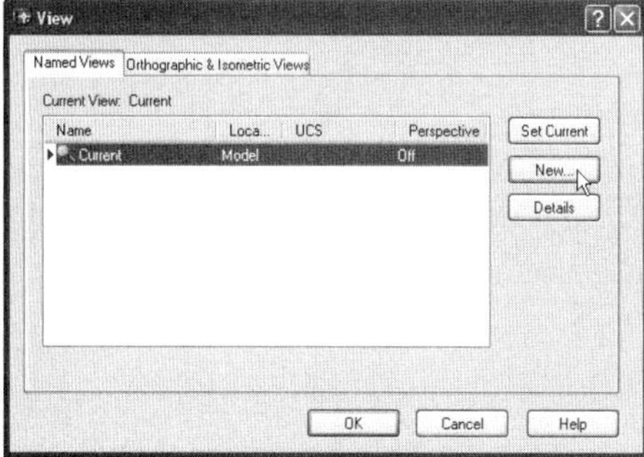

Figure 3–3

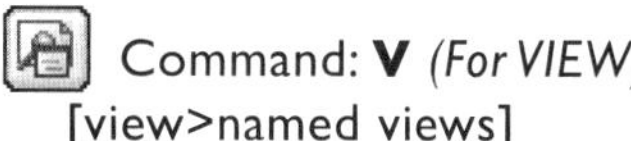

Command: **V** *(For VIEW)*
[view>named views]

You will first be creating a named view of the current drawing screen; in other words the house plan in its zoomed extents state. In the View dialog box, click on the New... button shown in Figure 3–3. This will activate the New View dialog box shown in Figure 3–4. Type the name **TOTAL VIEW** in the View name: edit box. (It is not significant to enter view names in all upper case letters.) Notice that the Current display radio button is selected. This means the named view will be based on the current screen display. Click the OK button to create the view. This will add the named view TOTAL PLAN to the current list of named views, as shown in Figure 3–5.

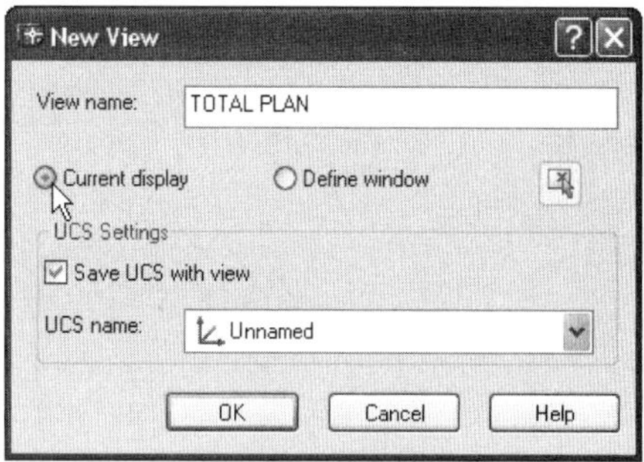

Figure 3–4

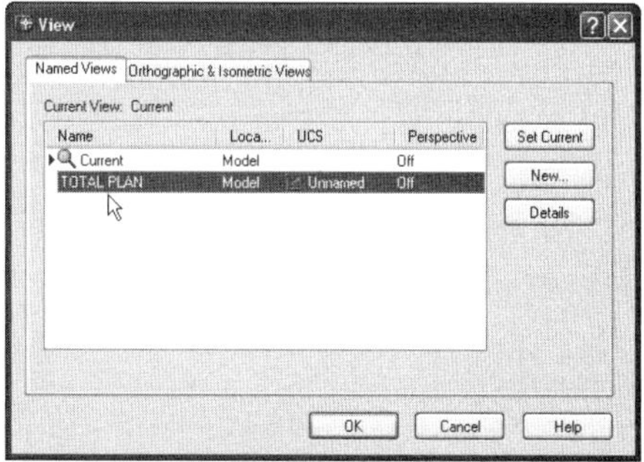

Figure 3–5

STEP 2

While still in the View dialog box, click on the New… button to display the New View dialog box. Type the name **BEDROOM 3** in the edit box as shown in Figure 3–6. You do not, however want to create this view based on the current display. In the New View dialog box, click on the Define window radio button to activate it. Then click on the Define View Window button, shown in Figure 3–7.

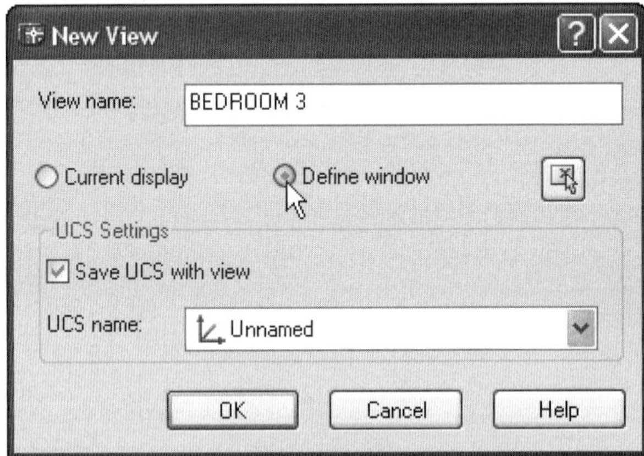

Figure 3–6

Figure 3–7

This will return you back to the AutoCAD editor and the house plan drawing. Construct a rectangular window from "A" to "B", as shown in Figure 3–8. This is the area you will be defining as part of the named view. When you return back to the New View dialog box, click the OK button. The named view BEDROOM 3 is added to the list of named views, as shown in Figure 3–9.

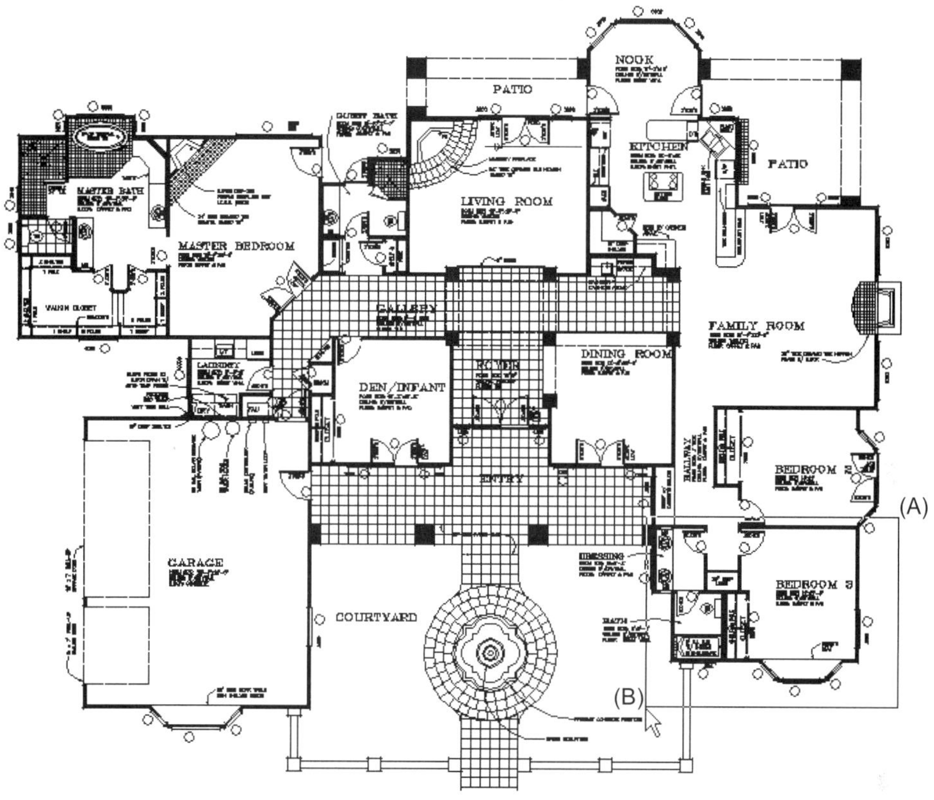

FLOOR PLAN
SCALE: 1/4" = 1'–0"

Figure 3–8

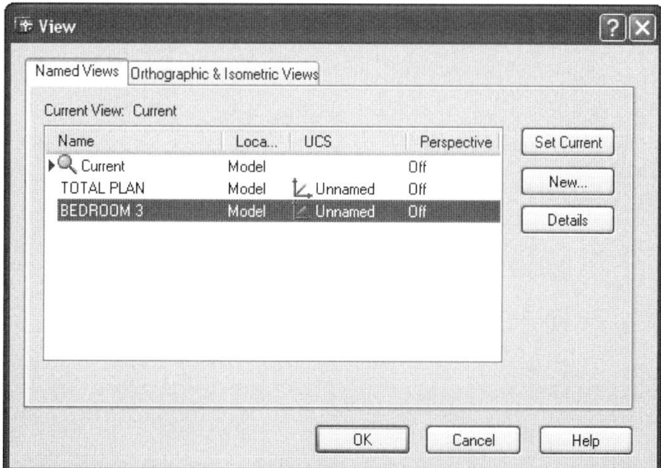

Figure 3–9

STEP 3

Click the OK button to dismiss the View dialog box. We will create one more view using a method different from the previous steps. Zoom into the NOOK area of the drawing. It is easily identified by the bay window at the top of the house plan. Your display should appear similar to Figure 3–10. Activate the View dialog box and click on the New… button to display the New View dialog box. Since you already zoomed into a portion of your drawing, you can again use the Current display button. Name this view **BREAKFAST NOOK** (see Figure 3–11). Click the OK button to add this new view to the list of existing named views.

The View dialog box should appear similar to Figure 3–12.

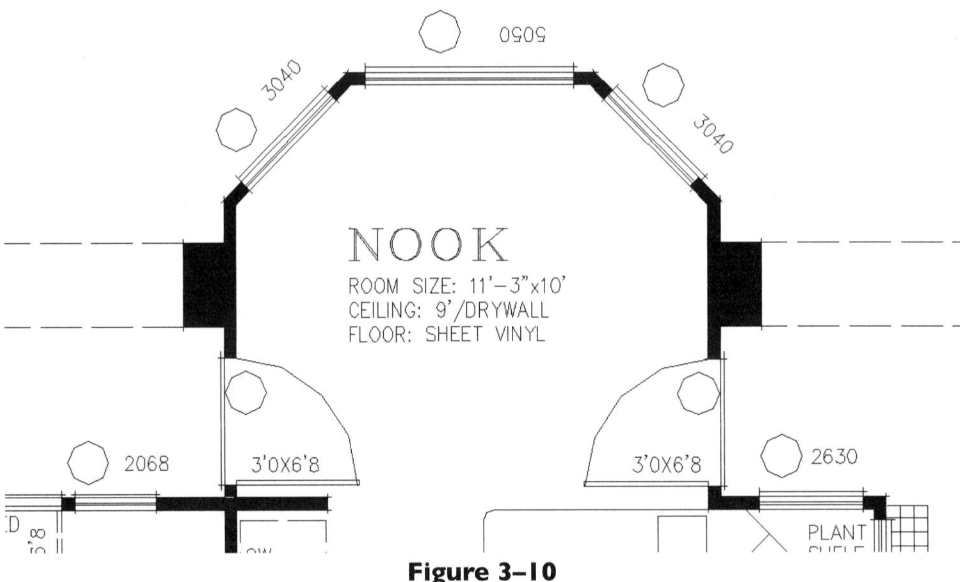

Figure 3–10

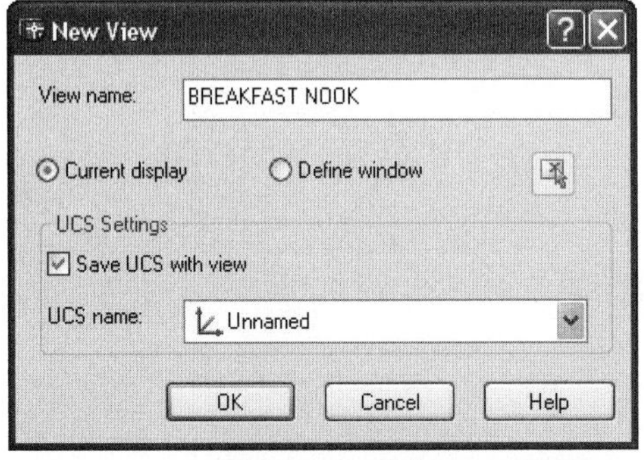

Figure 3–11

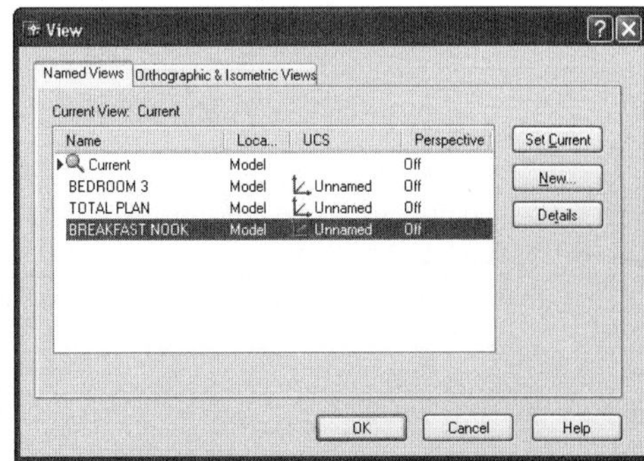

Figure 3–12

STEP 4

Next, test the named views out on your drawing. Activate the View dialog box, pick BEDROOM 3 from the list of named views, click on the Set Current button, and then click on the OK button as shown in Figure 3–13. Your display of BEDROOM 3 should appear similar to Figure 3–14.

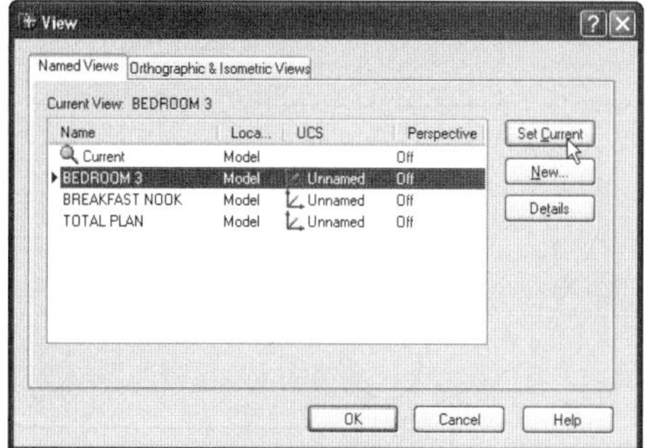

Figure 3–13

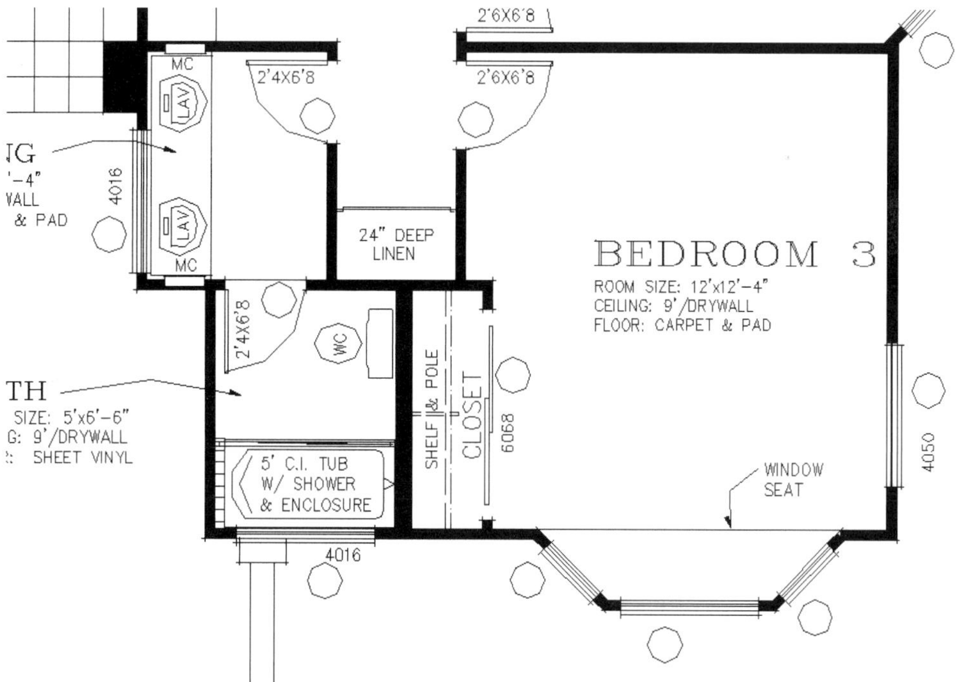

Figure 3–14

Note: An alternate method of making a view current is to double-click on its name and then click the OK button of the View dialog box to display the view.

STEP 5

To use a named view to return to the whole drawing, activate the View dialog box, double-click TOTAL PLAN from the list of named views, and click on the OK button, as shown in Figure 3–15. The whole house plan drawing should display.

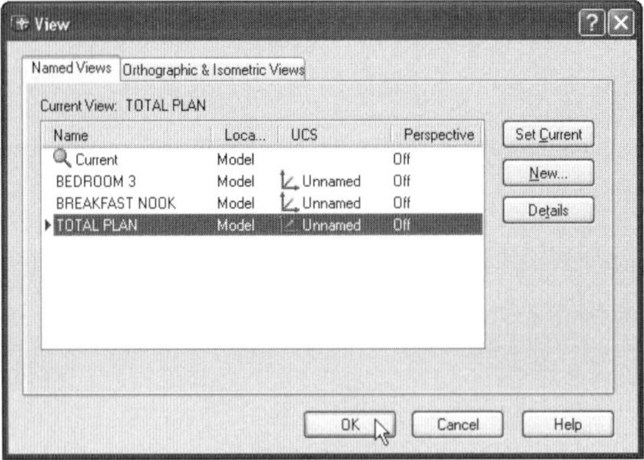

Figure 3–15

There is yet another feature of manipulating named views that is often overlooked, by even the most seasoned user. To illustrate this feature, first save this drawing under its original name of *03_HOUSE PLAN.DWG*. Then exit this drawing back to the AutoCAD neutral gray screen. Issue the OPEN command and select the 03_HOUSE PLAN drawing. Notice the check box labeled Select Initial View; place a check in this box (see Figure 3–16) and then click the Open button. Before the drawing displays, the Select Initial Views dialog box appears, as shown in Figure 3–17. This allows you to open the drawing and perform a zoom operation based on the named view you select. Select BREAKFAST NOOK from the list of named views and click the OK button.

Your drawing should appear similar to Figure 3–18.

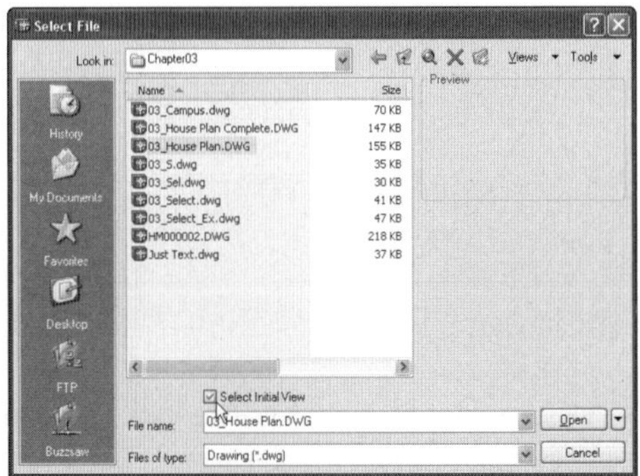

Figure 3–16

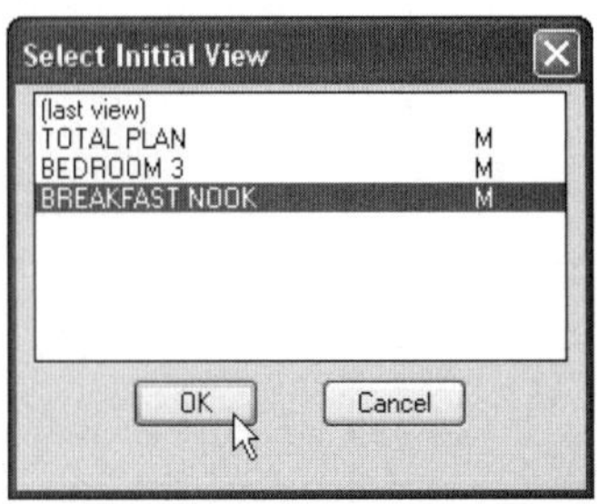

Figure 3–17

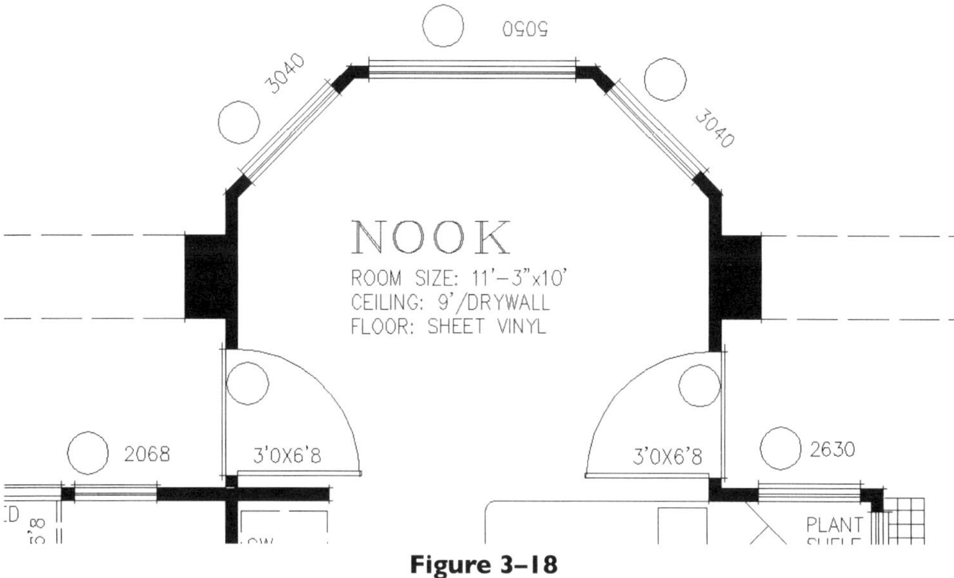

Figure 3–18

STEP 6

If you need to delete a named view, activate the View dialog box, select the named view, and right-click to display the menu shown in Figure 3–19. Click on the Delete button to remove the view from the list.

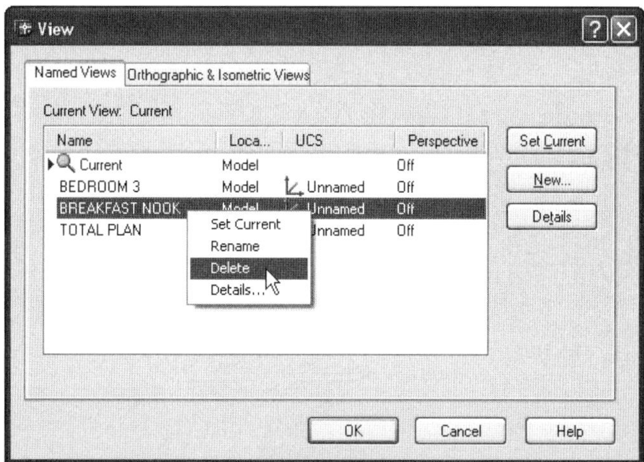

Figure 3–19

PROBLEM 3–1

Use the object selection tools to select the parts of the gear and place them on the correct layer. Open drawing *03_Gear.dwg*.

The necessary layers will need to be created so the objects can be changed into the proper layers.

Use the Change properties tool with the Select Objects button, to modify the gear pieces to match the following drawing and gear terminology. Remember, after selecting objects, a right mouse click or keyboard ENTER is necessary to move on to the next step.

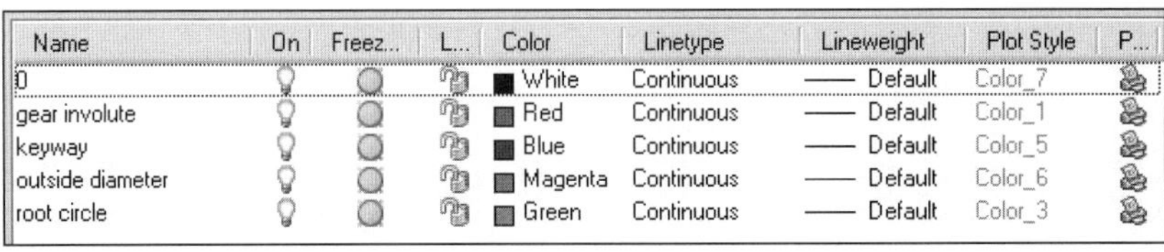

Name	On	Freez...	L...	Color	Linetype	Lineweight	Plot Style	P...
0				White	Continuous	—— Default	Color_7	
gear involute				Red	Continuous	—— Default	Color_1	
keyway				Blue	Continuous	—— Default	Color_5	
outside diameter				Magenta	Continuous	—— Default	Color_6	
root circle				Green	Continuous	—— Default	Color_3	

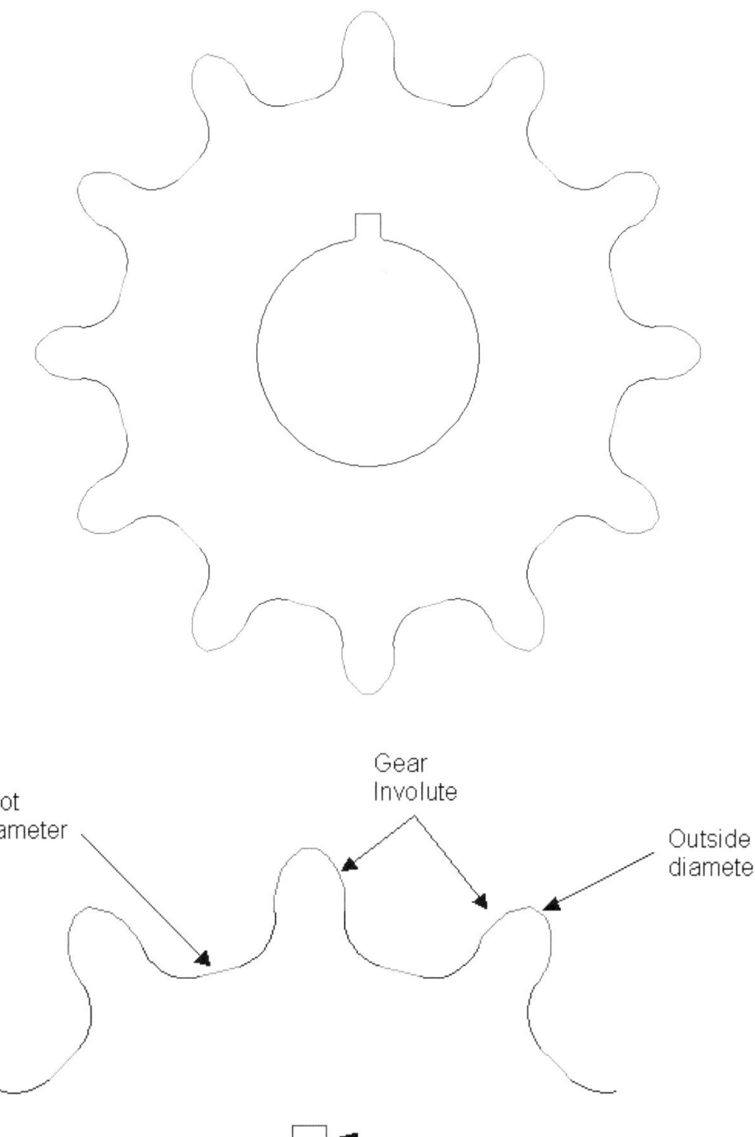
Gear
Involute
Root
Diameter
Outside
diameter
Key Way

Modify Commands

The exercises in this chapter introduce:

- Modification tools like copy, array, mirror and rotate
- Chamfer and fillet tools
- Offset objects
- Polyline editing
- The STRETCH command
- Trim and Extend tools

TUTORIAL EXERCISE: TILE.DWG

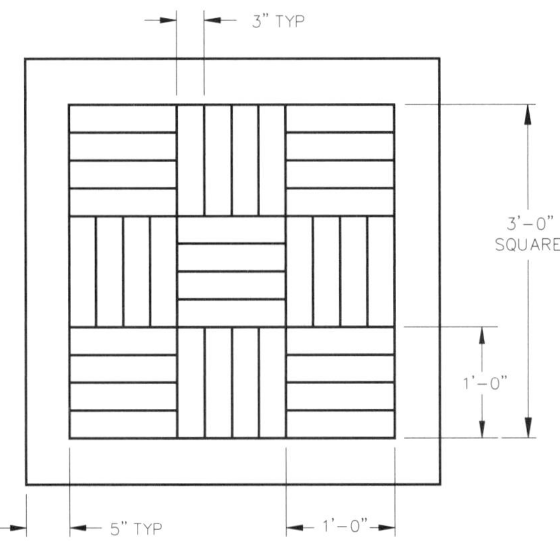

Figure 4–1

Purpose

This tutorial is designed to use the OFFSET and TRIM commands to complete the drawing of the floor tile shown in Figure 4–1.

System Settings

Use the Drawing Units dialog box and change the units of measure from decimal to architectural units. Keep the remaining default settings. Use the LIMITS command and change the limits of the drawing to (0,0) for the lower-left corner and (17',11') for the upper-right corner. Use the ZOOM command and the All option to fit the new drawing limits to the display screen. Use the SNAP command and change the value from 1/2" to 3". (If the GRID command is set to 0, the snap setting of 3" will also change the grid spacing to 3").

Check to see that the following Object Snap modes are currently set: Endpoint, Extension, Intersection, and Center.

Layers

Create the following layer with the format:

Name	Color	Linetype
Object	White	Continuous

Suggested Commands

Make the "Object" layer current. Use the LINE command to begin the inside square of the Tile. The ARRAY command is used to copy selected line segments in a rectangular pattern at a specified distance. The TRIM command is then used to form the inside tile patterns. The OFFSET command is used to copy the parallel lines that form the outer square and the FILLET command is used to form the corners of the square.

Whenever possible, substitute the appropriate command alias in place of the full AutoCAD command in each tutorial step. For example, use "CP" for the COPY command, "L" for the LINE command, and so on. The complete listing of all command aliases is located in Chapter 1, Table 1–2.

tutorial EXERCISE

STEP 1

Begin this exercise by using the LINE command and Polar Coordinate mode to draw a 3'0" square, as shown in Figure 4–2. The Direct Distance mode can also be used to construct the square.

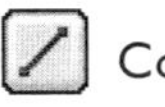 Command: **L** *(For LINE)* [draw>line]
Specify first point: **6',3'**
Specify next point or [Undo]: *(Move your cursor to the right and enter* **3'***)*
Specify next point or [Undo]: *(Move your cursor up and enter* **3'***)*
Specify next point or [Close/Undo]: *(Move your cursor to the left and enter* **3'***)*
Specify next point or [Close/Undo]: **C** *(To Close)*

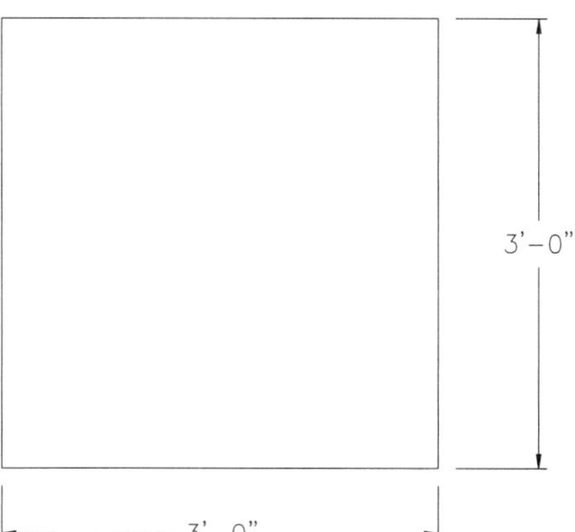

Figure 4–2

STEP 2

Use the Array dialog box to copy the bottom line in a rectangular pattern and have all lines spaced 3 units from each other. Select the bottom line as the object to array and perform a rectangular array consisting of twelve rows and one column (see Figure 4–3). Because the bottom line selected will be copied straight up, a positive distance can be entered to perform this operation. Once all values have been entered into the dialog box fields, click the Preview button observe the results. Another popular command that could be used here is OFFSET. However, because each line must be offset separately, the Array dialog box is the more efficient command to use.

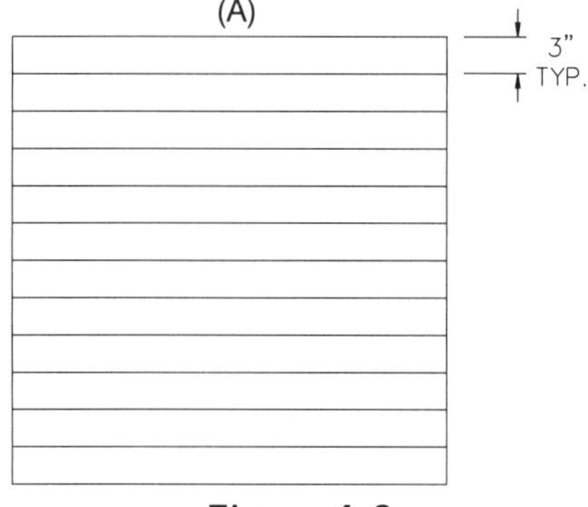

Figure 4–3

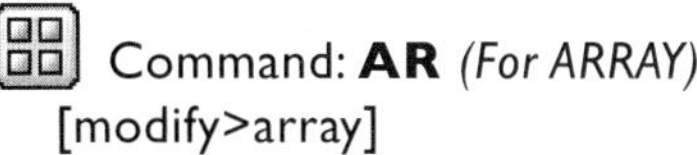 Command: **AR** *(For ARRAY)* [modify>array]

STEP 3

Again use the Array dialog box, but this time copy the left vertical line in a rectangular pattern and have all lines spaced 3 units away from each other. Turn off Running OSNAP before continuing. Select the left line as the object to array and perform a rectangular array consisting of one row and twelve columns (see Figure 4–4).

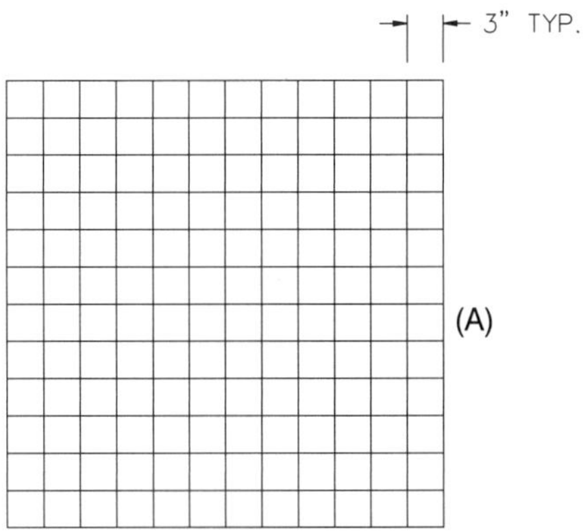

Figure 4–4

Because the left line selected will be copied to the right, you can enter a positive distance to perform this operation.

Command: **AR** *(For ARRAY)*
[modify>array]

STEP 4

Magnify the image by using the ZOOM command and a rectangular window. Use the TRIM command and select the two vertical dashed lines shown in Figure 4–5 as cutting edges. Use Figure 4–6 as a guide for determining which lines to trim out.

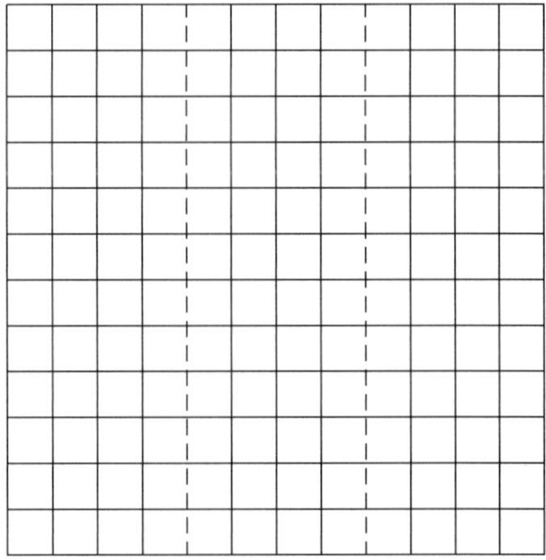
Figure 4–5

Command: **TR** *(For TRIM)*
[modify>trim]

Current settings: Projection=UCS Edge=None Select cutting edges ...
Select objects: *(Select the two dashed lines in Figure 4–5)*
Select objects: *(Press* ENTER *to continue)*
Select object to extend or shift-select to trim or [Project/Edge/Undo]:

(Continue to Step 5)

STEP 5

For the last prompt of the TRIM command in Step 4, select all horizontal lines in the areas marked "A," "B," "C," and "D" in Figure 4–6. When finished selecting the objects to trim, press ENTER to exit the commands selection mode.

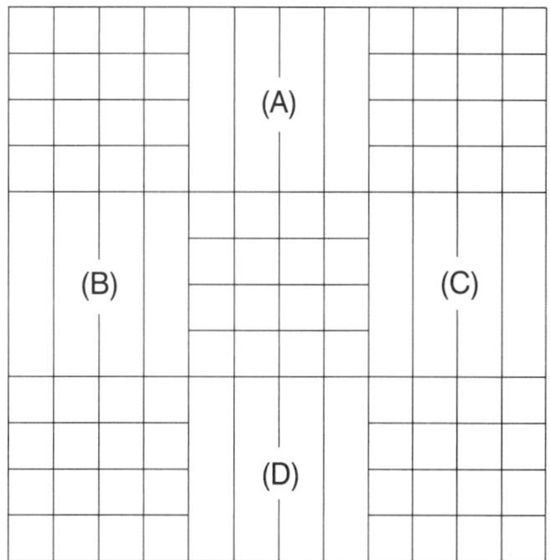

Figure 4–6

Select object to extend or shift-select to trim or [Project/Edge/Undo]: *(Select all horizontal lines in area "A")*
Select object to extend or shift-select to trim or [Project/Edge/Undo]: *(Select all horizontal lines in area "B")*
Select object to extend or shift-select to trim or [Project/Edge/Undo]: *(Select all horizontal lines in area "C")*
Select object to extend or shift-select to trim or [Project/Edge/Undo]: *(Select all horizontal lines in area "D")*
Select object to extend or shift-select to trim or [Project/Edge/Undo]: *(Press ENTER to exit this command)*

STEP 6

Use the TRIM command and select the two horizontal dashed lines as cutting edges in Figure 4–7.

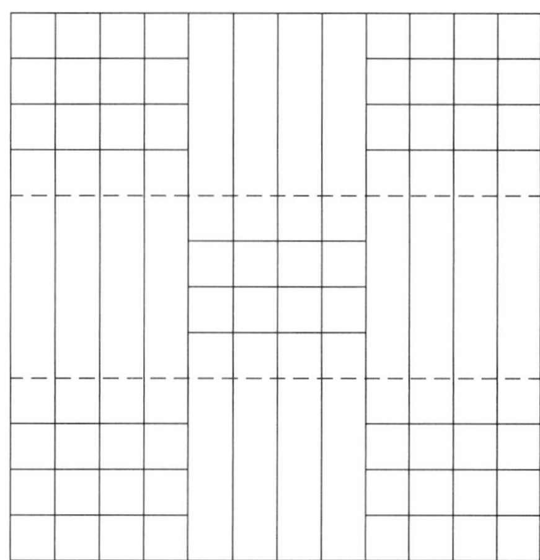

Figure 4–7

Command: **TR** *(For TRIM)* [modify>trim]
Current settings: Projection=UCS Edge=None Select cutting edges ...
Select objects: *(Select the two dashed lines in Figure 4–7)*
Select objects: *(Press ENTER to continue)*
Select object to extend or shift-select to trim or [Project/Edge/Undo]:

(Continue to Step 7)

STEP 7

For the last step of the TRIM command in Step 6, select all vertical lines in the areas marked "A," "B," "C," "D," and "E" in Figure 4–8. When finished selecting the objects to trim, press ENTER to exit the commands selection mode.

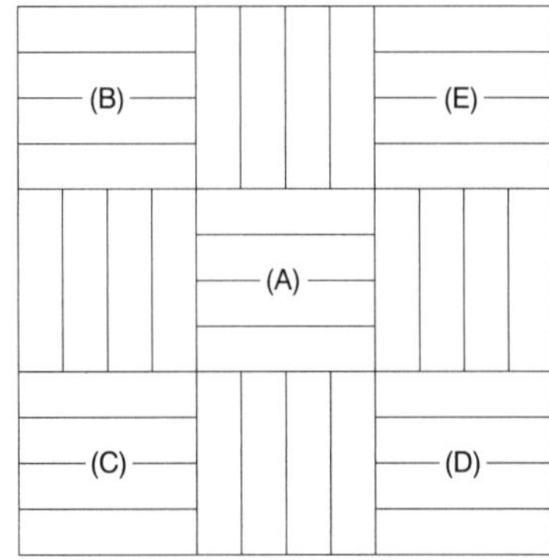

Figure 4–8

Select object to extend or shift-select to trim or [Project/Edge/Undo]: *(Select all vertical lines in area "A")*
Select object to extend or shift-select to trim or [Project/Edge/Undo]: *(Select all vertical lines in area "B")*
Select object to extend or shift-select to trim or [Project/Edge/Undo]: *(Select all vertical lines in area "C")*
Select object to extend or shift-select to trim or [Project/Edge/Undo]: *(Select all vertical lines in area "D")*
Select object to extend or shift-select to trim or [Project/Edge/Undo]: *(Select all vertical lines in area "E")*
Select object to extend or shift-select to trim or [Project/Edge/Undo]: *(Press* ENTER *to exit this command)*

STEP 8

Use the OFFSET command to offset the lines "A," "B," "C," and "D" five units in the directions shown in Figure 4–9.

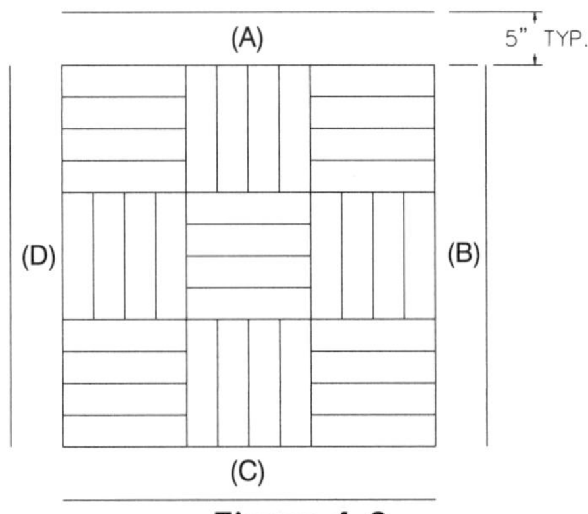

Figure 4–9

 Command: **O** *(For OFFSET)*
[modify>offset]
Specify offset distance or [Through] <0'-1">: **5**
Select object to offset or <exit>: *(Select the line at "A")*
Specify point on side to offset: *(Pick a point above the line)*
Select object to offset or <exit>: *(Select the line at "B")*
Specify point on side to offset: *(Pick a point to the right of the line)*
Select object to offset or <exit>: *(Select the line at "C")*
Specify point on side to offset: *(Pick a point below the line)*
Select object to offset or <exit>: *(Select the line at "D")*
Specify point on side to offset: *(Pick a point to the left of the line)*
Select object to offset or <exit>: *(Press* ENTER *to exit this command)*

STEP 9

Use the FILLET command, check that the radius is set to 0, and use the Multiple option to place corners at each intersection (see Figure 4–10).

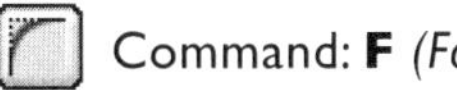

Command: **F** *(For FILLET)*

Current settings: Mode = TRIM, Radius = 0'-0"

Select first object or [Polyline/Radius/Trim/mUltiple]: **U** *(For Multiple)*

Select first object or [Polyline/Radius/Trim/mUltiple]: *(Select line "A")*

Select second object: *(Select line "B")*

Repeat this procedure for the other three corners.

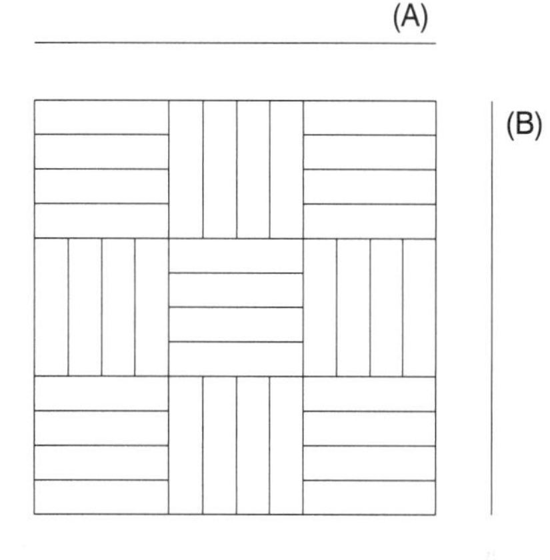

Figure 4–10

STEP 10

The completed tile drawing is shown in Figure 4–11. Use the ZOOM command and the Previous option to return to the previous display image. Follow on to Step 11 to add more tiles in a rectangular pattern using the ARRAY command.

Command: **Z** *(For ZOOM)* [view>zoom>previous]

Specify corner of window, enter a scale factor (nX or nXP), or [All/Center/Dynamic/Extents/Previous/Scale/Window] <real time>: **P** *(For Previous)*

Figure 4–11

STEP 11

As an alternate step, use the Array dialog box to copy the initial design in a rectangular pattern by row and column. Use Figure 4–12 as a guide for entering row, column and spacing distances.

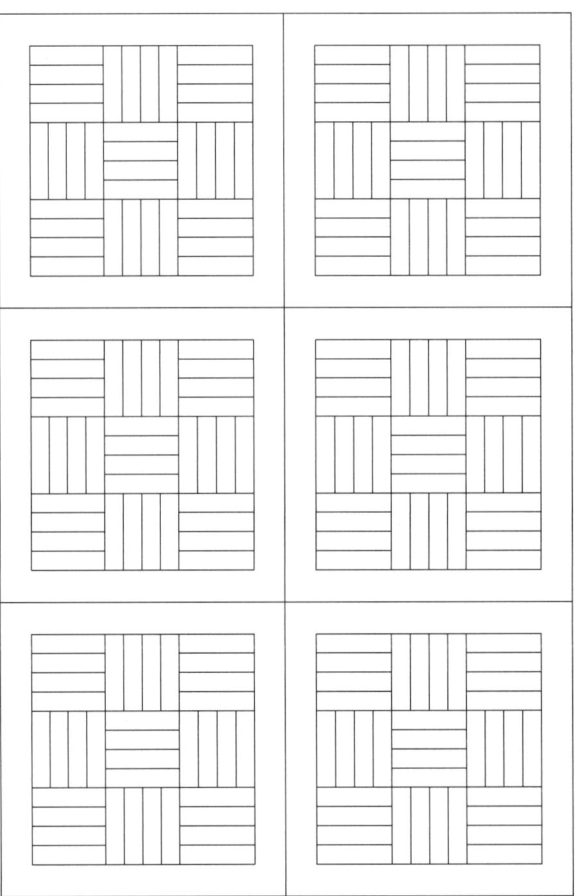

Figure 4–12

TUTORIAL EXERCISE: INLAY.DWG

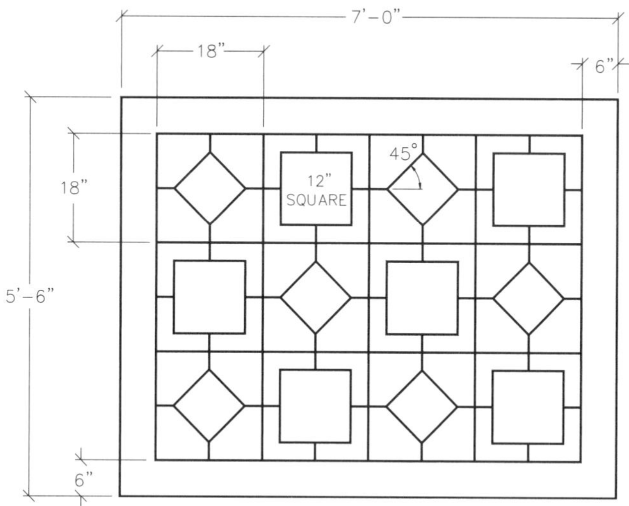

Figure 4–13

Purpose

This tutorial is designed to allow you to construct a drawing of the inlay shown in Figure 4–13 using the COPY command and the Multiple modifier.

System Settings

Use the Drawing Units dialog box and change the units of measure from decimal to architectural units. Keep the remaining default settings. Use the LIMITS command and change the limits of the drawing to (0,0) for the lower-left corner and (17',11') for the upper-right corner. Use the ZOOM command and the All option to fit the new drawing limits to the display screen. Use the SNAP command and change the value from 0' 0-1/2" to 3". (If the Grid is set to 0, the snap setting of 3" will also change the grid spacing to 3".) Check to see that the following Object Snap modes are currently set: Endpoint, Extension, Intersection, and Center.

Layers

Create the following layer with the format:

Name	Color	Linetypes
Object	White	Continuous

Suggested Commands

Make "Object" Layer current. Then begin this tutorial by drawing a 6'0" × 4'6" rectangle using the LINE command. Use the ARRAY command to copy designated lines in a rectangular pattern at a distance of 18". Then, using the 3" grid as a guide along with the Snap On, draw the diamond and square shapes. Use the COPY command along with the Multiple modifier to copy the diamond and square shapes numerous times to the designated areas.

Whenever possible, substitute the appropriate command alias in place of the full AutoCAD command in each tutorial step. For example, use "CP" for the COPY command, "L" for the LINE command, and so on. The complete listing of all command aliases is located in Chapter 1, Table 1–2.

STEP 1

Begin this exercise by using the LINE command and Polar coordinate mode to draw a rectangle 6'0" by 4'6", as shown in Figure 4–14. The Direct Distance mode may also be used to construct the rectangle.

Command: **L** *(For LINE)* [draw>line]
Specify first point: **3',2'**
Specify next point or [Undo]: **@6'<0**
Specify next point or [Undo]: **@4'6<90**
Specify next point or [Close/Undo]:
(Move your cursor over the intersection at "A" to acquire this point. Then move your cursor up until it locks with the horizontal snap position. Pick this point at "B").
Specify next point or [Close/Undo]: **C** *(To Close)*

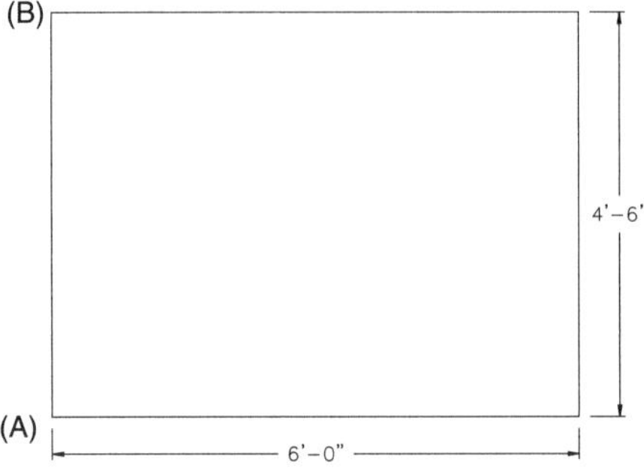

Figure 4–14

STEP 2

Use the ARRAY dialog box to copy the bottom line in a rectangular pattern and have all lines spaced 18 units away from each other. Select the bottom horizontal line as the object to array; the number of rows to create is 3. Since this line will be copied straight up, a value of 18 units for the spacing between rows will perform this operation. Repeat this command to copy the left vertical line three times to the right at a distance of 18 units. Enter a value of 18 units for the spacing in between columns since the copying is performed in the right direction as shown in Figure 4–15.

Command: **AR** *(For ARRAY)* [modify>array]

Repeat the preceding rectangular ARRAY command for the vertical line at "B." Use 1 for the number of rows, 4 for the number of columns, and -18 as the distance between columns.

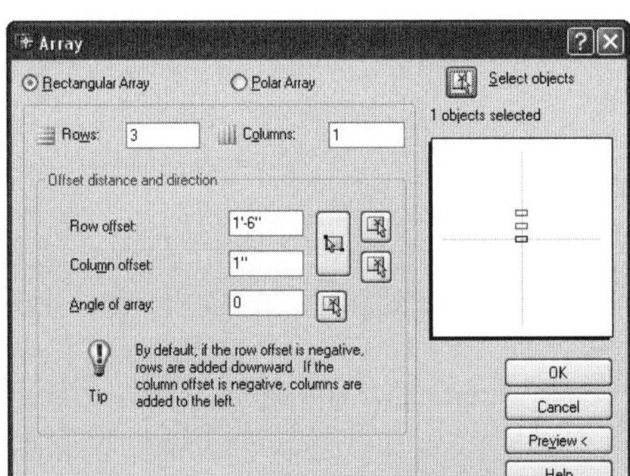

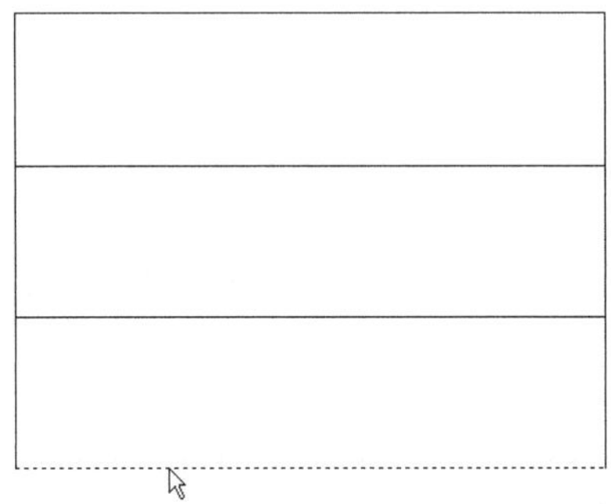

Figure 4–15

STEP 3

Draw one 12" × 12" diamond shape in the position shown in Figure 4–16. Use the ZOOM command with the Window option to magnify the area around the position of the diamond figure.

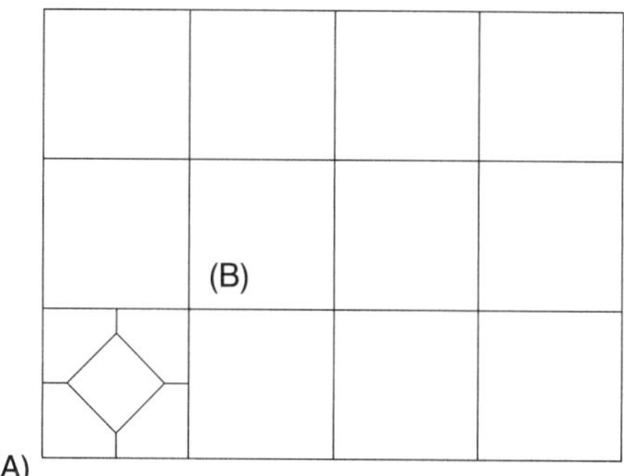

Figure 4–16

 Command: **Z** *(For ZOOM)* [view>zoom]

Specify corner of window, enter a scale factor (nX or nXP), or [All/Center/ Dynamic/Extents/Previous/ Scale/ Window] <real time>: *(Pick a point at "A")*
Specify opposite corner: *(Pick a point at "B")*

STEP 4

Turn OSNAP off by pressing F3. Be sure the Grid and Snap values are set to 3"; the Snap should already be turned on. Then use the LINE command and snap to the grid dots to draw the four lines shown in Figure 4–17. Use "A," "B," "C," and "D" as the starting points for the four lines.

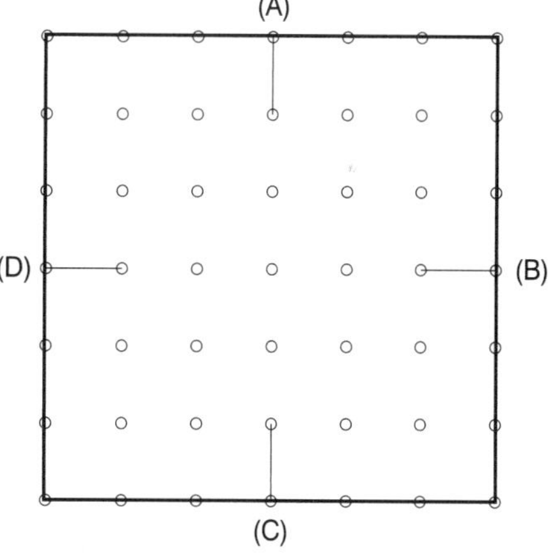

Figure 4–17

Command: *(Press F3 to turn OSNAP off)*

 Command: **L** *(For LINE)* [draw>line]

Specify first point: *(Pick a point at "A")*
Specify next point or [Undo]: *(Pick a point one grid space below "A". Or enter a polar coordinate of* **@3<270***)*
Specify next point or [Undo]: *(Press ENTER to exit this command)*

Use the same process for the other three lines, beginning at B, C, and D. The polar coordinate from "B" would be **@3<180**; from "C" **@3<90**; and from "D" **@3<0**.

STEP 5

Turn OSNAP on by pressing F3. Use the LINE command to draw the diamond shaped figure shown in Figure 4–18. Be sure Ortho mode is turned off before constructing the line segments. When finished, perform a ZOOM PREVIOUS operation; turn the snap off by pressing F9.

Command: *(Press F3 to turn OSNAP back on)*

Command: **L** *(For LINE)* [draw>line]

Specify first point: *(Pick the endpoint at "A")*

Specify next point or [Undo]: *(Pick the endpoint at "B")*

Specify next point or [Undo]: *(Pick the endpoint at "C")*

Specify next point or [Close/Undo]: *(Pick the endpoint at "D")*

Specify next point or [Close/Undo]: **C** *(To Close)*

Command: **Z** *(For ZOOM)*

Specify corner of window, enter a scale factor (nX or nXP), or [All/Center/Dynamic/Extents/Previous/Scale/Window] <real time>: **P** *(For Previous)*

Command: *(Press F9 to turn the SNAP off)*

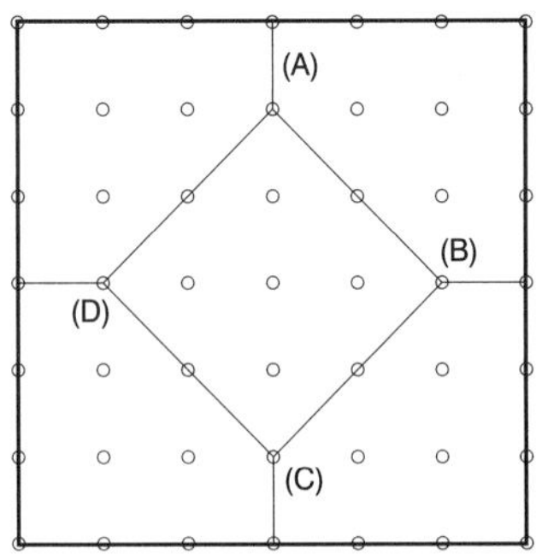

Figure 4–18

STEP 6

Use the COPY command and the Multiple option to repeat the diamond-shaped pattern shown in Figure 4–19. The OSNAP Intersection and Endpoint modes should already be enabled.

Command: **CP** *(For COPY)* [modify>copy]

Select objects: *(Select all dashed lines in Figure 4–19)*

Select objects: *(Press ENTER to continue)*

Specify base point or displacement, or [Multiple]: **M** *(For Multiple)*

Specify base point: *(Select the endpoint at "A")*

Specify second point of displacement or <use first point as displacement>: *(Select the endpoint at "B")*

Specify second point of displacement or <use first point as displacement>: *(Select the intersection at "C")*

Specify second point of displacement or <use first point as displacement>: *(Select the intersection at "D")*

Specify second point of displacement or <use first point as displacement>: *(Select the endpoint at "E")*

Specify second point of displacement or <use first point as displacement>: *(Select the intersection at "F")*

Specify second point of displacement or <use first point as displacement>: *(Press ENTER to exit this command)*

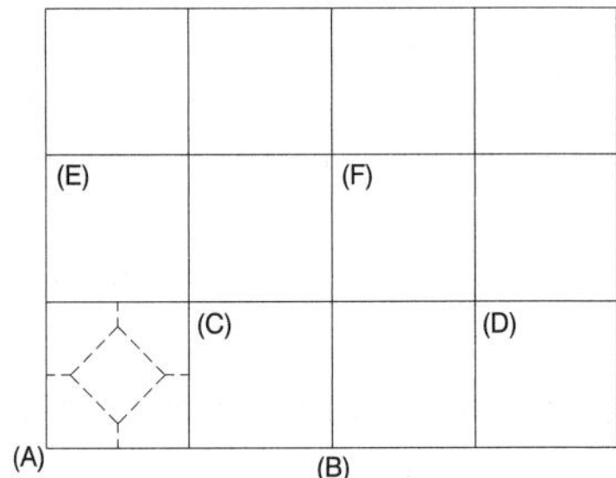

Figure 4–19

STEP 7

The diamond pattern of the Inlay floor tile should be similar to Figure 4–20.

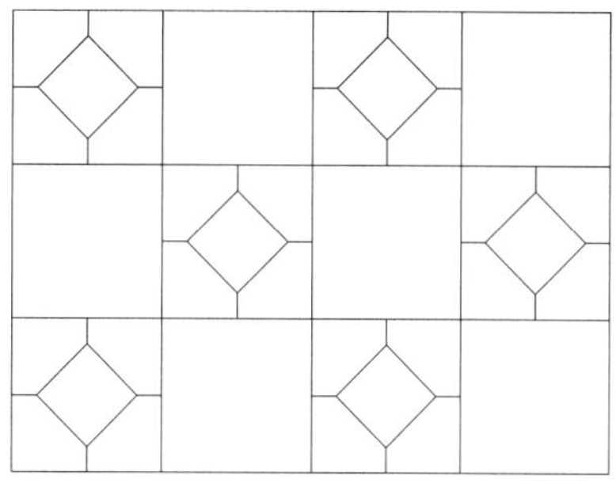

Figure 4–20

STEP 8

Draw one 12" × 12" square shape in the position shown in Figure 4–21. Use the ZOOM command along with the Window option to magnify the area around the position of the square figure.

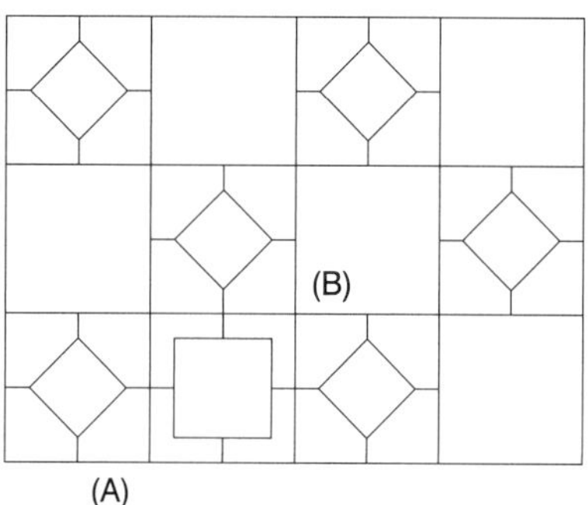

Figure 4–21

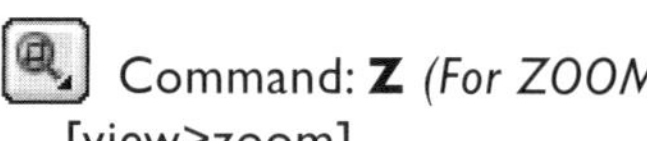

Command: **Z** *(For ZOOM)* [view>zoom]

Specify corner of window, enter a scale factor (nX or nXP), or [All/Center/Dynamic/ Extents/Previous/ Scale/Window] <real time>: *(Pick a point at "A")*
Specify opposite corner: *(Pick a point at "B")*

STEP 9

Be sure the Grid and Snap values are set to 3" and turn the snap back on by pressing F9. Also, turn off Running OSNAP mode by either clicking OSNAP in the status bar or pressing F3. Then use the LINE command to draw the four lines shown in Figure 4–22. Use "A," "B," "C," and "D" as the starting points for the four lines.

Command: *(Press F3 to turn OSNAP off)*
Command: *(Press F9 to turn SNAP on)*

Command: **L** *(For LINE)* [draw>line]
Specify first point: *(Pick a point at "A")*
Specify next point or [Undo]: *(Pick a point at "B")*
Specify next point or [Undo]: *(Press ENTER to exit this command)*

Repeat this procedure for the other three lines at "C," "D," and "E."

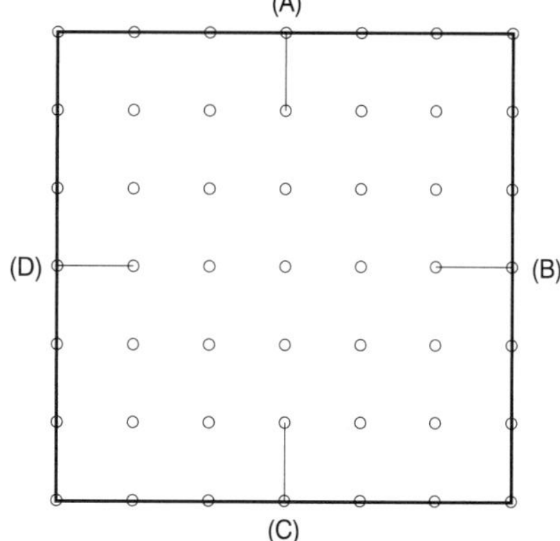

Figure 4–22

STEP 10

Use the RECTANG command to construct the square-shaped object in Figure 4–23. Be sure Object Snap mode is turned on.

Command: *(Press F3 to turn OSNAP back on)*

Command: **REC** *(For RECTANGLE)* [draw>rectangle]

Specify first corner point or [Chamfer/ Elevation/Fillet/Thickness/Width]: *(Pick a point at "A")*

Specify other corner point: *(Pick a point at "B")*

Command: *(Press F7 to turn the GRID off and F9 to turn the SNAP off)*

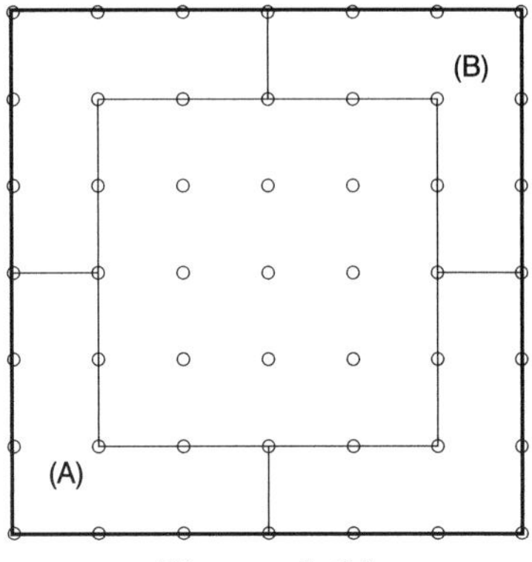

Figure 4–23

STEP 11

Return to the previous display using the ZOOM command and the Previous option. Use the COPY command and the Multiple option to repeat the square-shaped pattern in Figure 4–24.

Command: **Z** *(For ZOOM)* [view>zoom>previous]

Specify corner of window, enter a scale factor (nX or nXP), or [All/Center/ Dynamic/Extents/Previous/Scale/ Window] <real time>: **P** *(For Previous)*

Command: **CP** *(For COPY)* [modify>copy]

Select objects: Specify opposite corner: *(Select all dashed lines in Figure 4–24)*

Select objects: *(Press ENTER to exit the commands selection mode)*

Specify base point or displacement, or [Multiple]: **M** *(For Multiple)*

Specify base point: *(Select the endpoint at "A")*

Specify second point of displacement or <use first point as displacement>: *(Select the endpoint at "B")*

Specify second point of displacement or <use first point as displacement>: *(Select the endpoint at "C")*

Specify second point of displacement or <use first point as displacement>: *(Select the intersection at "D")*

Specify second point of displacement or <use first point as displacement>: *(Select the intersection at "E")*

Specify second point of displacement or <use first point as displacement>: *(Select the intersection at "F")*

Specify second point of displacement or <use first point as displacement>: *(Press ENTER to exit this command)*

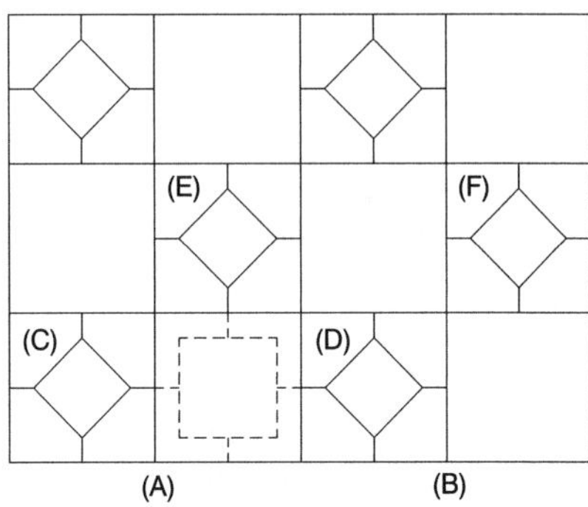

Figure 4–24

STEP 12

The drawing of the Inlay should appear similar to Figure 4–25.

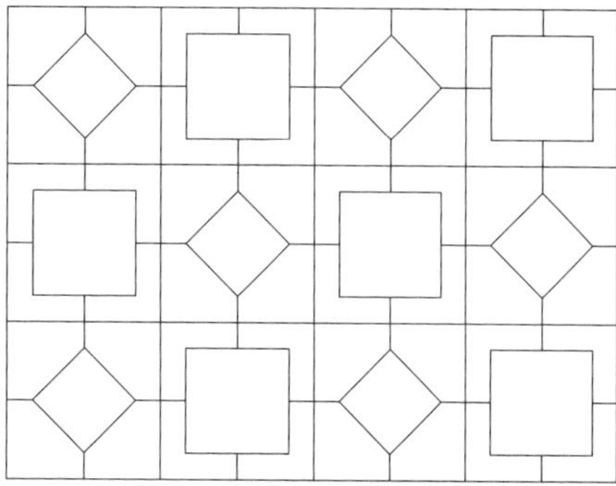

Figure 4–25

STEP 13

Use the OFFSET command to copy the outline of the Inlay outward at a distance of six units, as shown in Figure 4–26.

 Command: **O** *(For OFFSET)* [modify>offset]

Specify offset distance or [Through] <Through>: **6**

Select object to offset or <exit>: *(Select line "A")*

Specify point on side to offset: *(Pick a point above the line)*

Select object to offset or <exit>: *(Select line "B")*

Specify point on side to offset: *(Pick a point to the left of the line)*

Select object to offset or <exit>: *(Select line "C")*

Specify point on side to offset: *(Pick a point below the line)*

Select object to offset or <exit>: *(Select line "D")*

Specify point on side to offset: *(Pick a point to the right of the line)*

Select object to offset or <exit>: *(Press* ENTER *to exit this command)*

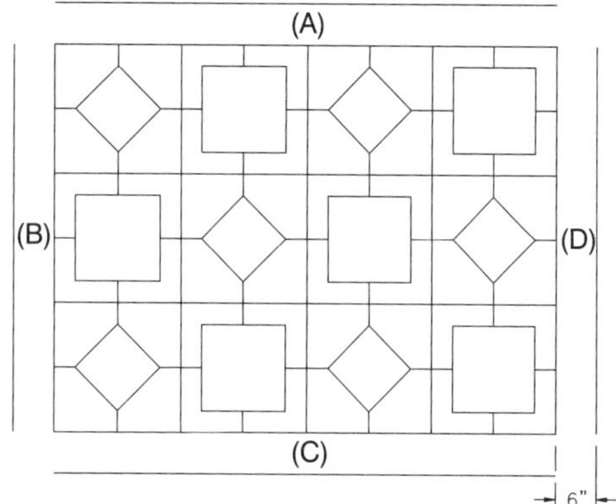

Figure 4–26

STEP 14

Use the FILLET command set to a radius of "0" to place a corner at the intersection of lines "A" and "B." See Figure 4–27.

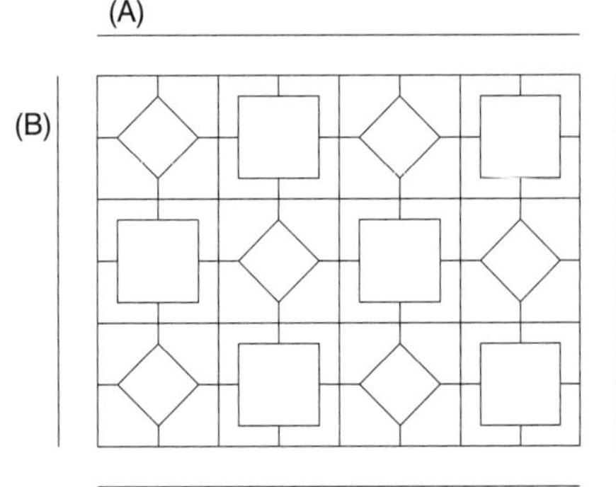

Figure 4–27

Command: **F** *(For FILLET)*
[modify>fillet]

Current settings: Mode = TRIM, Radius = 0'-0 1/2" Select first object or [Polyline/Radius/ Trim]: **R** *(For Radius)*
Specify fillet radius <0'-0 1/2">: **0**
Select first object or [Polyline/Radius/ Trim]: *(Select line "A" in Figure 4–27)*
Select second object: *(Select line "B")*

Repeat this procedure for the remaining three corners.

STEP 15

Perform a ZOOM ALL. The completed drawing is shown in Figure 4–28. Dimensions may be added upon the request of your instructor.

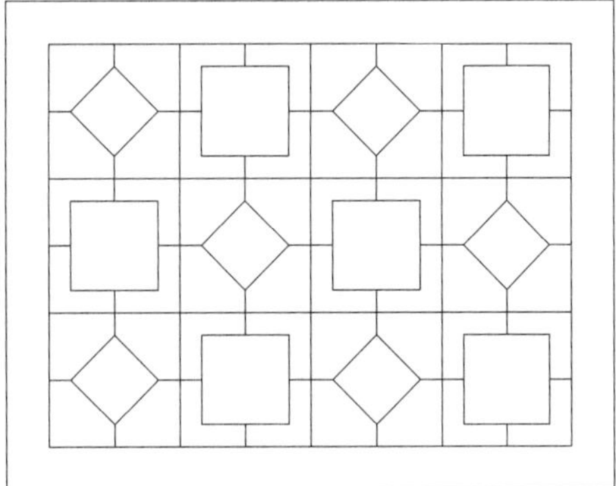
Figure 4–28

 Command: **Z** *(For ZOOM)*
[view>zoom>all]

Specify corner of window, enter a scale factor (nX or nXP), or [All/Center/ Dynamic/Extents/Previous/ Scale/ Window] <real time>: **A** *(For All)*

Beginning Level Drawings
Directions for Problems 4–1 through 4–15:

Construct each one-view drawing using the appropriate coordinate mode or Direct Distance mode. Utilize advanced commands such as ARRAY and MIRROR whenever possible.

PROBLEM 4–1

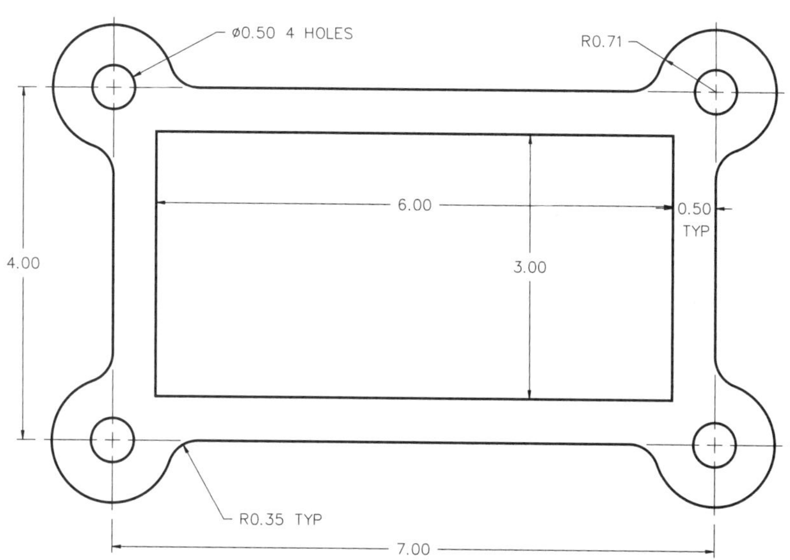

PROBLEM 4–2

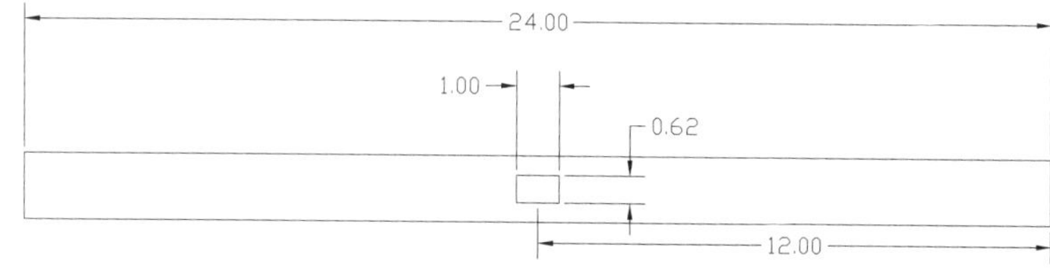

Ø2.50
Ø2.00
8.00
8.00
4.00
1.50
0.37
1.75
3.50
0.25

problem EXERCISES

PROBLEM 4–3

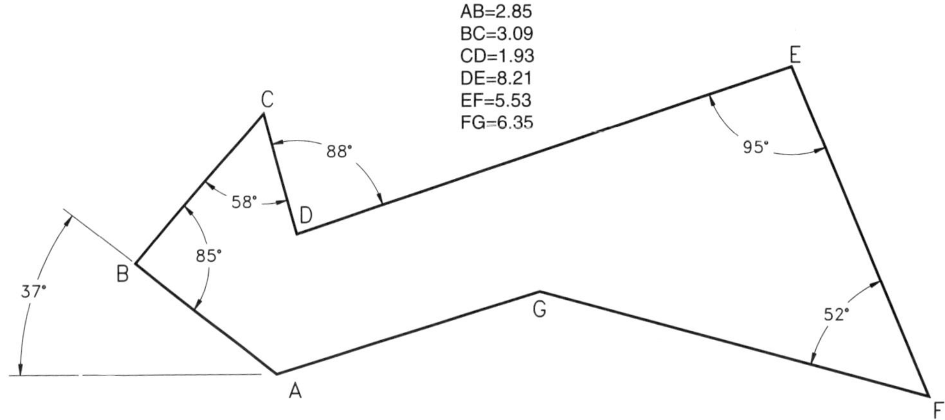

PROBLEM 4–4

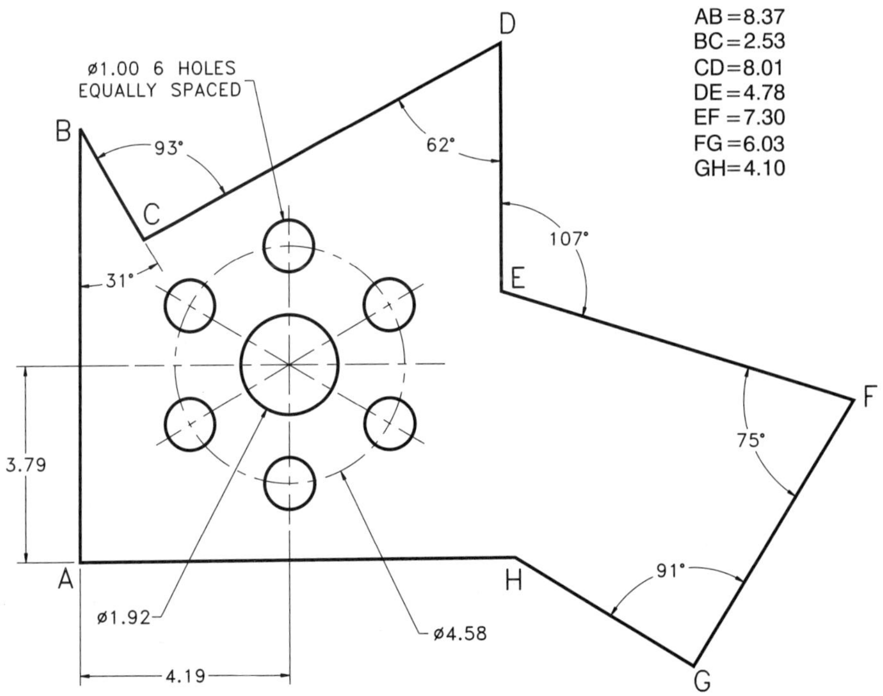

PROBLEM 4–5

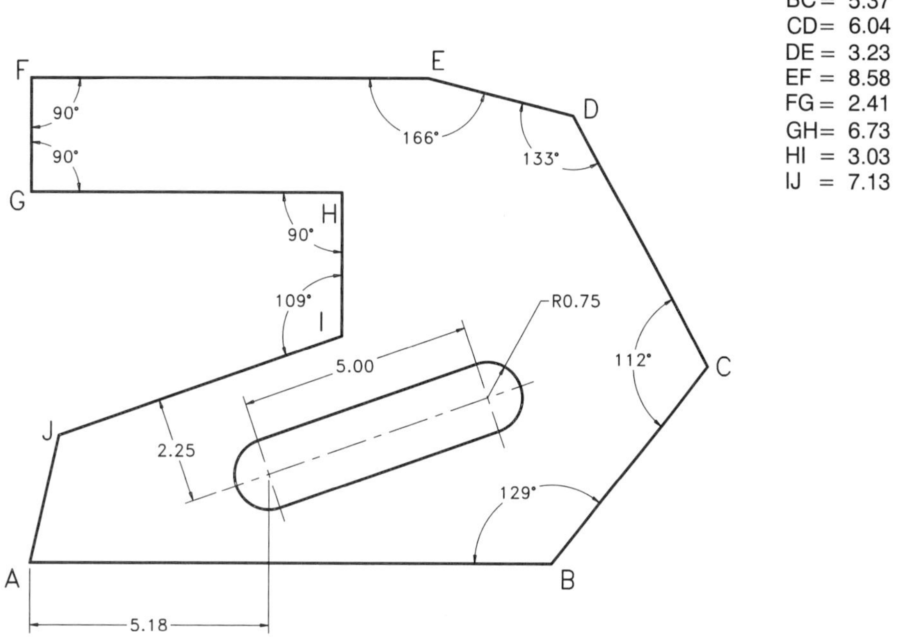

PROBLEM 4–6

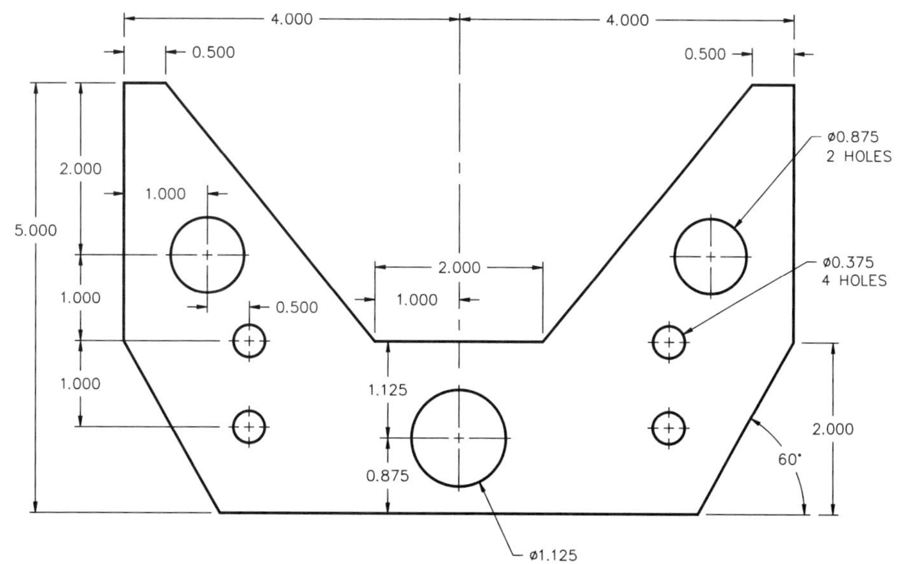

PROBLEM 4–7

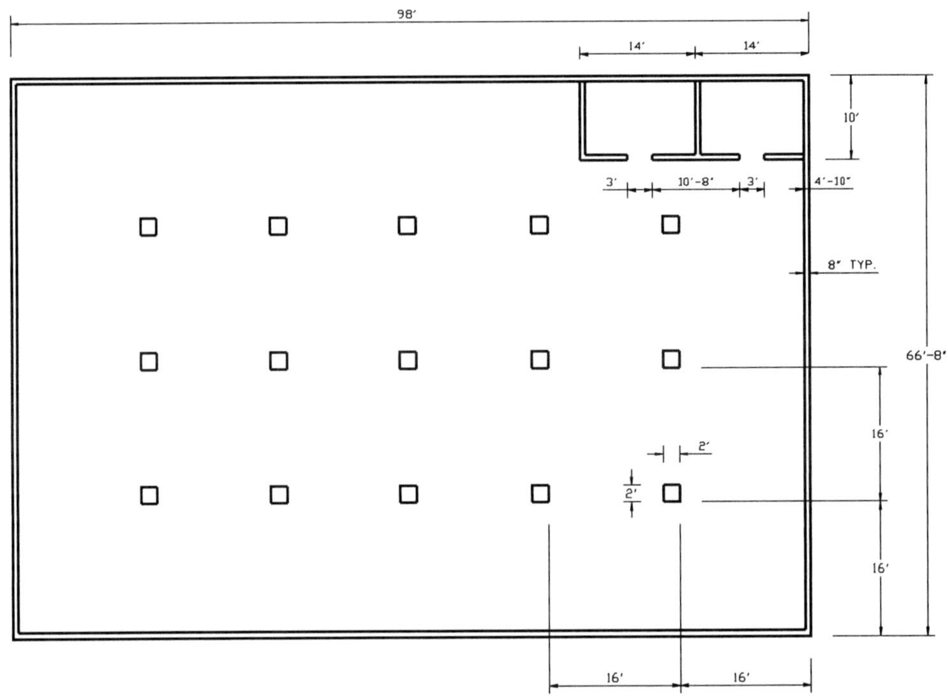

PROBLEM 4–8

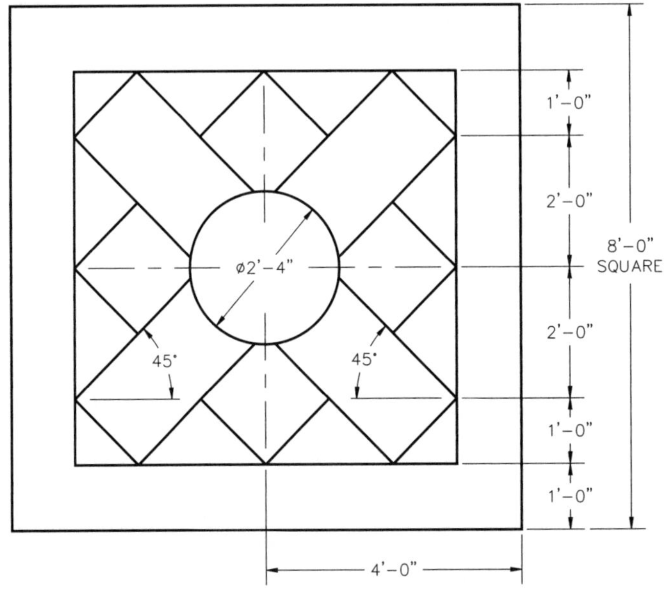

Intermediate Level Drawings

PROBLEM 4–9

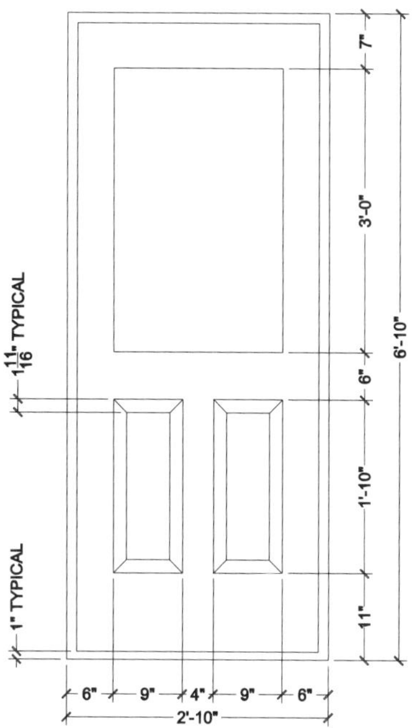

PROBLEM 4–10

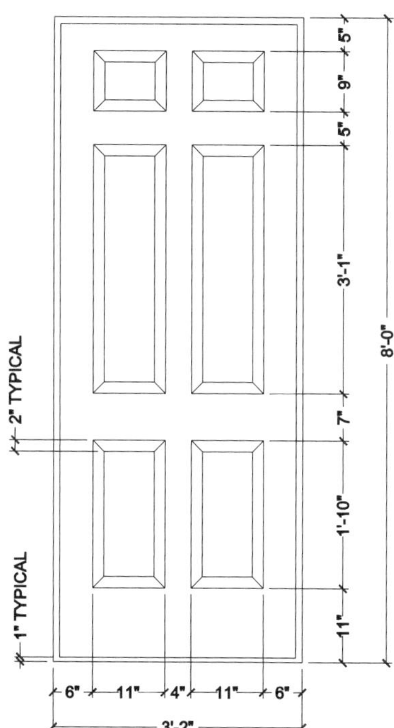

PROBLEM 4–11

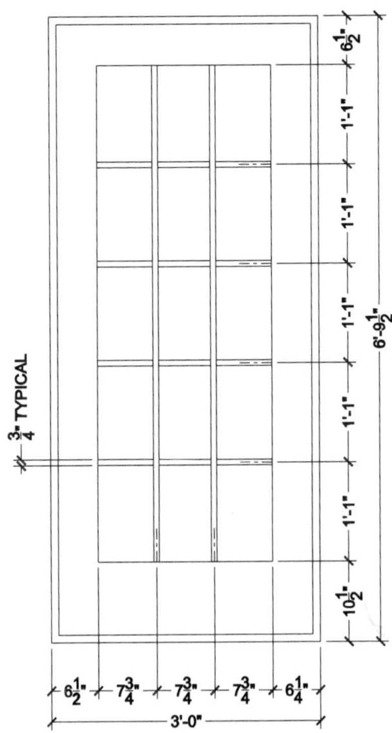

PROBLEM 4–12

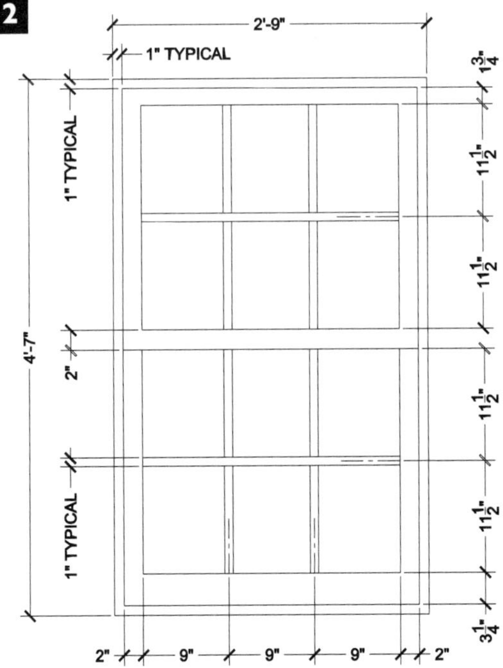

PROBLEM 4–13

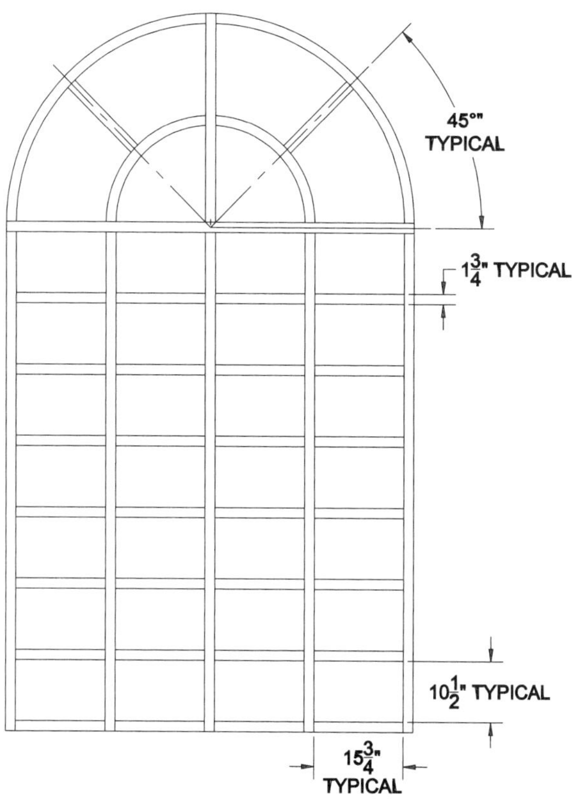

Advanced Level Drawings

PROBLEM 4–14

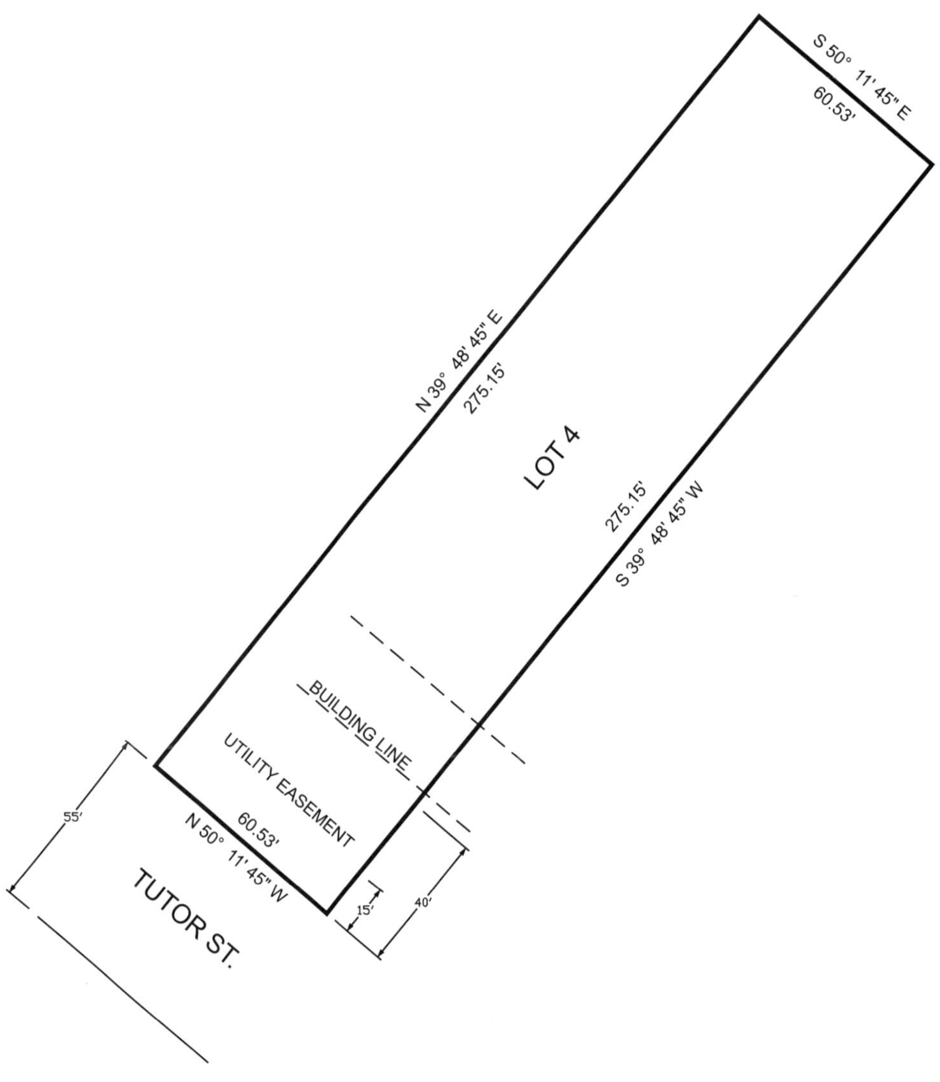

PROBLEM 4–15

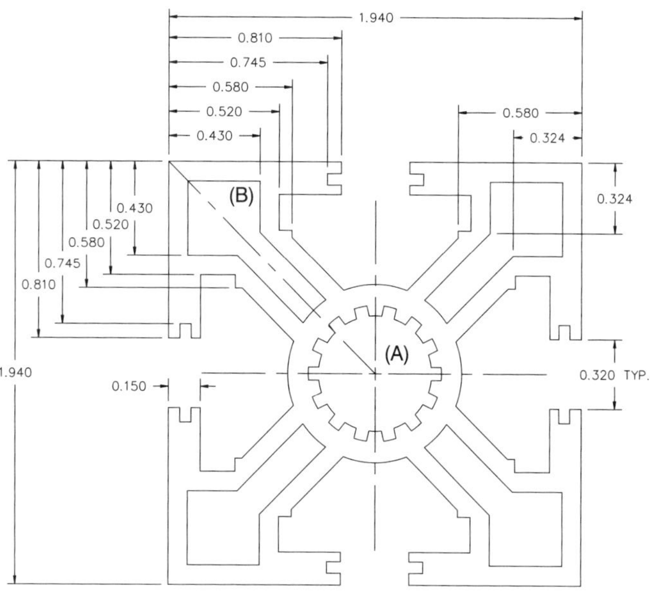
1.940
0.810
0.745
0.580
0.520
0.430
0.580
0.324
0.324
(B)
0.430
0.520
0.580
0.745
0.810
1.940
(A)
0.150
0.320 TYP.

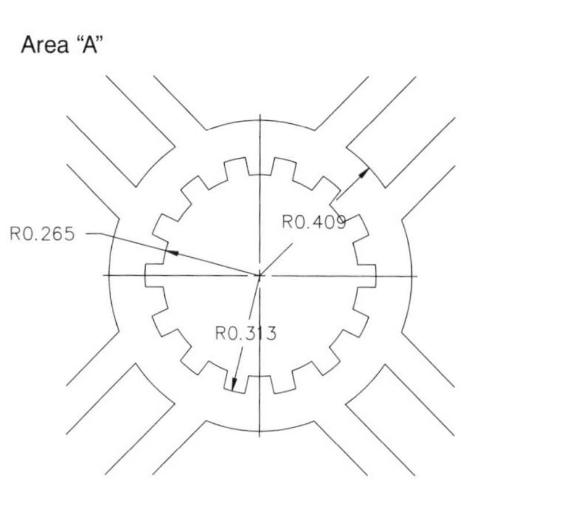
Area "A"
R0.265
R0.409
R0.313

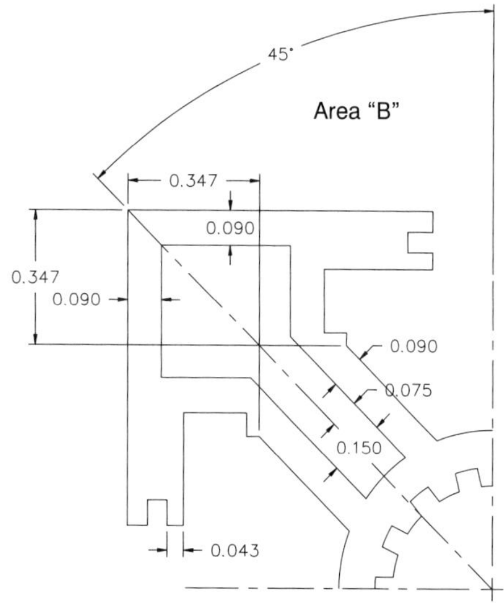
45°
Area "B"
0.347
0.090
0.347
0.090
0.090
0.075
0.150
0.043

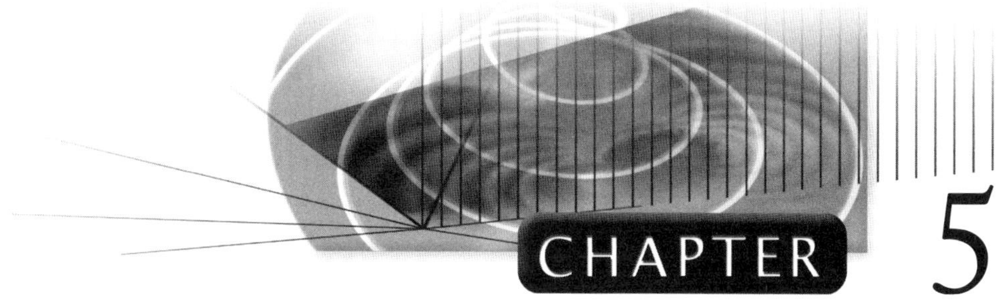

Performing Geometric Constructions

The exercises in this chapter introduce:

- Creating arcs
- Using circles to create tangent arcs
- Creating donuts and ellipses
- Drawing multilines
- Using Divide and Measure
- Multi-sided shape development
- Ogee curve creation

TUTORIAL EXERCISE: PATTERN1.DWG

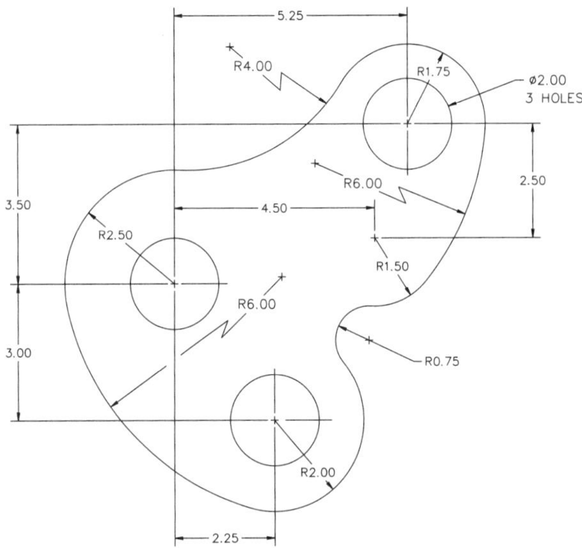

Figure 5–1

Purpose:

This tutorial is designed to use various Draw commands to construct a one-view drawing of Pattern1, as shown in Figure 5–1. Refer to the following special system settings and suggested command sequences.

System Settings:

Use the Drawing Units dialog box to change the number of decimal places past the zero from four to two. Keep the remaining default unit values. Using the LIMITS command, keep (0,0) for the lower-left corner and change the upper-right corner from (12,9) to (21.00,16.00). Perform a ZOOM-ALL after changing the drawing limits. Check to see that the following Object Snap modes are already set: Endpoint, Extension, Intersection, and Center.

Setting OSNAP Mode would also be helpful with completing this exercise.

Layers:

Create the following layers with the format:

Name	Color	Linetype
Object	White	Continuous
Center	Yellow	Center
Dimension	Yellow	Continuous

Suggested Commands:

Begin a new drawing called "Pattern1." Begin constructing this object by first laying out four points, which will be used as centers for circles. Use the CIRCLE/TTR command/option to construct tangent arcs to the circles already drawn. Use the TRIM command to clean up and partially delete circles to obtain the outline of the pattern. Then add the 2.00-diameter holes followed by the center markers, using the L command.

Whenever possible, substitute the appropriate command alias in place of the full AutoCAD command in each tutorial step. For example, use "CP" for the COPY command, "L" for the L command, and so on. The complete listing of all command aliases is located in Chapter 1, Table 1–2.

STEP 1

Check that the current layer is set to "Object." Use the Layer Control box in Figure 5–2 to accomplish this task.

Change the default point appearance from a "dot" to the "plus", using the Point Style dialog box (see Figure 5–3), which you can open by choosing Point Style from the Format pull-down menu.

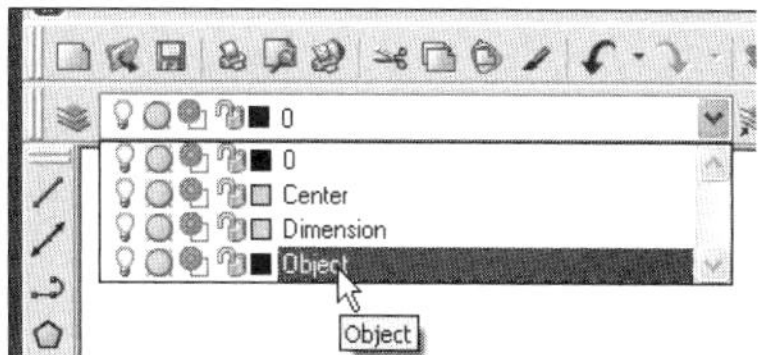

Figure 5–2

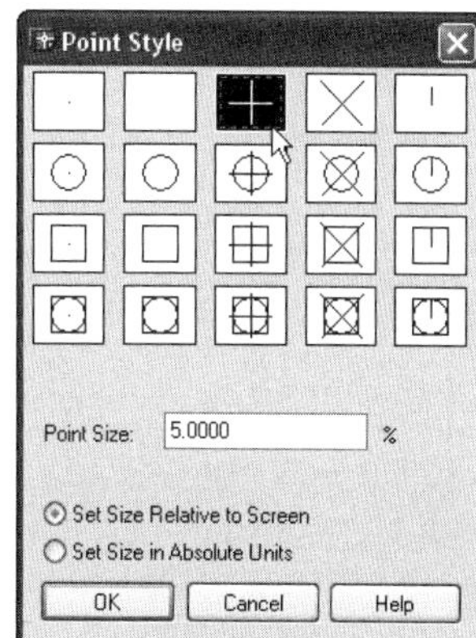

Figure 5–3

STEP 2

Locate one point at absolute coordinate 7.50,7.50. Then use the COPY command and the dimensions in Figure 5–4 as a guide for duplicating the remaining points.

Command: **PO** *(For Point)*

Current point modes: PDMODE=2 PDSIZE=0.00 Specify a point: **7.50,7.50** *(Locates the point at "A" in Figure 5–4)*

Command: **CP** *(For COPY)*

Select objects: **L** *(This should select the point, this option selected the last point)*

Select objects: *(Press* ENTER *to continue, pressing enter terminates the commands selection mode)*

Specify base point or displacement, or [Multiple]: **M** *(For Multiple)*

Specify base point: **Nod**

of *(Select the point at "A" in Figure 5–77)*

Specify second point of displacement or <use first point as displacement>: **@2.25,-3.00** *(To locate point "B")*

Specify second point of displacement or <use first point as displacement>: **@4.5,1.00** *(To locate point "C")*

Specify second point of displacement or <use first point as displacement>: **@5.25,3.50** *(To locate point "D")*

Specify second point of displacement or <use first point as displacement>: *(Press* ENTER *to exit this command)*

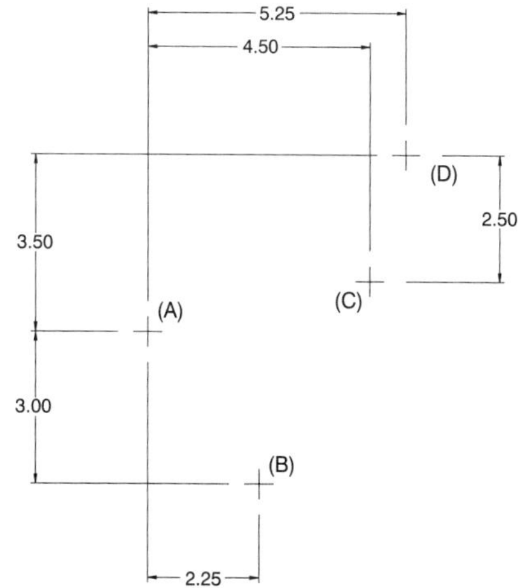

Figure 5–4

STEP 3

Use the CIRCLE command to place four circles of different sizes from points located at "A," "B," "C," and "D", as shown in Figure 5–5. When you have completed drawing the four circles, use the ERASE command to erase points "A," "B," "C," and "D."

 Command: **C** *(For CIRCLE)*
Specify center point for circle or [3P/2P/
Ttr (tan tan radius)]: **Nod**
of *(Select the point at "A")*
Specify radius of circle or [Diameter]:
2.50

 Command: **C** *(For CIRCLE)*
Specify center point for circle or [3P/2P/
Ttr (tan tan radius)]: **Nod**
of *(Select the point at "B")*
Specify radius of circle or [Diameter]
<2.50>: **2.00**

 Command: **C** *(For CIRCLE)*
Specify center point for circle or [3P/2P/
Ttr (tan tan radius)]: **Nod**
of *(Select the point at "C")*
Specify radius of circle or [Diameter]
<2.00>: **1.50**

 Command: **C** *(For CIRCLE)*

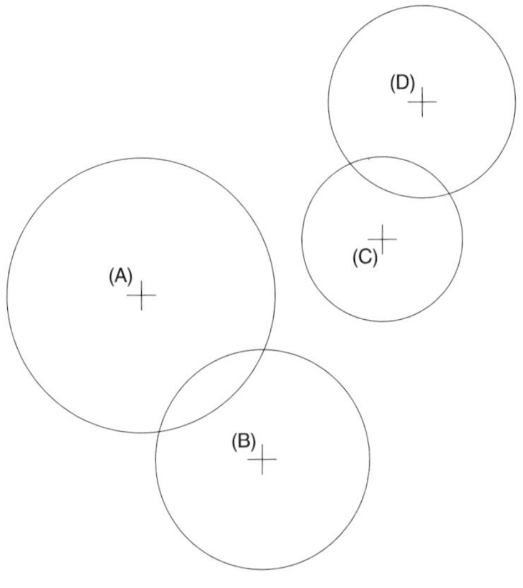

Figure 5–5

Specify center point for circle or [3P/2P/
Ttr (tan tan radius)]: **Nod**
of *(Select the point at "D")*
Specify radius of circle or [Diameter]
<1.50>: **1.75**

 Command: **E** *(For ERASE)*
Select objects: *(Pick the four points labeled "A", "B", "C", and "D")*
Select objects: *(Press ENTER to execute this command)*

STEP 4

Use the CIRCLE/TTR command/option to construct a 4.00-radius circle tangent to the two dashed circles in Figure 5–6. Then use the TRIM command to trim away part of circle "C."

Command: **C** *(For CIRCLE)*
Specify center point for circle or [3P/2P/
Ttr (tan tan radius)]: **T** *(For TTR)*
Specify point on object for first tangent of circle: *(Select the dashed circle at "A")*
Specify point on object for second tangent of circle: *(Select the dashed circle at "B")*
Specify radius of circle <1.75>: **4.00**

 Command: **TR** *(For TRIM)*

Current settings: Projection=UCS
Edge=None Select cutting edges ...
Select objects: *(Select the two dashed circles shown in Figure 5–6)*
Select objects: *(Press* ENTER *to continue)*
Select object to trim or shift-select to extend or [Project/Edge/Undo]: *(Select the large circle at "C")*
Select object to trim or shift-select to extend or [Project/Edge/Undo]: *(Press* ENTER *to exit this command)*

If, during the trimming process, you trim the wrong segment by mistake, you can use a built-in Undo to trim the correct item.

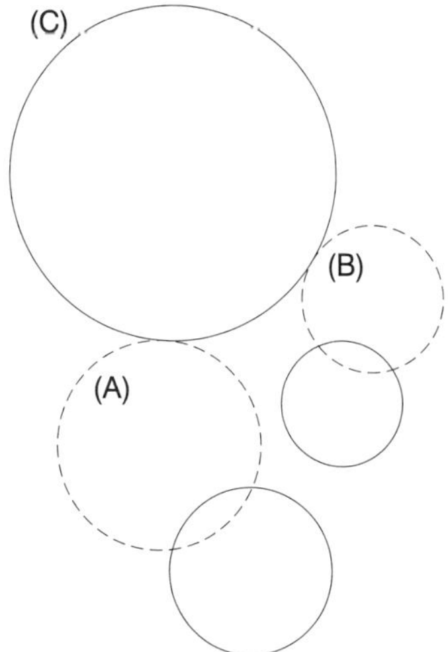

Figure 5–6

STEP 5

Use the CIRCLE/TTR command/option to construct a 6.00-radius circle tangent to the two dashed circles in Figure 5–7. Then use the TRIM command to trim away part of circle "C."

Command: **C** *(For CIRCLE)*
Specify center point for circle or [3P/2P/Ttr (tan tan radius)]: **T** *(For TTR)*
Specify point on object for first tangent of circle: *(Select the dashed circle at "A")*
Specify point on object for second tangent of circle: *(Select the dashed circle at "B")*
Specify radius of circle <4.00>: **6.00**

Command: **TR** *(For TRIM)*
Current settings: Projection=UCS
Edge=None Select cutting edges ...
Select objects: *(Select the two dashed circles "A" and "B" in Figure 5–7)*
Select objects: *(Press* ENTER *to continue)*

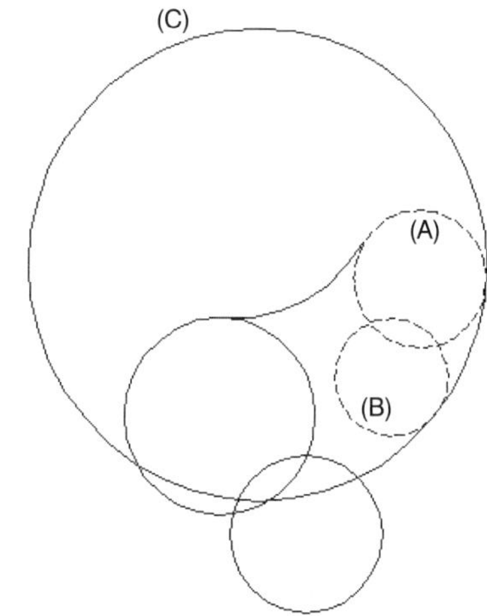

Figure 5–7

Select object to trim or shift-select to extend or [Project/Edge/Undo]: *(Select the large circle at "C")*
Select object to trim or shift-select to extend or [Project/Edge/Undo]: *(Press* ENTER *to exit this command)*

STEP 6

Use the CIRCLE/TTR command/option to construct a 6.00-radius circle tangent to the two dashed circles in Figure 5–8. Then use the TRIM command to trim away part of circle "C."

 Command: **C** *(For CIRCLE)*

Specify center point for circle or [3P/2P/Ttr (tan tan radius)]: **T** *(For TTR)*

Specify point on object for first tangent of circle: *(Select the dashed circle at "A")*

Specify point on object for second tangent of circle: *(Select the dashed circle at "B")*

Specify radius of circle <6.00>: *(Press ENTER to accept this default value)*

 Command: **TR** *(For TRIM)*

Current settings: Projection=UCS Edge=None Select cutting edges ...

Select objects: *(Select the two dashed circles "A" and "B" in Figure 5–8)*

Select objects: *(Press ENTER to continue)*

Select object to trim or shift-select to extend or [Project/Edge/Undo]: *(Select the large circle at "C")*

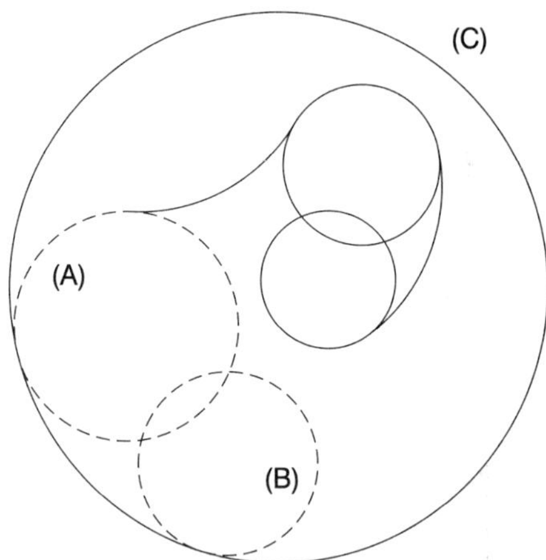

Figure 5–8

Select object to trim or shift-select to extend or [Project/Edge/Undo]: *(Press ENTER to exit this command)*

Select objects: *(Press ENTER to continue)*

Select object to trim or shift-select to extend or [Project/Edge/Undo]: *(Select the circle at "C")*

Select object to trim or shift-select to extend or [Project/Edge/Undo]: *(Press ENTER to exit this command)*

STEP 7

Use the CIRCLE/TTR command/option to construct a 0.75-radius circle tangent to the two dashed circles in Figure 5–9. Then use the TRIM command to trim away part of circle "C."

 Command: **C** *(For CIRCLE)*

Specify center point for circle or [3P/2P/Ttr (tan tan radius)]: **T** *(For TTR)*

Specify point on object for first tangent of circle: *(Select the dashed circle at "A")*

Specify point on object for second tangent of circle: *(Select the dashed circle at "B")*

Specify radius of circle <6.00>: **0.75**

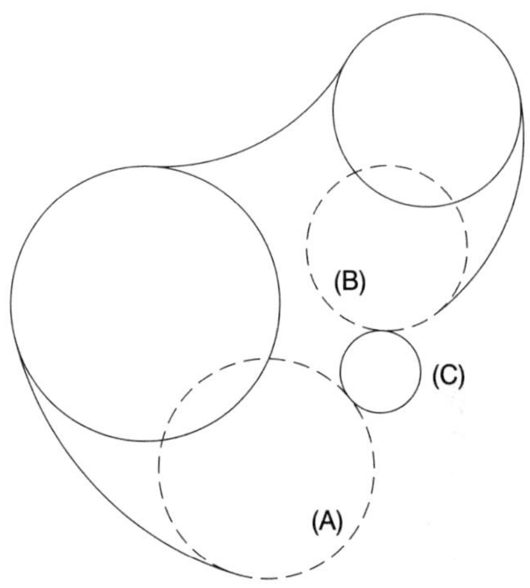

Figure 5–9

Command: **TR** *(For TRIM)*
Current settings: Projection=UCS Edge=None Select cutting edges ...
Select objects: *(Select the two dashed circles "A" and "B" in Figure 5–9)*
Select objects: *(Press* ENTER *to continue)*
Select object to trim or shift-select to extend or [Project/Edge/Undo]: *(Select the circle at "C)*
Select object to trim or shift-select to extend or [Project/Edge/Undo]: *(Press* ENTER *to exit this command)*

STEP 8

Use the TRIM command, select all dashed arcs in Figure 5–10 as cutting edges, and trim away the circular segments to form the outline of the Pattern1 drawing.

 Command: **TR** *(For TRIM)*
Current settings: Projection=UCS Edge=None Select cutting edges ...
Select objects: *(Select the four dashed arcs in Figure 5–10)*
Select objects: *(Press* ENTER *to continue)*
Select object to trim or shift-select to extend or [Project/Edge/Undo]: *(Select the circle at "A")*
Select object to trim or shift-select to extend or [Project/Edge/Undo]: *(Select the circle at "B")*
Select object to trim or shift-select to extend or [Project/Edge/Undo]: *(Select the circle at "C")*

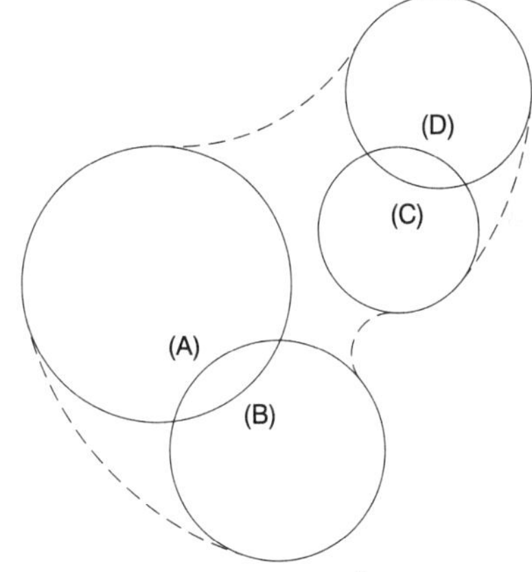

Figure 5–10

Select object to trim or shift-select to extend or [Project/Edge/Undo]: *(Select the circle at "D")*
Select object to trim or shift-select to extend or [Project/Edge/Undo]: *(Press* ENTER *to exit this command)*

STEP 9

Your drawing should be similar to the illustration in Figure 5–11.

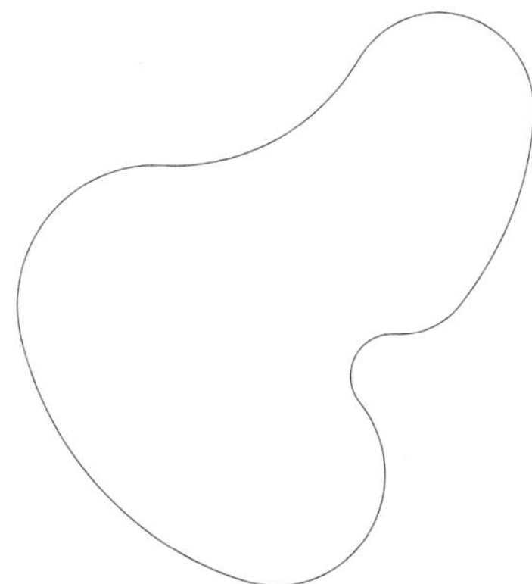

Figure 5–11

STEP 10

Use the CIRCLE command to place a circle of 2.00-unit diameter at the center of arc "A." Then use the COPY command to duplicate the circle at the center of arcs "B" and "C" (the OSNAP-Center mode should already be running). See Figure 5–12.

 Command: **C** *(For CIRCLE)*

Specify center point for circle or [3P/2P/ Ttr (tan tan radius)]: *(Select the edge of arc "A")*

Specify radius of circle or [Diameter] <0.7500>: **D** *(For Diameter)*

Specify diameter of circle <1.50>: **2.00**

Command: **CP** *(For COPY)*

Select objects: **L** *(For Last)*

Select objects: *(Press* ENTER *to continue)*

Specify base point or displacement, or [Multiple]: **M** *(For Multiple)*

Specify base point: *(Select the edge of the arc at "A")*

Specify second point of displacement or <use first point as displacement>: *(Select the edge of the arc at "B")*

Specify second point of displacement or <use first point as displacement>: *(Select the edge of the arc at "C")*

Specify second point of displacement or <use first point as displacement>: *(Press* ENTER *to exit this command)*

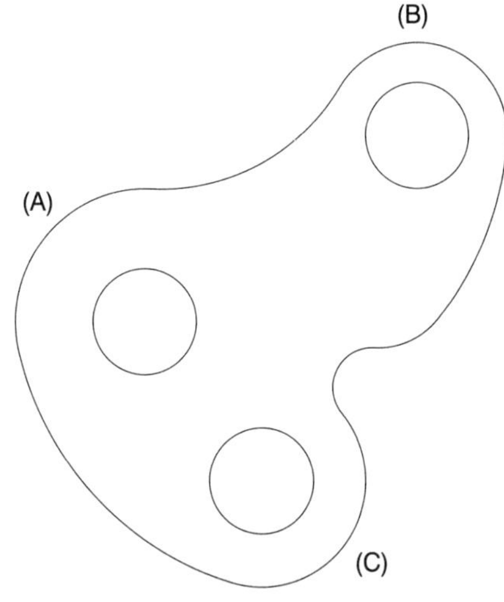

Figure 5–12

STEP 11

Use the Layer Control box to set the current layer to "Center" (see Figure 5–13).

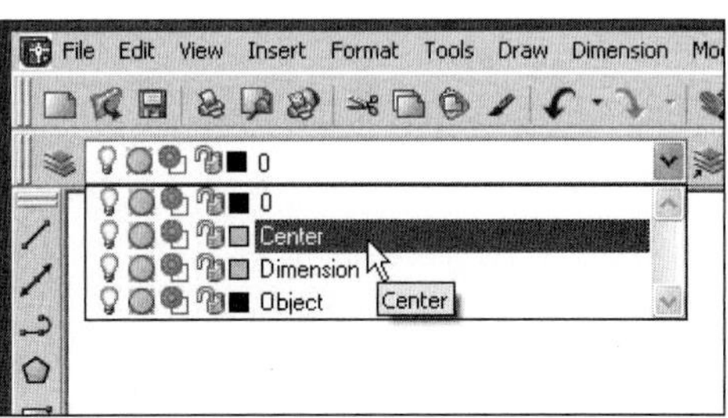

Figure 5–13

STEP 12

Change the DIMCEN variable from a value of 0.09 to -0.12. This will place the center marker identifying the centers of circles and arcs when the DIMCENTER command is used. Also set the linetype scale to 2 using the LTSCALE command. See Figure 5–14.

Command: **LTS** *(For LTSCALE)*
Enter new linetype scale factor <1.00>: **2**
Command: **DIMCEN**
Enter new value for DIMCEN <0.09>: **-0.12**

Command: **DCE** *(For DIMCENTER)*
Select arc or circle: *(Select the arc at "A")*

Command: **DCE** *(For DIMCENTER)*
Select arc or circle: *(Select the arc at "B")*

Command: **DCE** *(For DIMCENTER)*
Select arc or circle: *(Select the arc at "C")*

Command: **DCE** *(For DIMCENTER)*
Select arc or circle: *(Select the arc at "D")*

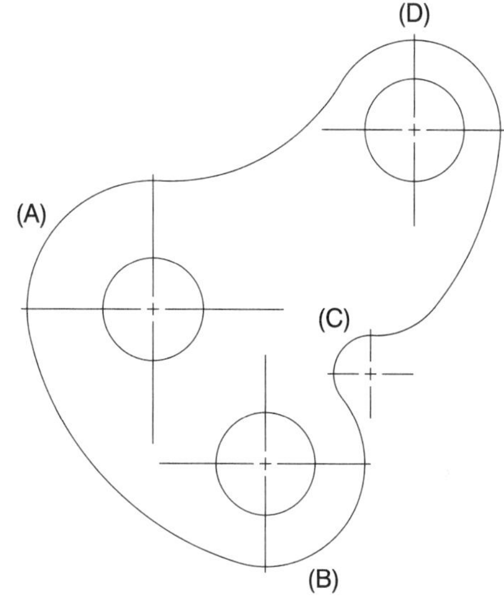

Figure 5–14

STEP 13

Dimensions may be added to this drawing at a later time. Place them on the Dimension layer as shown in Figure 5–15.

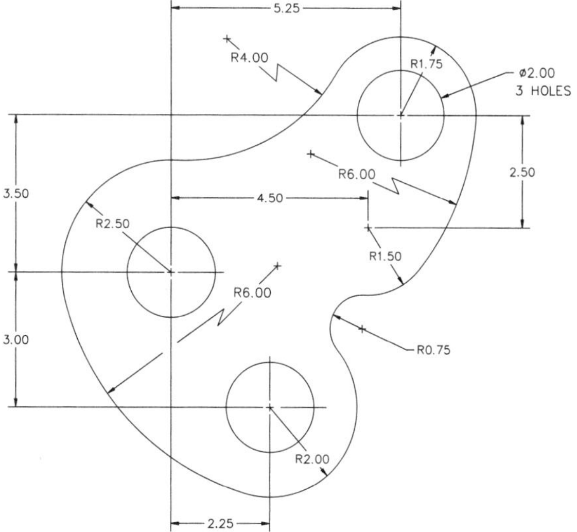

Figure 5–15

Beginning Level Drawings
Directions for Problems 5–1 through 5–27:

Construct these geometric construction figures using existing AutoCAD commands.

PROBLEM 5–1

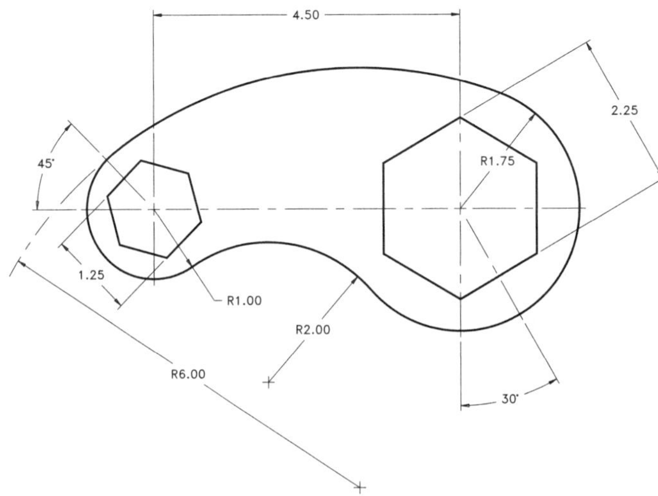

PROBLEM 5–2

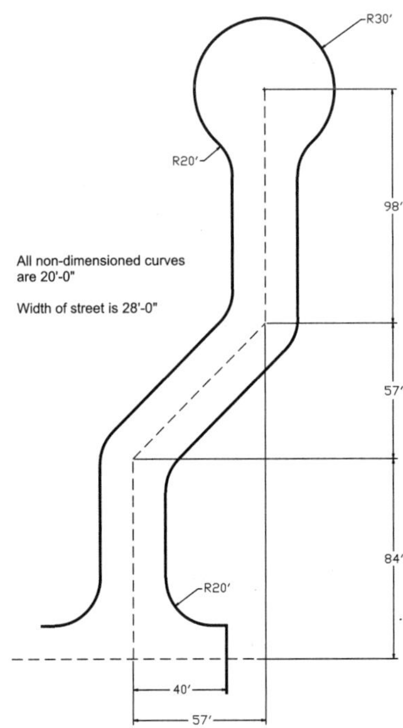

PROBLEM 5–3

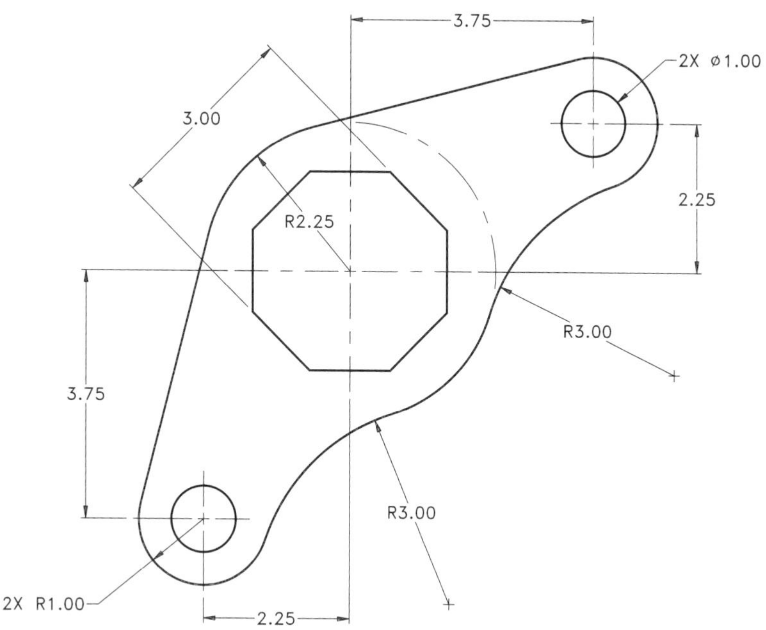

PROBLEM 5–4

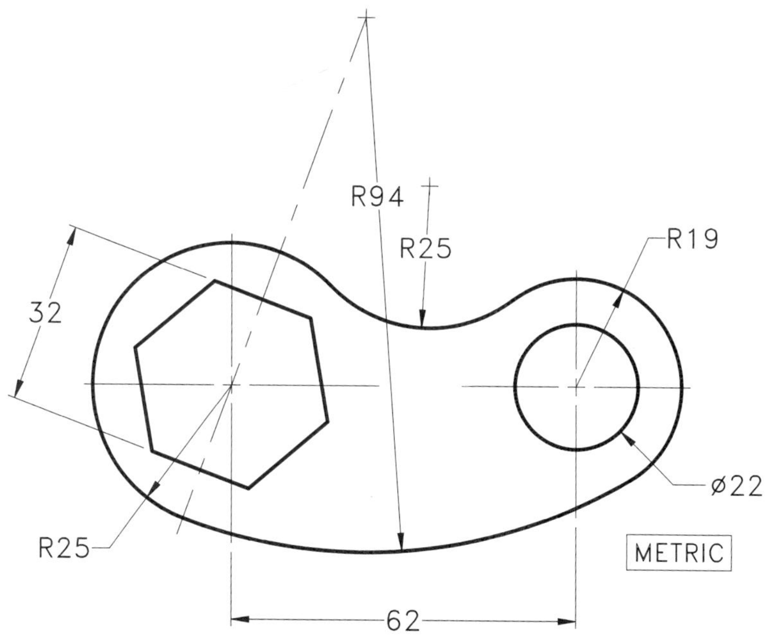

PROBLEM 5–5

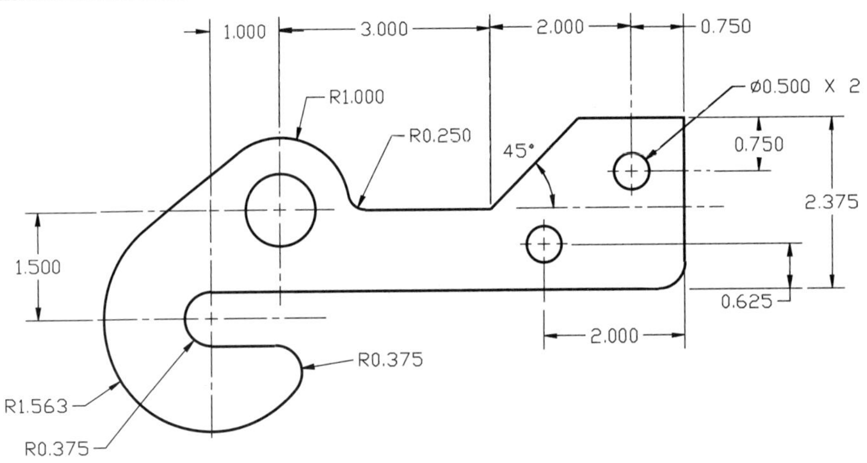

PROBLEM 5–6

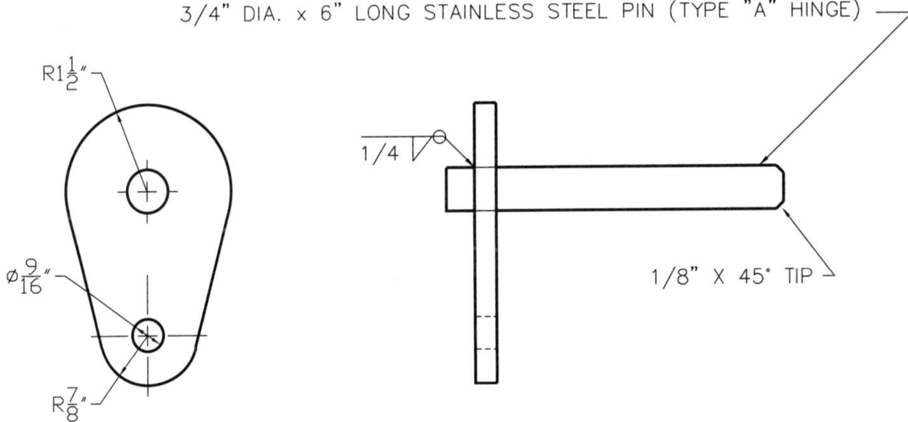

PROBLEM 5–7

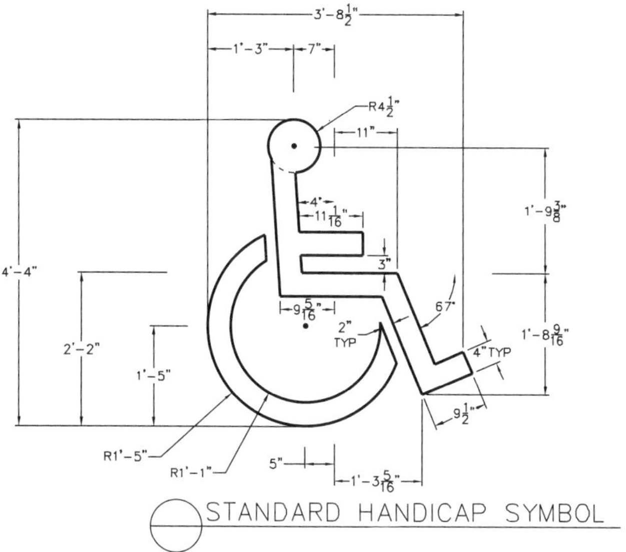

Intermediate Level Drawings

PROBLEM 5–8

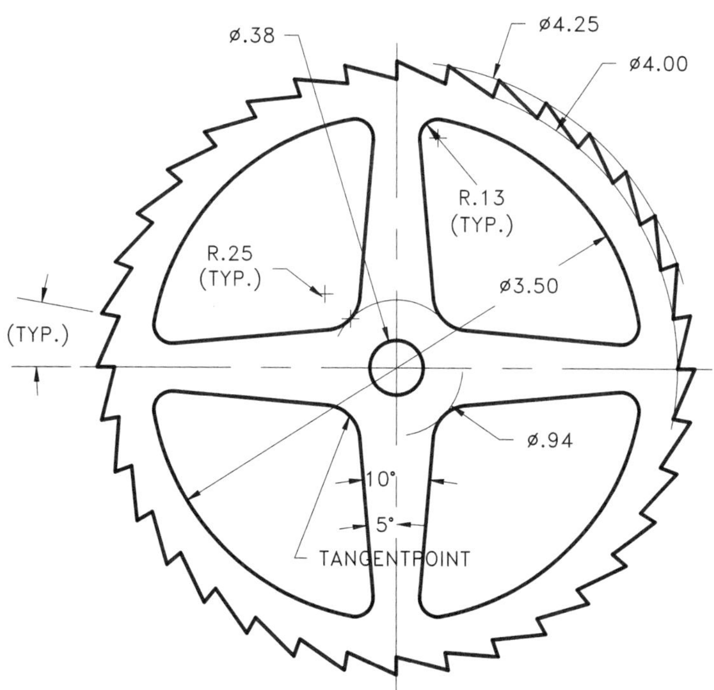

PROBLEM 5–9

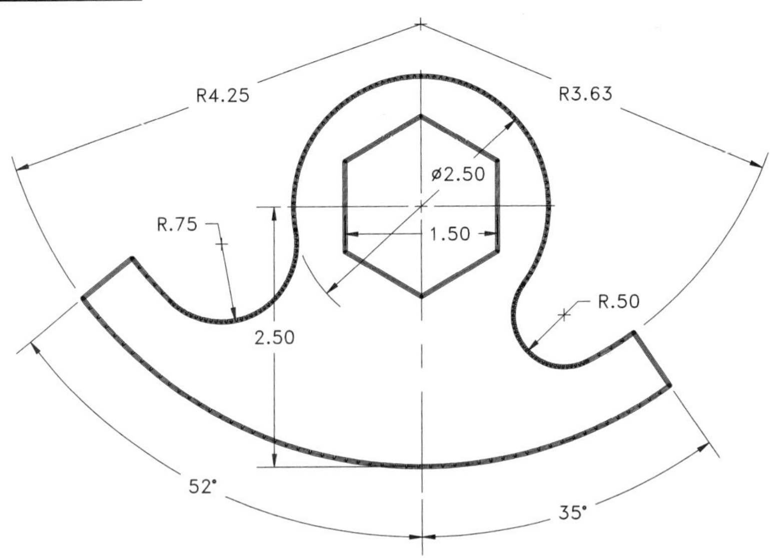

PROBLEM 5–10

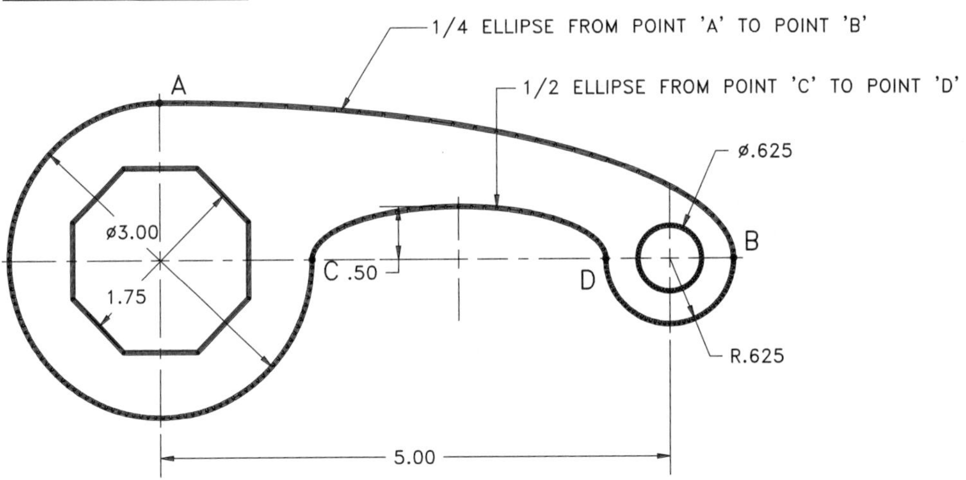

PROBLEM 5–11

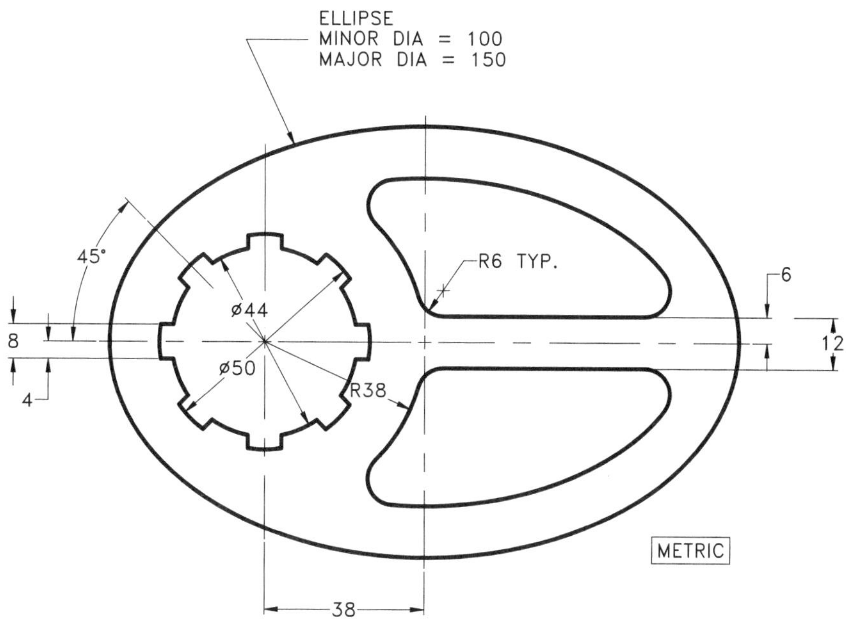

PROBLEM 5–12

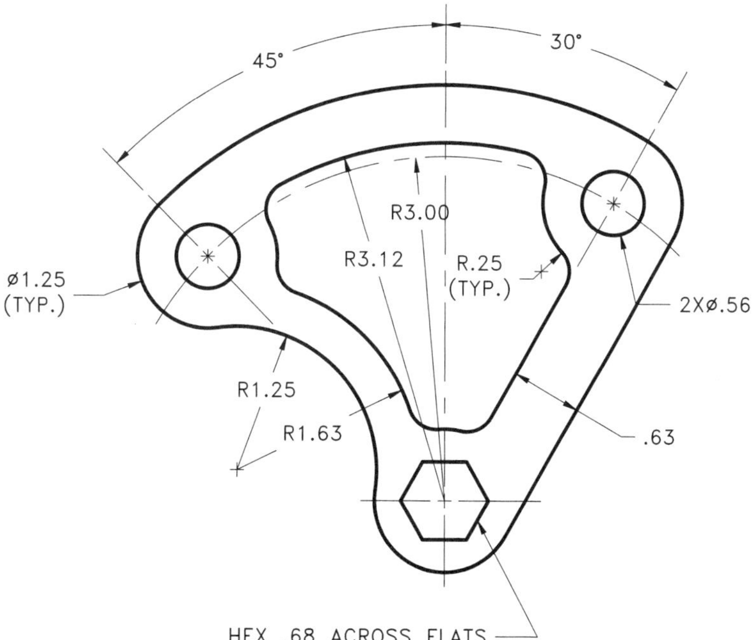

PROBLEM 5–13

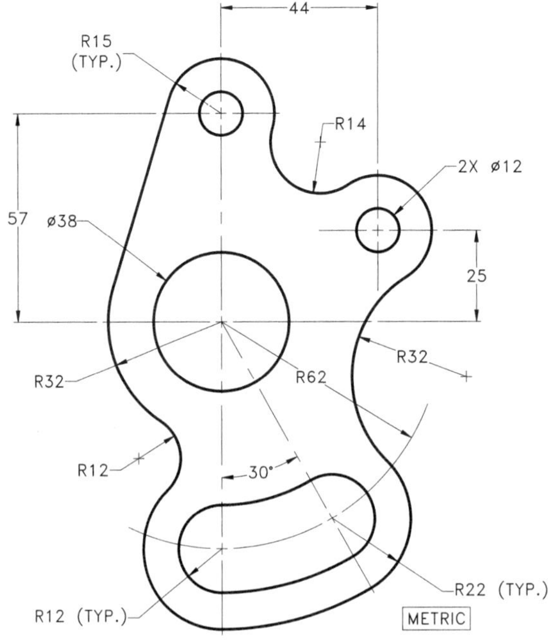

PROBLEM 5–14

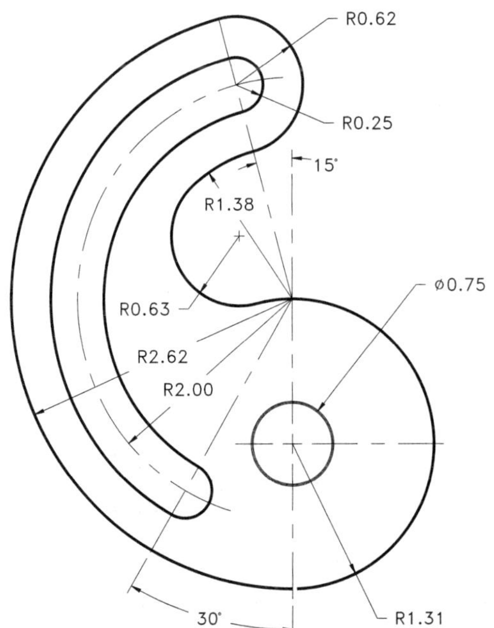

PROBLEM 5–15

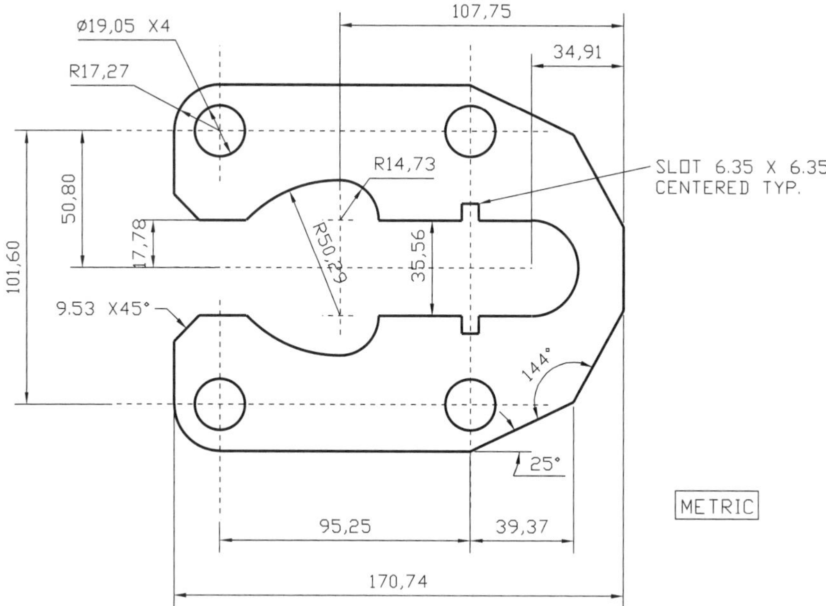

PROBLEM 5–16

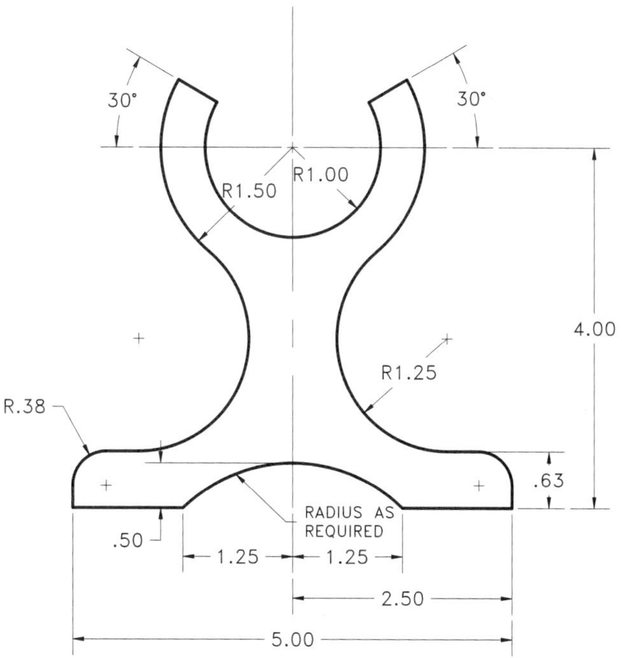

PROBLEM 5–17

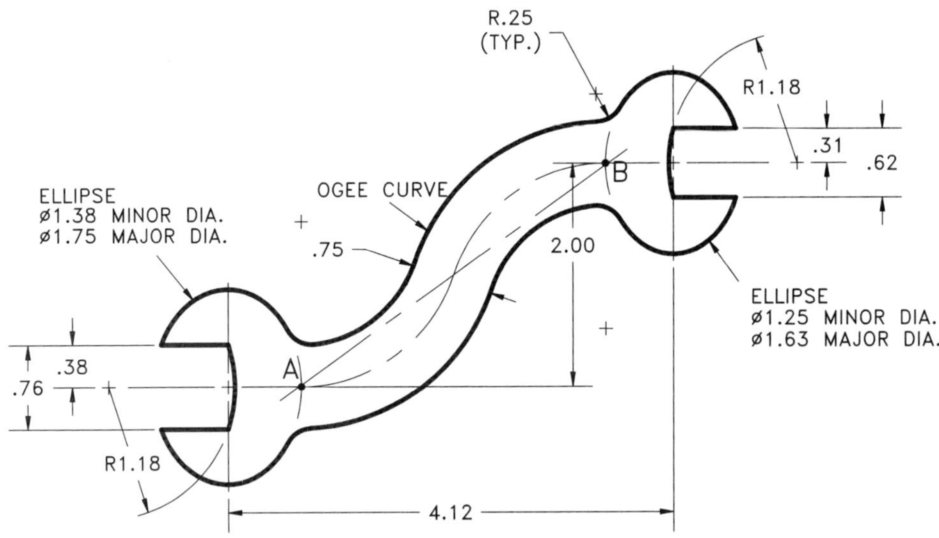

PROBLEM 5–18

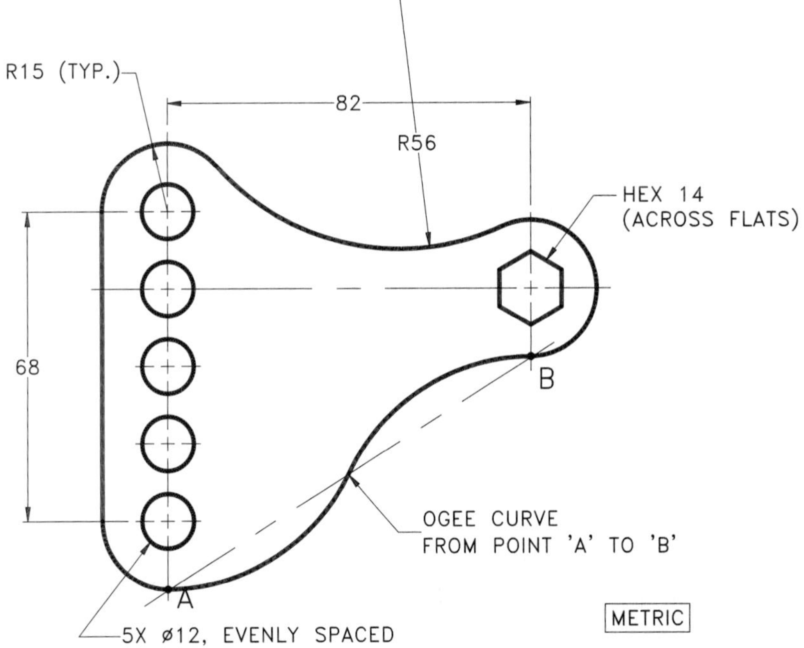

PROBLEM 5–19

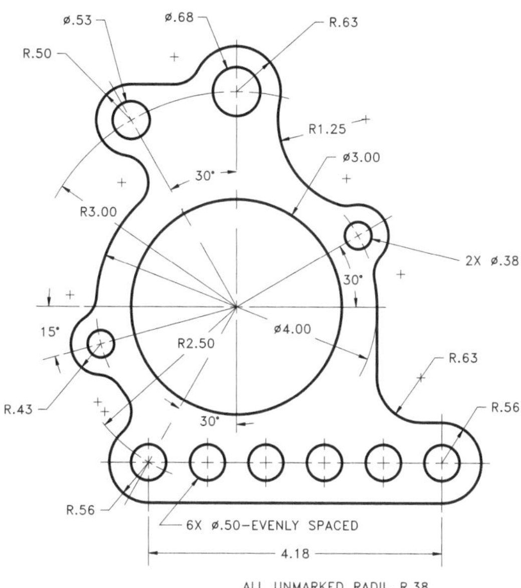

PROBLEM 5–20

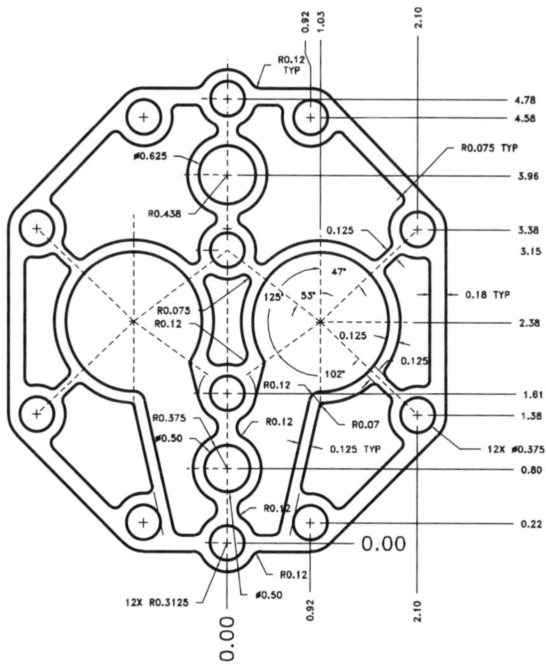

Advanced Level Drawings

PROBLEM 5–21

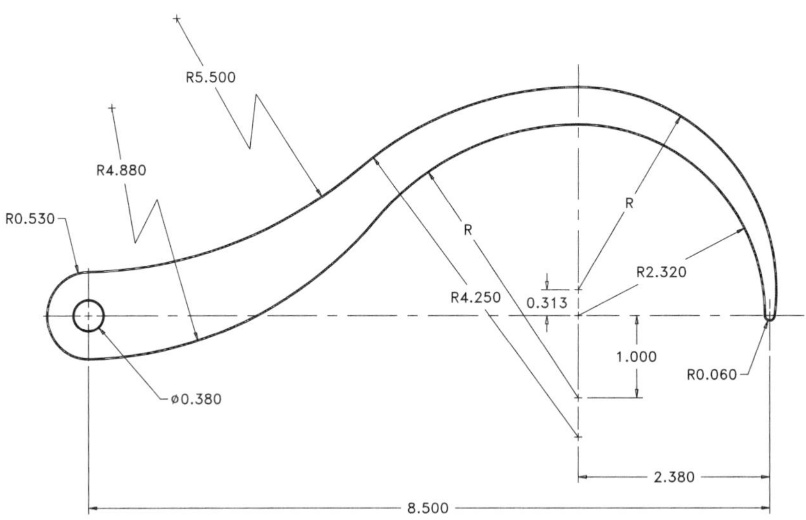

PROBLEM 5–22

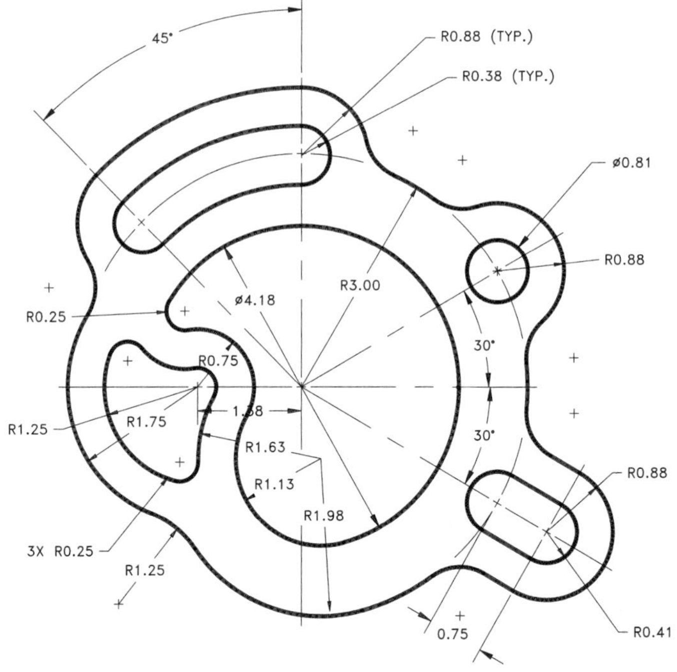

PROBLEM 5–23

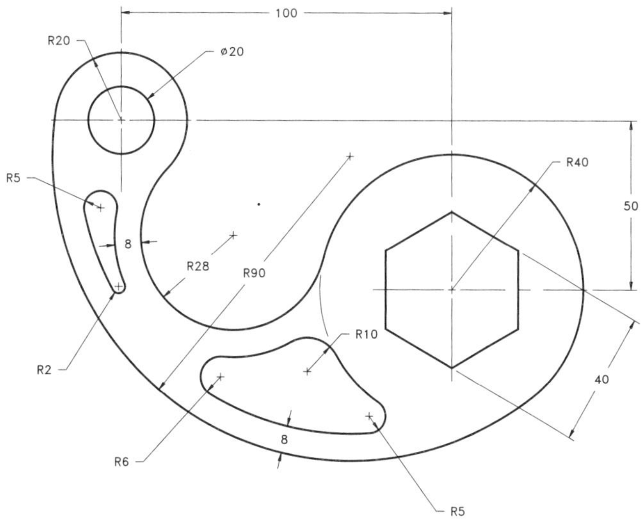

PROBLEM 5–24

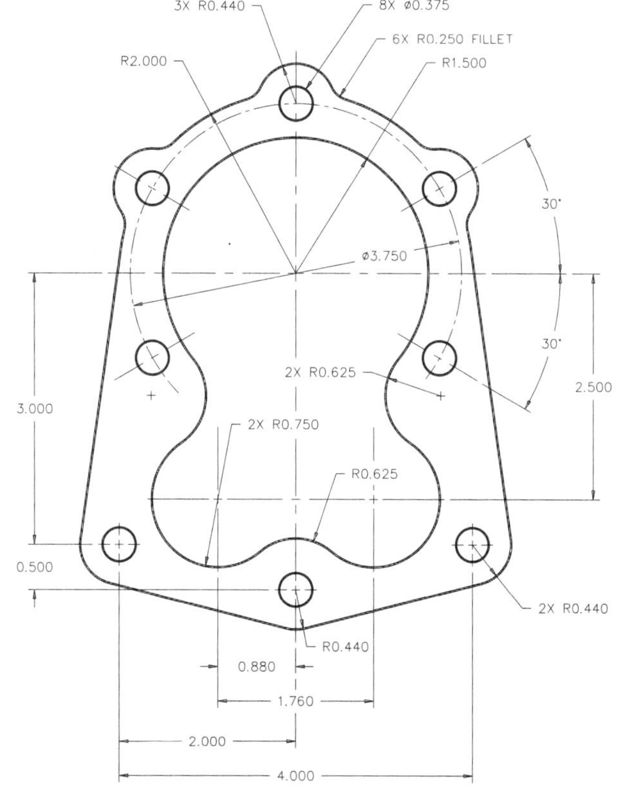

PROBLEM 5–25

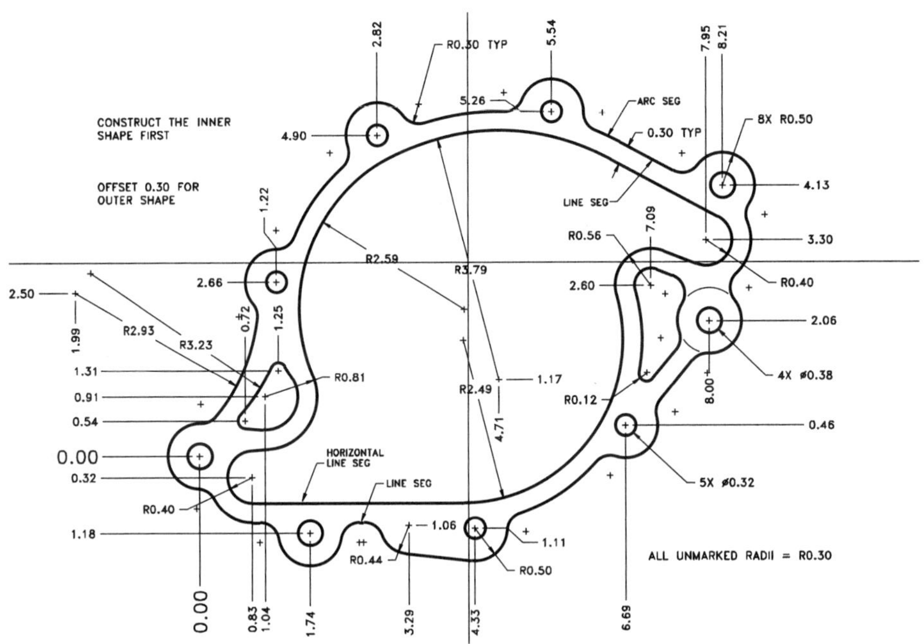

PROBLEM 5–26

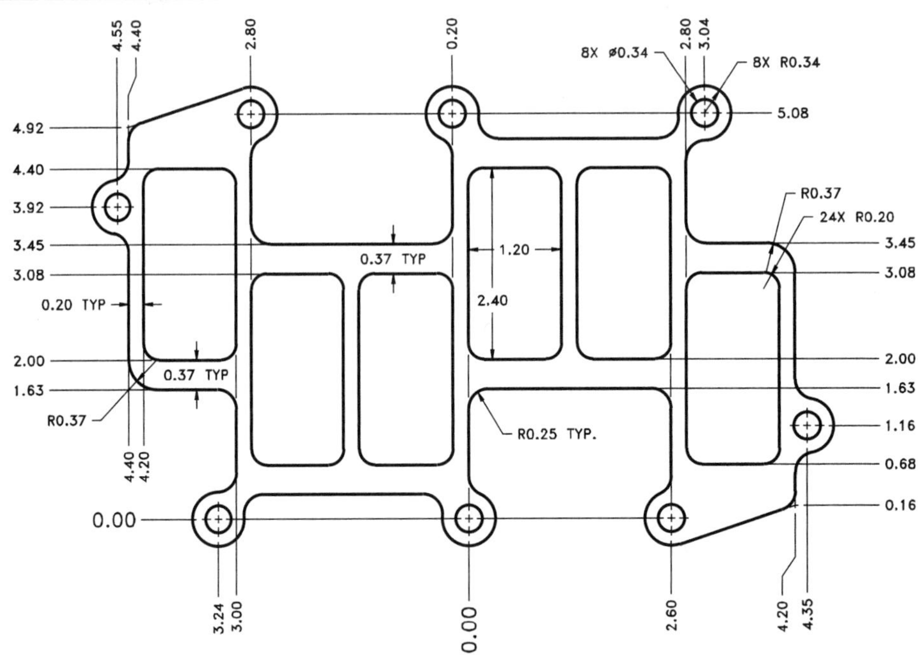

PROBLEM 5–27

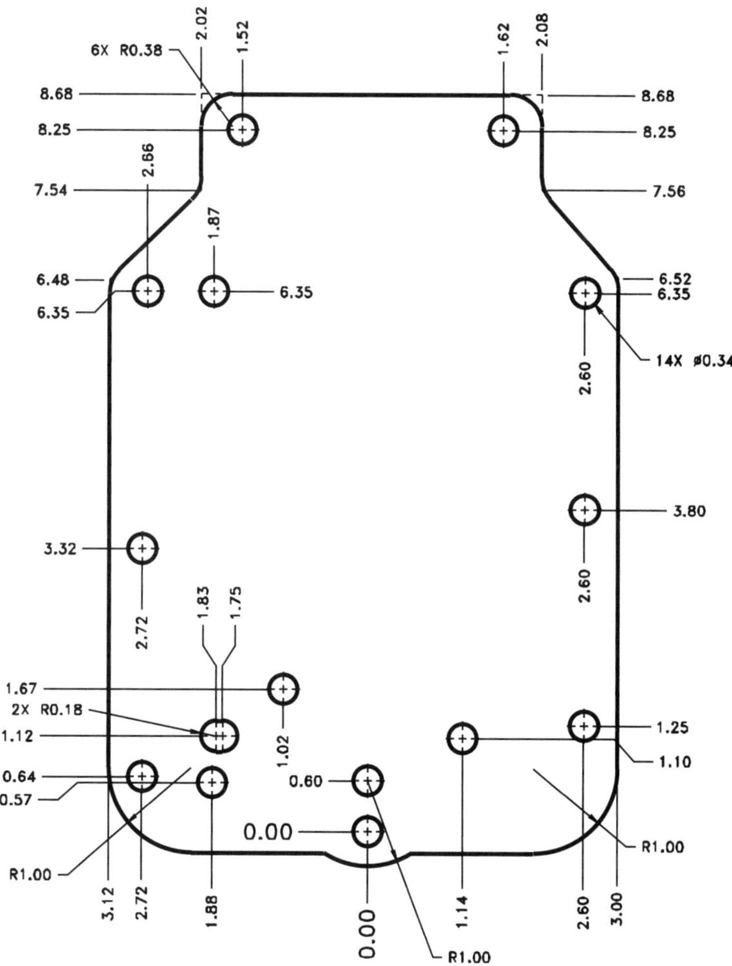
6X R0.38
2.02
1.52
1.62
2.08
8.68
8.25
8.68
8.25
2.66
7.54
7.56
1.87
6.48
6.35
6.35
6.52
6.35
14X ⌀0.34
2.60
3.80
3.32
2.60
2.72
1.83
1.75
1.67
2X R0.18
1.12
1.02
1.25
1.10
0.64
0.57
0.60
0.00
R1.00
R1.00
3.12
2.72
1.88
0.00
1.14
2.60
3.00
R1.00

PROBLEM 5–28

Directions for Problem 5–28

Develop the lot plan as shown. Set up the units to follow surveyor's units. Allow the lines to intersect and then add the corner pins (circles). Use text to detail the bearing and distance on each of the lot segments.

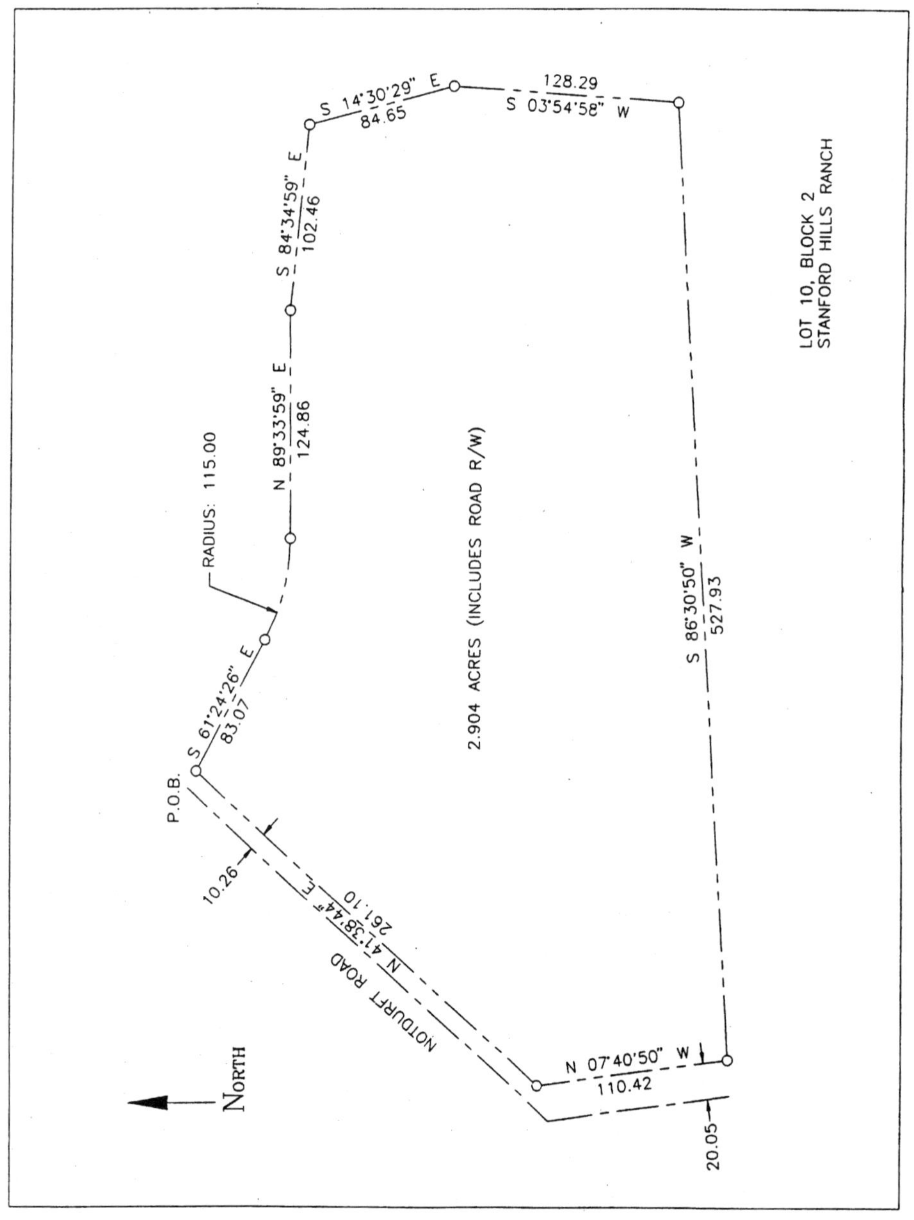

Adding Text to Your Drawing

The exercises in this chapter introduce:

- Using the MTEXT editor
- Applying special characters to text
- Spell checking drawing text
- Using the character mapping capabilities
- Modifying text

TUTORIAL EXERCISE: 06_BASE MAP.DWG

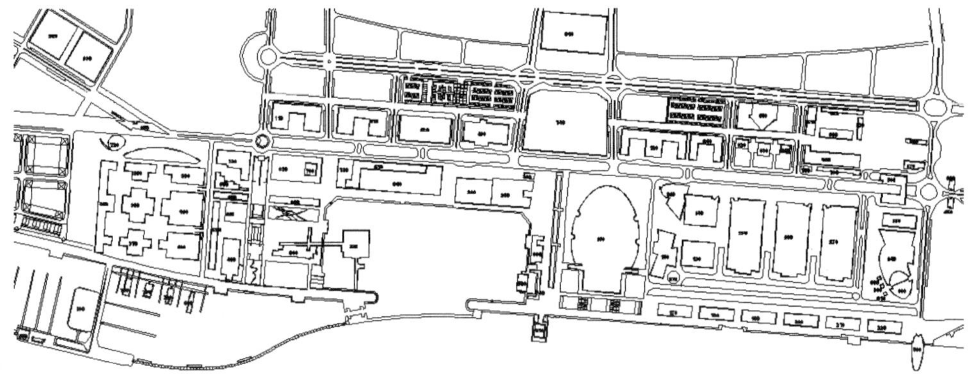

Figure 6–1

Purpose:

This tutorial teaches how to use the Zoom To feature of the Find and Replace dialog box on the base map, shown in Figure 6–1.

System Settings:

Changes to the drawing settings will not be necessary for this tutorial exercise.

Layers:

Layers have already been created for this tutorial exercise.

Suggested Commands:

Whenever possible, substitute the appropriate command alias for the full AutoCAD command in each tutorial step. For example, use "CP" for the COPY command, "L" for the LINE command, and so on. The complete listing of all command aliases is located in Chapter 1, Table 1–2.

STEP 1

Zoom into various portions of the base map and observe the building numbering system. In some cases, building numbers run in succession of each other (340 to 400), as shown in Figure 6–2. In other cases, the buildings are numbered randomly. The purpose of this exercise is to use the Find and Replace dialog box to not only locate the building number, but to also zoom into the building.

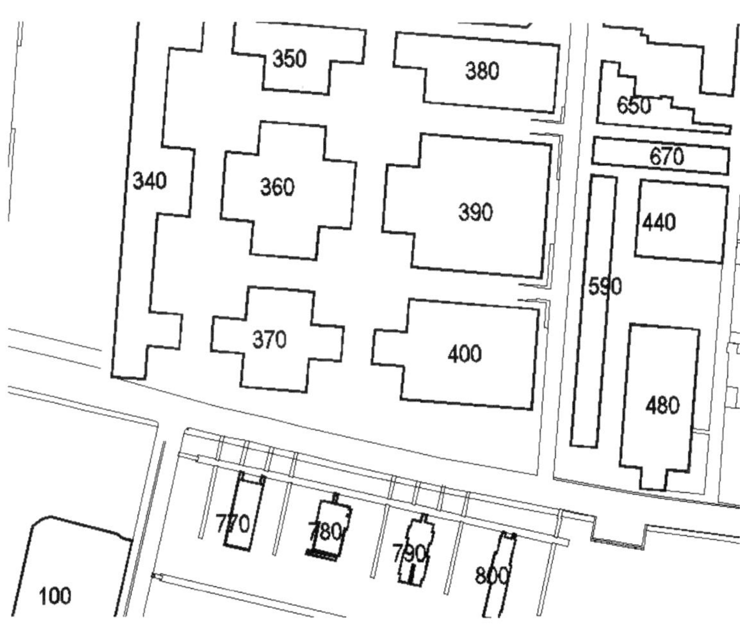

Figure 6–2

STEP 2

Activate the Text toolbar and click on the Find and Replace tool, as shown in Figure 6–3.

Figure 6–3

STEP 3

When the Find and Replace dialog box appears, type the building number to search for. In the dialog box, type the number **870** in the Find text string edit box (see Figure 6–4) and then press ENTER. This adds the number "870" to the Search results area of the dialog box. Click the Zoom to button at the right side of the dialog box.

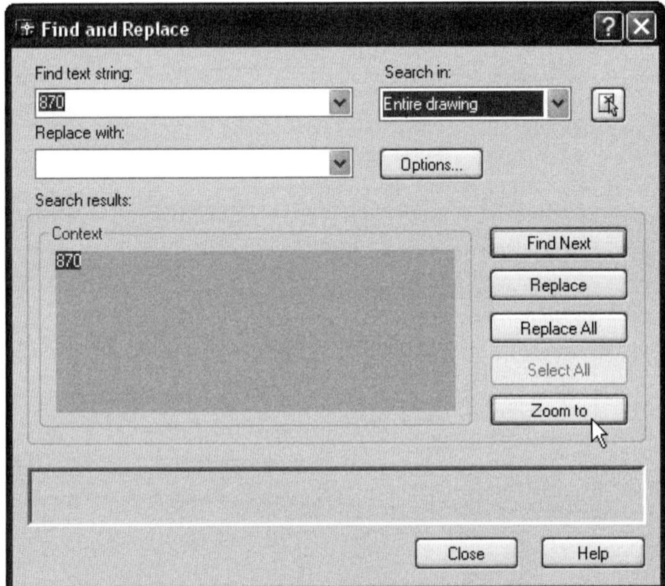

Figure 6–4

STEP 4

Your screen will change by automatically zooming into the text that you are searching for. However, initially you will not see the text; this is because the Find and Replace dialog box repositions itself directly on top of the text you were searching for. Simply move the dialog box to one side of the display screen by positioning the cursor over the dialog box's title bar, hold down the left mouse button and drag it aside. Once the window is in the desired location, simply release the mouse button. Figure 6–5 shows how the search text will now be visible.

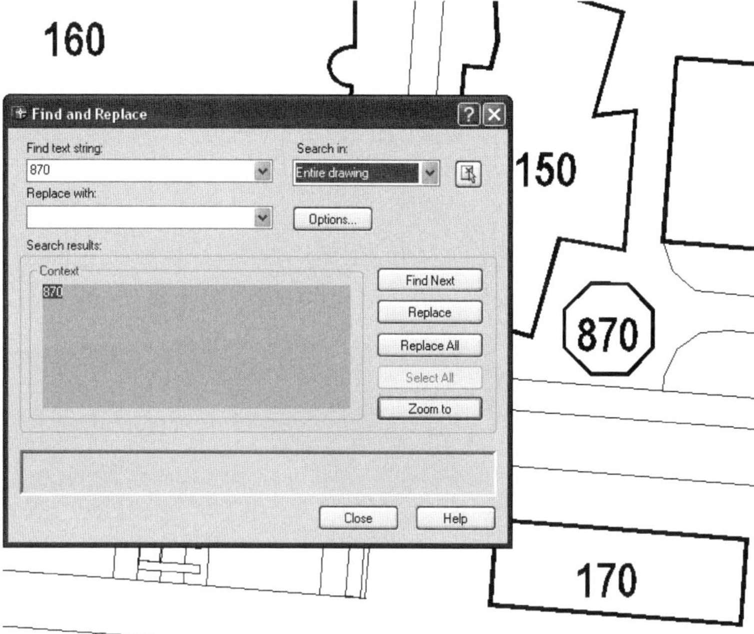

Figure 6–5

STEP 5

Next, try to locate building number "290". Activate the Find and Replace dialog box, type **290** into the Find text string edit box, press ENTER, and click on the Zoom to button. After your screen zooms to the desired text, move the dialog box to one side of the screen and the text will be visible, as shown in Figure 6–6.

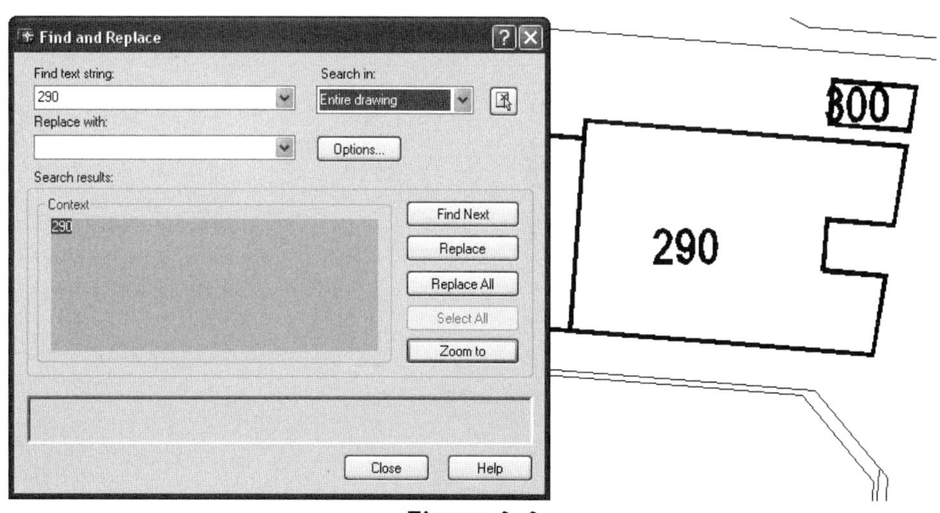

Figure 6–6

Beginning Level Drawing

PROBLEM 6–1

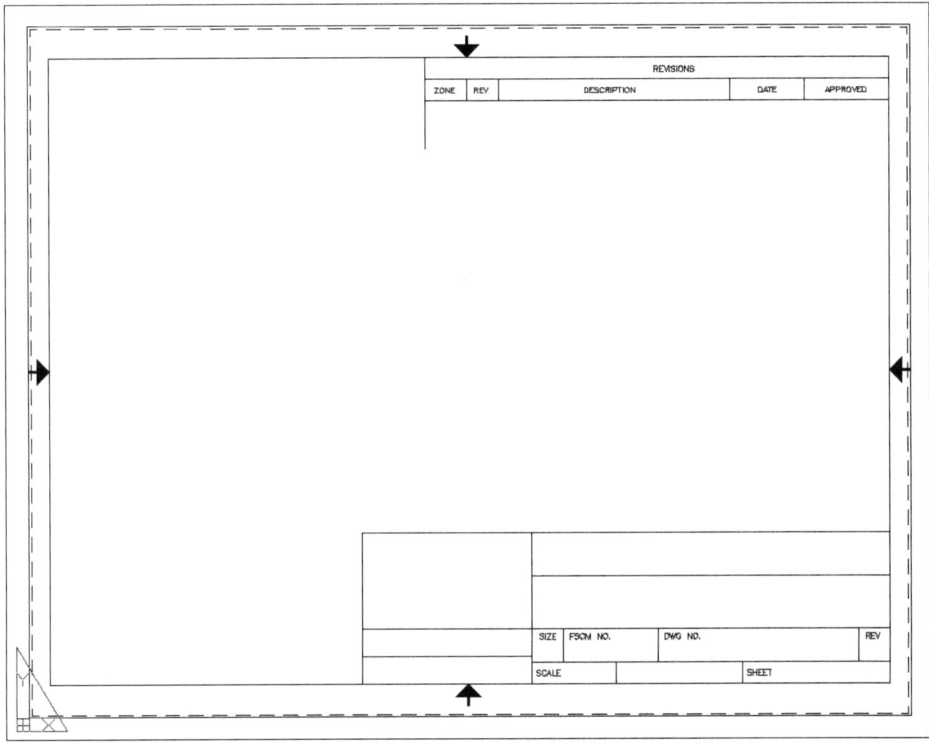

A

B

C

J

SIZE D

FSCM NO. E

DWG NO. F

REV

SCALE G

H

SHEET I

Using a standard Autodesk title block, fill in the necessary information about your drawing in fields A-J

A. Create a business logo for this field

B. Company name and address in bold lettering

C. The name of the drawing

D. Plotting sheet paper size

E. Federal supply code for manufacturers (FSCM) – Used for the area of the government that the design is created for. (optional)

F. Drawing number that is assigned

G. Scale of the drawing

H. Date the drawing originated (drawing revision dates are placed in the revision table)

I. Sheet number if there are more than 1 drawing sheet included in the drawings packet

J. General tolerance information used in the drawing

Object Grips and Changing the Properties of Objects

The exercises in this chapter introduce:

- Grips
- Grip configuration tools
- Applying grips to objects
- Dimensional grips
- Grip editing several items
- Using grips as a selection tool (for editing and properties)

TUTORIAL EXERCISE: 07_LUG.DWG

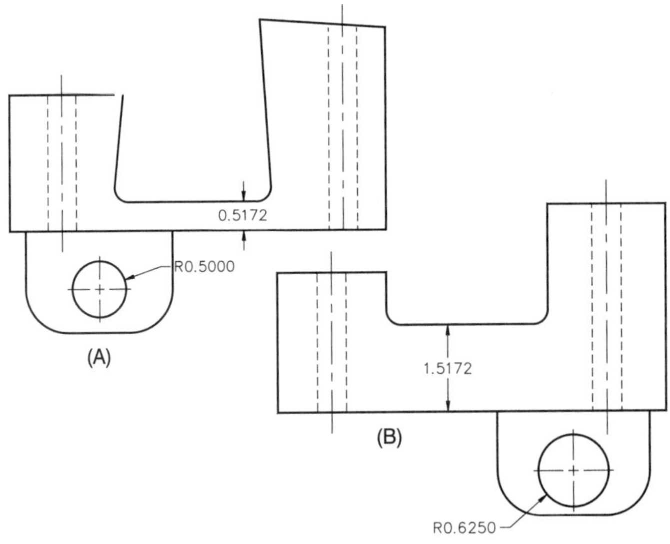

Figure 7–1

Purpose

This tutorial teaches how to use object grips to edit the drawing of the lug shown in Figure 7–1, at "A", until it appears like the drawing at "B."

System Settings

Since this drawing is provided on the CD, open the existing drawing file named *07_Lug.dwg*. Follow the steps in this tutorial for using object grips to edit and make changes to the Lug.

Layers

The following layers have already been created with the format:

Name	Color	Linetype
Object	White	Continuous
Center	Yellow	Center
Hidden	Red	Hidden
Dim	Yellow	Continuous
Defpoints	White	Continuous

Suggested Commands

Be sure grip mode is enabled through the Selection tab of the Options dialog box.

Then use the STRETCH, MOVE, ROTATE, SCALE, and MIRROR modes of object grips to make changes to the existing drawing file.

Whenever possible, substitute the appropriate command alias in place of the full AutoCAD command in each tutorial step. For example, use "CP" for the COPY command, "L" for the LINE command, and so on. The complete listing of all command aliases is located in Chapter 1, Table 1–2.

STEP 1

Be sure that Object Grip mode is enabled by choosing Options... from the Tools pull-down menu, which activates the Options dialog box. Click on the Selection tab to determine that grips are turned on (the default setting). Your dialog box should be similar to Figure 7–2.

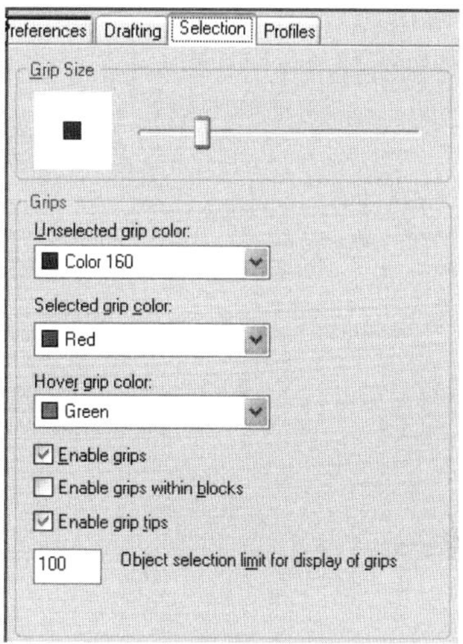

Figure 7–2

STEP 2

Begin by turning ORTHO off by clicking on ORTHO in the Status bar. Next, at the command prompt, use the grip cursor to select the inclined line "A" shown in Figure 7–3. Notice the appearance of the grips at the endpoints and midpoint of the line.

Command: **ORTHO**
Enter mode [ON/OFF] <ON>: **OFF**
Command: *(Select the inclined line "A")*

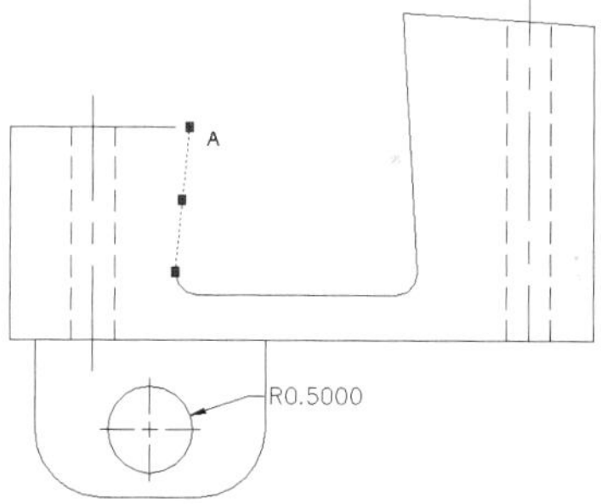

Figure 7–3

STEP 3

While still at the command prompt, select the grip shown in Figure 7–4 at "A." This grip becomes the current base point for the following editing option. Use the STRETCH option to reposition the endpoint of the highlighted line at the endpoint of the horizontal line shown in Figure 7–4 (OSNAP-Endpoint should already be enabled).

When this operation is complete, press ESC twice at the command prompt to remove the object highlight and grips from the display screen.

Command: *(Select the warm grip at the endpoint of the line at "A" to make it hot)*
STRETCH Specify stretch point or [Base point/Copy/Undo/eXit]: *(Select the endpoint of the horizontal line at "B")*

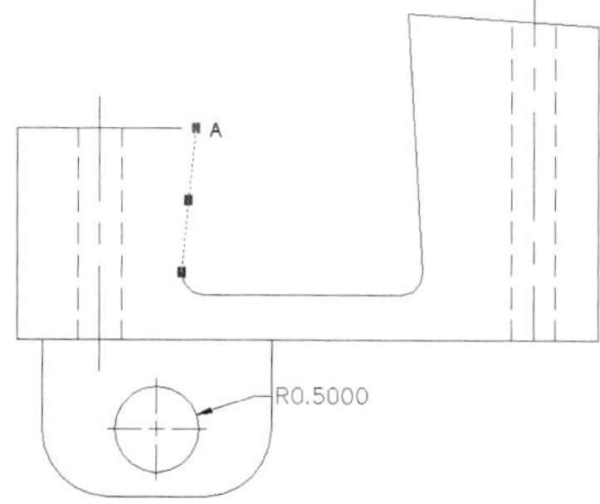

Figure 7–4

Command: *(Press ESC to remove the object highlight and grips)*

STEP 4

At the command prompt, use the grip cursor to select the two inclined lines "A" and "B" shown in Figure 7–5. Notice the appearance of the grips at the endpoints and midpoints of the lines.

Command: *(Select the inclined lines "A" and "B")*

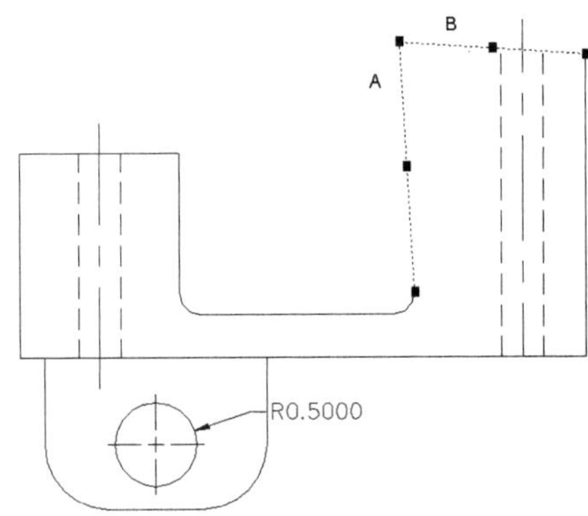

Figure 7–5

STEP 5

While still at the command prompt, select the grip shown in Figure 7–6 at "A." This grip becomes the current base point for the following editing options. Use the STRETCH option in combination with OSNAP-Tracking mode to reposition the corner of the highlighted lines to form a 90° corner. When this operation is complete, press ESC to remove the object highlight and grips from the display screen.

Command: *(Select the warm grip at the endpoint of the line at "A" to make it hot)*
** STRETCH ** Specify stretch point or [Base point/Copy/Undo/eXit]: **TK** *(To enable Tracking)*
First tracking point: *(Select the grip at "B" in Figure 7–6)*
Next point *(Press* ENTER *to end tracking)*: *(Select the grip at "C" in Figure 7–6)*
Next point *(Press* ENTER *to end tracking)*: *(Press* ENTER *to end tracking and perform the operation)*
Command: *(Press* ESC *to remove the object highlight and grips)*

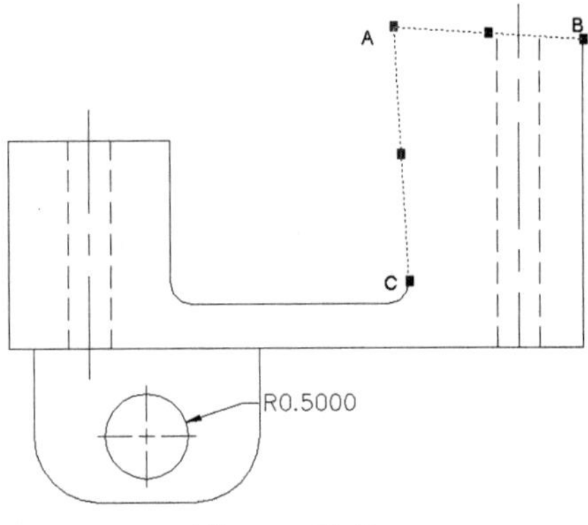

Figure 7–6

STEP 6

At the command prompt, use the grip cursor to select the two vertical lines, horizontal line, and both filleted corners as shown in Figure 7–7. Notice the appearance of the grips at the endpoints and midpoints of the lines and arcs.

Command: *(Select the two vertical lines, two arcs representing fillets, and the horizontal line)*

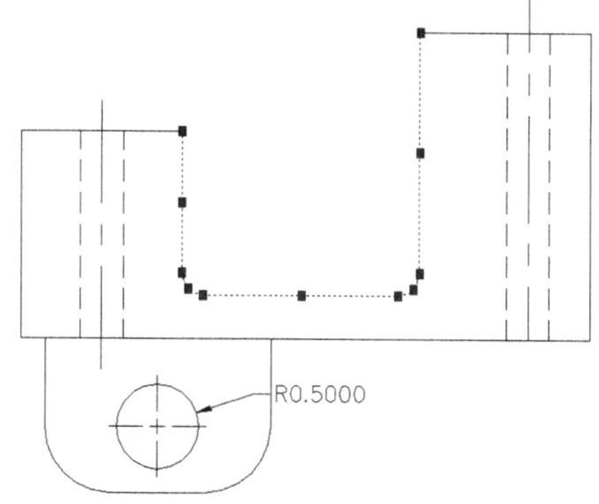

Figure 7–7

STEP 7

While still at the command prompt, select all the grips shown in Figure 7–8 by holding down SHIFT as you select the grips.

Release SHIFT and pick the selected grip at "A" again. This grip becomes the current base point for the following editing options. Use the STRETCH option in combination with a polar coordinate to stretch the objects a distance of one unit in the 90° direction (you could also use the Direct Distance mode to accomplish this operation).

When this operation is complete, press ESC at the command prompt to remove the object highlight and grips from the display screen.

Command: *(Hold down the SHIFT key and select all warm grips in Figure 7–8 to make them hot. Release the SHIFT key and pick the grip at "A" again.)*
STRETCH Specify stretch point or [Base point/Copy/Undo/eXit]: **@1.00<90**
Command: *(Press ESC to remove the object highlight and grips)*

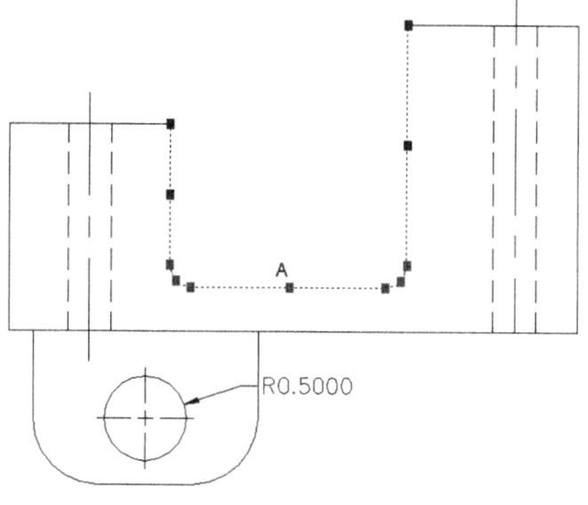

Figure 7–8

STEP 8

At the command prompt, use the grip cursor to select the circle and radius dimension shown in Figure 7–9. Notice the appearance of the grips at the quadrants and center of the circle and center, starting point, and text location of the radius dimension.

Command: *(Select the circle and the radius dimension)*

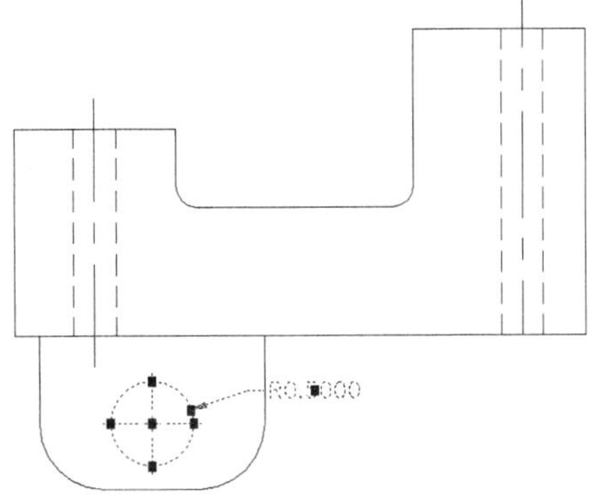

Figure 7–9

STEP 9

While still at the command prompt, select the grip at the center of the circle shown in Figure 7–10. This grip becomes the current base point for the following editing options. Use the SCALE option to increase the size of the circle using the REFERENCE option. Notice that this will also affect the value of the dimension. When this operation is complete, press ESC to remove the object highlight and grips from the display screen.

Command: *(Select the warm grip at the center of the circle to make it hot. Press the* SPACEBAR *until the SCALE mode appears at the bottom of the prompt line or press the right mouse button to activate the grip cursor menu to select Scale)*
** SCALE ** Specify scale factor or [Base point/Copy/ Undo/Reference/eXit]: **R** *(For Reference)*
Specify reference length <1.0000>: **0.50**
** SCALE ** Specify new length or [Base point/Copy/ Undo/Reference/eXit]: **0.625**
Command: *(Press* ESC *to remove the object highlight and grips)*

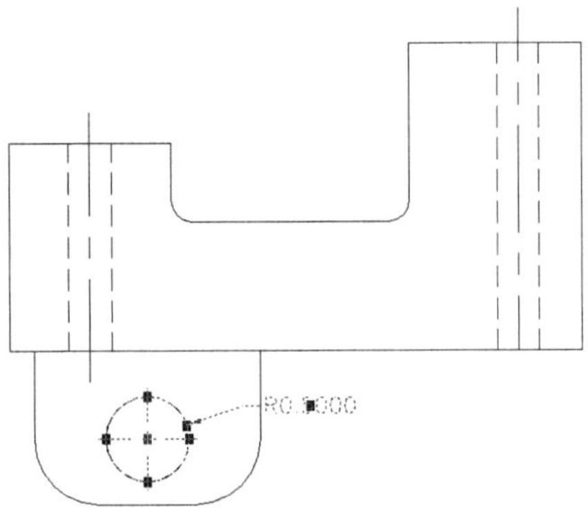

Figure 7–10

STEP 10

At the command prompt, use the grip cursor to select the 0.625 radius dimension shown in Figure 7–11. Notice the appearance of the grips at the center, starting point, and text location of the radius dimension. The dimension text will be relocated with grips.

Command: *(Select the dimension in Figure 7–11)*

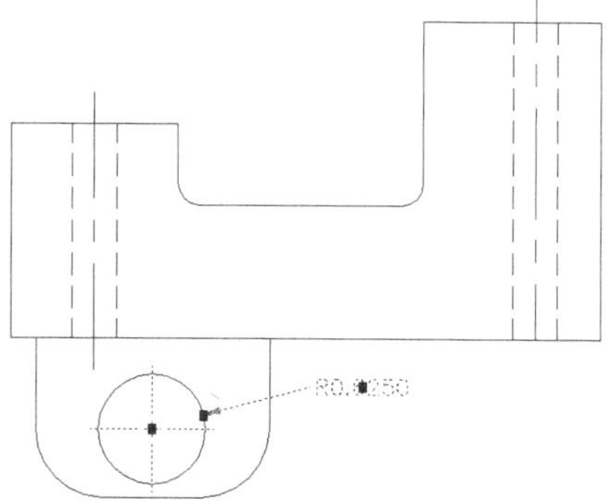

Figure 7–11

STEP 11

Still at the command prompt, select the text grip shown in Figure 7–12. This grip becomes the current base point for the following editing options. Use the STRETCH option to reposition the dimension text to a better location. When this operation is complete, press ESC to remove the object highlight and grips from the display screen.

Command: *(Select the warm grip at the text location to make it hot).*
** STRETCH ** Specify stretch point or [Base point/ Copy/Undo/eXit]: *(Pick a convenient location on the display screen to relocate the radius dimension)*
Command: *(Press ESC to remove the object highlight and grips)*

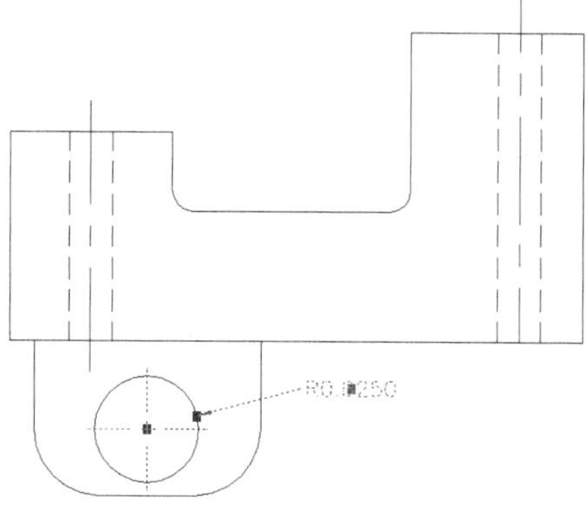

Figure 7–12

Note: To get the tail to re-appear on the leader line, drag the grip selected leader text to a steeper angle placement. It may be necessary to select the Modify Dimension Style dialog box and the Fit tab to change the Fit Options to both text and arrows and the Text Placement to beside the dimension line. This will be covered in greater detail in the Chapter 12, The Dimension Style Manager.

STEP 12

At the command prompt, use the grip cursor to select the centerline shown in Figure 7–13. Notice the appearance of the grips at the endpoints and midpoint of the centerline.

Command: *(Select the centerline shown in Figure 7–13)*

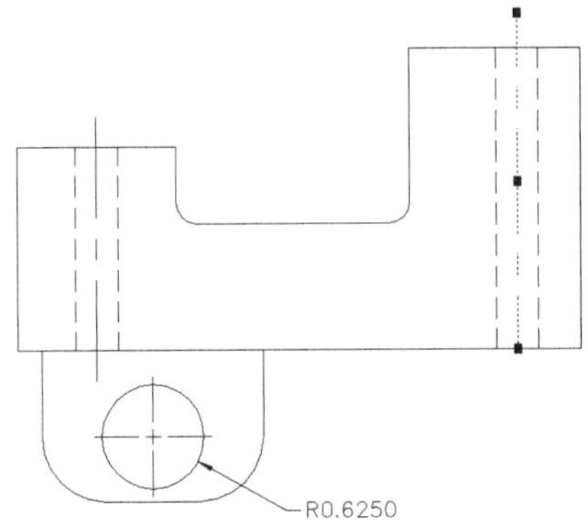

Figure 7–13

STEP 13

While still at the command prompt, select the grip shown in Figure 7–14 at "A." This grip becomes the current base point for the following editing options.

Use the STRETCH mode and BASE option to extend the centerline from a new base point at "B" to the intersection at "C." When this operation is complete, press ESC to remove the object highlight and grips from the display screen.

Command: *(Select the warm grip at the bottom of the centerline at "A" to make it hot).*
** STRETCH ** Specify stretch point or [Base point/Copy/Undo/eXit]: **B** *(For Base)*
Specify base point: *(Select the grip at "B")*
** STRETCH ** Specify stretch point or [Base point/Copy/Undo/eXit]: *(Select the intersection of the centerline and horizontal line at "C")*
Command: *(Press ESC to remove the object highlight and grips)*

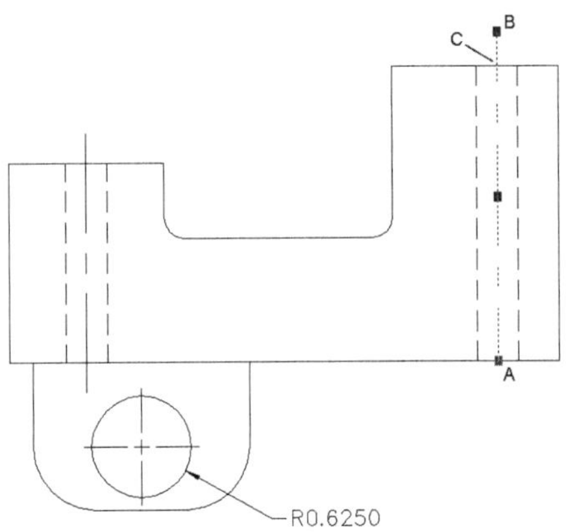

Figure 7–14

STEP 14

At the command prompt, use the grip cursor to select the two hidden lines and centerline shown in Figure 7–15. Notice the appearance of the cold grips at the endpoints and midpoints of the lines.

Command: *(Select the two hidden lines and the centerline)*

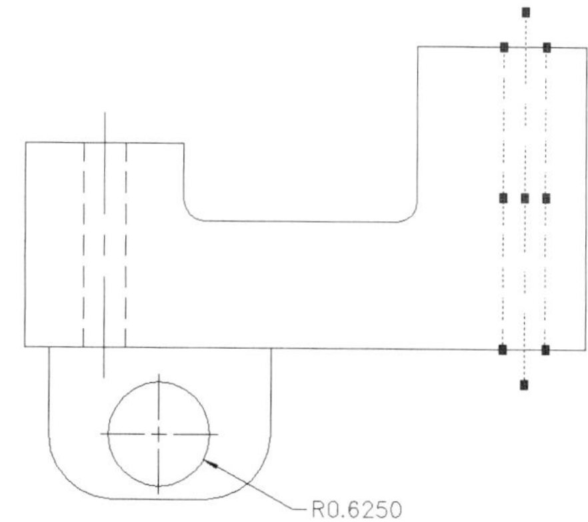

Figure 7–15

STEP 15

While still at the command prompt, select the three grips at the midpoints of the center and hidden lines by holding down SHIFT as they are selected, shown in Figure 7–16. These grips become the current base point for the following editing options. Release SHIFT and pick the selected middle grip again. Use the MOVE option to center the hidden and centerlines to the middle of the horizontal line. When this operation is complete, press ESC to remove the object highlight and grips from the display screen.

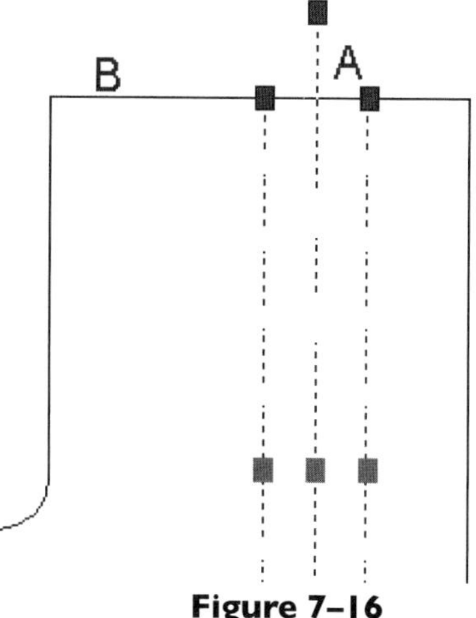

Figure 7–16

Command: *(Use* SHIFT *to individually select the three warm grips at the middle of the three lines to make them hot. Release* SHIFT *and pick the middle grip again. Press the* SPACEBAR *until* MOVE *mode appears at the bottom of the prompt line)*
** MOVE ** Specify move point or [Base point/Copy/Undo/eXit]: **B** *(For Base)*
Specify base point: *(Pick the intersection of the center and horizontal lines at "A")*
** MOVE ** Specify move point or [Base point/Copy/Undo/eXit]: **Mid**
of *(Select the horizontal line at "B")*
Command: *(Press* ESC *to remove the object highlight and grips)*

STEP 16

At the command prompt, use the grip cursor to select all of the objects shown in Figure 7–17. Notice the appearance of the grips at the various key points of these objects.

Command: *(Select all objects shown in Figure 7–17. You can accomplish this easily by using the Automatic Window mode and marking points at "X" and "Y")*

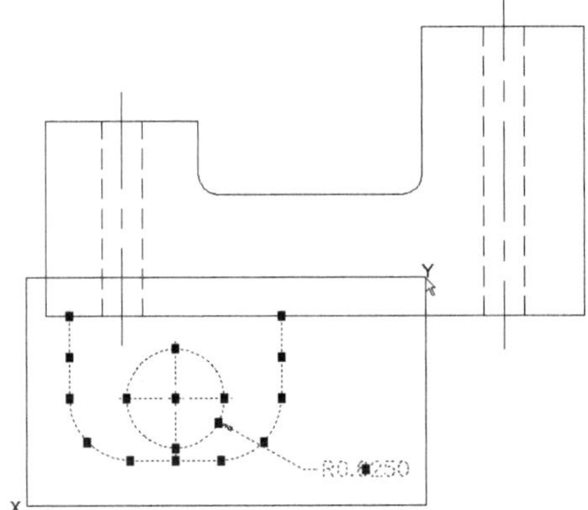

Figure 7–17

STEP 17

While still at the command prompt, pick the grip at "A" and press the SPACEBAR until MIRROR mode appears at the bottom of the prompt line. Mirror the selected objects using the midpoint of the line at "B" shown in Figure 7–18.

Command: *(Pick the warm grip at "A" to make it hot. Press the* SPACEBAR *until* MIRROR *mode appears at the bottom of the prompt line)*
** MIRROR** Specify second point or [Base point/Copy/Undo/eXit]: **B** *(For Base)*
New base point: **Mid**
of *(Select the midpoint of the horizontal line at "B")*
** MIRROR** Specify second point or [Base point/Copy/Undo/eXit]: **@1<90**
Command: *(Press* ESC *to remove the object highlight and grips)*

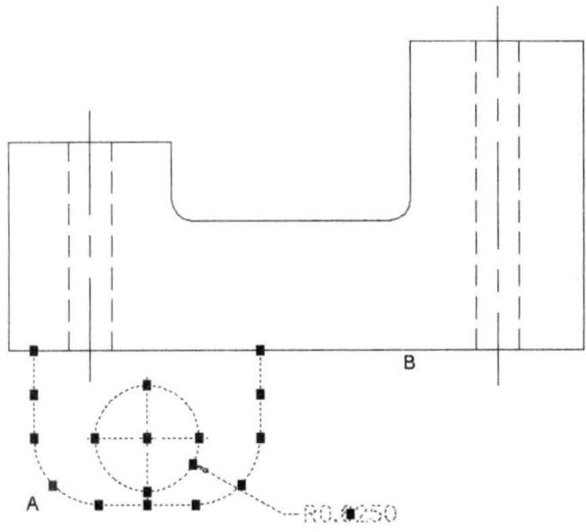

Figure 7–18

The completed object is illustrated in Figure 7–19.

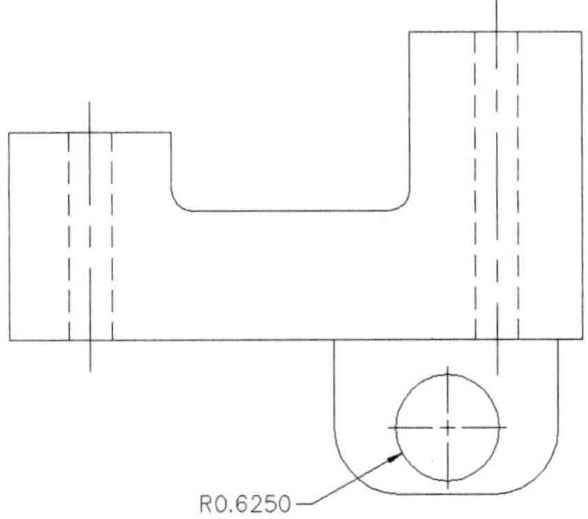

Figure 7–19

Beginning Level Drawing

PROBLEM 7–1

Directions for Problem 1

Using circles and text, develop a drawing of a clock. Try to use only the grip tools in the placement of each of the numerals (equal distance around the clock) and the hands on the clock. The exterior of the clock should be ornate.

Create and apply the following layers through grip selection: Numbers, Hands, Clock Exterior, and Construction Lines. Below is a partially finished clock project to help get you started. This project used grips to mirror the exterior shape, rotate and copy the number placements, and copy the numerals on the clock face.

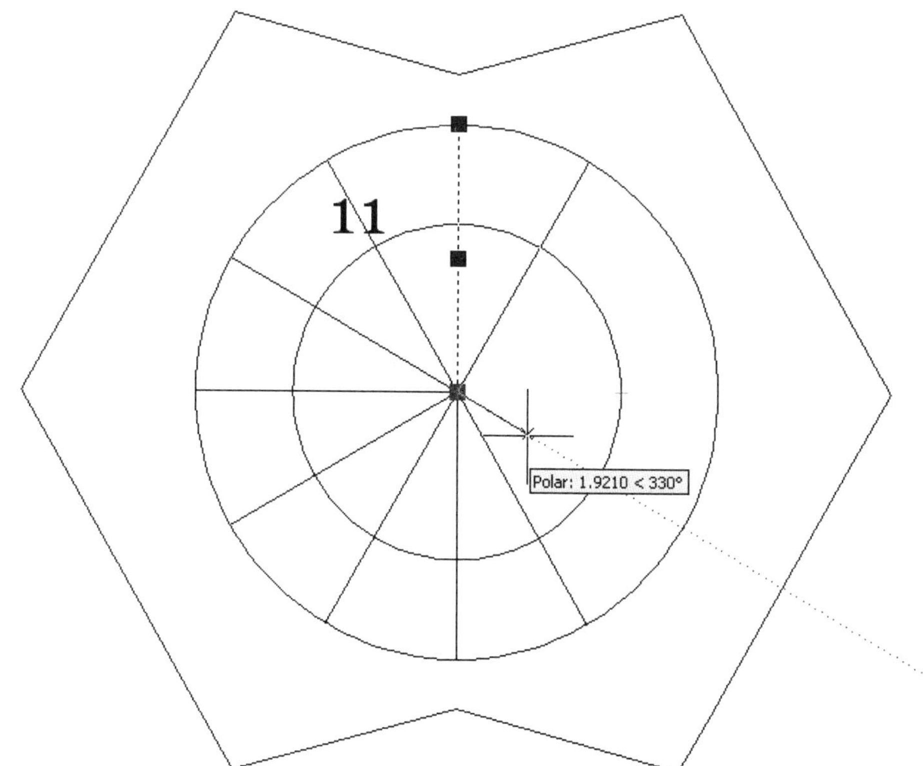

problem EXERCISE

Shape Description/Multiview Projection

The exercises in this chapter introduce:

- Single view drawings
- Primary views
- Projection between views
- Hidden line generation
- Centerline application

TUTORIAL EXERCISE: GAGE.DWG

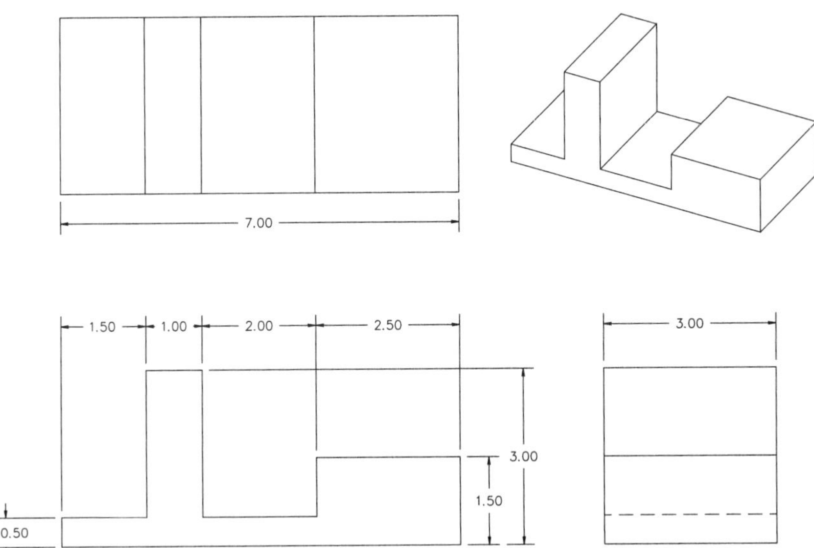

Figure 8–1

Purpose

This tutorial teaches you how to construct a three-view drawing of the Gage, with the aid of temporary tracking points. See Figure 8–1.

System Settings

Begin a new drawing called *Gage.dwg*. Use the Drawing Units dialog box to change the precision from four to two decimal places. Keep the remaining default unit values.

Using the LIMITS command, keep (0,0) for the lower-left corner and change the upper-right corner from (12,9) to (15.50,9.50).

Use the GRID command to change the grid spacing from 0.50 to 0.25 units. Do not turn the Snap or Ortho modes on. It is very important to check and see that the following Object Snap modes are already set: Endpoint, Extension, Intersection, and Center.

Layers

Create the following layers with the format:

Name	Color	Linetype
Object	Green	Continuous
Hidden	Red	Hidden

Suggested Commands

Begin this tutorial by laying out the three primary views using the LINE and OFFSET commands.

Use the TRIM command to clean up any excess line segments. As an alternate projection method, use temporary tracking points in combination with Object Snap options to add features in other views.

Whenever possible, substitute the appropriate command alias in place of the full AutoCAD command in each tutorial step; for example, use "CP" for the COPY command, "L" for the LINE command, and so on. The complete listing of all command aliases is located in Chapter 1, Table 1–2.

STEP 1

Begin by making the Object layer current. Then construct the Front view of the Gage by using absolute coordinates and the Direct Distance mode shown in Figure 8–2. Start the Front view at coordinate 1.50,1.00.

Command: **L** *(For LINE)* [draw>line]

Specify first point: **1.50,1.00** *(At "A")*

Specify next point or [Undo]: *(Move your cursor to the right of the last point and enter* **7.00***...for "B")*

Specify next point or [Undo]: *(Move your cursor up from the last point and enter* **1.50** *... for "C")*

Specify next point or [Close/Undo]: *(Move your cursor to the left of the last point and enter* **2.50** *... for "D")*

Specify next point or [Close/Undo]: *(Move your cursor down from the last point and enter* **1.00** *... for "E")*

Specify next point or [Close/Undo]: *(Move your cursor to the left of the last point and enter* **2.00** *... for "F")*

Specify next point or [Close/Undo]: *(Move your cursor up from the last point and enter* **2.50** *... for "G")*

Specify next point or [Close/Undo]: *(Move your cursor to the left of the last point and enter* **1.00** *... for "H")*

Specify next point or [Close/Undo]: *(Move your cursor to down from the last point and enter* **2.50** *... for "I")*

Specify next point or [Close/Undo]: *(Move your cursor to the left of the last point and enter* **1.50** *... for "J")*

Specify next point or [Close/Undo]: **C** *(To close the shape back to "A" and exit the command)*

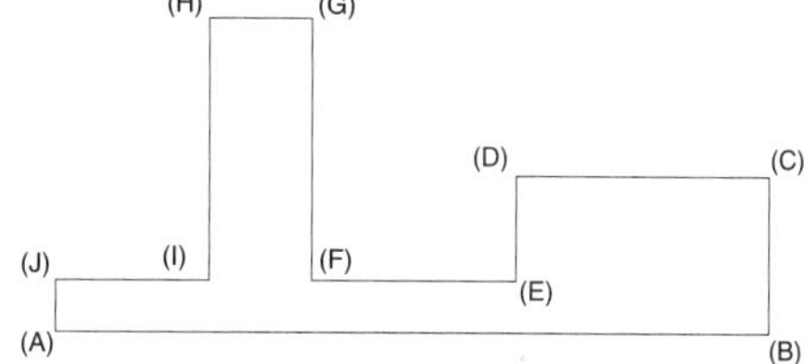

Figure 8–2

STEP 2

Begin the construction of the Top view by locating the lower-left corner at coordinate 1.50,5.50, shown in Figure 8–3.

 Command: **L** *(For LINE)* [draw>line]

Specify first point: **1.50,5.50** *(At "A")*

Specify next point or [Undo]: *(Move your cursor to the right of the last point and enter* **7.00** *...for "B")*

Specify next point or [Undo]: *(Move your cursor up from the last point and enter* **3.00** *... for "C")*

Specify next point or [Close/Undo]: *(Move your cursor to the left of the last point and enter* **7.00** *... for "D")*

Specify next point or [Close/Undo]: **C** *(To close the shape back to "A" and exit the command)*

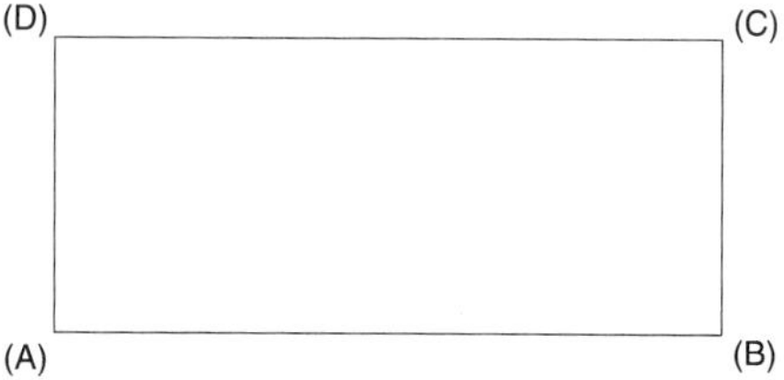

Figure 8–3

STEP 3

Perform a ZOOM ALL command before beginning this step. Begin the construction of the Right Side view by locating the lower-left corner at coordinate 10.00,1.00, as shown in Figure 8–4.

Command: **L** *(For LINE)* [draw>line]
Specify first point: **10.00,1.00** *(At "A")*
Specify next point or [Undo]: *(Move your cursor to the right of the last point and enter* **3.00** *… for "B")*
Specify next point or [Undo]: *(Move your cursor to up from the last point and enter* **3.00** *… for "C")*
Specify next point or [Close/Undo]: *(Move your cursor to the left of the last point and enter* **3.00** *… for "D")*
Specify next point or [Close/Undo]: **C** *(To close the shape back to "A" and exit the command)*

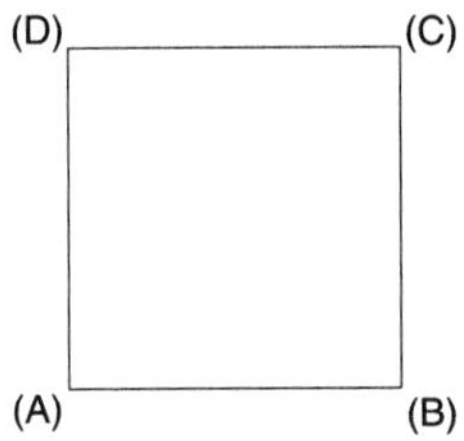

Figure 8–4

STEP 4

Your display should appear similar to the illustration in Figure 8–5, with the placement of the Top view above the Front view and the Right Side view directly to the right of the Front view. A new method of using a temporary snap point will be used to complete the Top and Right Side views.

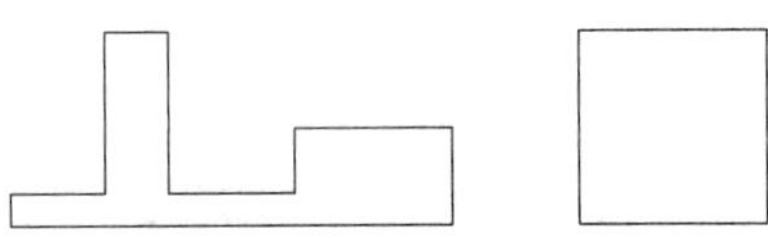

Figure 8–5

STEP 5

Zoom in to the drawing so the Front and Top views appear similar to the illustration in Figure 8–7. Use the following steps to create a temporary snap point in conjunction with the LINE command for completing the Top view. With POLAR, OSNAP, and OTRACK (Object Tracking) all turned on, begin to draw a line in the Top view using the intersection at "A" as a temporary tracking point. The temporary tracking point allows you to jump the gap between views.

 Command: **L** *(For LINE)* [draw>line]
Specify first point: **TT** *(For Temporary Tracking point; or pick the Temporary Tracking Point button in the Standard toolbar in Figure 8–6)*

Specify temporary OTRACK point: *(Pick the intersection at "A" as the temporary tracking point in Figure 8–7)*
Specify first point: *(Pick the intersection at "B")*
Specify next point or [Undo]: *(Pick the Polar intersection at "C")*
Specify next point or [Undo]: *(Press ENTER to exit this command)*

Figure 8–6

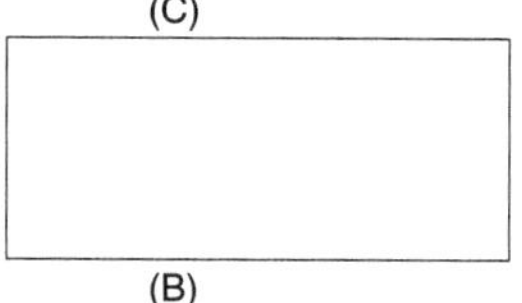

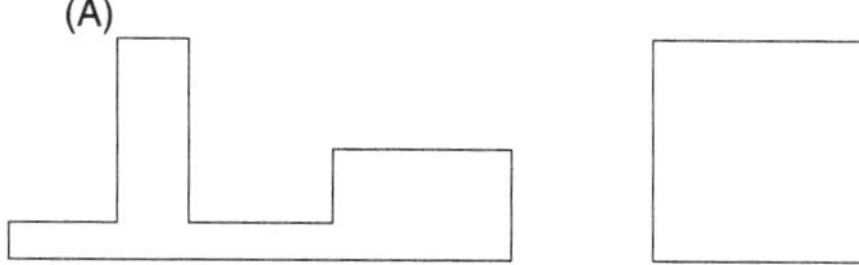

Figure 8–7

STEP 6

Your display should appear similar to the illustration in Figure 8–8. Rather than use a temporary tracking point for the next line, simply duplicate the last line drawn using the COPY command.

 Command: **CP** *(For COPY)* [modify>copy]
Select objects: **L** *(For Last; this should select the last line drawn)*
Select objects: *(Press ENTER to continue)*
Specify base point or displacement, or [Multiple]: *(Pick the endpoint of the line at "A")*
Specify second point of displacement or <use first point as displacement>: *(Pick the endpoint of the line at "B")*

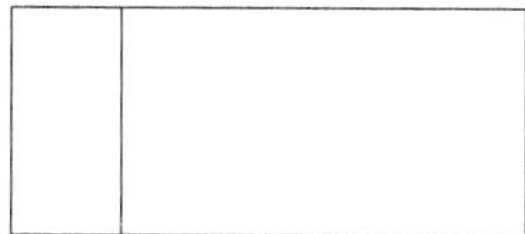

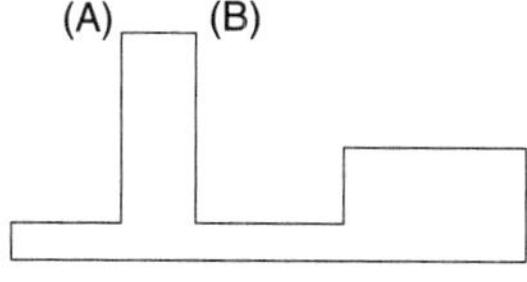

Figure 8–8

STEP 7

After you perform the COPY command, your display should appear similar to the illustration in Figure 8–9. Use a temporary tracking point to project the last object line in the Top view from the Front view. Use the procedure from Step 5 as a guide for performing this operation.

 Command: **L** *(For LINE)* [draw>line]

Specify first point: **TT** *(For Temporary Tracking point; or pick the Temporary Tracking Point button in the Standard toolbar in Figure 8–6)*

Specify temporary OTRACK point: *(Pick the intersection at "A" as the temporary tracking point)*

Specify first point: *(Pick the intersection at "B")*

Specify next point or [Undo]: *(Pick the Polar intersection at "C")*

Specify next point or [Undo]: *(Press* ENTER *to exit this command)*

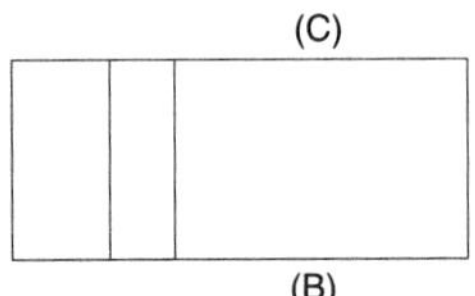

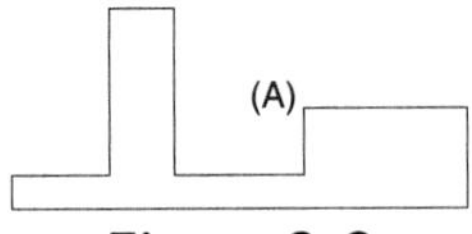

Figure 8–9

STEP 8

The complete Front and Top views are illustrated in Figure 8–10. Use the same procedure with the temporary tracking point for completing the lines located in the Right Side view.

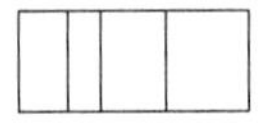

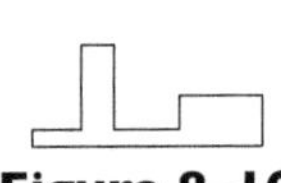

Figure 8–10

STEP 9

Before adding the missing lines to the Right Side view, first use the ZOOM command to magnify the area illustrated in Figure 8–11.

 Command: **Z** *(For ZOOM)* [view>zoom]

Specify corner of window, enter a scale factor *(nX or nXP)*, or [All/Center/Dynamic/Extents/Previous/Scale/Window] <real time>: *(Pick a point approximately at "A")*

Specify opposite corner: *(Pick a point approximately at "B")*

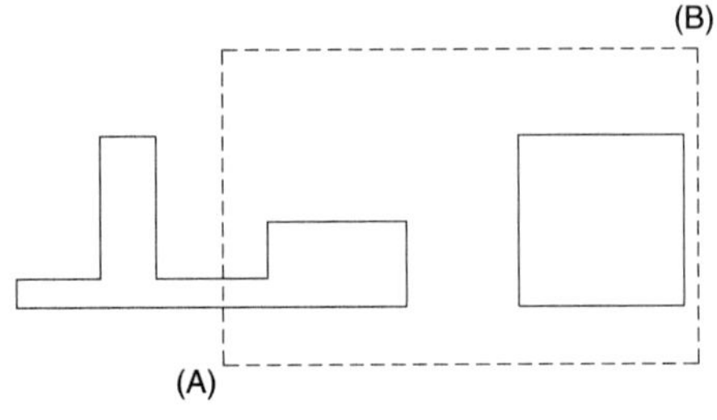

Figure 8–11

STEP 10

Use the procedure from Step 7 to construct the line in the Right Side view using a Temporary Tracking point (see Figure 8–12).

 Command: **L** *(For LINE)* [draw>line]

Specify first point: **TT** *(For Temporary Tracking point; or click the Temporary Tracking Point button in the Standard toolbar in Figure 8–6)*

Specify temporary OTRACK point: *(Pick the intersection at "A" as the temporary tracking point)*

Specify first point: *(Pick the intersection at "B")*

Specify next point or [Undo]: *(Pick the Polar intersection at "C")*

Specify next point or [Undo]: *(Press* ENTER *to exit this command)*

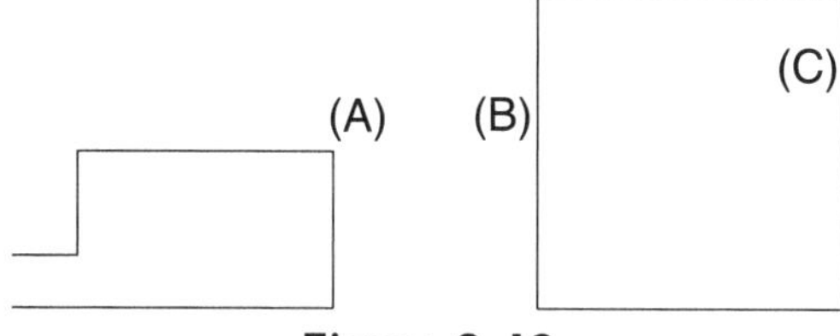

Figure 8–12

STEP 11

Use the procedure from Step 10 to construct the line in the Right Side view using a Temporary Tracking point in Figure 8–13.

 Command: **L** *(For LINE)* [draw>line]

Specify first point: **TT** *(For Temporary Tracking point; Pick the Temporary Tracking Point button in the Standard toolbar in Figure 8–6)*

Specify temporary OTRACK point: *(Pick the intersection at "A" as the temporary tracking point)*

Specify first point: *(Pick the intersection at "B")*

Specify next point or [Undo]: *(Pick the Polar intersection at "C")*

Specify next point or [Undo]: *(Press* ENTER *to exit this command)*

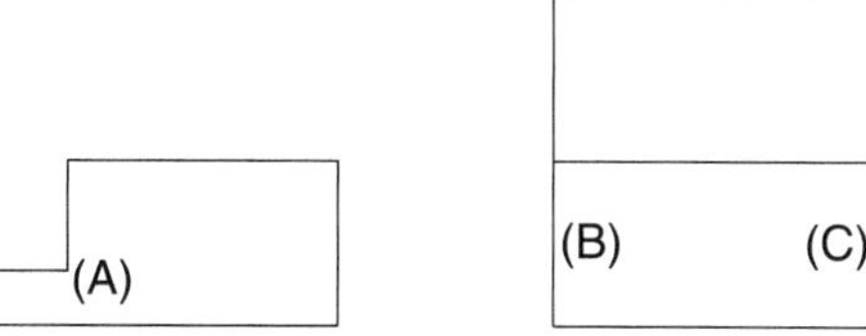

Figure 8–13

STEP 12

Use the Layer Control box options, shown in Figure 8–14, to change the layer of the last line drawn from **Object** to **Hidden**. When finished, press ESC twice to remove the object highlight and the grips.

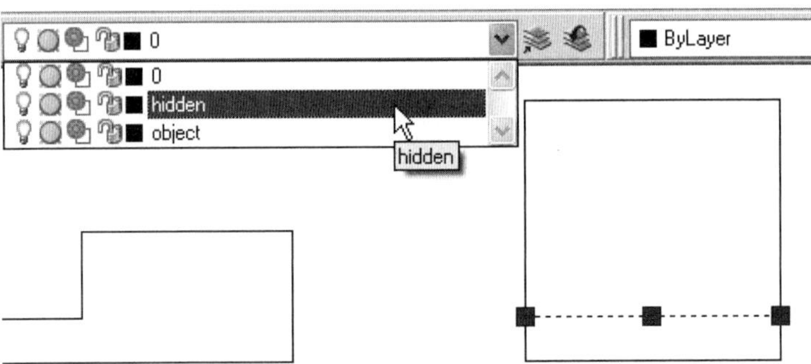

Figure 8–14

STEP 13

The completed Right Side view (illustrated in Figure 8–15) is complete with visible lines and hidden lines that represent invisible surfaces. Use the ZOOM-Previous or ZOOM-All options to demagnify your display and show all three views of the Gage.

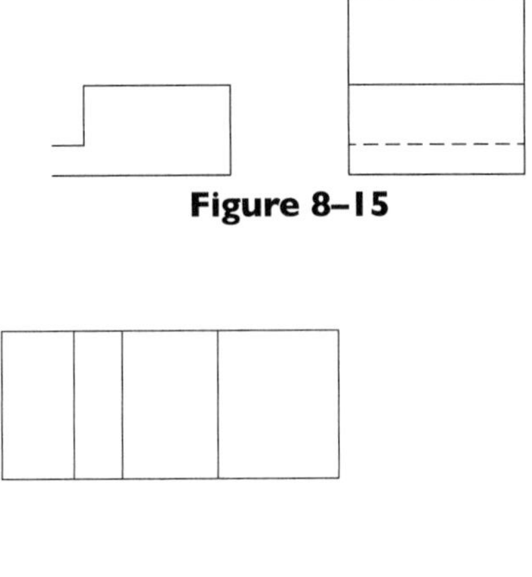

Figure 8–15

The next step for the completed drawing with tree views, illustrated in Figure 8–16, would be to add dimensions to the views for manufacturing purposes. This topic will be discussed in a later chapter.

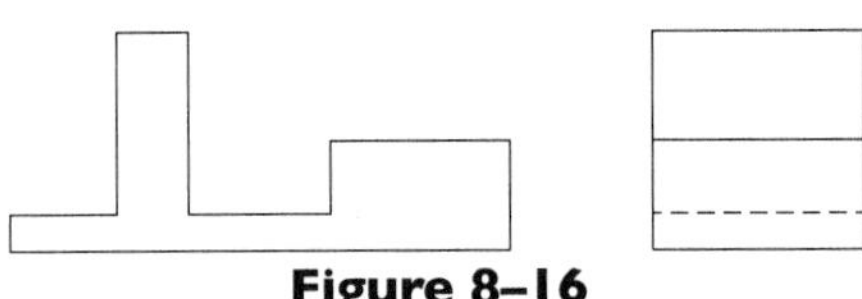

Figure 8–16

TUTORIAL EXERCISE: ORTHOGRAPHIC BLOCK.DWG

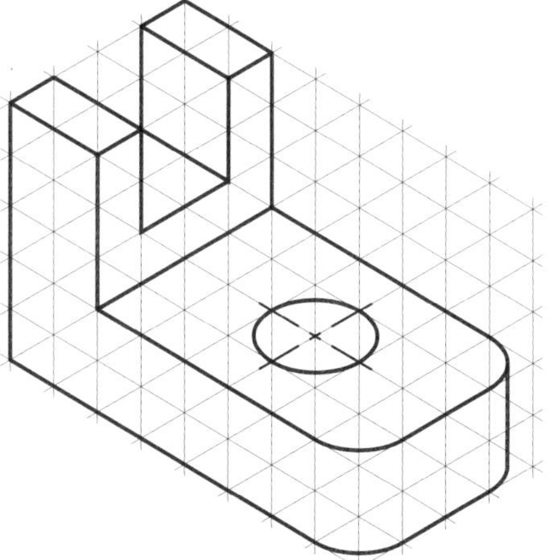

Figure 8–17

Purpose

Using the isometric blocks shown in Figure 8–17, construct the Front, Top, and Right Side views of the Orthographic Block.

System Settings

Begin a new drawing called *Orthographic Block.dwg* Use the Drawing Units dialog box to change the precision from four to two decimal places. Keep the remaining default unit values.

Using the LIMITS command, keep (0,0) for the lower-left corner and change the upper-right corner from (12,9) to (15.50,9.50).

Use the GRID command to change the grid spacing from 1.00 to 0.25 units. Do not turn the Snap or Ortho modes on. It is very important to check and see that the following Object Snap modes are already set: Endpoint, Extension, Intersection, and Center.

Layers

Create the following layers with the format:

Name	Color	Linetype
Object	Green	Continuous
Hidden	Red	Hidden

Suggested Commands

Begin this tutorial by laying out the three primary views using the LINE and OFFSET commands.

Use the TRIM command to clean up any excess line segments. As an alternative projection method, use temporary tracking points in combination with Object Snap options, to add features in other views.

Whenever possible, substitute the appropriate command alias in place of the full AutoCAD command in each tutorial step; for example, use "CP" for the COPY command, "L" for the LINE command, and so on. The complete listing of all command aliases is located in Chapter 1, Table 1–2.

tutorial EXERCISE

STEP 1

Begin constructing the Orthographic Block by laying out the Front, Top, and Right Side views using only the overall dimensions. Do not be concerned about details such as holes or slots; these will be added to the views at a later step. The length of the object shown in Figure 8–18 is 8 grid units. The height of the object is 5 grid units; the depth of the object is 4 units. The distance between views is 5 grid units.

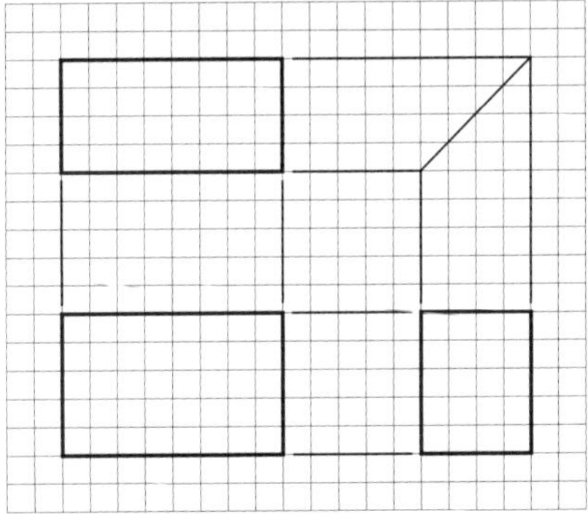

Figure 8–18

STEP 2

Once the overall dimensions have been used to lay out the Front, Top, and Right Side views, begin adding visible details to the views. The "L" shape is added to the Front view; the hole and corner fillets are added to the Top view; the rectangular slot is added to the Right Side view (see Figure 8–19).

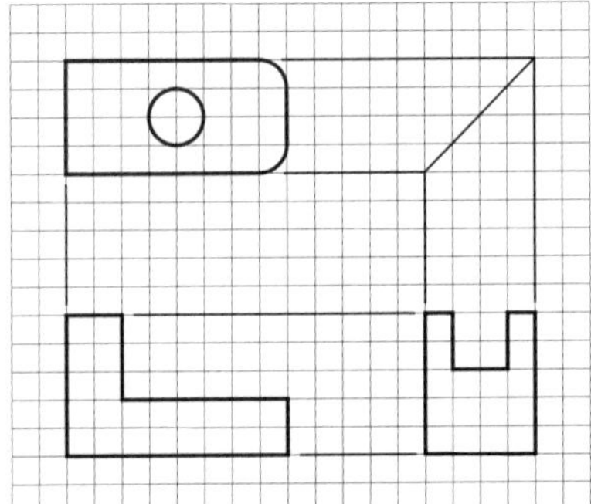

Figure 8–19

STEP 3

Visible edges from hole and slot features are projected onto other views. Slot information is added to the Top view and Height information is projected onto the Right Side view from the Front view. At this point, only add visible information to other views where required (see Figure 8–20).

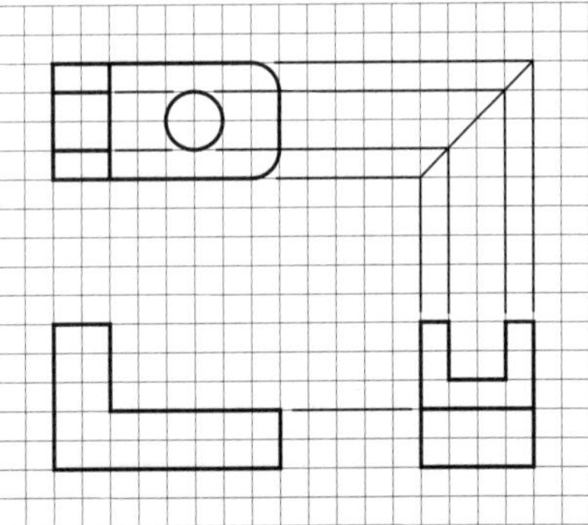

Figure 8–20

STEP 4

Now project hidden features to the views. The hole projection is hidden in the Front view along with the slot visible in the Right Side view. The hole is also hidden in the Right Side view (see Figure 8–21).

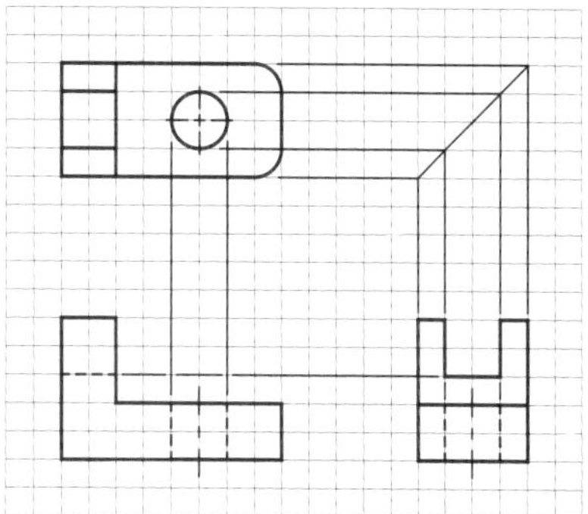

Figure 8–21

STEP 5

The completed multiview drawing solution is illustrated in Figure 8–22. Dimensions are added (see Figure 8–23) for use when the object is to be constructed. Proper placement of dimensions will be discussed in a later chapter.

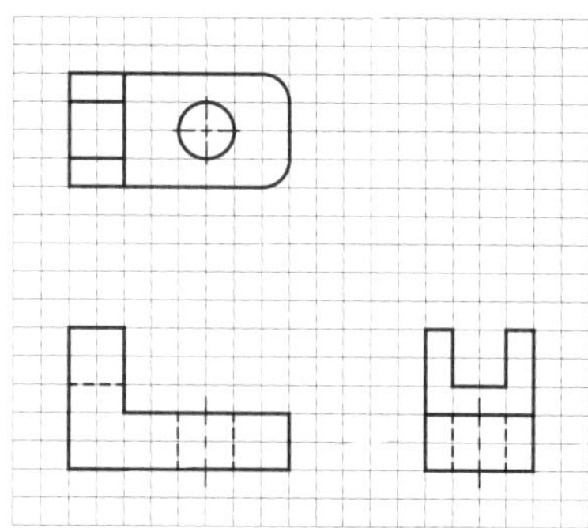

Figure 8–22

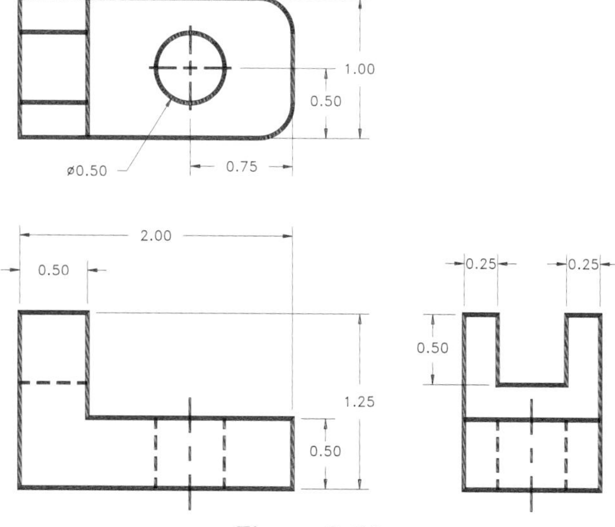

Figure 8–23

Beginning Level Drawings

Directions for Problems 8–1 and 8–2:

Find the missing lines in these problems and sketch the correct solution.

PROBLEM 8–1

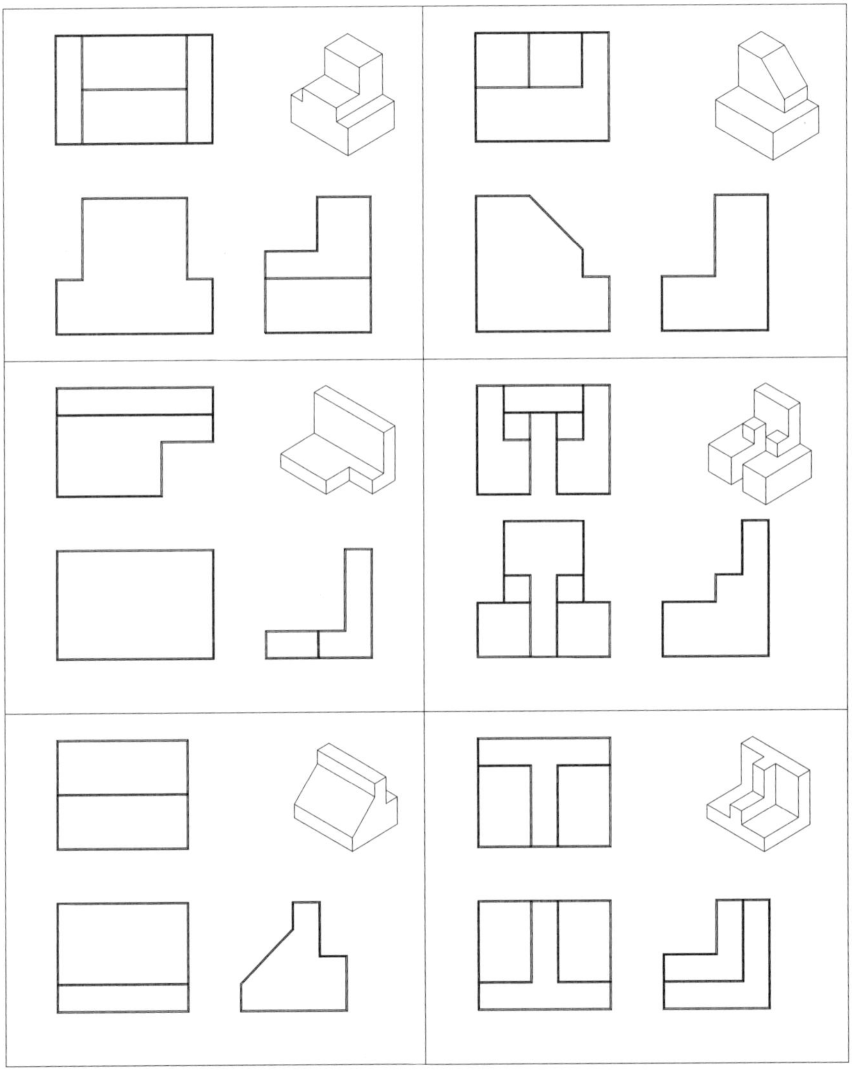

PROBLEM 8–2

Directions for Problems 8–3 through 8–25:

Constructs a multiview drawing by sketching the Front, Top, and Right Side views. Use a grid of 0.25 units to assist in the construction of the sketches. These drawings may also be constructed directly in the CAD system through AutoCAD commands.

PROBLEM 8–3

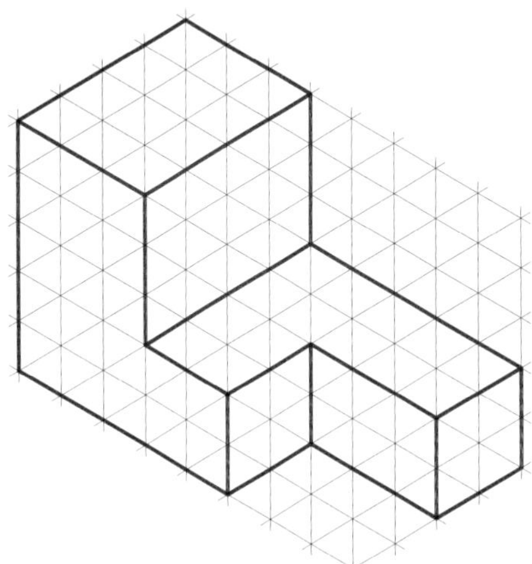

PROBLEM 8–5

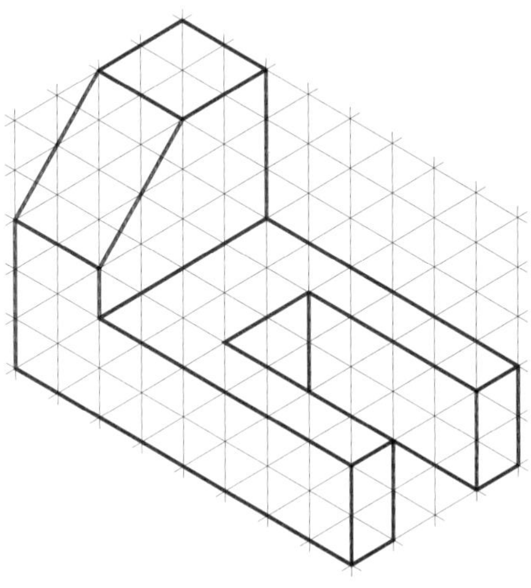

PROBLEM 8–4

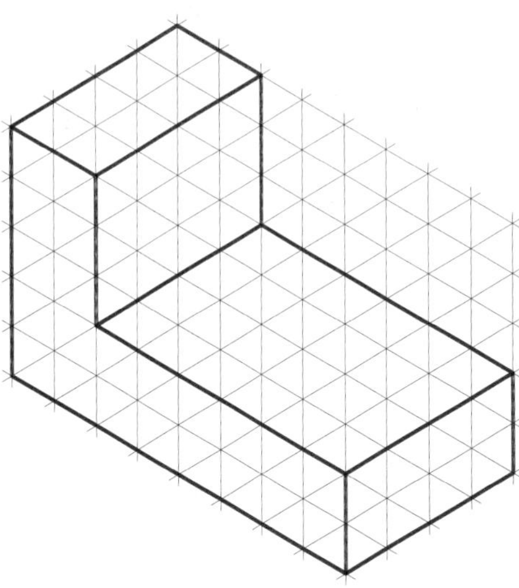

PROBLEM 8–6

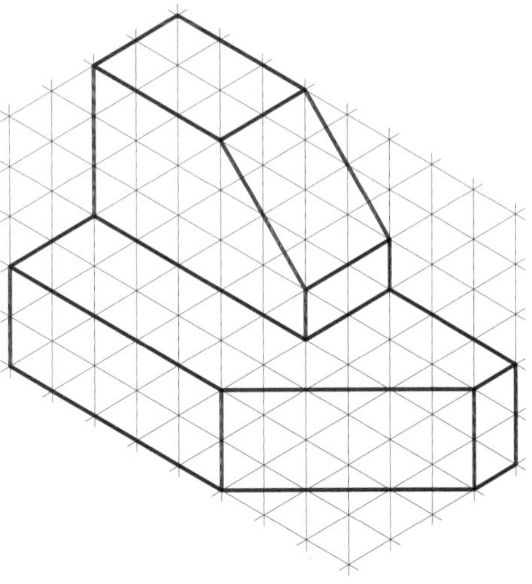

PROBLEM 8–7

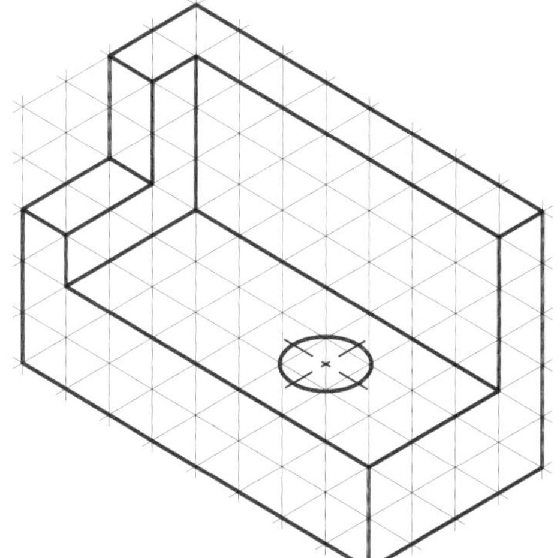

PROBLEM 8–9

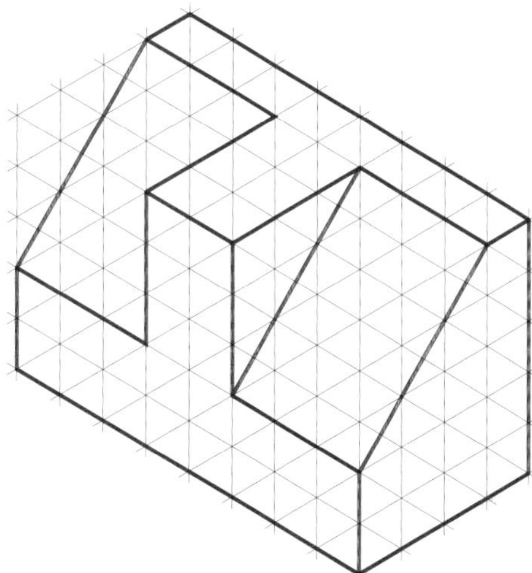

PROBLEM 8–8

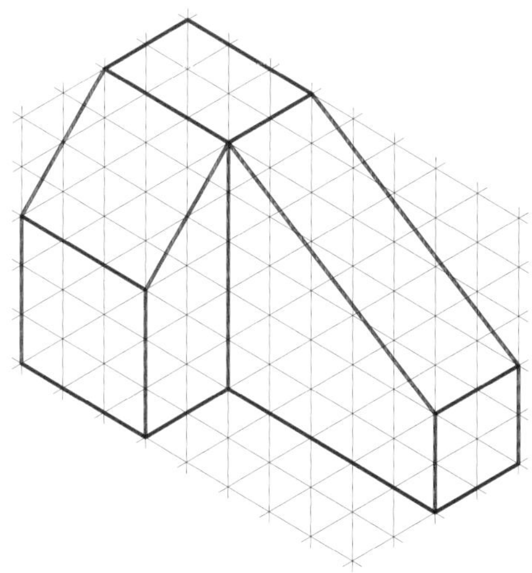

PROBLEM 8–10

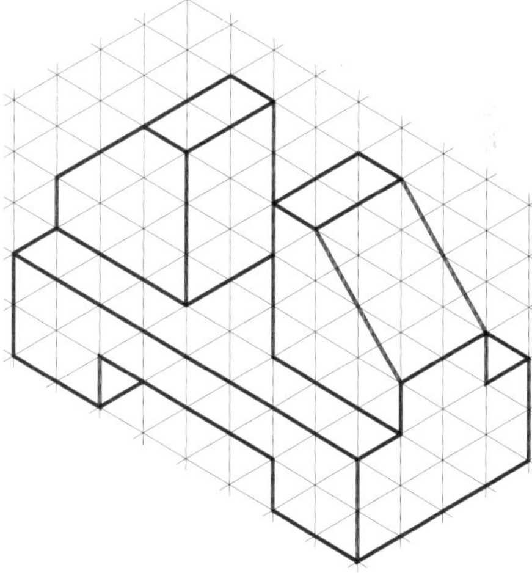

PROBLEM 8–11

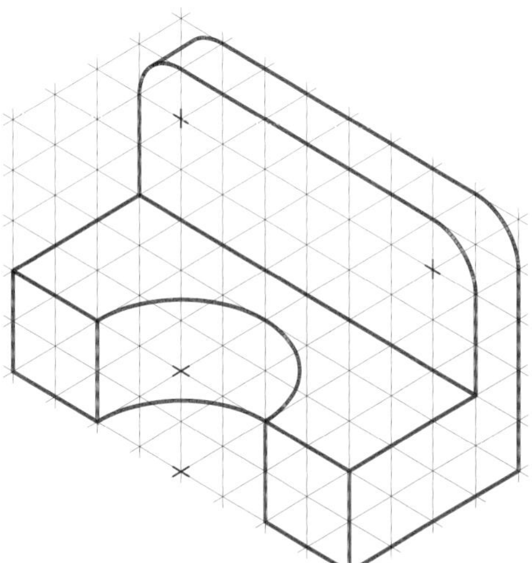

PROBLEM 8–13

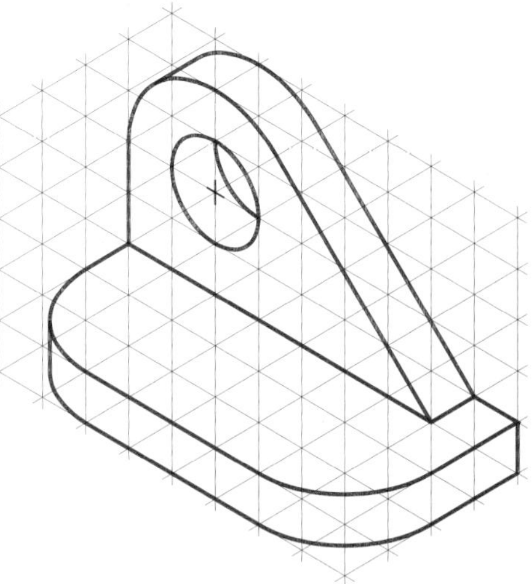

PROBLEM 8–12

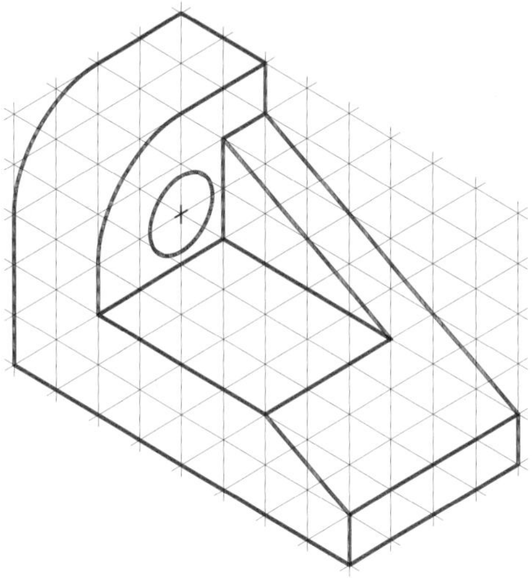

PROBLEM 8–14

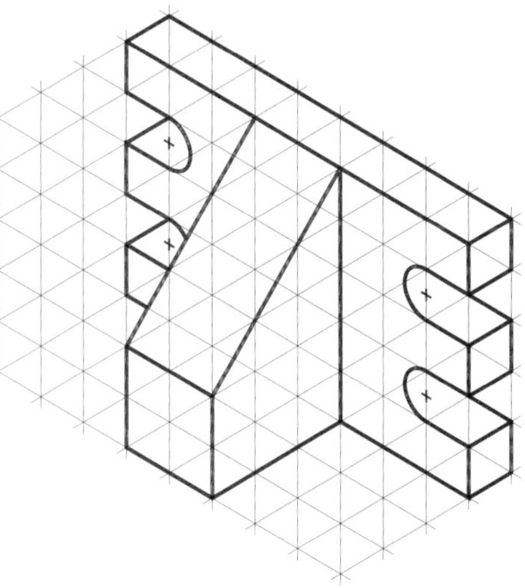

PROBLEM 8–15

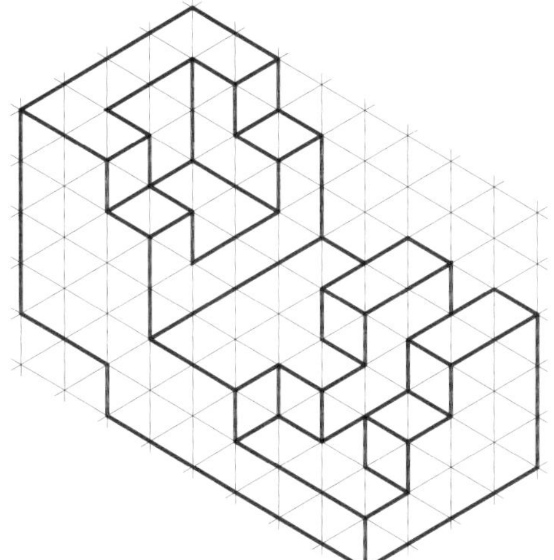

PROBLEM 8–17

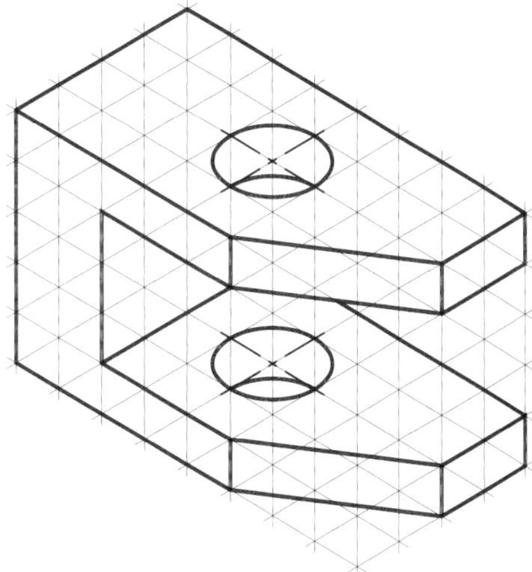

PROBLEM 8–16

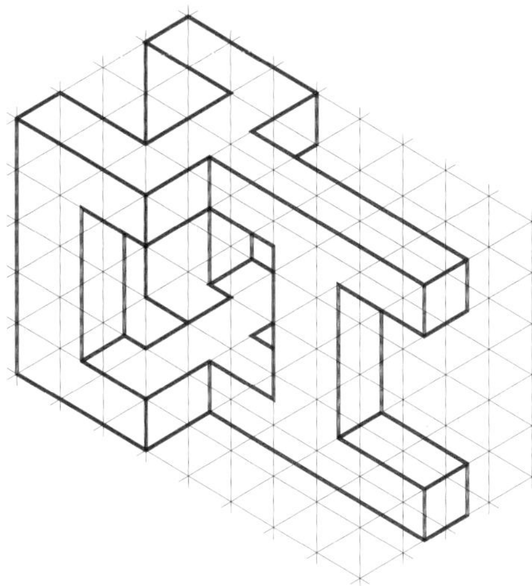

PROBLEM 8–18

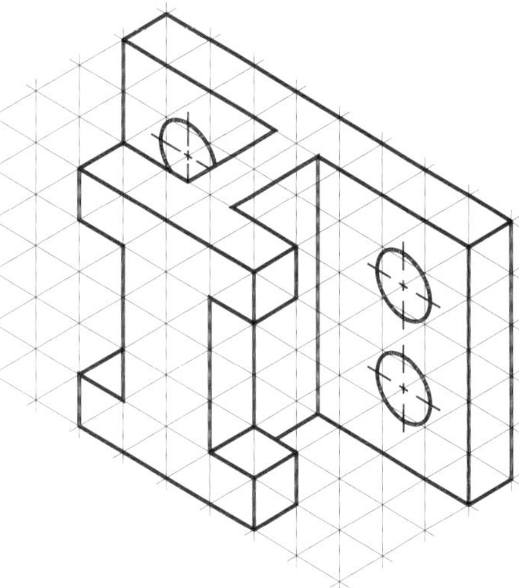

PROBLEM 8–19

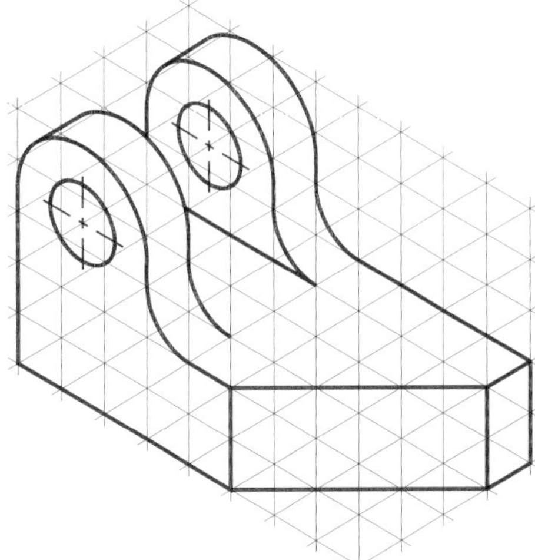

PROBLEM 8–21

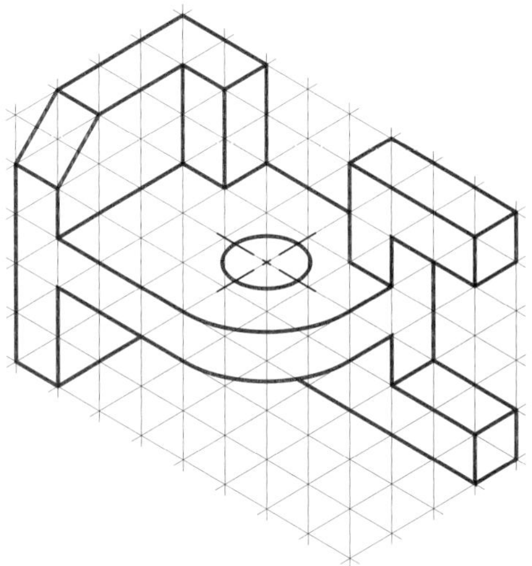

PROBLEM 8–20

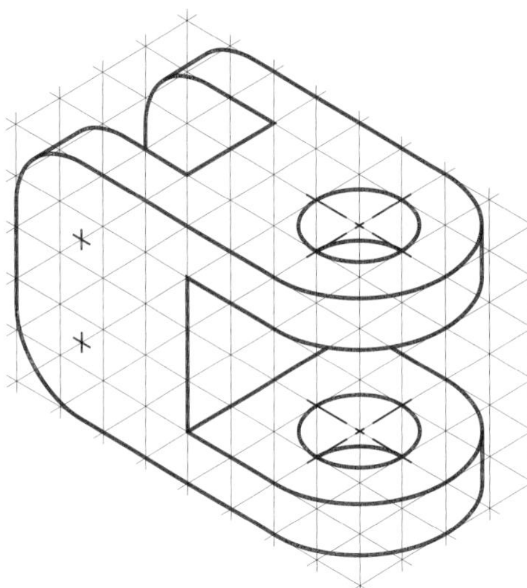

PROBLEM 8–22

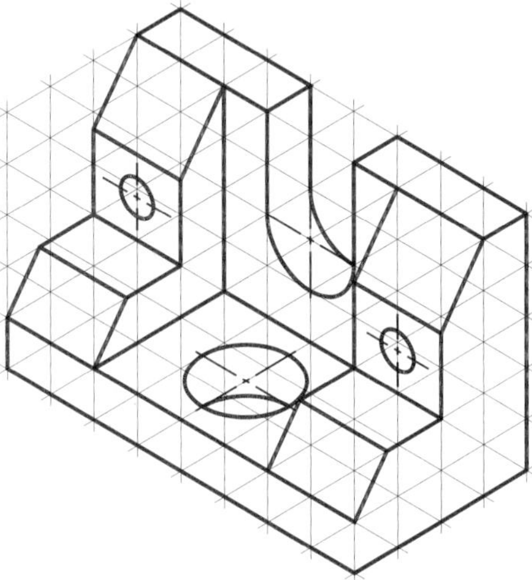

PROBLEM 8–23

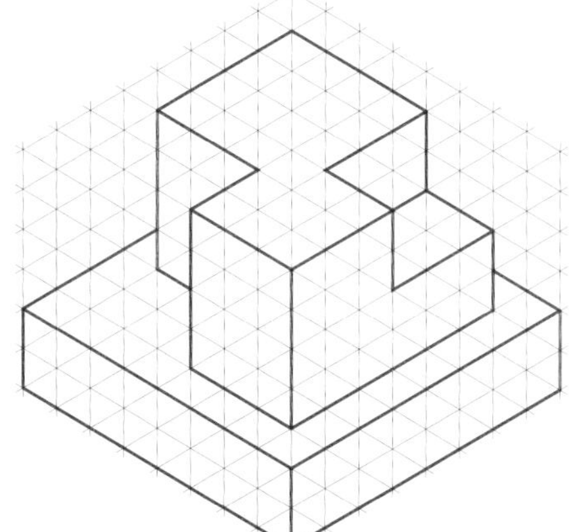

PROBLEM 8–24

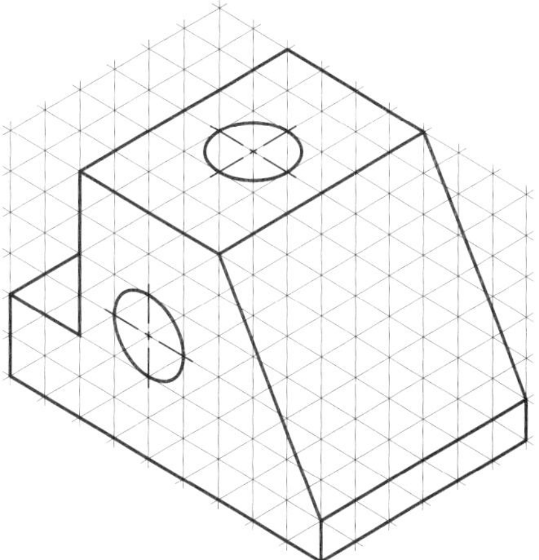

PROBLEM 8–25

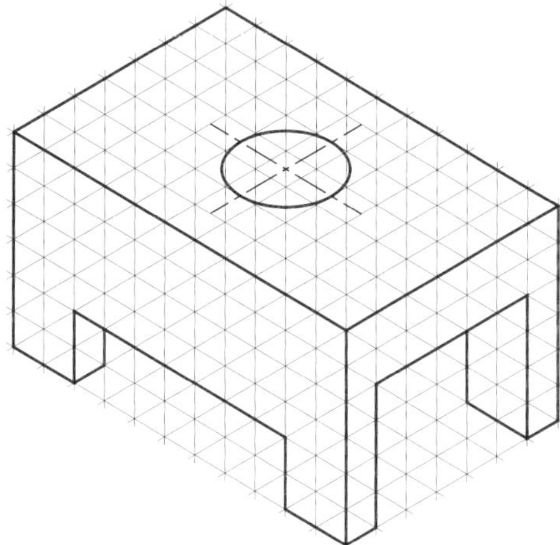

Directions for Problems 8–26 through 8–47:

Construct a multiview drawing of each object. Be sure to construct only those views that accurately describe the object.

PROBLEM 8–26

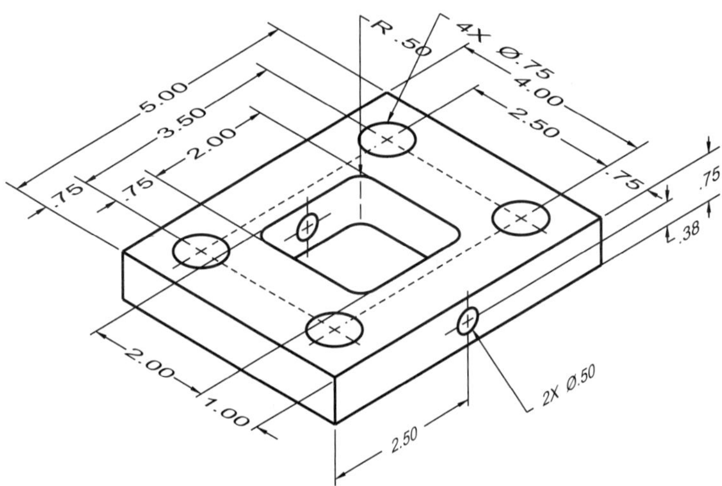

PROBLEM 8–27

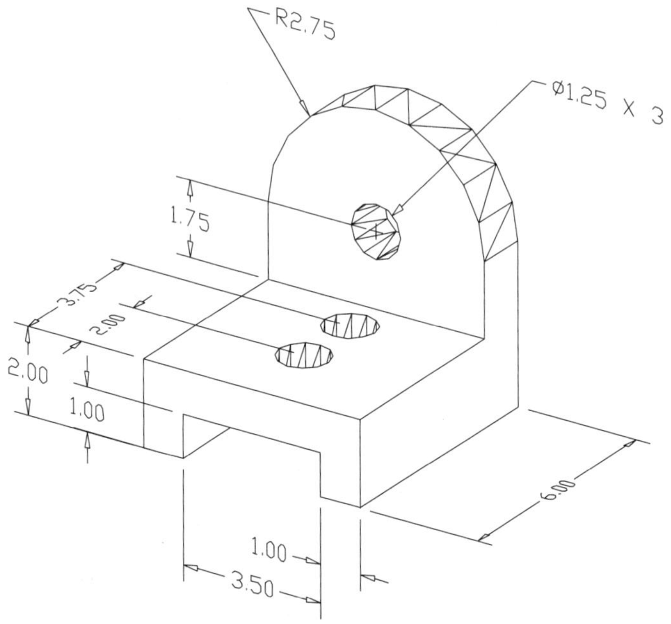

PROBLEM 8–28

Intermediate Level Drawings

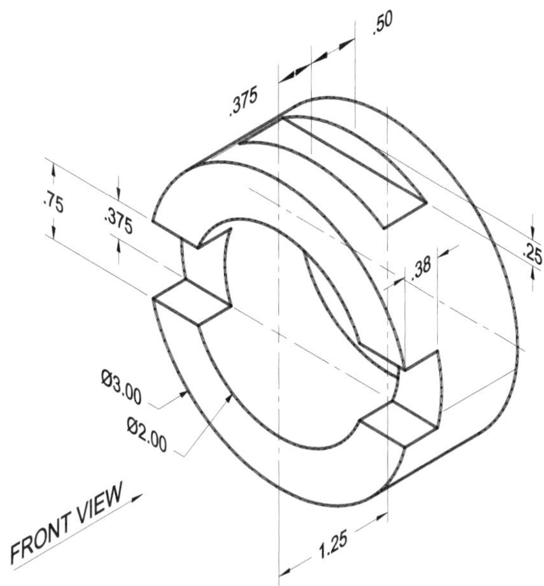

PROBLEM 8–29

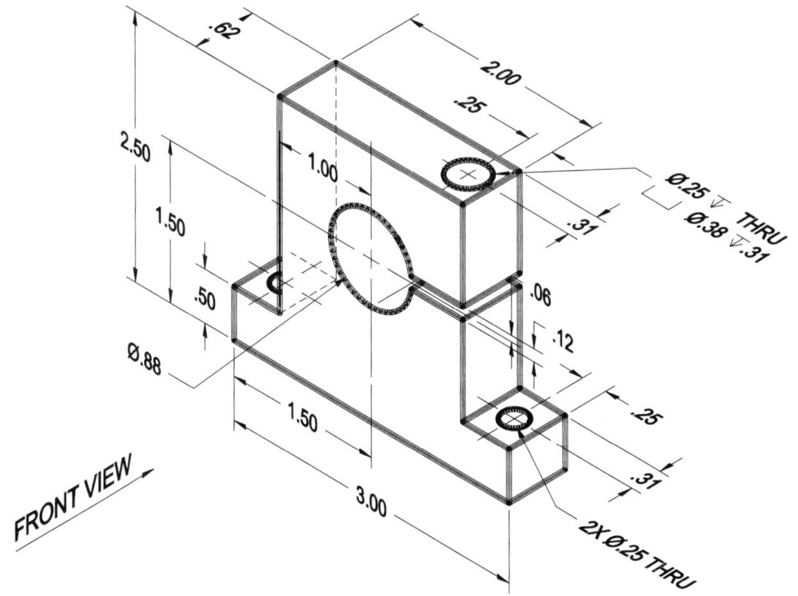

PROBLEM 8–30

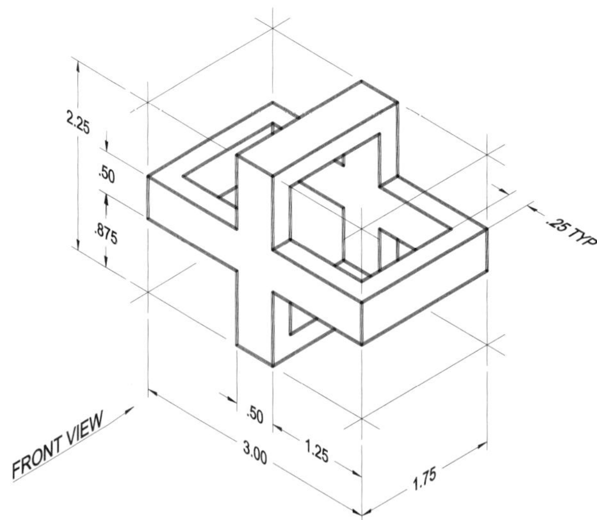

PROBLEM 8–31

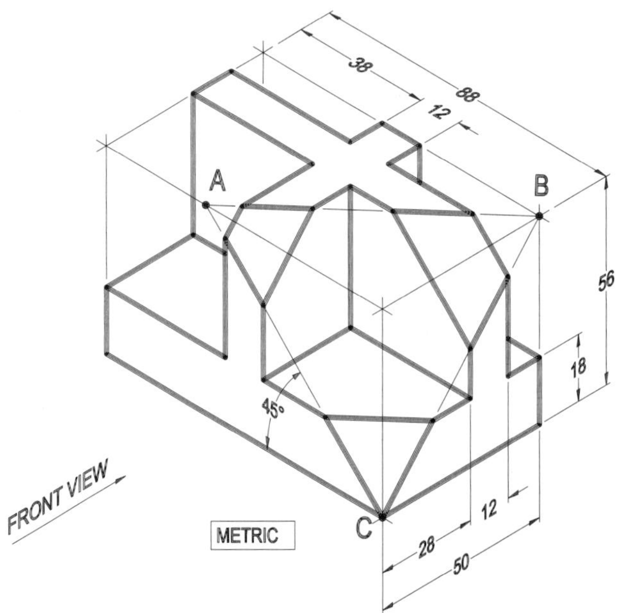

PROBLEM 8–32

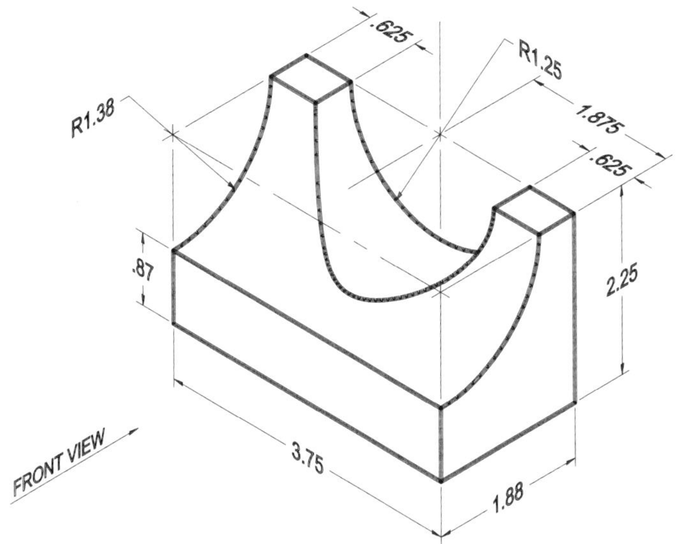

PROBLEM 8–33

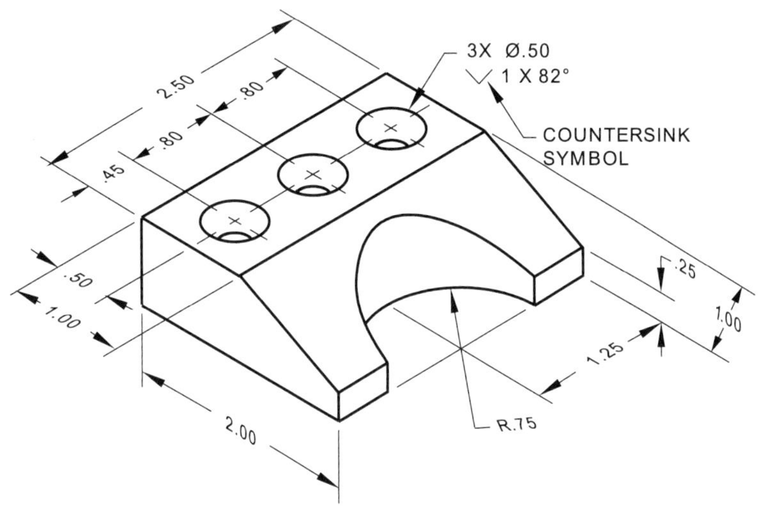

PROBLEM 8–34

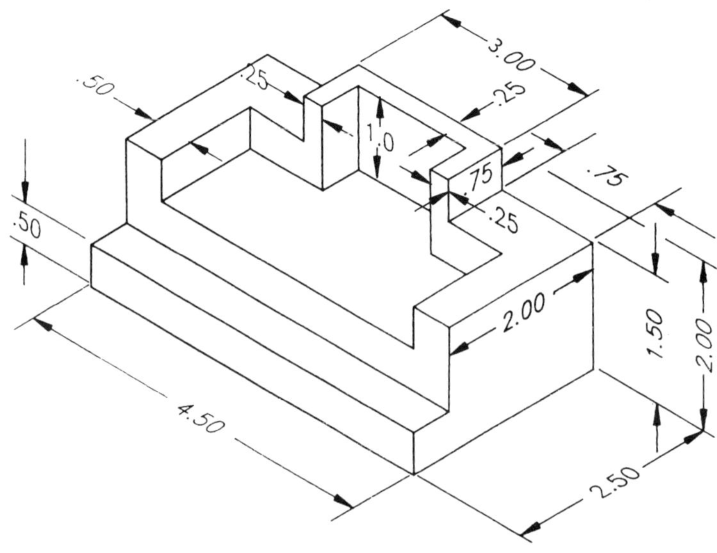

PROBLEM 8–35

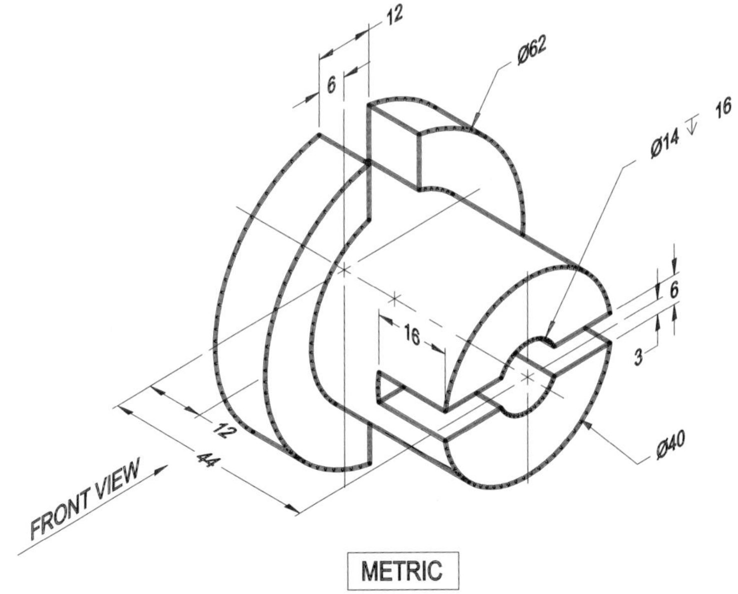

PROBLEM 8–36

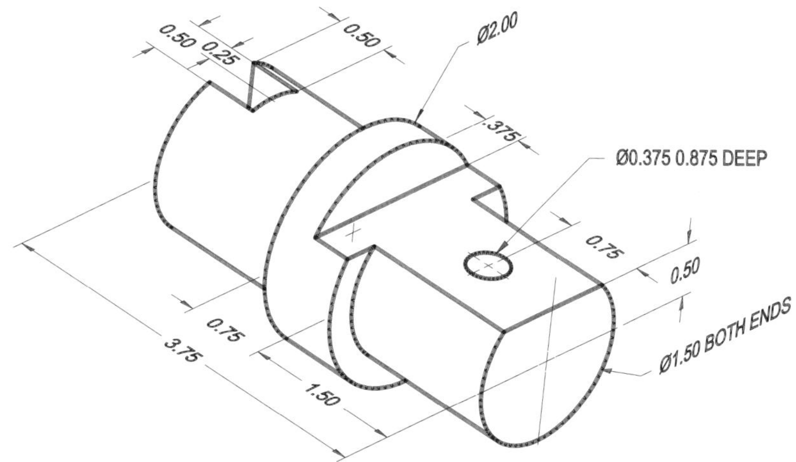

PROBLEM 8–37

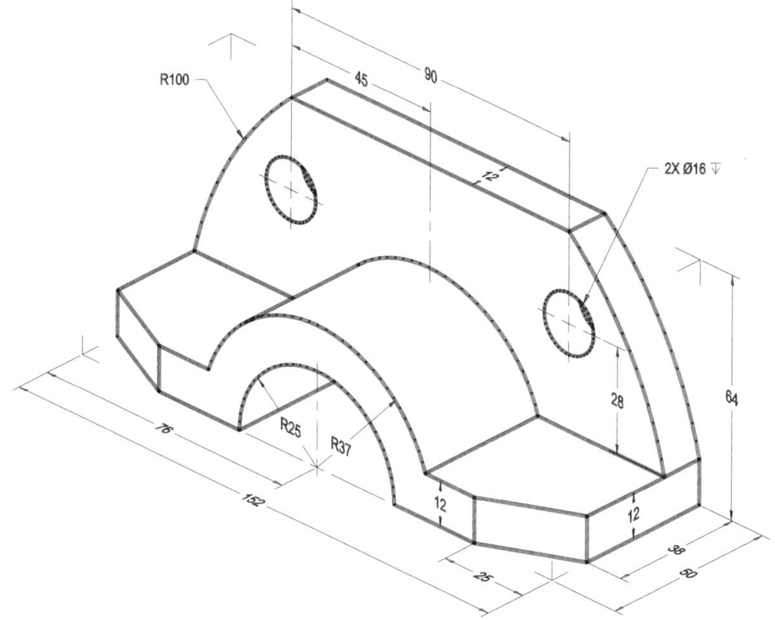

PROBLEM 8–38

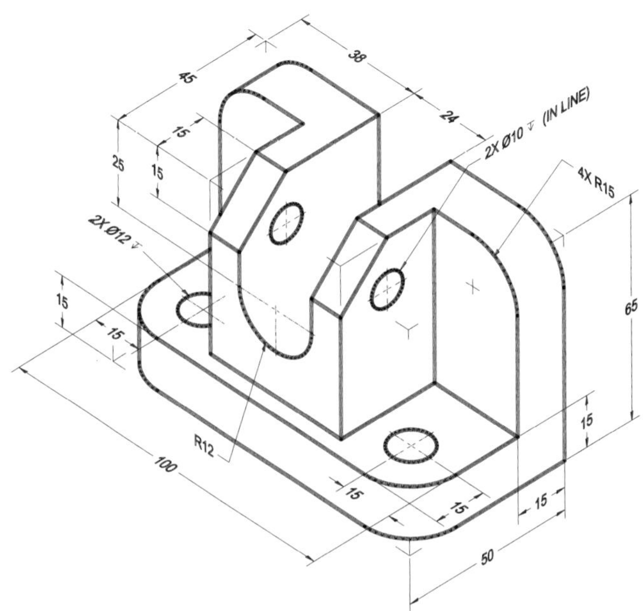

PROBLEM 8–39

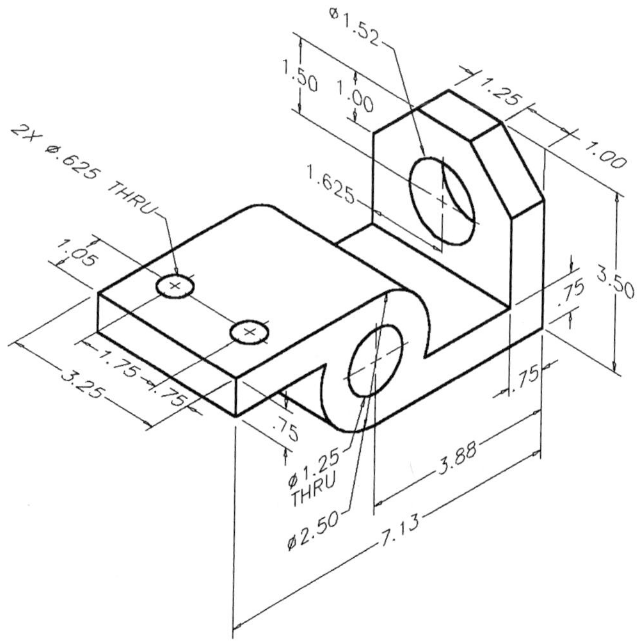

Advanced Level Drawings

PROBLEM 8–40

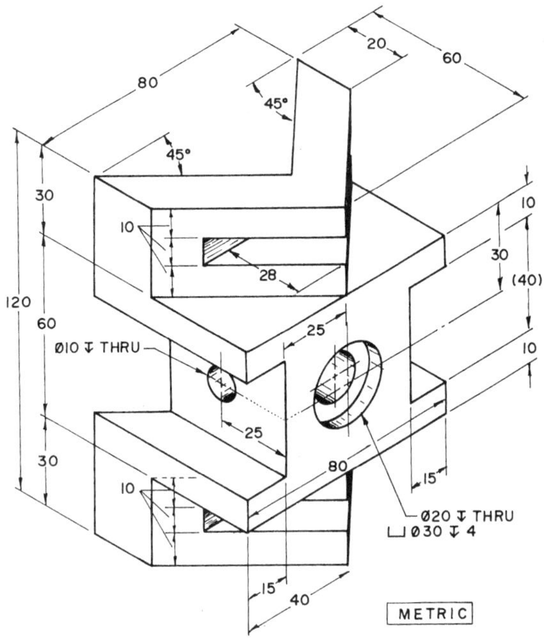

PROBLEM 8–41

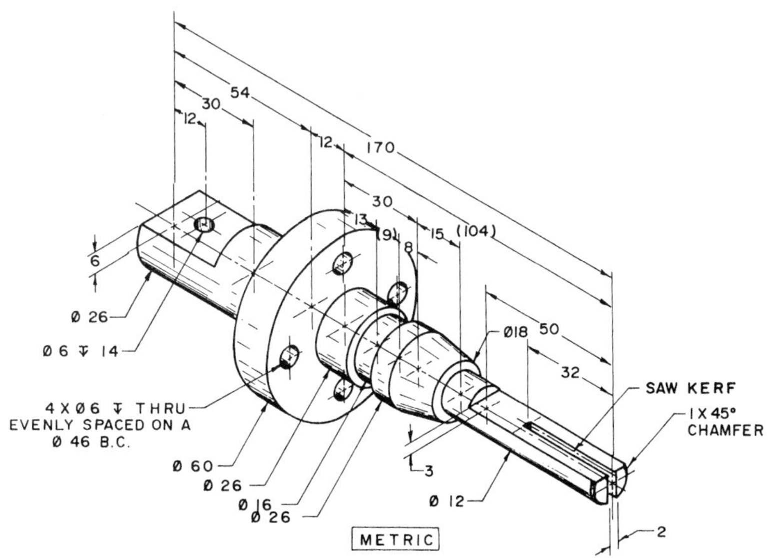

PROBLEM 8–42

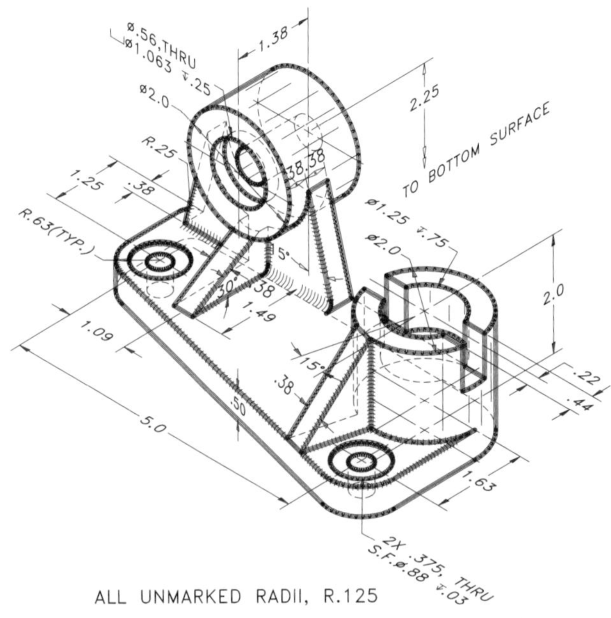

ALL UNMARKED RADII, R.125

PROBLEM 8–43

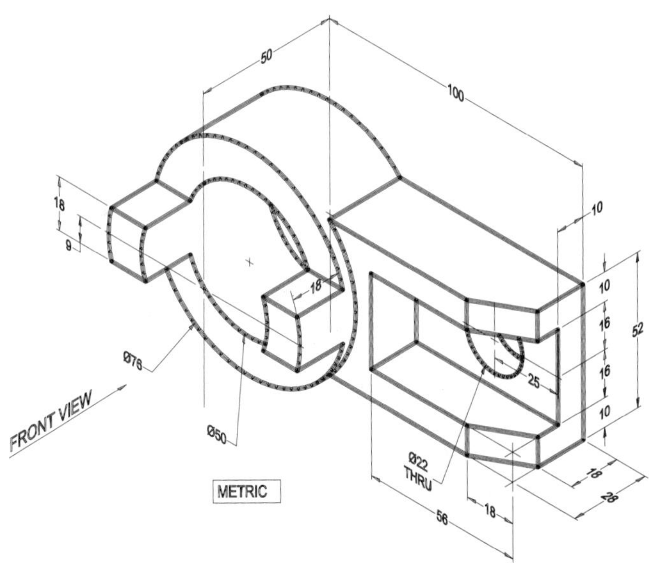

PROBLEM 8–44

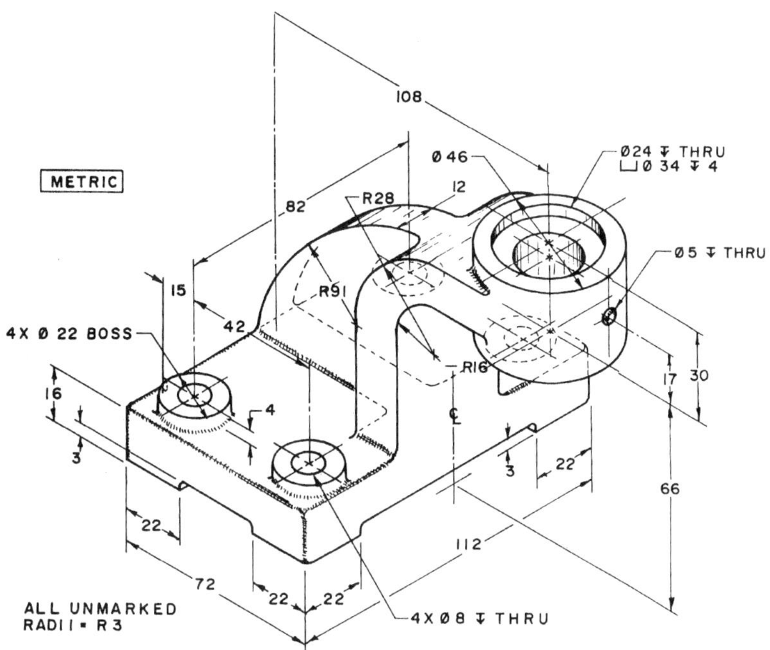

PROBLEM 8–45

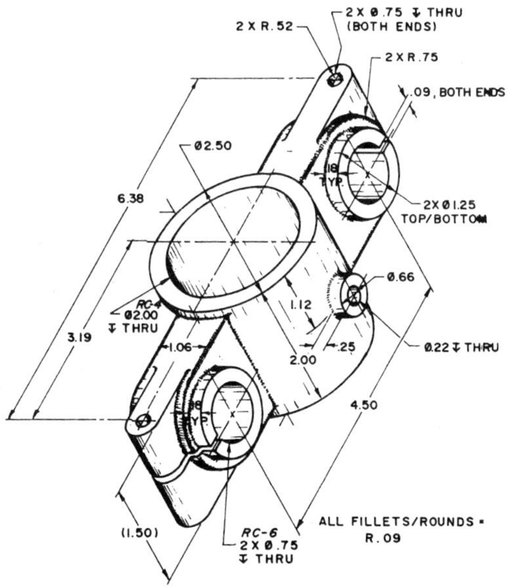

PROBLEM 8–46

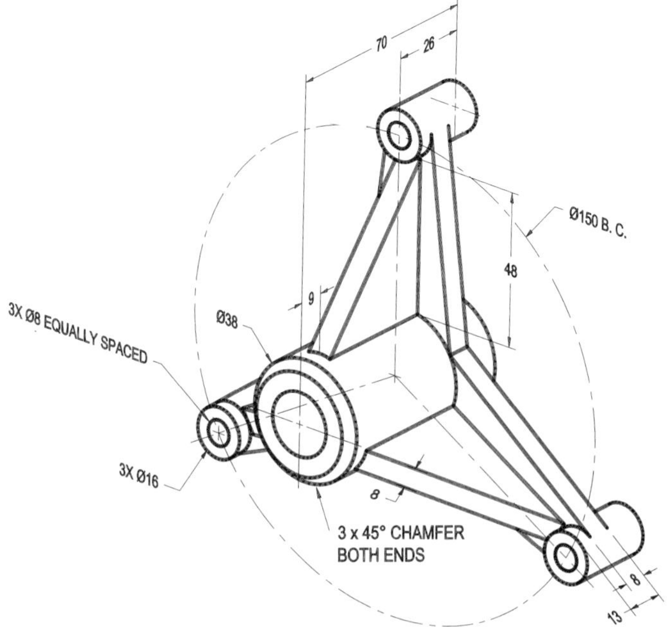

PROBLEM 8–47

Create the 2 elevation drawings for the garage.

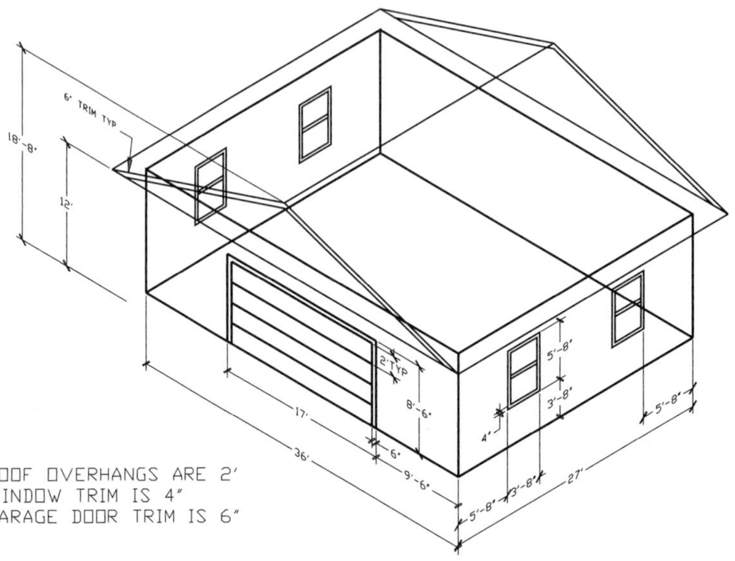

An Introduction to Drawing Layouts

The exercises in this chapter introduce:

- Model space and Layout space
- Viewport creation
- Using layouts for mechanical, architectural and metric drawings
- Creating multiple drawing layouts

TUTORIAL EXERCISE: 09_CENTER_GUIDE.DWG

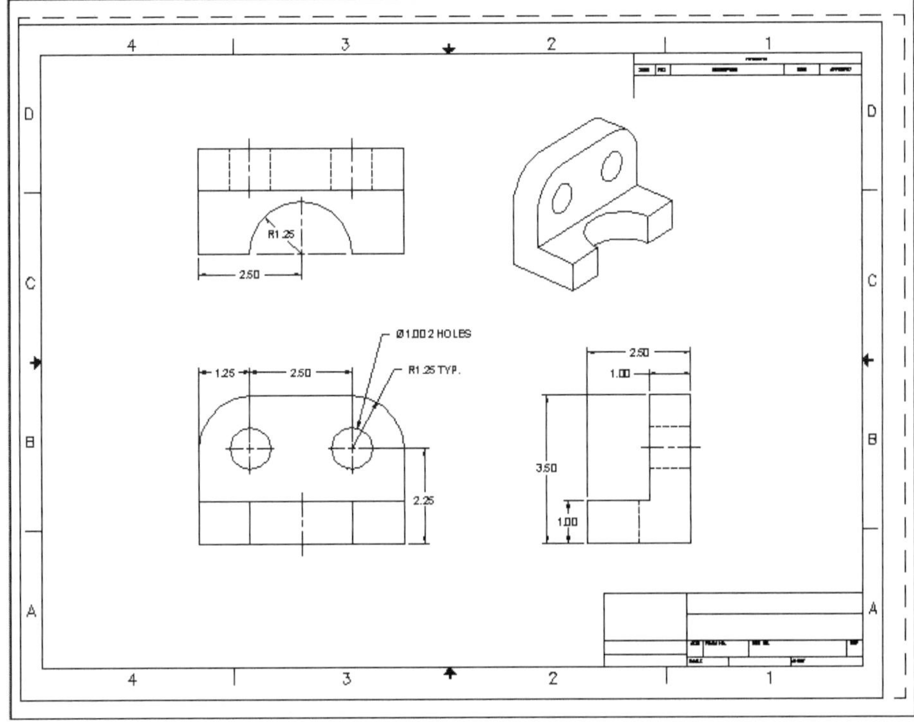

Figure 9–1

Purpose

This tutorial is designed to create a layout of the four-view drawing in Figure 9–1 in Paper Space.

System Settings

All unit and limit values have been pre-set. In order for this tutorial to function properly, the HP DesignJet 750C Plus C4708B must be configured. If this not the case, please perform the first tutorial exercise in Chapter 10 to configure this plotter.

Layers

In addition to the current layers of the drawing, create the following layer:

Name	Color	Linetype
Viewport	Green	Continuous

Suggested Commands

Begin by opening the drawing file *09_Center_Guide.Dwg*. Use the Create Layout wizard to perform the initial work of creating the layout for the Center Guide. When the layout is created and properly scaled, change the viewport object to the Viewport layer. Then, turn off this layer in preparation for plotting the drawing at a scale of 1:1.

Whenever possible, substitute the appropriate command alias in place of the full AutoCAD command in each tutorial step; for example, use "CP" for the COPY command, "L" for the LINE command, and so on. The complete listing of all command aliases is located in Chapter 1, Table 1–2.

STEP 1

Illustrated in Figure 9–2 is a three-view orthographic drawing of a Center Guide. An isometric view is also provided to better interpret the three views. The drawing was created in Model Space. This is easily identified by the appearance of the User Coordinate System icon in the lower-left corner. Use this as the starting point for arranging the model in Paper Space to prepare the drawing to be plotted out at a scale of 1=1.

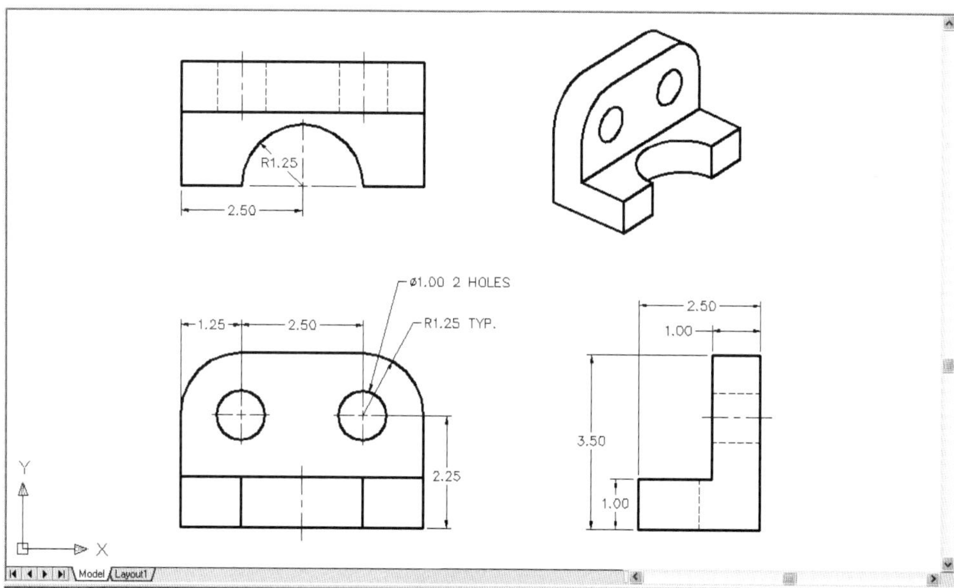

Figure 9–2

STEP 2

Choose Wizards from the Tools pull-down menu. Choosing the Wizards>Create Layout... menu option shown in Figure 9–3 displays the Create Layout dialog box in Figure 9–4. Begin the process of creating this new layout by typing **Four Views** as the name of the layout. When finished, click on the Next> button.

Figure 9–3

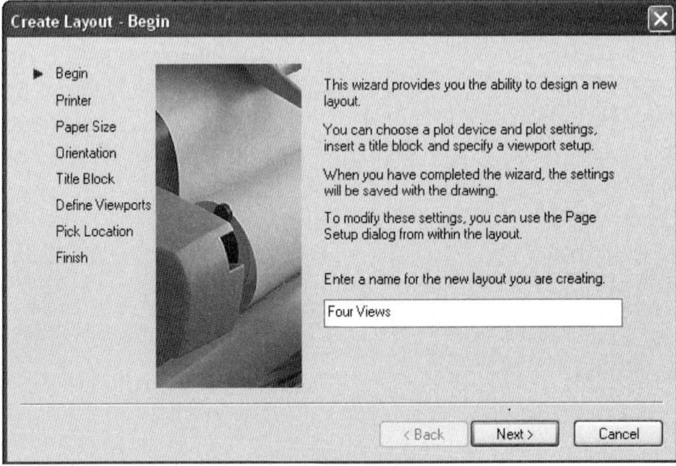

Figure 9–4

STEP 3

In the Printer dialog, you must select a printer that has already been configured. From the list of printers shown in Figure 9–5, click on the DesignJet 750C Plus C4708B.pc3 as the device. (Note: If this printer device is not listed, stop this tutorial and see Chapter 10 of the AutoCad 2004 Tutor for Engineering Graphics; follow the steps for configuring this device and begin this exercise again.) When finished, click the Next> button.

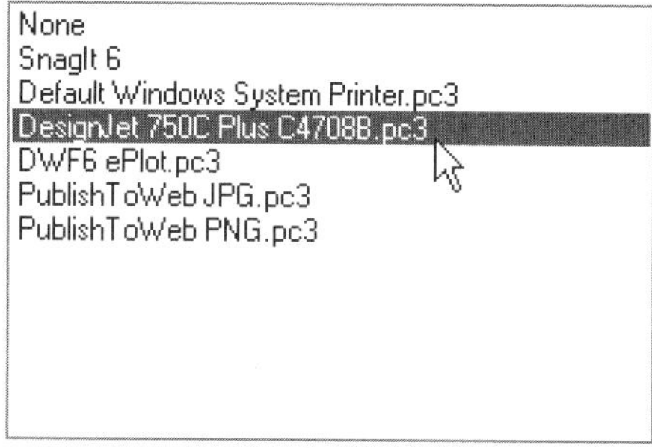

Figure 9–5

STEP 4

In the Paper dialog of the Create Layout wizard, select the paper size to be used. In the edit box shown in Figure 9–6, choose the ANSI expand C (22.00 x 17.00 Inches) sheet size. Keep all other default settings. When finished, click on the Next> button.

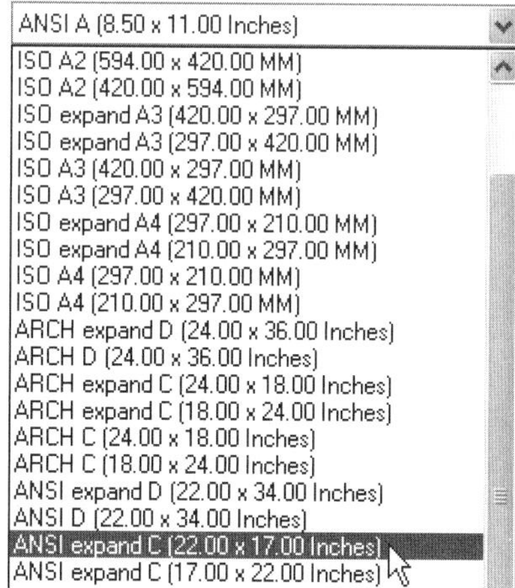

Figure 9–6

STEP 5

In the Orientation dialog of the Create Layout wizard, select the Landscape radio button as shown in Figure 9–7, to ensure proper orientation of the drawing on the paper. When finished, click the Next > button.

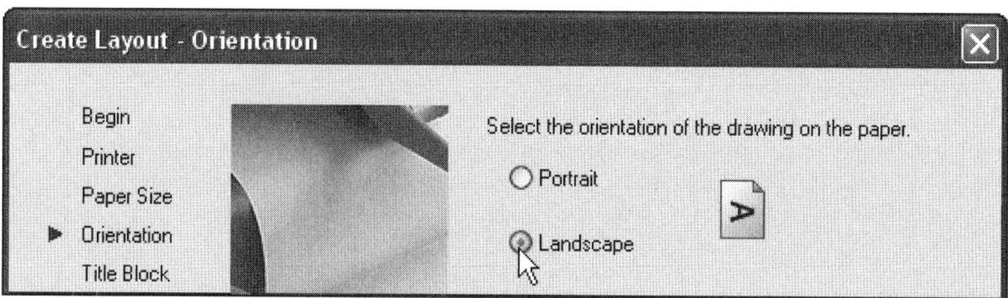

Figure 9–7

STEP 6

In the Title Block dialog, choose the *ANSI C title block.dwg* shown in Figure 9–8. This is the title block that will automatically be inserted in the newly created layout. When finished, click the Next > button.

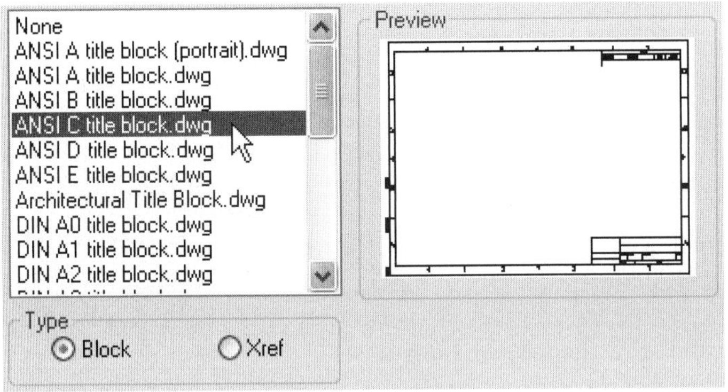

Figure 9–8

STEP 7

In the Define Viewports dialog, be sure the Viewport setup is set to Single, as shown in Figure 9–9. You also have the opportunity to pre-scale the drawing to paper space units. The drawing will be scaled inside the viewport to 1:1; select this value, as shown in Figure 9–9. When finished, click the Next > button.

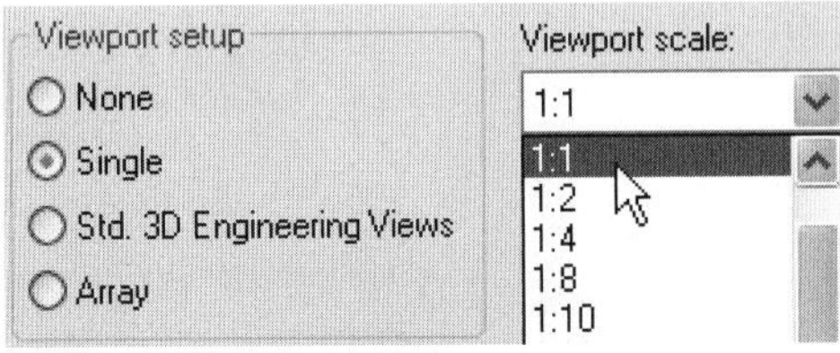

Figure 9–9

STEP 8

In the Pick Location dialog, you can define the size of the viewport that will be created in the drawing layout. If you choose not to click the Select location< button, a viewport is created to match the printable area—the area inside the dashed margins in the layout. Click on the Select location< button to create a smaller viewport, as shown in Figure 9–10. This returns you to the drawing, where the sheet size and title block are displayed as shown in Figure 9–11. Pick a point at "A" and "B" to define the new viewport. After specifying the viewport, you are returned to the Create Layout wizard. In the last dialog box, click on the Finish button.

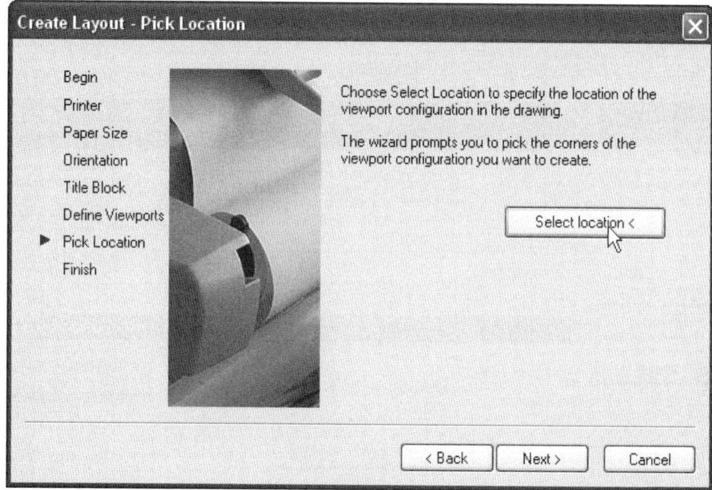

Figure 9–10

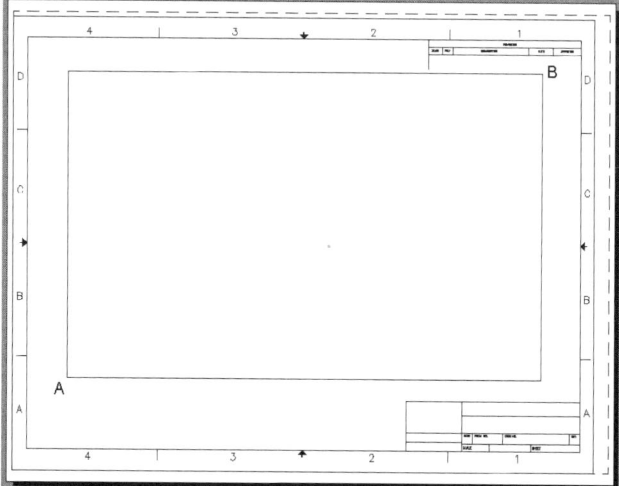

Figure 9–11

STEP 9

The completed layout is shown in Figure 9–12. Notice that the following items are already created in the layout: Layout title (Four Views), Viewport containing the four-view drawing, and C-size title block already inserted. The Paper Space icon is present, alerting you to the layout mode. If your image doesn't exactly fit inside of the viewport or is not centered in relation to the title block, use grips to stretch the title block to a better size and then move the viewport containing the image to a better location inside the title block.

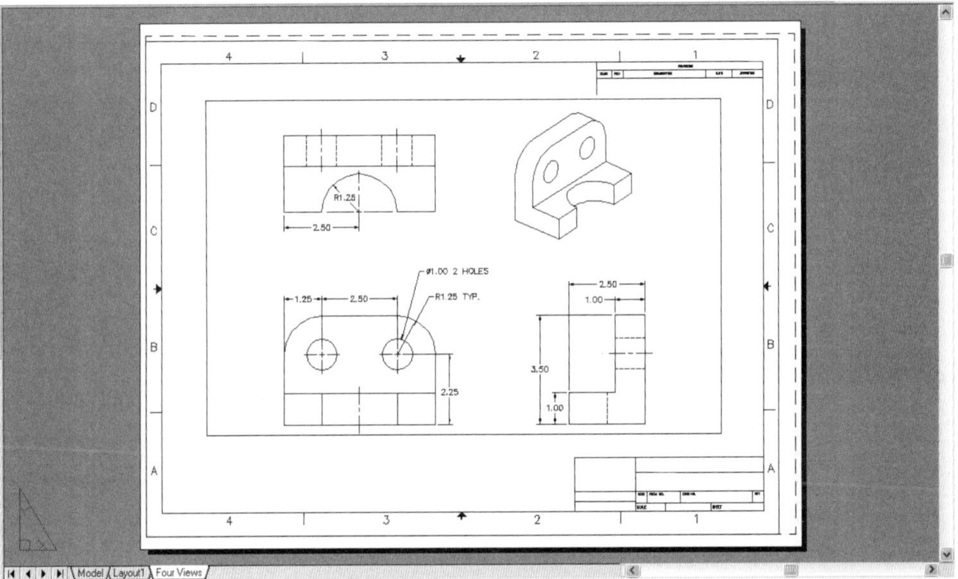

Figure 9–12

STEP 10

When the Layout Wizard created the viewport holding the drawing information, it did not place the viewport on the correct layer. To accomplish this, first, click on the viewport; it is highlighted and grips appear. Click in the Layer Control box and change the viewport to the Viewport layer, as shown in Figure 9–13. When finished, press ESC to remove the object highlight and remove the grips.

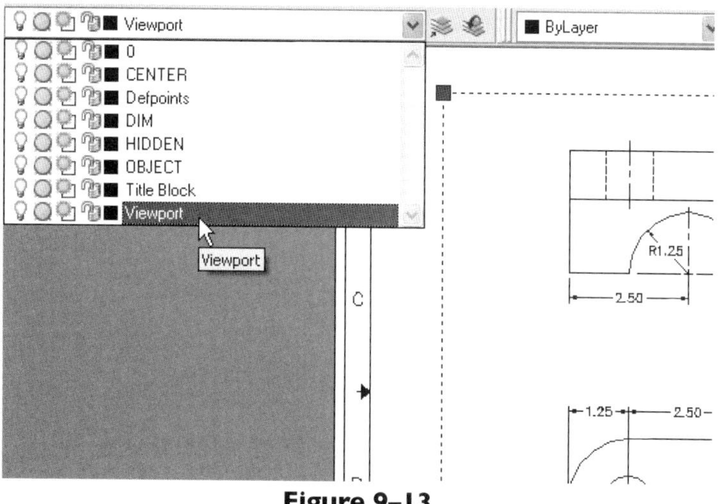

Figure 9–13

STEP 11

Verify the correct Paper Space scale by double-clicking in the viewport at "A." This places you in floating Model Space, with the appearance of the User Coordinate System icon as in Figure 9–14. Activate the Viewports toolbar and check that the correct scale of 1:1 is reflected in the edit box. When finished verifying, dismiss the Viewports toolbar and double click at "B" to return to Paper Space.

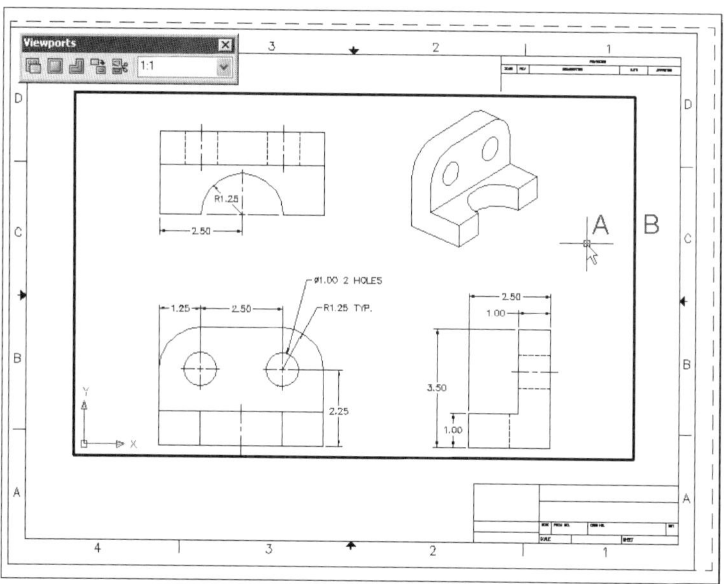

Figure 9–14

STEP 12

Your display should appear similar to Figure 9–15.

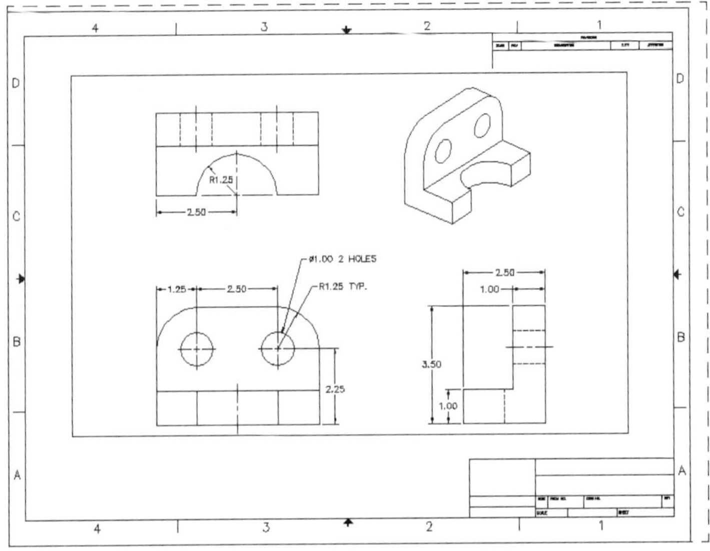

Figure 9–15

STEP 13

Notice that the Layout1 tab is still present at the bottom of your display screen. Since the Layout Wizard was used to create the Four Views tab, Layout1 still remains. Right-clicking on the Layout1 tab displays the context menu shown in Figure 9–16. Click on the Delete option to remove Layout1 from the drawing database. An AutoCAD alert box will appear (see Figure 9–17), telling you that the layout will be permanently deleted. Also, it reminds you that the Model tab cannot be deleted.

Click the OK button to delete Layout1.

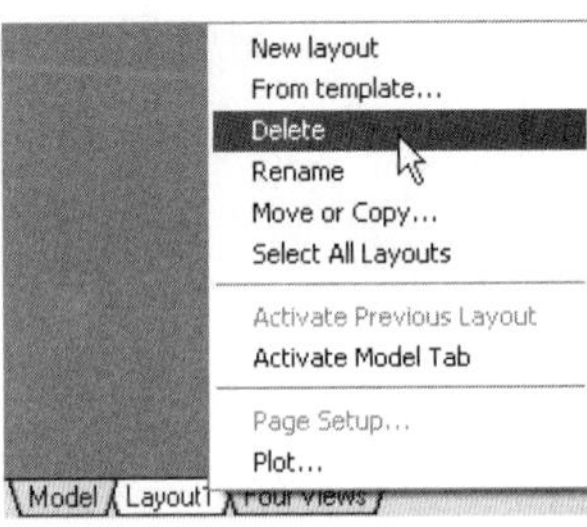

Figure 9–16

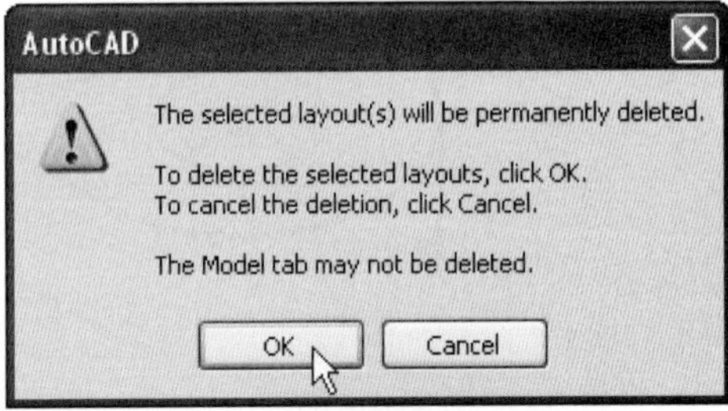

Figure 9–17

STEP 14

Before plotting, turn off the Viewport layer in the Layer Control box, as shown in Figure 9–18.

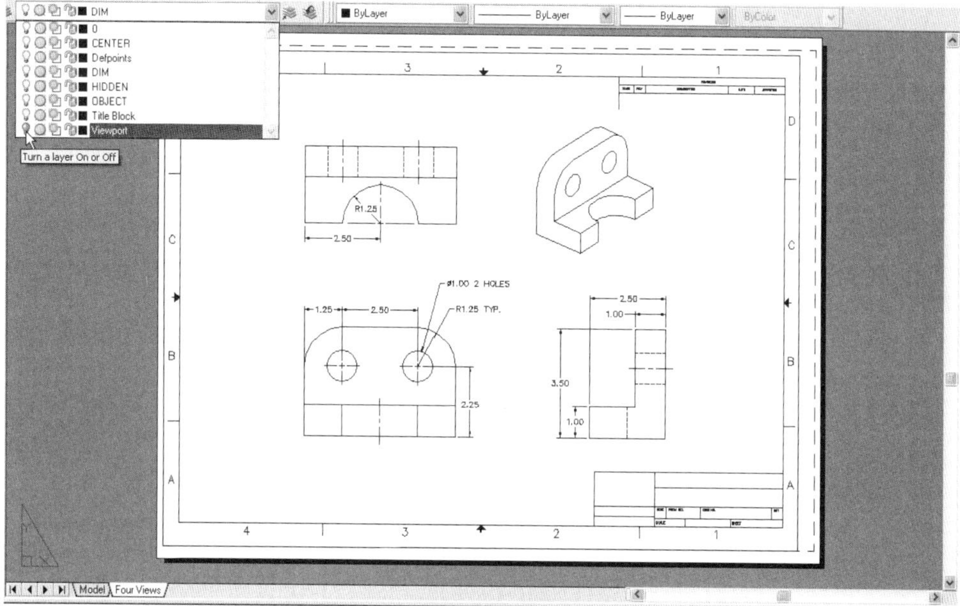

Figure 9–18

STEP 15

If at any point you need to change the drawing layout or plotter device, you can accomplish this by choosing Page Setup... from the File pull-down menu, as shown in Figure 9–19. The Page Setup dialog box, shown in Figure 9–20 will appear. Changes made here will be reflected in the current layout.

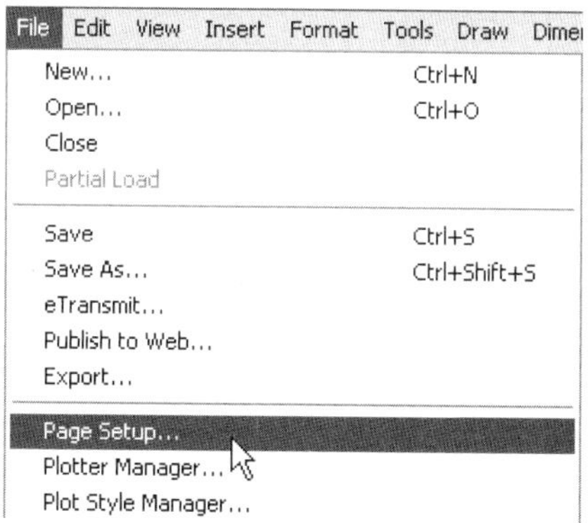

Figure 9–19

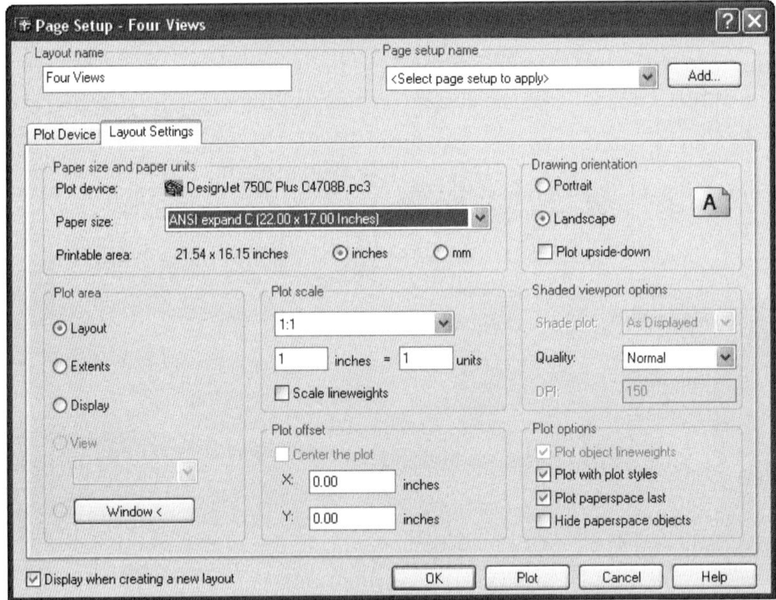

Figure 9–20

STEP 16

The completed layout is shown in Figure 9–21. In this final step, save the drawing file. The layout will also be saved, since it is part of the drawing.

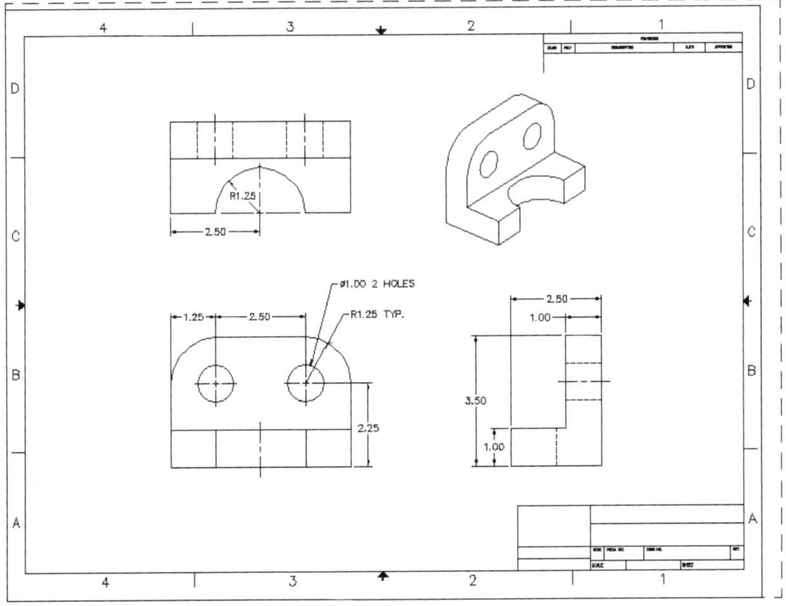

Figure 9–21

Viewport Toolbar Scaling Tools in Layout Space

Using the viewport toolbar to scale the drawing images in layouts for plotting has really simplified the drawing to plotting process. The toolbar shown in Figure 9–22 can be used to create a viewport in a layout and apply the scale factor to the layout.

Figure 9–22

Try It! – Open the *drawing 09_utah dip.dwg*. The ability to scale viewports in a layout eliminates the need to plot the whole drawing to a specific scale factor since a scale factor has been applied within each viewport created in a layout. To accomplish the task of applying a scale to a specific viewport, first turn on the viewports toolbar so that it is displayed on the screen in the layout space. If a viewport is not present, then create a viewport by using the single viewport button. (See Figure 9–23.)

Figure 9–23

Once the single viewport is created, select it using the grip selection technique (see Figure 9–24). Notice that the "Scale to fit" selection in the Viewports tool bar list has been replaced by numbers. This value is the current scale factor.

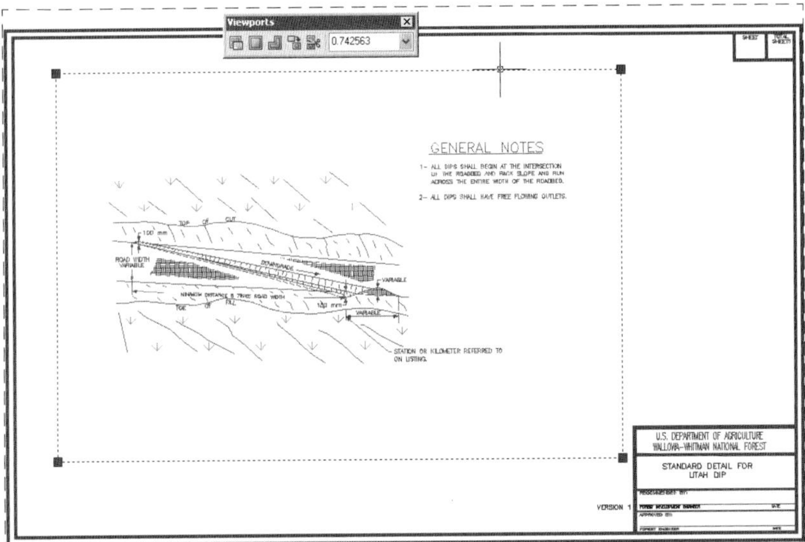

Figure 9–24

To change the scale of the drawing image, displayed in the viewport, use the drop down list on the Viewports toolbar shown in Figure 9–25. This will display a variety of typical scale factors applied on mechanical, civil, and architectural drawings. Select the appropriate scale factor from the dropdown list. Notice the split in scale factors on the bar. The scale factors that have been used recently are shown above the double line for quick re-use.

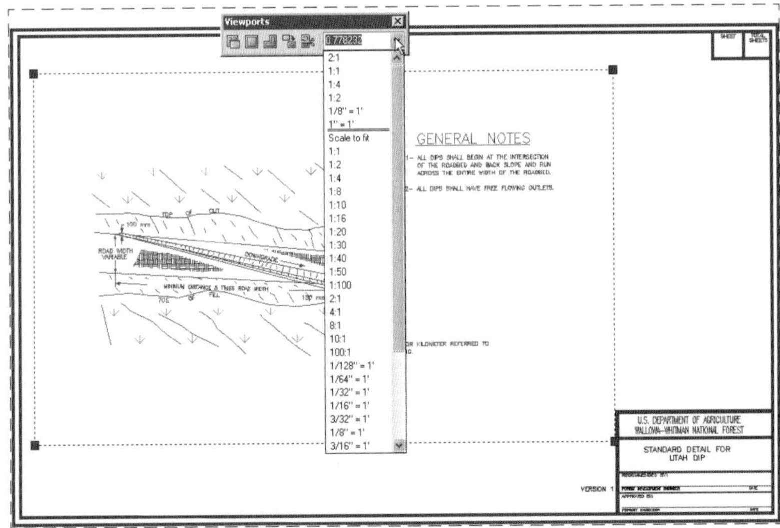

Figure 9–25

There may be some trial attempts at selecting the best scale for the drawing. There is no wasted paper on these attempts! Select different scale factors for the drawing viewport. Try selecting 1/8"=1', 1"=1', and 1:1. Which selection fits the image on the drawing sheet best? See Figure 9–26 for the answer.

If the image is very close to fitting, but cut off by one edge of a viewport there are two options to resolve the problem. One way is to stretch the viewport (using grips or the STRETCH command) the other way is to pan the image. A combination of both methods usually proves to be the most successful method. Notice the viewport size change in Figure 9–26.

To pan the image, double-click inside the viewport to start a floating model space. Notice the thick borders on the viewport. Using the PAN command or holding down the wheel on a wheel mouse, move the image to fit the viewport. Once panned into position, double-click outside of the viewport on the drawing surface (usually the area surrounding the drawing sheet) to return to the layout space.

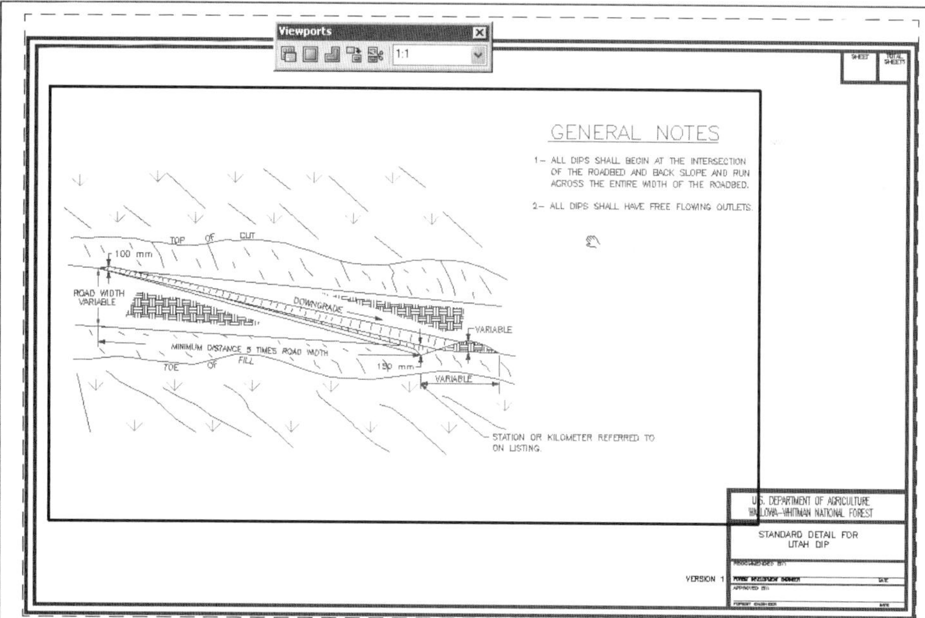

Figure 9–26

Beginning Level Drawings

PROBLEM 9–1

Develop the layout shown in the figure below. Open the drawing *09_traffic barrier.dwg* from the CD. On Layout1 create a viewport and apply a scale of ¾"=1' to it. Plot it to a B size sheet.

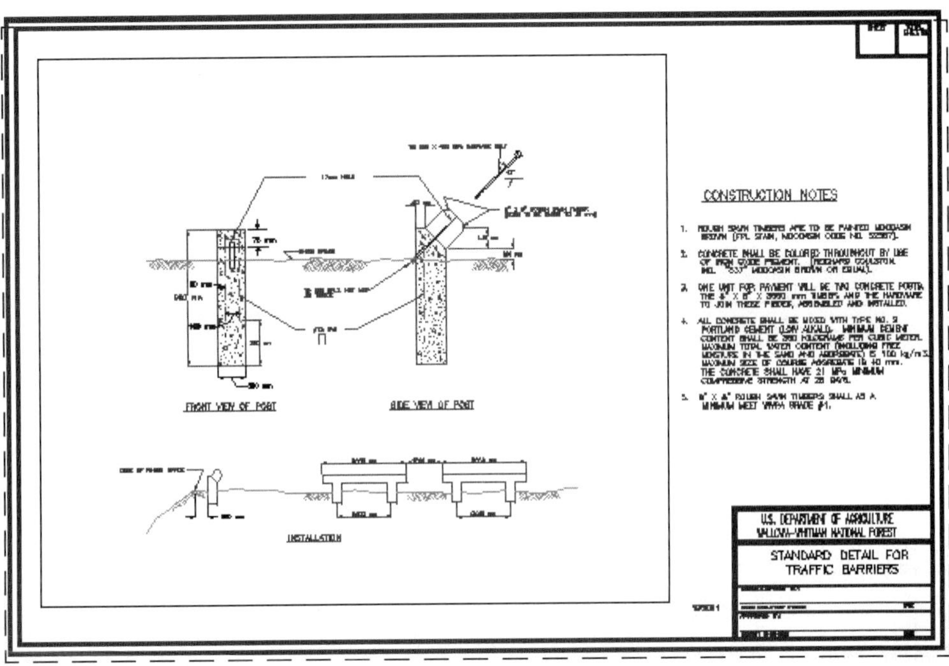

PROBLEM 9–2

Arrange the following drawings from Chapter 5, "Performing Geometric Constructions." Use the figure below (showing Problem 5–1) as an example of how the layout will appear when finished.

For Problems 5–1, 5–7, 5–8, 5–9, 5–11, 5–13, 5–15, 5–16, 5–18, 5–20, 5–21 use *ANSI_B.Dwg* as the title block inside Paper Space. This title block can be found in the *Template* folder of AutoCAD 2004. Scale the image inside floating Model Space at a scale of 1:1 using the viewports toolbar.

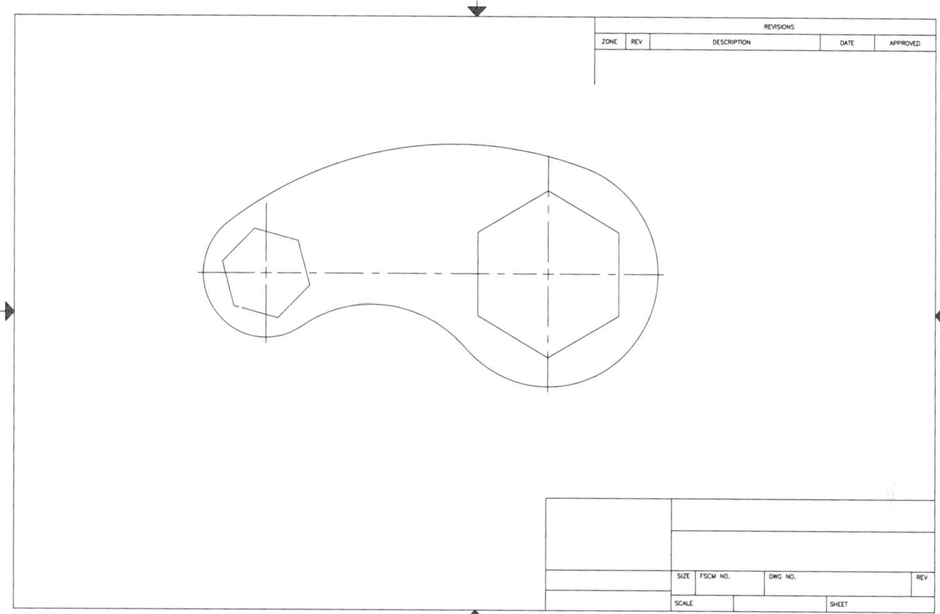

PROBLEM 9–3

Arrange the following drawings from Chapter 5, "Performing Geometric Constructions." Use the figure below (showing Problem 5–3) as an example of how the layout will appear when finished.

For Problems 5–3, 5–23 use *ANSI_C.Dwg* as the title block inside Paper Space. This title block can be found in the *Template* folder of AutoCAD 2004. Scale the image inside floating Model Space at a scale of 1:1 using the viewports toolbar.

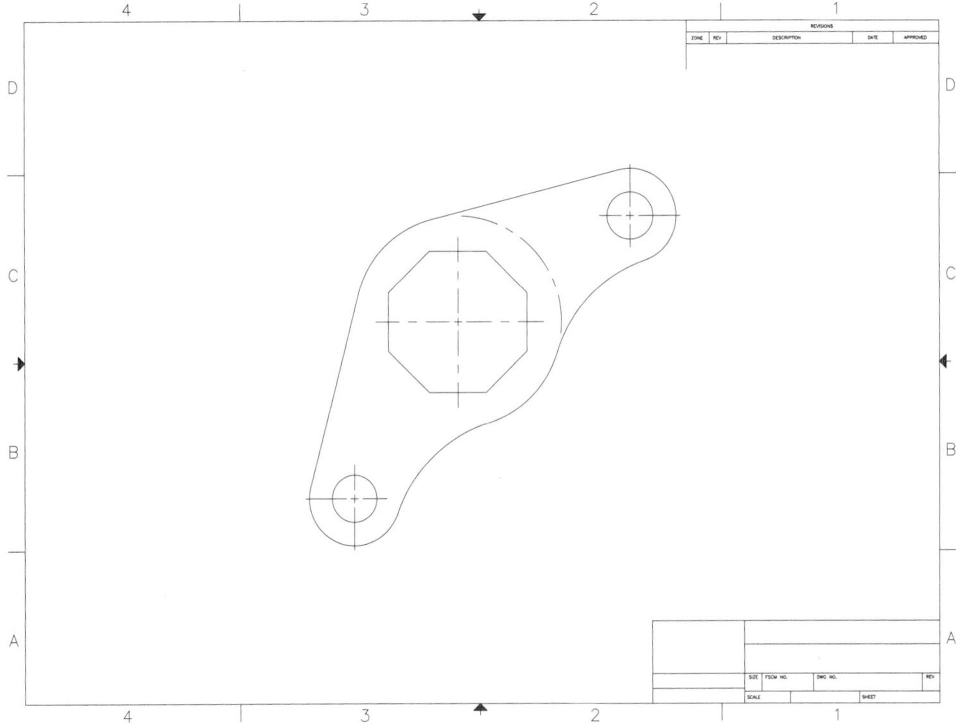

PROBLEM 9–4

Arrange the following metric drawings from Chapter 5, "Performing Geometric Constructions." Use the figure below (showing Problem 5–9) as an example of how the layout will appear when finished.

For Problems 5–4, 5–10, 5–12, 5–17, 5–22 use *ISO_A3.Dwg* as the title block inside Paper Space. This title block can be found in the *Template* folder of AutoCAD 2004. Scale the image inside floating Model Space at a scale of 1:1 using the viewports toolbar.

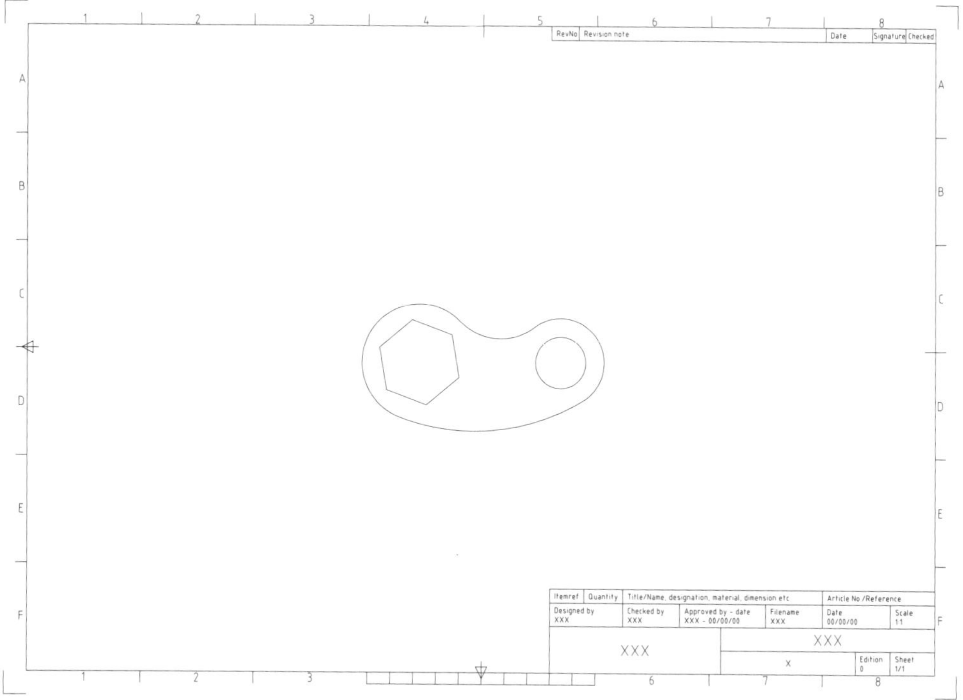

Intermediate Level Drawings

PROBLEM 9–5

For Problems 9–5 & 9–6, create the drawing and use layouts for scaling and plotting of the drawing. Optional: use an Autodesk default or customized template drawing to add a title block into the layout.

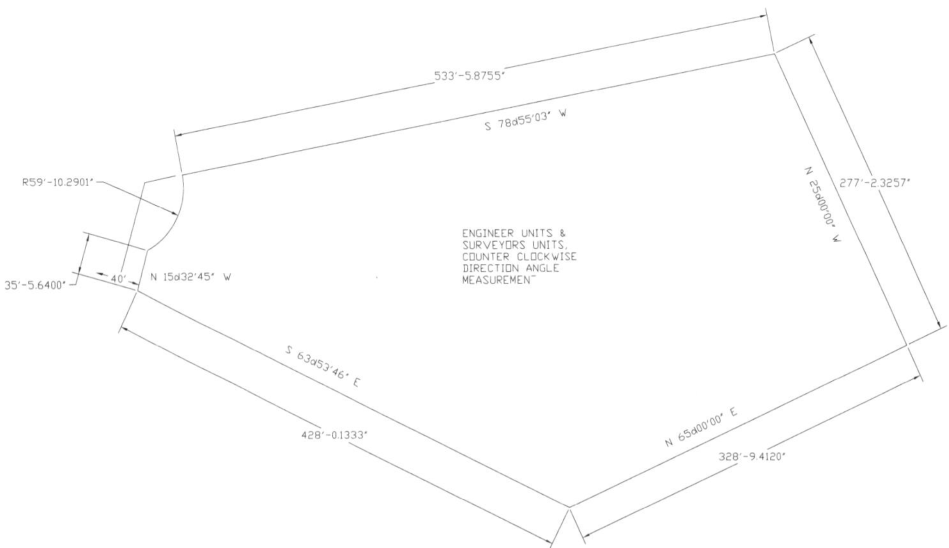

PROBLEM 9–6

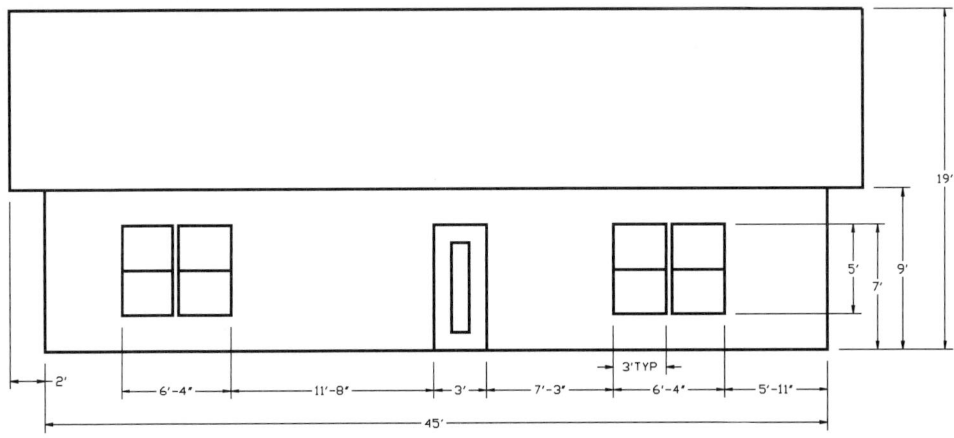

Plotting Your Drawings

The exercises in this chapter introduce:

- Plotter configuration
- Plotting from layouts
- Plotting from model space
- Creating plot styles
- Publishing to the web

TUTORIAL EXERCISE: CREATING A NAMED PLOT

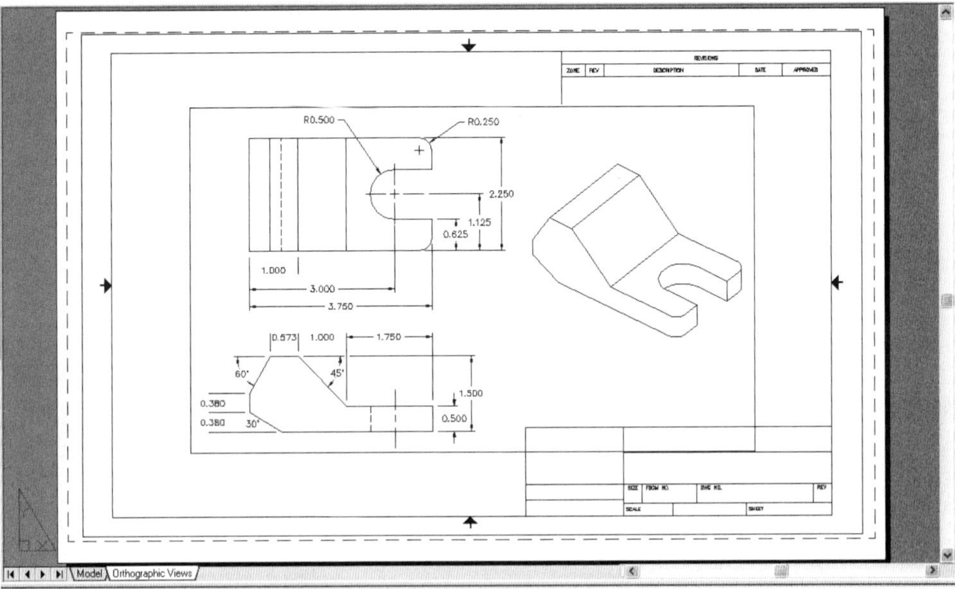

Figure 10–1

What if you want to assign different lineweights to the same color in a drawing, and what if these colors are assigned to different layers? This is where you would create a named plot style table and attach the plot style to an individual layer. To see how this can be accomplished, open the drawing called *10_Named_R-Guide.Dwg* shown in Figure 10–1. Notice that the entire drawing is red; yet if you examine the Layer Properties Manager dialog box, you will notice the drawing is organized in layers. We want all red objects to plot in black. We also want to assign thin, medium, thick, and heavy lineweights to certain layers as follows:

Layer	Plot Style
Hidden	Medium
Center	Thin
Viewports	Thin
Text	Medium
Title Block	Heavy
Dimensions	Thin
Object	Thick

Follow the next series of steps for creating a named plot style with the extension of *.STB* (Plot Style Table).

STEP 1

As with the creation of a color-dependent plot style (see Chapter 10 in The AutoCAD 2004 Tutor for Engineering Graphics), begin the process of creating a named plot style table by choosing Plot Style Manager from the File pull-down menu or type **_stylesmanager** at the AutoCAD command prompt.

When the dialog box appears, double-click on Add-A-Plot Style Table Wizard to start the Add Plot Style Table wizard.

Click the Next> button in the Add Plot Style Table dialog box. On the Begin dialog of the wizard turn on the Start from scratch option button.

Clicking the Next> button takes you to the Table Type dialog box of the wizard shown in Figure 10–2. Click the Named Plot Style Table radio button and then click the Next> button to display the next dialog box.

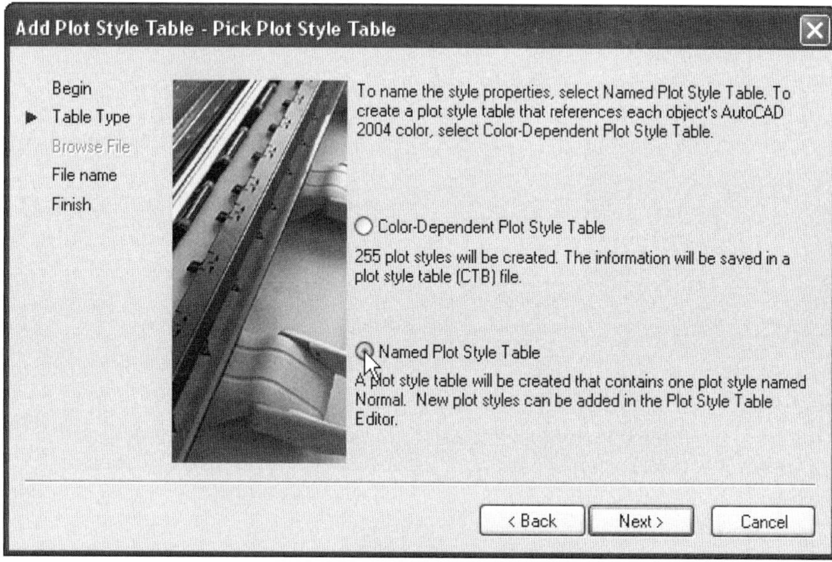

Figure 10–2

STEP 2

In the File name dialog box of the wizard, shown in Figure 10–3, type the name **Ortho_Drawings** for the name of this plot style. Do not be concerned that this name is identical to the one just used for the color-dependent table. The named plot style extension is *.STB*, which is quite different from the color-dependent extension of *.CTB*. Click the Next> button.

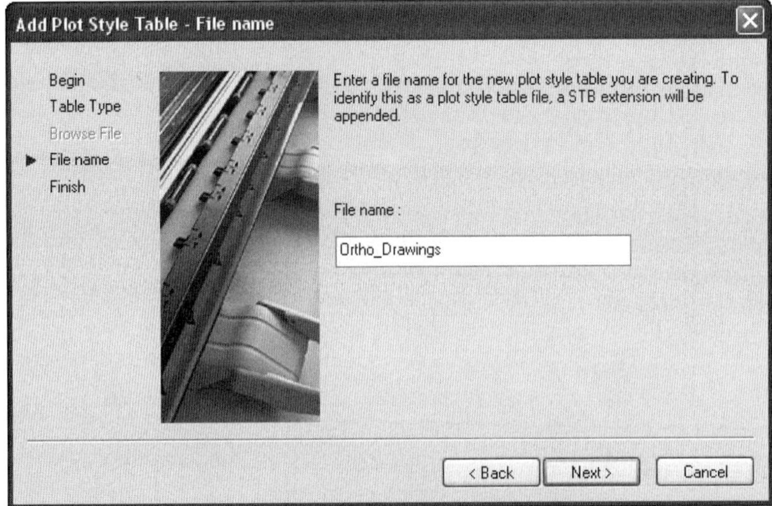

Figure 10–3

STEP 3

The Finish dialog box appears. Click on the Plot Style Table Editor... button, shown in Figure 10–4, to make changes to the named plot style properties.

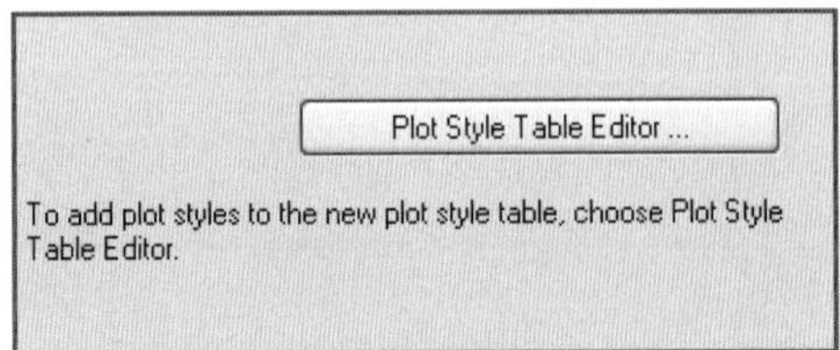

Figure 10–4

STEP 4

In the Plot Style Table Editor dialog box shown in Figure 10–5, notice the name of the plot style table you are editing is displayed in the title bar of the dialog box (*Ortho_Drawings.stb*). Click on the Form View tab and notice the plot style "Normal" is listed. By default, when you use a named plot style, "Normal" is always present. Four new plot styles need to be added to the Plot styles list to handle thin, medium, thick, and heavy lines. We will use these same names when creating each style. Click on the Add Style button near the bottom of the dialog box.

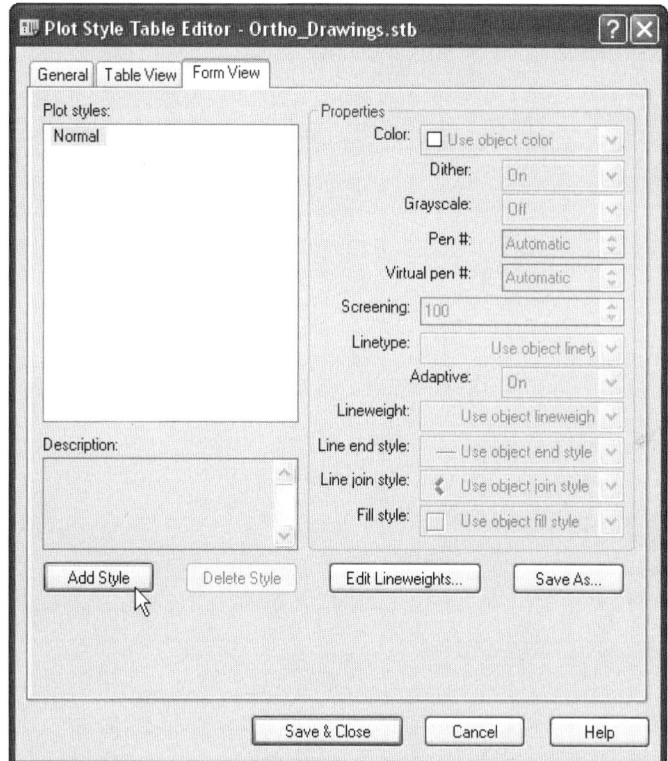

Figure 10–5

STEP 5

In the Add Plot Style dialog box shown in Figure 10–6, enter the name **Thin** as the new plot style name. Click the OK button to return to the Plot Style Table Editor.

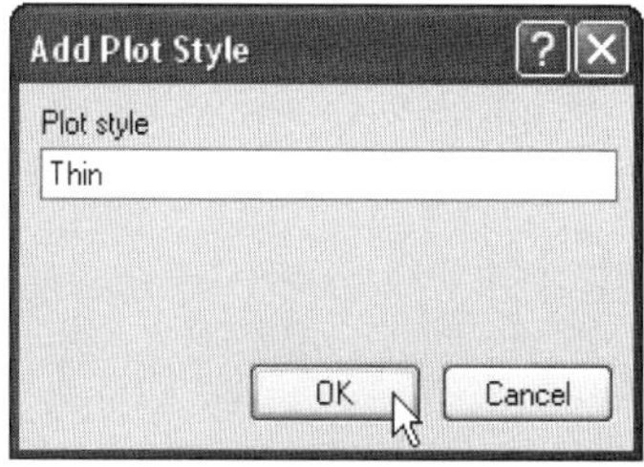

Figure 10–6

STEP 6

Back in the Form View tab of the Plot Style Table Editor dialog box shown in Figure 10–7, change the Color property to Black and the Lineweight to 0.10 mm. Under the Plot styles section there is a Description textbox for you to add a note regarding the purpose of the "Thin" style. Insert the text shown, to remind you that objects on the center, viewport, and dimension layers will plot out with thin, black lines. Click on the Add Style button to create another plot style called "Medium."

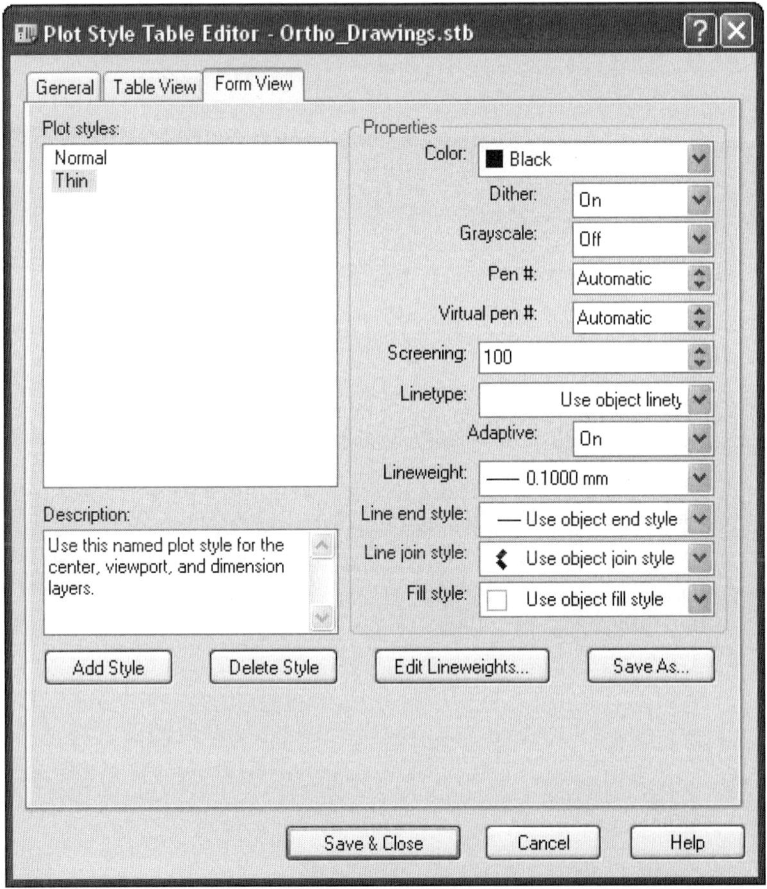

Figure 10–7

STEP 7

Back in the Form View tab shown in Figure 10–8, change the Color property to Black and the Lineweight to 0.30 mm for the "Medium" plot style. Add a note in the Description area to remind you that objects on the hidden and text layers will plot out with medium, black lines. Click on the Add Style button to create another plot style called "Thick."

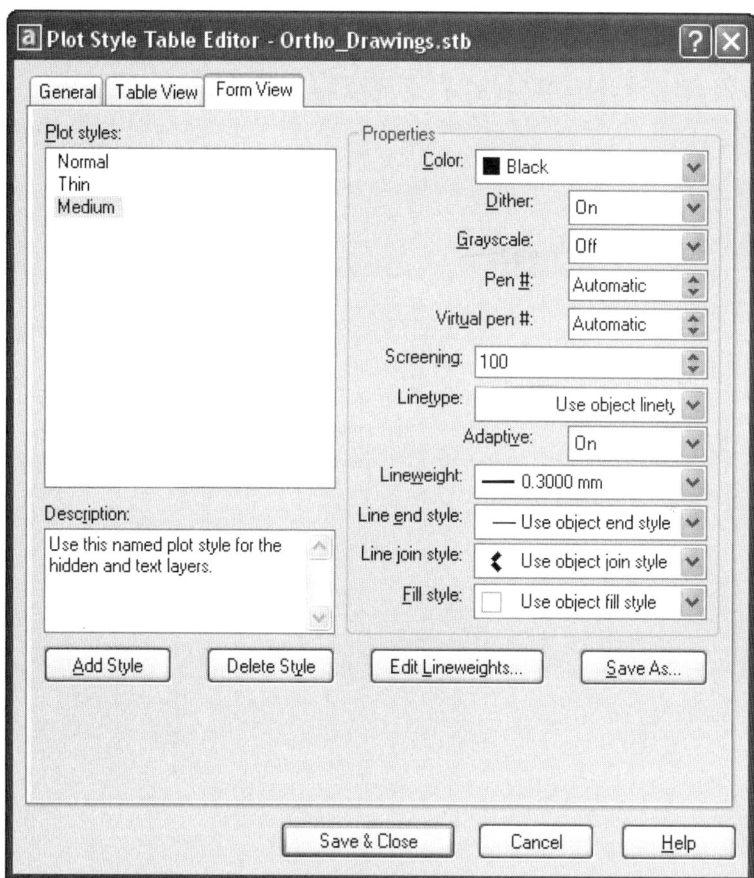

Figure 10–8

STEP 8

Back in the Form View tab shown in Figure 10–9, change the Color property to Black and the Lineweight to 0.70 mm for the "Thick" plot style. Add a note in the Description area to remind you that the object layer will plot out with thick, black lines. Click on the Add Style button to create the final plot style called "Heavy."

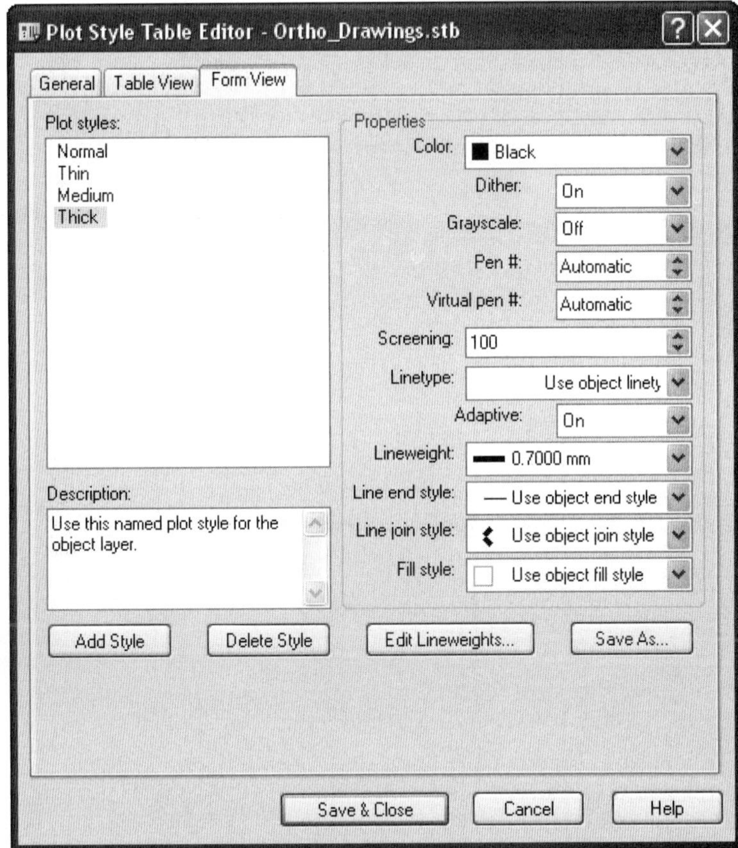

Figure 10–9

STEP 9

Back in the Form View tab shown in Figure 10–10, change the Color property to Black and the Lineweight to 0.80 mm for the "Heavy" plot style. Add a note in the Description area to remind you that the title block layer will plot out with heavy, black lines. Click on the Save & Close button at the bottom of the dialog box to save these changes to the *Ortho_Drawings.stb* file. When the Finish dialog box displays, click on the Finish button. Finally, close the dialog box that started the Add-A-Plot Style Table Wizard from and return to your drawing.

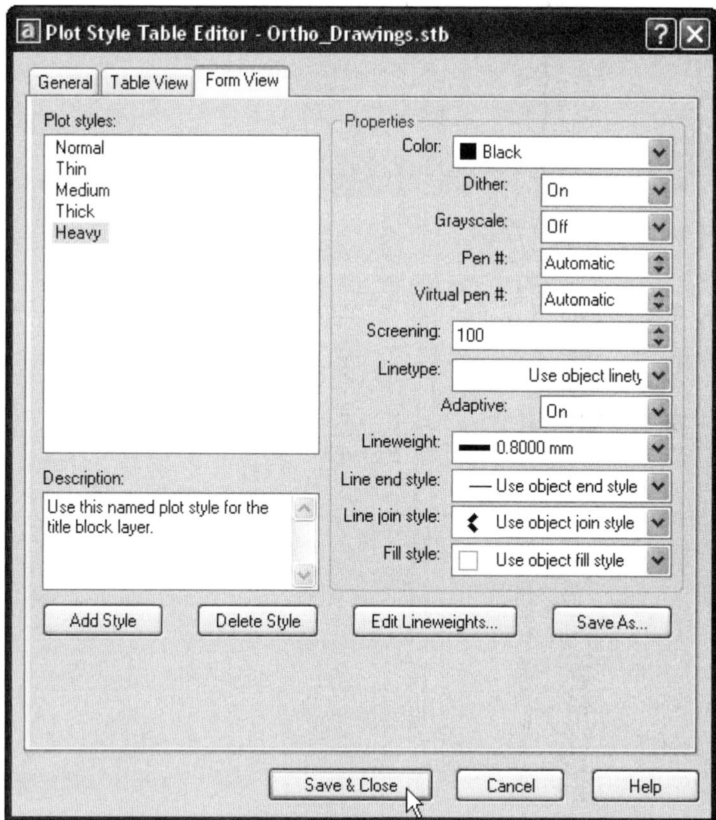

Figure 10–10

STEP 10

Before you plot this drawing, a few more actions need to be performed in preparation for the plot. First, activate the Page Setup dialog box, shown in Figure 10–11, and change Name: to *Ortho_Drawings.stb* in the Plot style table (pen assignments) area. Also, in this same area, make sure that the Display plot styles option is selected so you can view the results when previewing the plot. When finished, click the OK button, which will return you to your drawing.

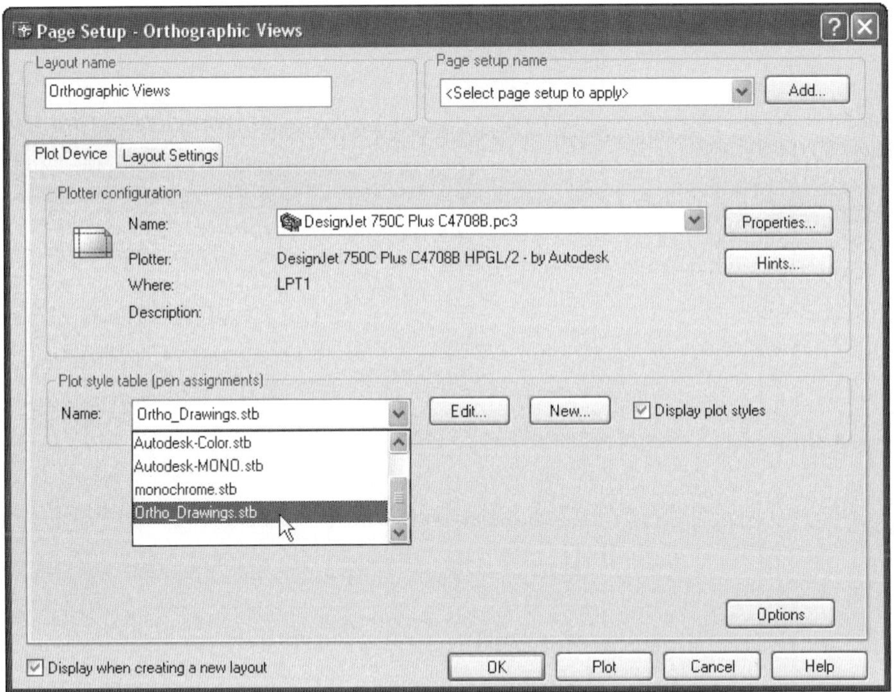

Figure 10–11

STEP 11

The purpose of the named plot style is to assign the plot style to objects and not to rely on colors. To illustrate, click on the Layer Properties Manager dialog box. As shown in Figure 10–12, click on the three layers identified as Center, Dimension, and Viewports. Click on the name "Normal" in the Plot Style heading of any of these three layers. The Select Plot Style dialog box appears with all defined plot styles that belong to the plot style table *Ortho_Drawings.stb*. Click on "Thin" and then the OK button. This will change the "Normal" plot style of these three layers to "Thin."

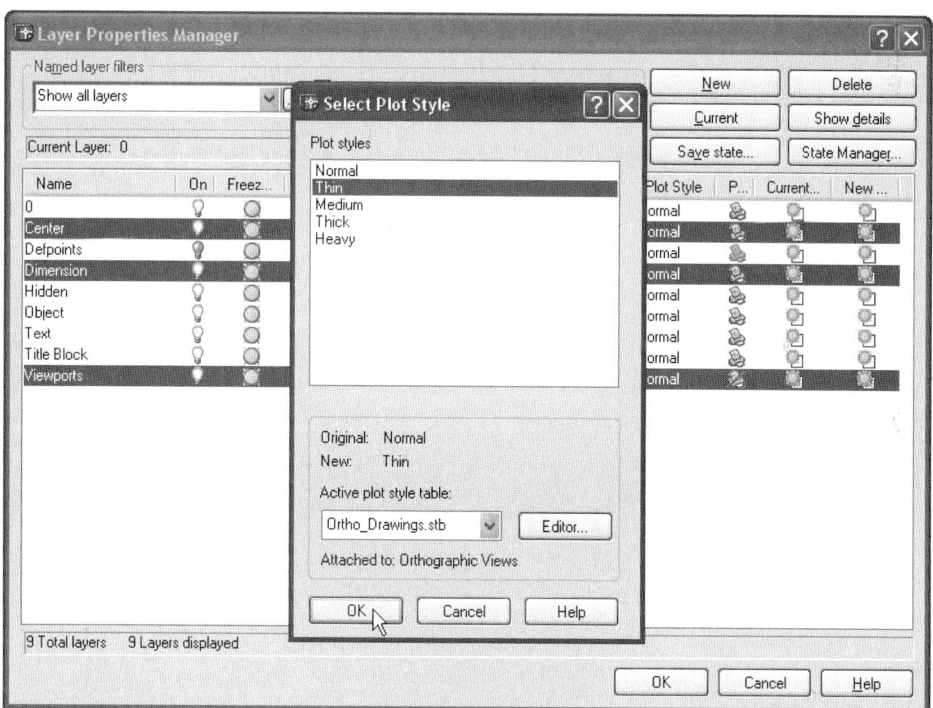

Figure 10–12

STEP 12

Repeat this procedure to change the plot style of the Hidden and Text layers to "Medium," the Object layer to "Thick" and the Title Block layer to "Heavy." Click on the Viewports and click the printer icon under Plot. This will make the viewport layer visible but not included in the final plot of the drawing. Your display should appear similar to Figure 10–13. Click on the OK button.

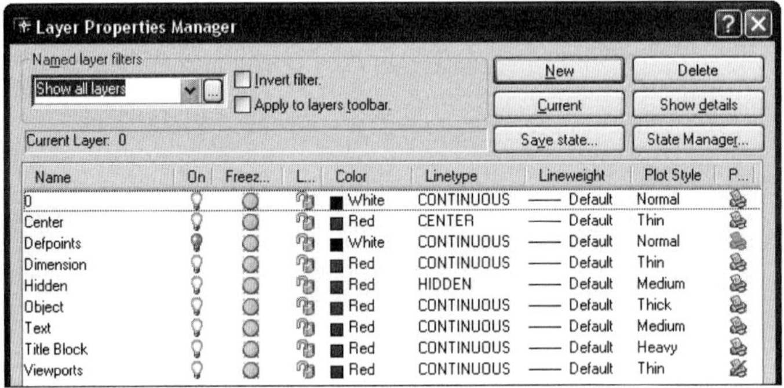

Figure 10–13

STEP 13

Activate the Plot dialog box, shown in Figure 10–14. Verify that the current plot device is the DesignJet 750C Plus and that the current plot style table name is *Ortho_Drawings.stb*. Click on the Full Preview… Button.

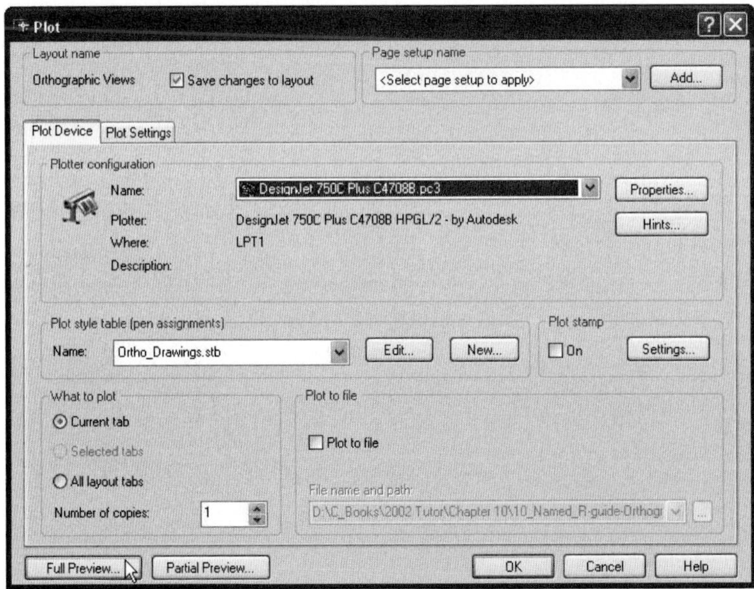

Figure 10–14

STEP 14

The results of performing a full preview are illustrated in Figure 10–15. Notice that all lines are black, even though they appear in color in the drawing file. This is the result of using a named plot style table on this drawing. This plot style table file can also be attached to other drawings that share the same layer names.

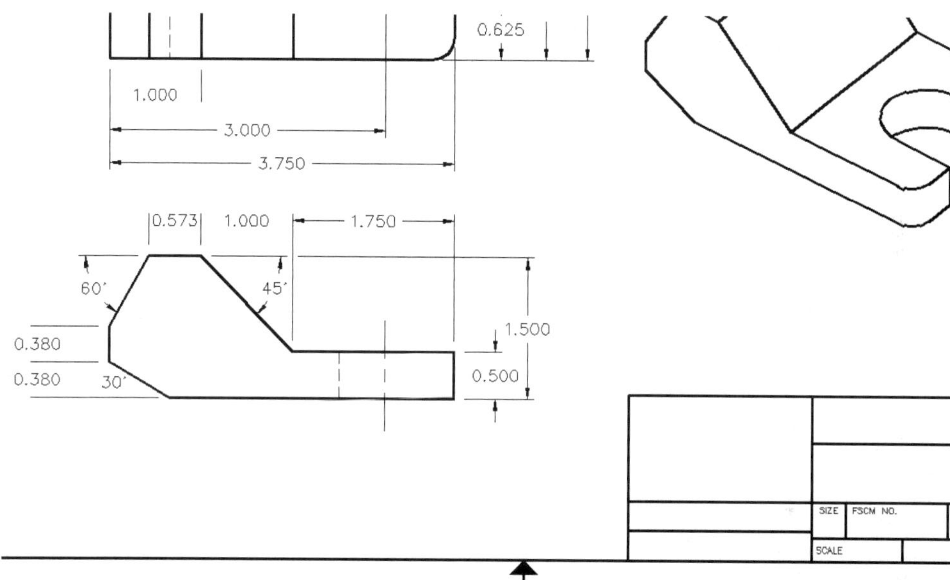

Figure 10–15

TUTORIAL EXERCISE: ELECTRONIC PLOTTING TO A DWF FILE

In addition to plotting out drawings on sheets of paper, drawings can be plotted out electronically to a file, for either printing or viewing purposes. These files are stored in a special Drawing Web Format with a file extension of *.DWF*. Once created, these DWF files can be opened, viewed, and plotted using the Autodesk Express Viewer. This method of viewing files over the Internet is very popular with individuals who do not own a copy of AutoCAD software.

STEP 1

Begin by opening the drawing file *10_Roof_Plan.dwg* To prepare this drawing to be plotted electronically, you must first select the proper plotter configuration, which is the DWF6 ePlot device, as shown in Figure 10–19. Notice that the Plot to file area is now activate. It is here that you will type the name and destination location of the DWF file. You can also attach a plot style table and click on the Plot Settings tab to change other plotter settings before pressing OK.

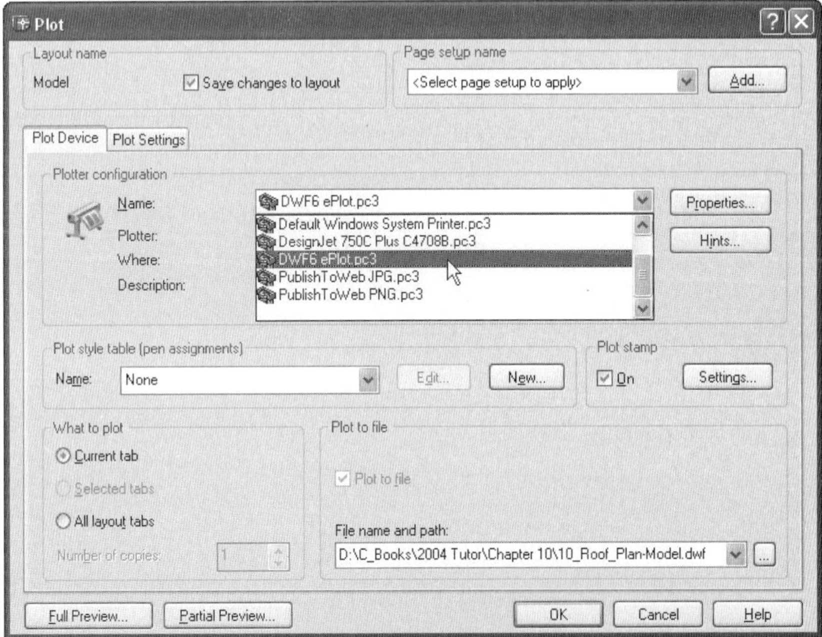

Figure 10–19

STEP 2

When the electronic plotting utility creates the file, it can now be viewed with Autodesk Express Viewer which is provided with AutoCAD 2004. First, start the Autodesk Express Viewer application. Next, choose Open from the File pull-down menu and search for the file in the Open File dialog box, shown in Figure 10–20. Finally, select the file, click the Open button and the drawing displays, as shown in Figure 10–21.

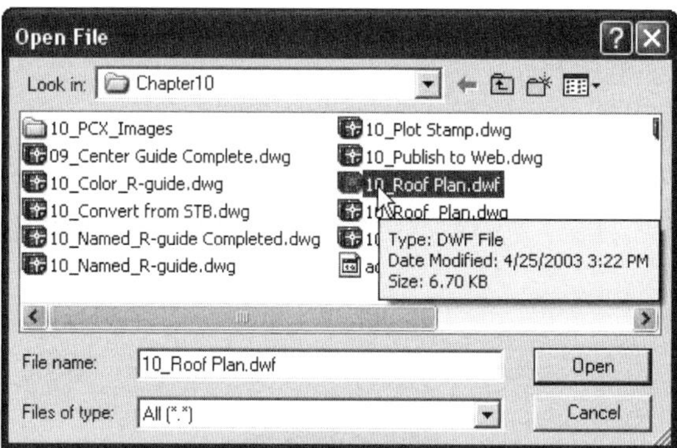

Figure 10–20

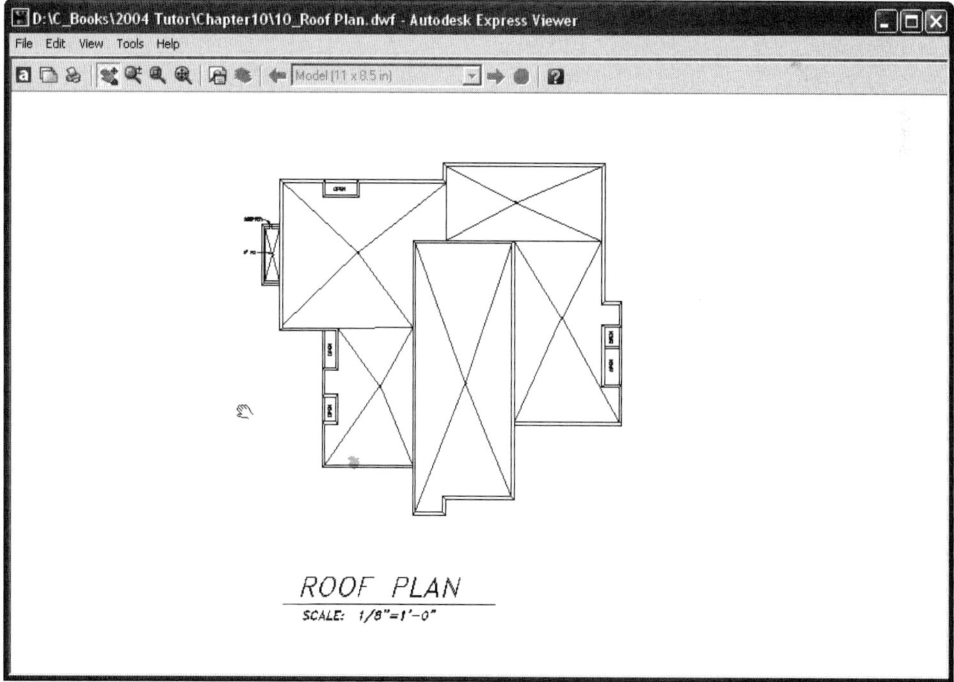

Figure 10–21

STEP 3

Right-click the mouse button, with the mouse in the drawing area, to display the cursor menu, shown in Figure 10–22. This menu contains various tools used to view the DWF image.

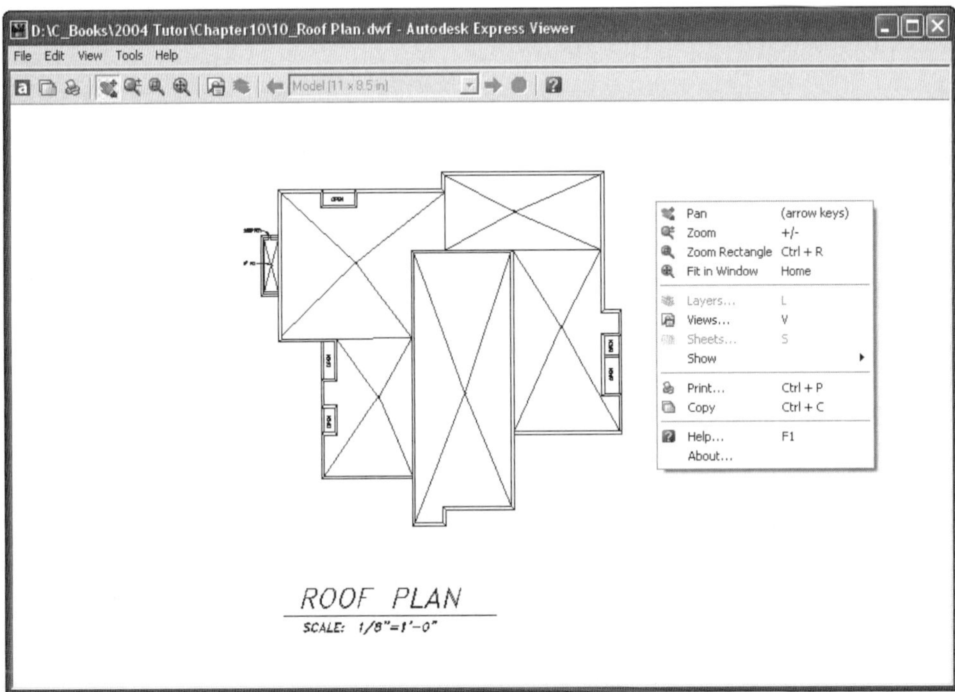

Figure 10–22

PROBLEM 10–1

Beginning Level Drawing

Open any existing drawing that has been created. Choose Plot… Using the Full Preview button, try different plotting combinations. And answer the following questions.

1. What are the differences between Limit, Display, Extent and Window Plots?
2. The Plot offset option affects the plot in what way?
3. How can the Scaled to fit option be used as a guide in selecting measurable plotting scales?

PROBLEM 10–2

Intermediate Level Drawing

Develop an architectural floor plan and create several layouts within the single drawing. The separate layouts should be used for detailing specific rooms of the floor plan.

Use the Publish to Web wizard to create a website that is based around your drawing. View your design in your favorite web browser.

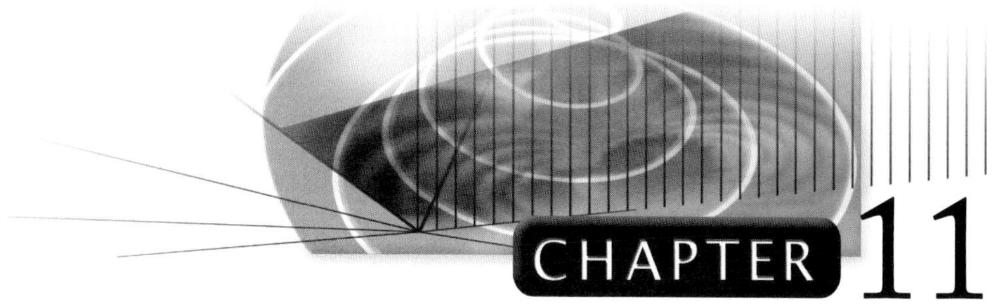

Dimensioning Basics

The exercises in this chapter introduce:

- Dimensioning principles
- Dimension application to a drawing
- Styles of dimensioning (ordinate, unidirectional, and aligned)
- Leaders
- General tolerancing
- GDT - Geometric Dimensioning and Tolerancing

TUTORIAL EXERCISE: 11_BAS-PLAT.DWG

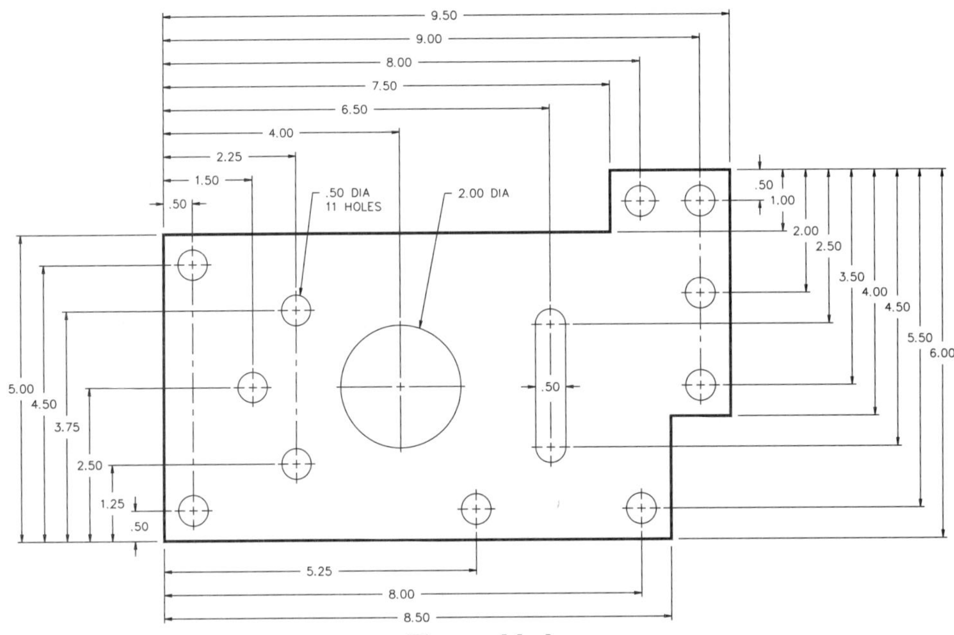

Figure 11–1

Purpose

The purpose of this tutorial is to annotate the drawing in *Bas-plat.dwg* (Baseplate) file, using ordinate dimensioning.

System Settings

The drawing, shown in Figure 11–1, is already constructed and has been dimensioned using the ordinate style. Open AutoCAD and follow the steps in this tutorial for adding dimensions in Ordinate mode. Be sure the following Object Snap modes are currently set: Endpoint, Center, Intersect, and Extension.

Layers

All layers have already been created:

Name	Color	Linetype
Cen	Yellow	Center
Defpoints	White	Continuous
Dim	Yellow	Continuous
Object	White	Continuous

Suggested Commands

Use the DIMORDINATE command for placing ordinate dimensions throughout this tutorial.

Whenever possible, substitute the appropriate command alias in place of the full AutoCAD command in each tutorial step; for example, use "CP" for the COPY command, "L" for the LINE command, and so on. The complete listing of all command aliases is located in Table 1–2.

STEP 1

Open the drawing file *11_Bas-Plat.dwg* All ordinate dimensions make reference to the current 0,0 location identified by the position of the User Coordinate System (see Figure 11–2). Because this icon is located in the lower-left corner of the display screen by default, the coordinate system must be moved to a point on the object where all ordinate dimensions will be referenced from. First, use the UCS command to define a new coordinate system with the origin at the lower-left corner of the object. Then, use the UCSICON command to force the icon to display at the new origin.

Command: **UCS** [tools>ucs>origin]
Current ucs name: *NO NAME*
Enter an option [New/Move/orthoGraphic/Prev/Restore/Save/Del/Apply/?/World]
<World>: **O** *(For Origin)*
Specify new origin point <0,0,0>: *(Pick the intersection at "A")*

Command: **UCSICON** [view>display>ucsicon]
Enter an option [ON/OFF/All/Noorigin/ORigin] <ON>: **OR** *(For Origin)*

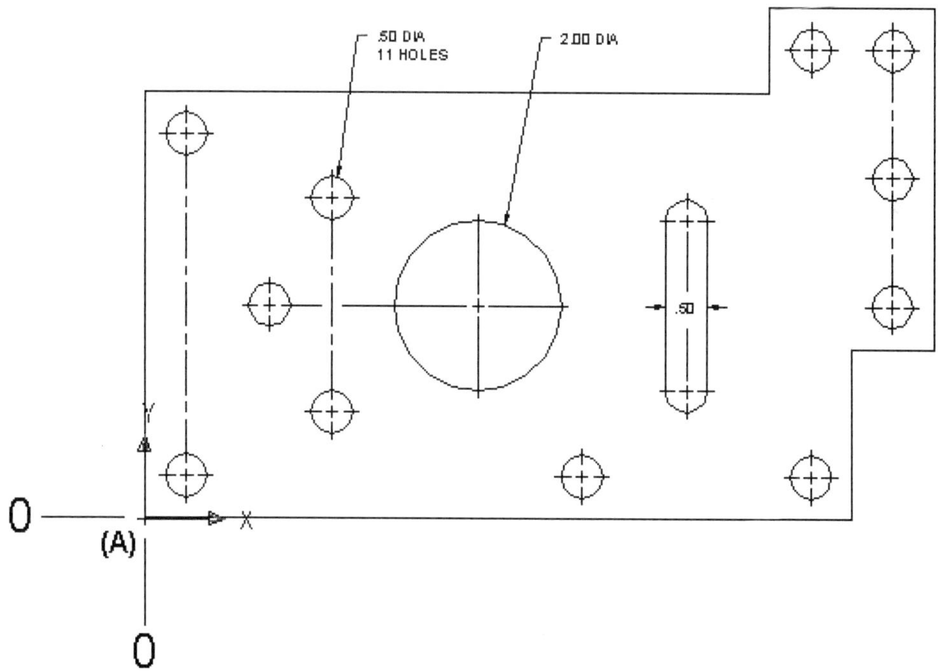

Figure 11–2

STEP 2

Use the ZOOM-CENTER command to magnify the screen, similar to the illustration in Figure 11–3.

Command: **Z** *(For ZOOM)*
[view>display>center]
Specify corner of window, enter a scale factor (nX or nXP), or
[All/Center/Dynamic/Extents/Previous/Scale/Window] <real time>: **C** *(For Center)*
Specify center point: **0,0**
Enter magnification or height <11.25>: **4**

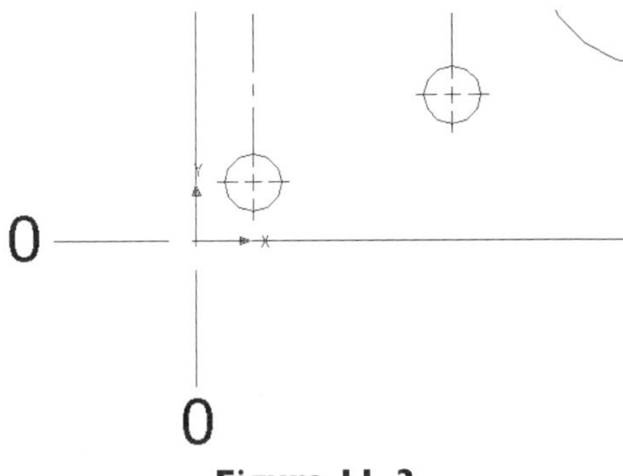

Figure 11–3

STEP 3

Before continuing with this next step, be sure that the grid is turned on. Use F7 to do this. Next, begin to place the first ordinate dimension using the DIMORDINATE command, or choose the DIMORDINATE command button from the Dimension toolbar. Select the endpoint of the centerline as a feature. With the Snap and Ortho modes turned on, locate a point two snap distances below the object to identify the leader endpoint. AutoCAD will automatically determine if the dimension is Xdatum or Ydatum. See Figure 11–4.

Command: **DOR** *(For DIMORDINATE)*
[dimension>ordinate]
Specify feature location: *(Pick the endpoint of the center line near point "A")*
Specify leader endpoint or [Xdatum/Ydatum/Mtext/Text/Angle]: *(Pick a point two snap points below the edge of the object at "B")*
Dimension text = 0.50

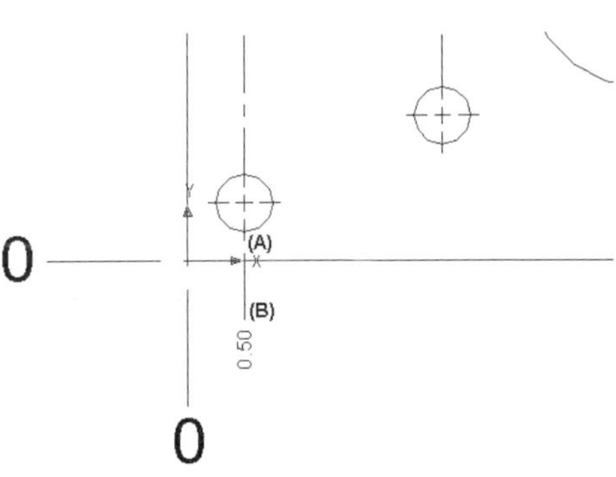

Figure 11–4

STEP 4

Perform a ZOOM-PREVIOUS operation to display the overall drawing of the base plate, similar to the illustration in Figure 11–5. Repeat the procedure in Step 3 to place ordinate dimensions at locations "A" through "G." Be sure Ortho mode is on and that the leader location is two snap distances below the bottom edge of the object. The OSNAP-Endpoint mode should be used on the endpoints of each centerline to satisfy the prompt "Select feature."

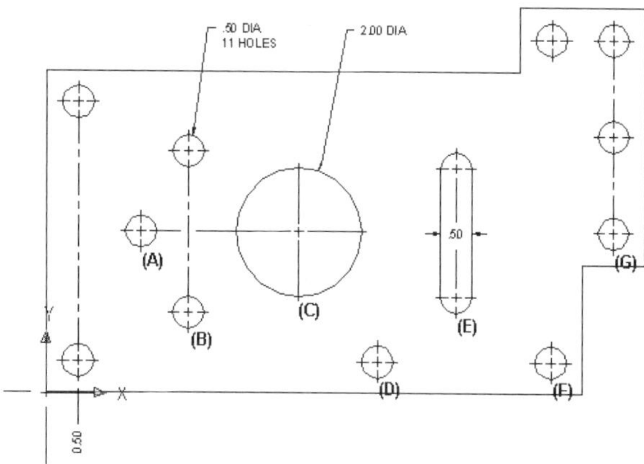

Figure 11–5

STEP 5

Continue placing ordinate dimensions similar to the procedure used in Step 3. Use the OSNAP-Endpoint mode to select features at "A" and "B" in Figure 11–6. Have Ortho mode on and identify the leader endpoint two grid dots below the bottom edge of the object.

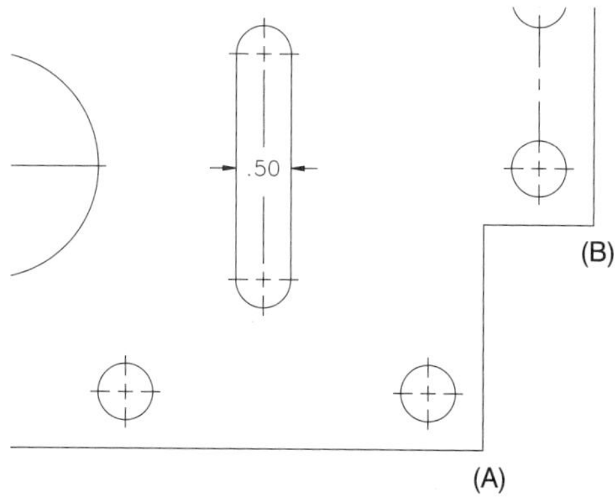

Figure 11–6

STEP 6

Use the UCSICON command to turn off the display of the icon. This will still keep the 0,0 origin at the lower-left corner of the object. After you turn off the icon, your display should appear similar to Figure 11–7.

Command: **UCSICON**
[view>display>ucsicon]
Enter an option [ON/OFF/All/Noorigin/ORigin] <ON>: **Off**

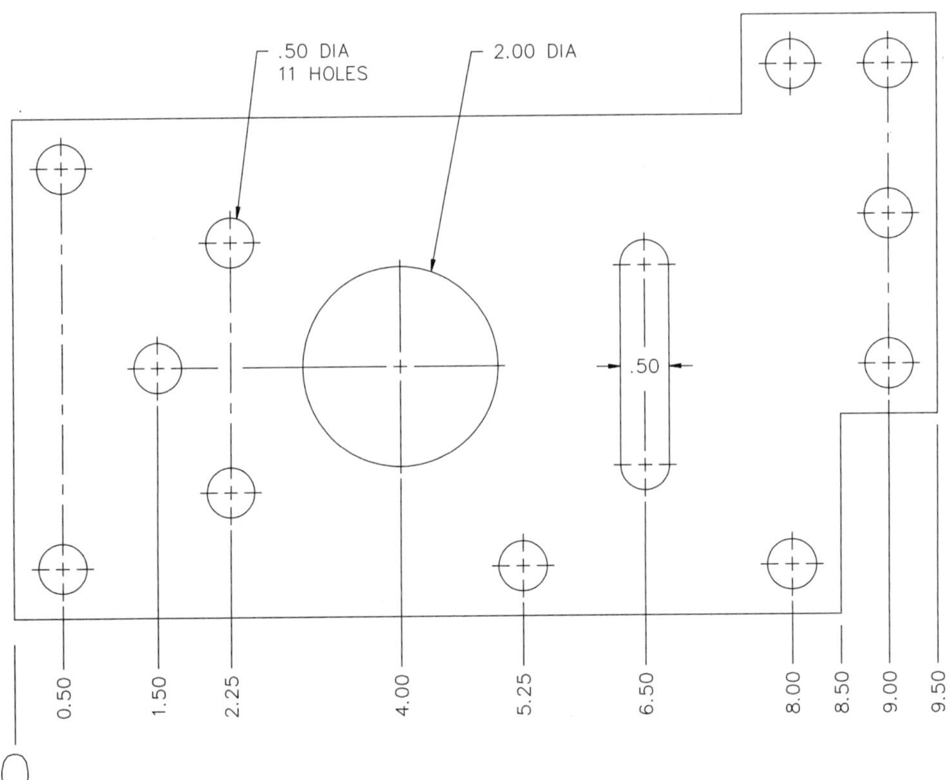

Figure 11–7

STEP 7

With Snap on and Ortho on, begin placing the first vertical ordinate dimension using the DIMORDINATE command. The procedure and prompts are identical to those for placing a horizontal ordinate dimension. See Figure 11–8 and the following prompts to perform this operation. (OSNAP-Endpoint should still be running.)

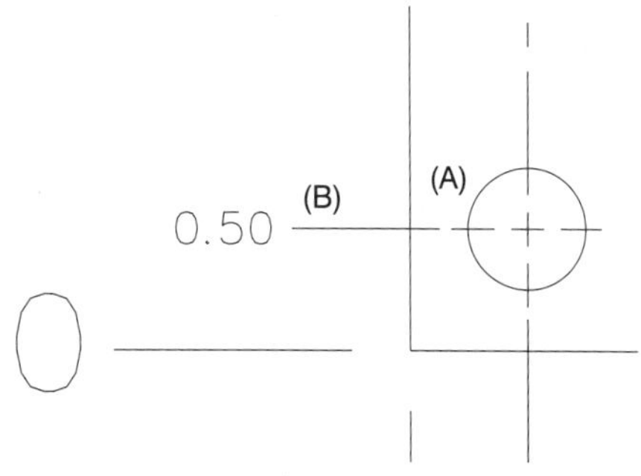

Figure 11–8

Command: **DOR** *(For DIMORDINATE)*
[dimension>ordinate]
Specify feature location: *(Select the endpoint of the center line at "A")*
Specify leader endpoint or [Xdatum/Ydatum/Mtext/Text/Angle]: *(Pick a point two snap distances to the left of the object at "B")*
Dimension text = 0.50

STEP 8

Follow the procedure in the previous step to complete the vertical ordinate dimensions along this edge of the object, as in Figure 11–9. Use the OSNAP-Quadrant mode for "A" through "D" and OSNAP-Endpoint for "E." Again, have Ortho on and Snap on. For the leader endpoint, count two grid dots to the left of the object and place the dimensions.

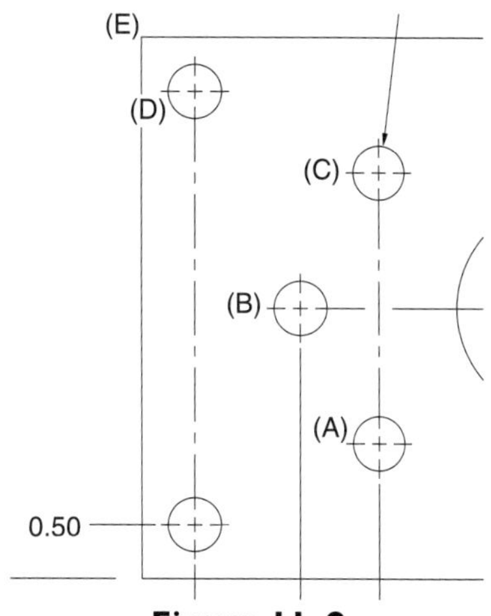

Figure 11–9

STEP 9

Your display should appear similar to the illustration in Figure 11–10.

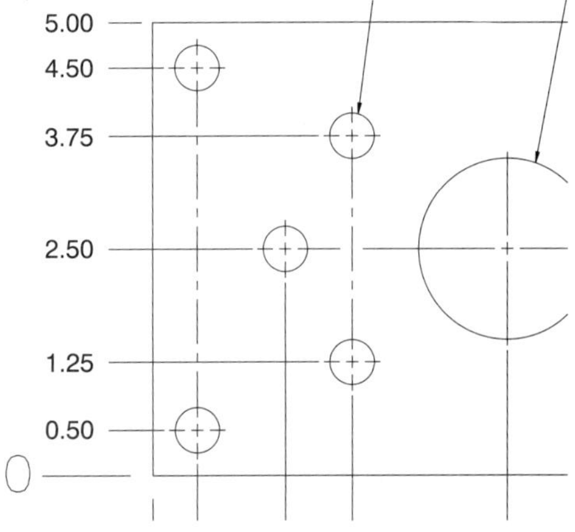

Figure 11–10

STEP 10

Magnify the right portion of the object using the ZOOM-WINDOW command. Use ordinate dimensions and OSNAP-Endpoint to place vertical ordinate dimensions from "A" to "I," as shown in Figure 11–11.

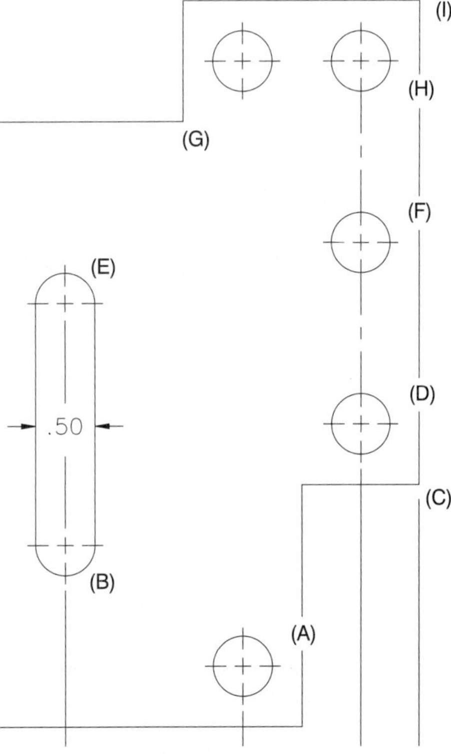

Figure 11–11

STEP 11

Your display should appear similar to the illustration in Figure 11–12.

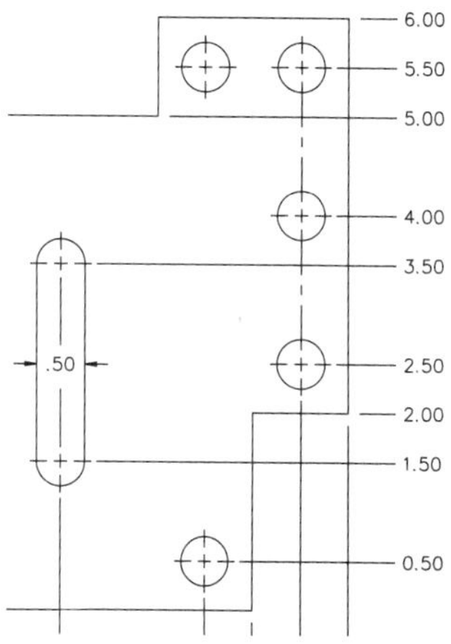

Figure 11–12

STEP 12

Complete the dimensioning by placing two horizontal ordinate dimensions at the locations illustrated in Figure 11–13. As in the earlier steps, have Ortho and Snap modes on, and use OSNAP-Endpoint to select the features. Place the leader endpoint two grid dots above the top line of the object. Use the ZOOM-ALL option to return the entire object to your display. See Figure 11–14.

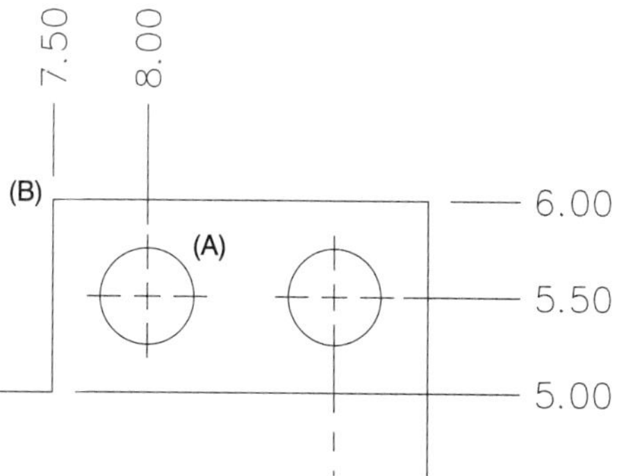

Figure 11–13

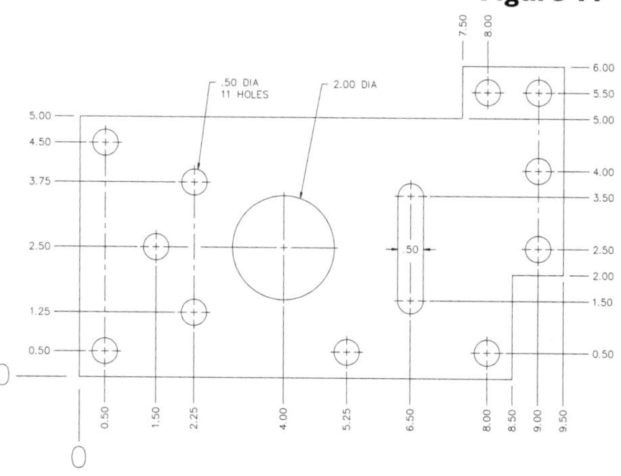

Figure 11–14

TUTORIAL EXERCISE: 11_DIMENSION VIEWS.DWG

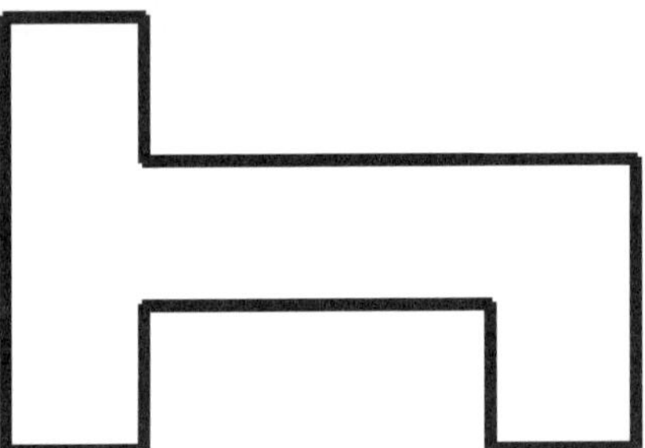

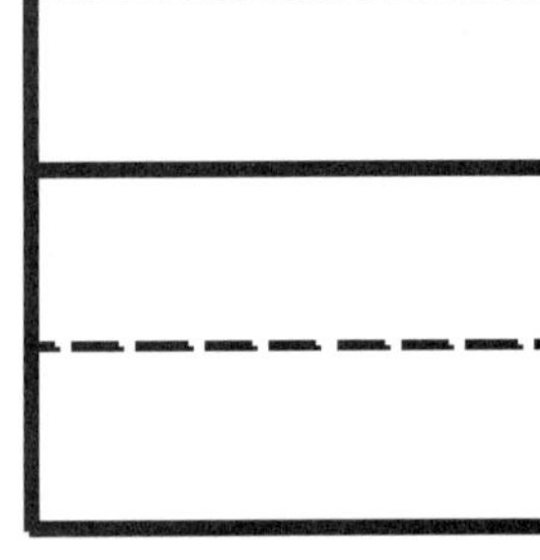

Figure 11–15

Purpose

The purpose of this tutorial is to add dimensions to the three-view drawing named *11_Dimension Views.dwg.*

System Settings

The drawing in Figure 11–15 is already constructed. Dimensions must be added to various views to call out overall distances, in addition to features such as cuts and slots. Be sure the following Object Snap modes are currently set: Endpoint, Center, Intersect, and Extension.

Layers

All layers have already been created:

Name	Color	Linetype
Cen	Yellow	Center
Defpoints	White	Continuous
Dim	Yellow	Continuous
Object	White	Continuous

Suggested Commands

Use the DIMLINEAR command for horizontal and vertical on different views in this drawing.

Whenever possible, substitute the appropriate command alias in place of the full AutoCAD command in each tutorial step; for example, use "CP" for the COPY command, "L" for the LINE command, and so on. The complete listing of all command aliases is located in Table 1–2.

STEP 1

Open the drawing file *11_Dimension Views.dwg*. Linear dimensions are placed identifying the overall length, width, and depth dimensions. The DIMLINEAR command is used to perform this task. The overall depth shown in Figure 11–16 is placed above the Right Side view. It could also be placed in the Top view as shown in Figure 11–17.

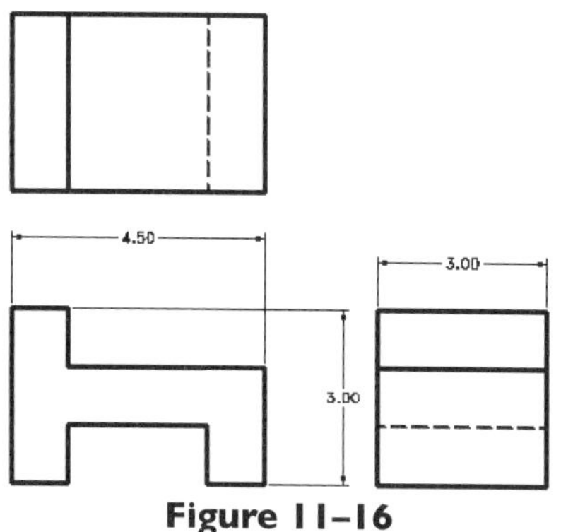

Figure 11–16

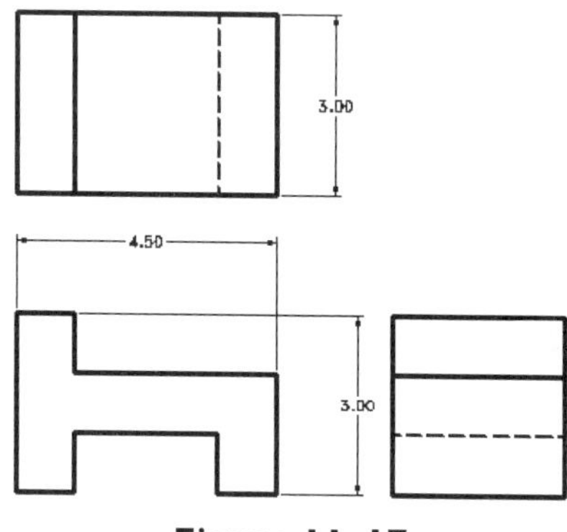

Figure 11–17

STEP 2

Detail dimensions that identify cuts and slots are placed. Because these cuts are visible in the Front view, the dimensions are placed there. See Figure 11–18. Use Dimension Continue (DCO) for the second 1.00 dimension.

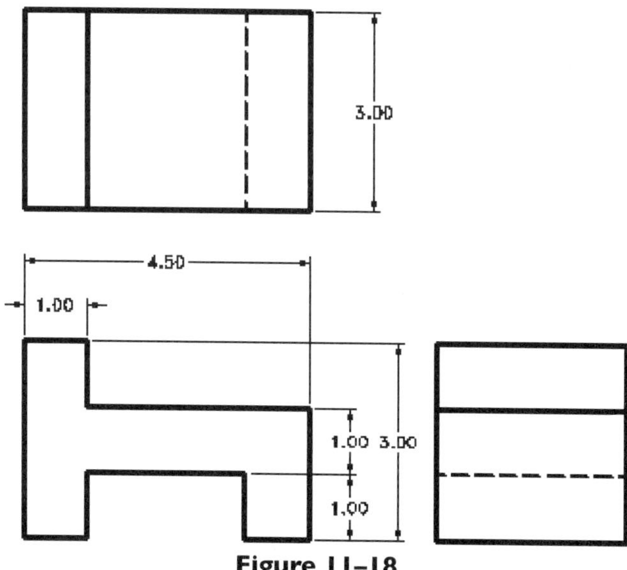

Figure 11–18

STEP 3

Once the spaces between the Front, Top, and Right Side views are used up by dimensions, the outer areas are used for placing additional dimensions such as the two horizontal dimensions in Figure 11–19. Again, use Dimension Continue (DCO) for the second dimension. Use Grips to adjust dimension text locations as necessary.

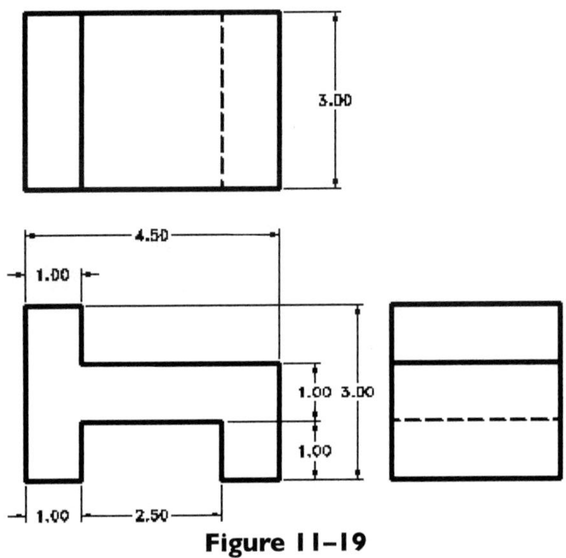

Figure 11–19

STEP 4

All dimensions in Steps 1 through 3 were placed with the DIMLINEAR command. Figure 11–20 shows the effects of using the QDIM command and the Baseline mode for placing the three vertical dimensions located in the Front view.

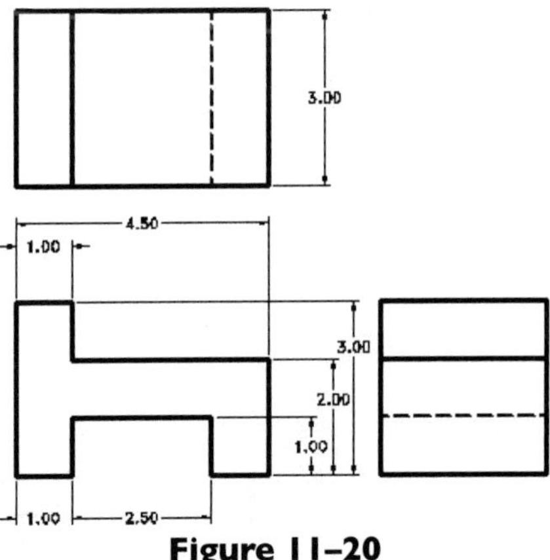

Figure 11–20

Beginning Level Drawings

Problems 11–1 through 11–19

1. Open each drawing from the CD
2. Use proper techniques to dimension each drawing.
3. Follow the steps used in completing the 11_Dimension Views tutorial.

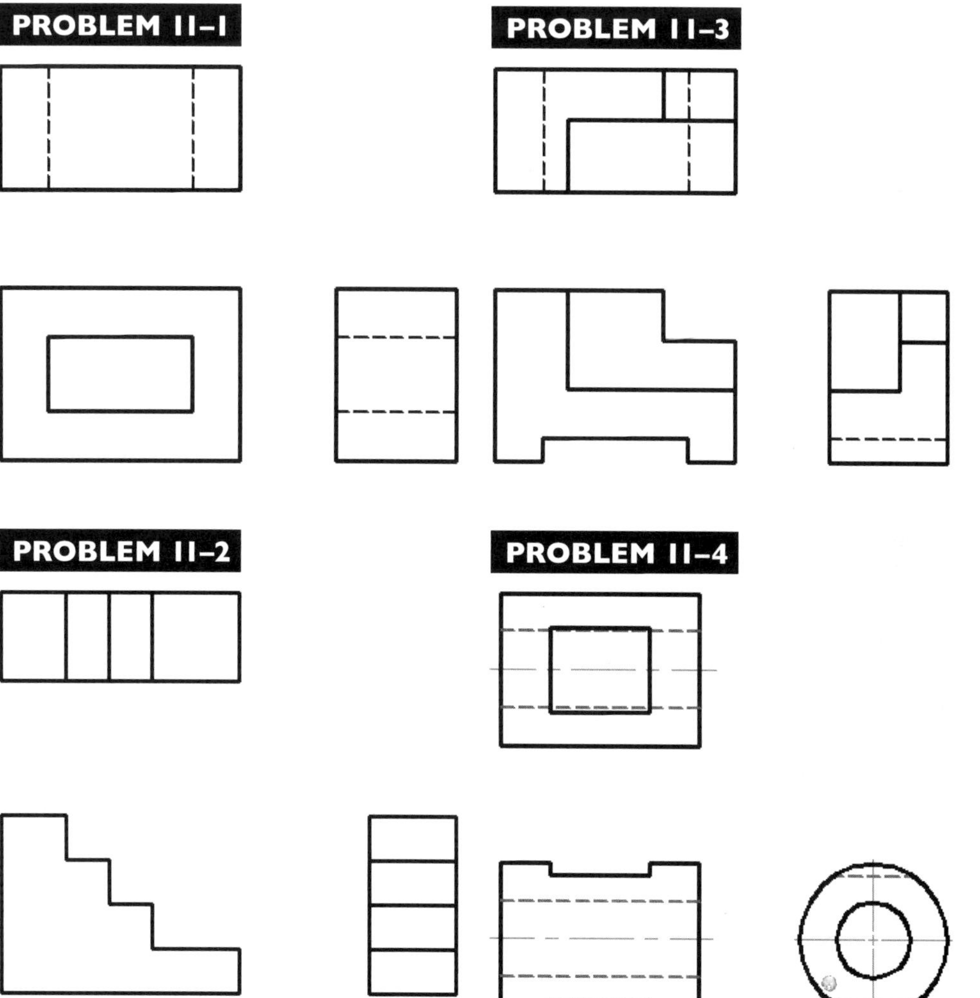

PROBLEM 11–5

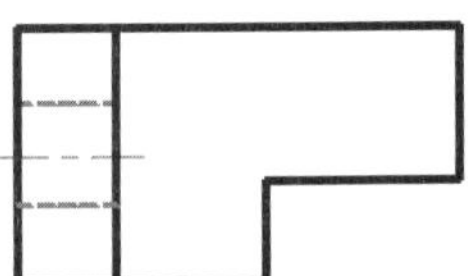

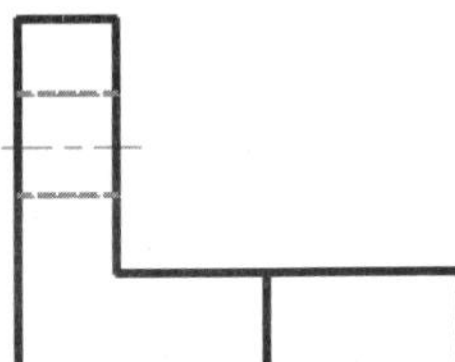

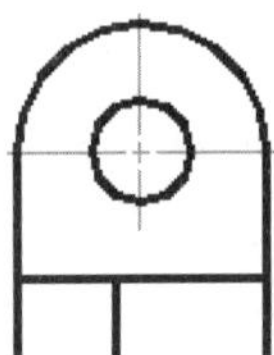

PROBLEM 11–7

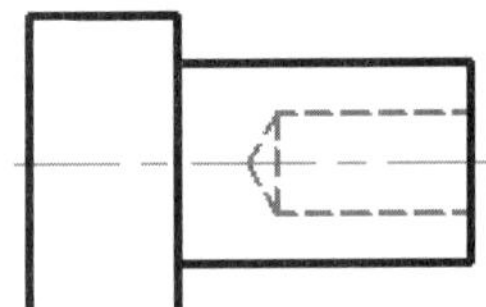

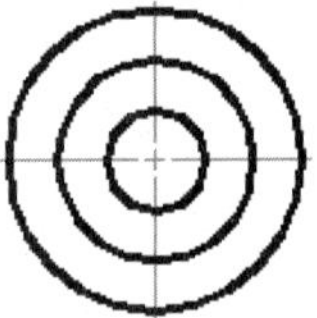

PROBLEM 11–6

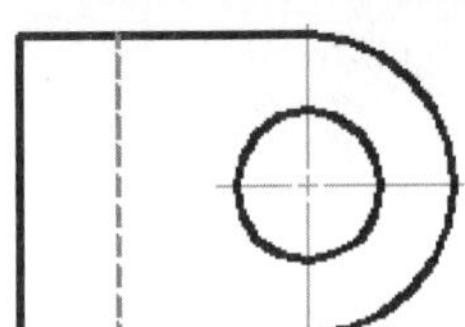

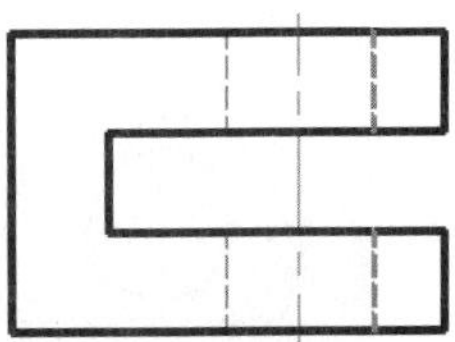

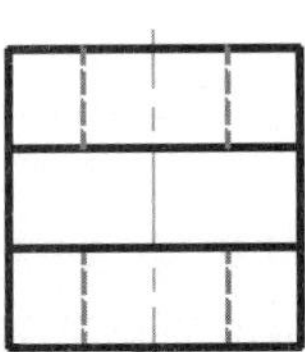

PROBLEM 11–8

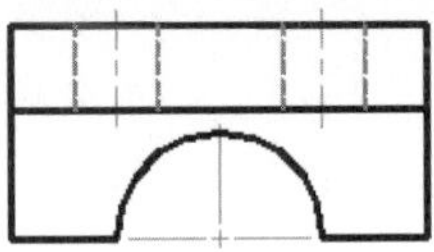

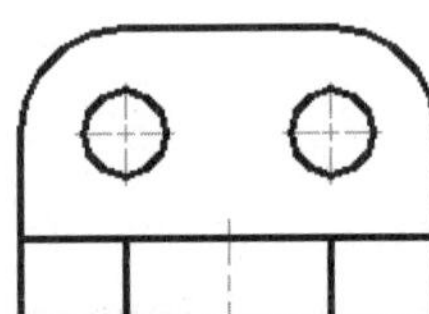

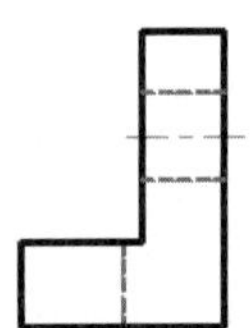

PROBLEM 11–9

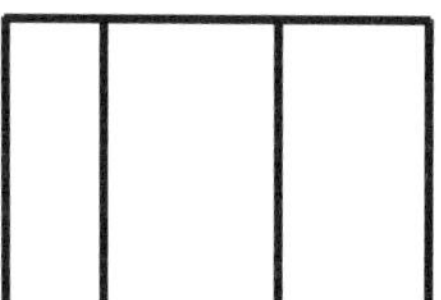

PROBLEM 11–11

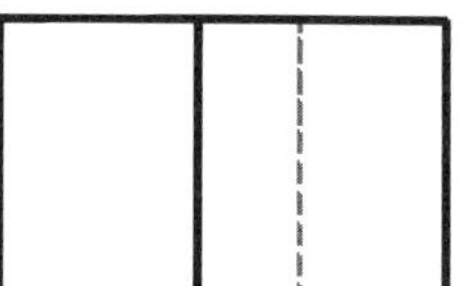

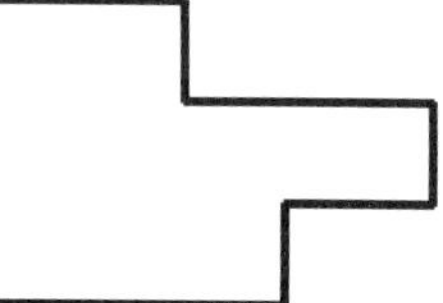

PROBLEM 11–10

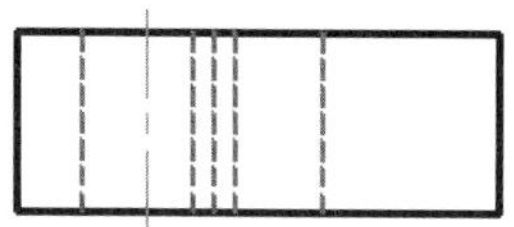

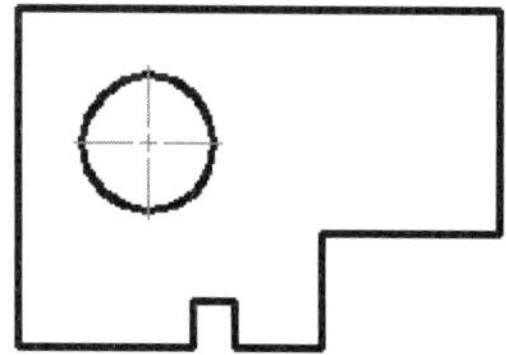

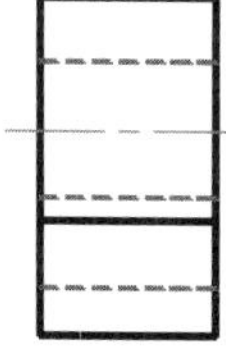

PROBLEM 11–12

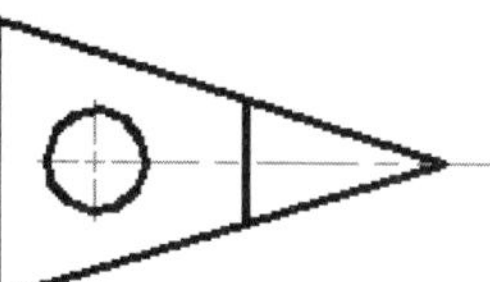

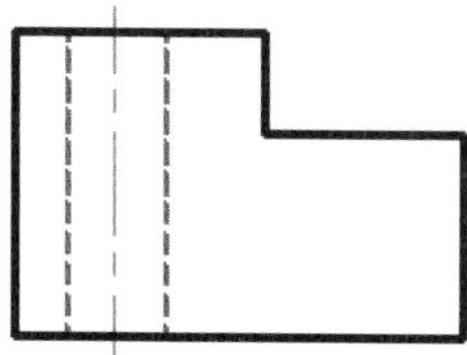

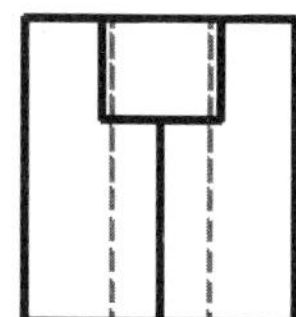

PROBLEM 11–15

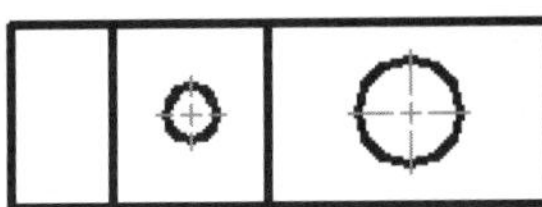

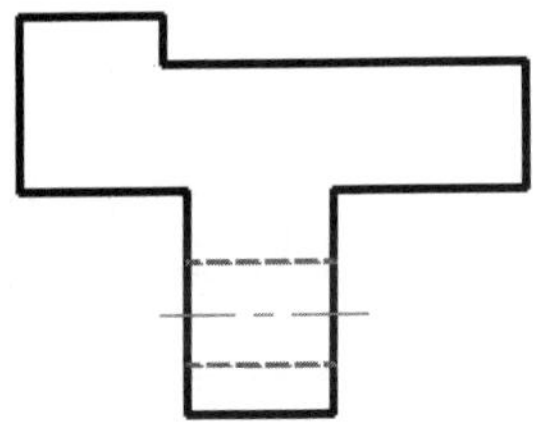

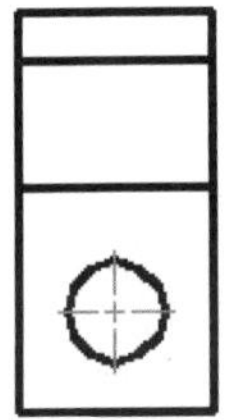

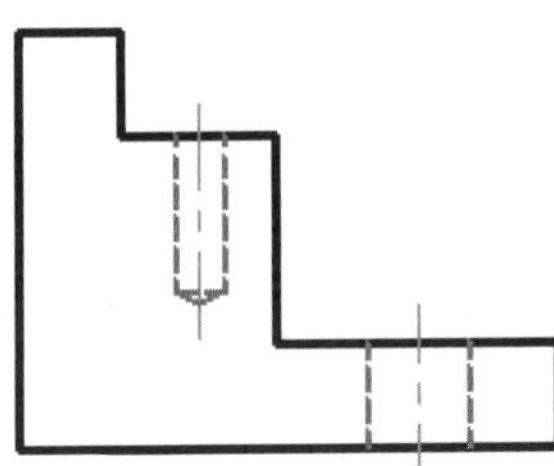

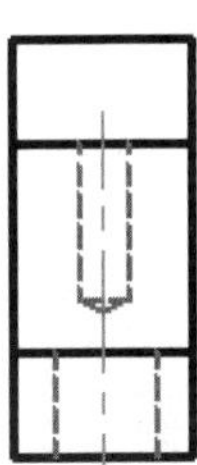

PROBLEM 11–14

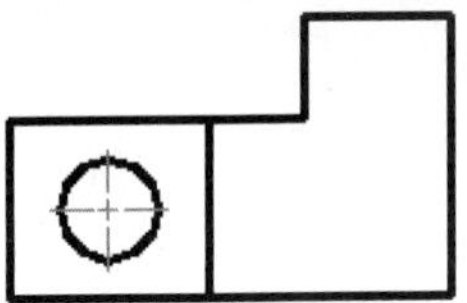

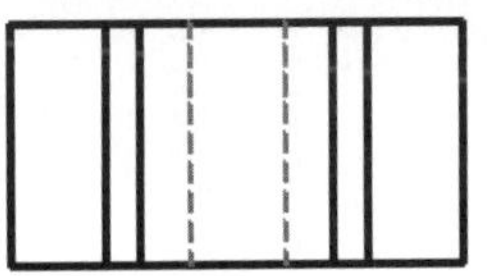

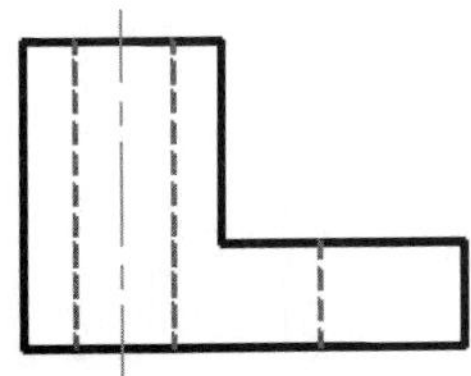

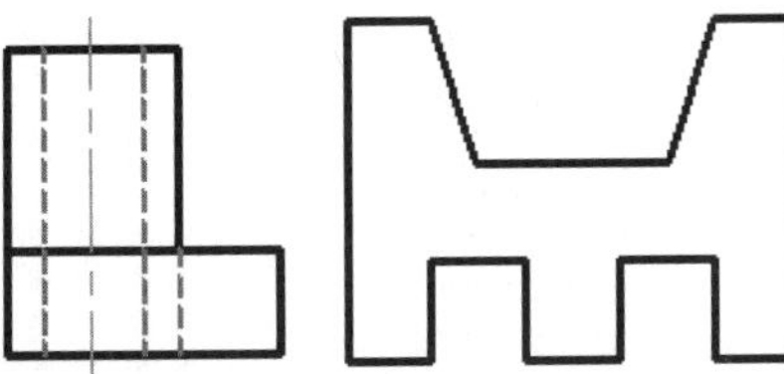

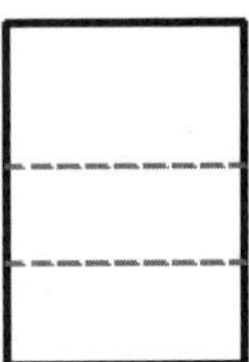

PROBLEM 11–17

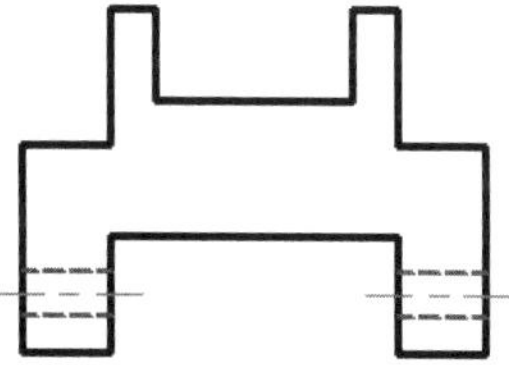

PROBLEM 11–19

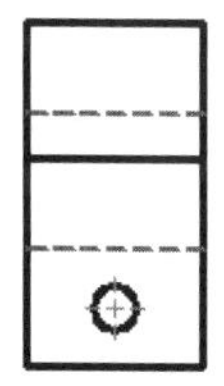

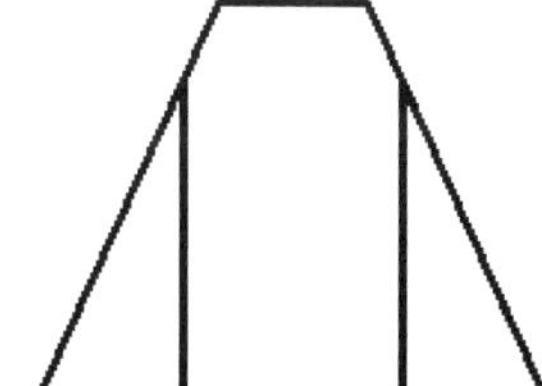

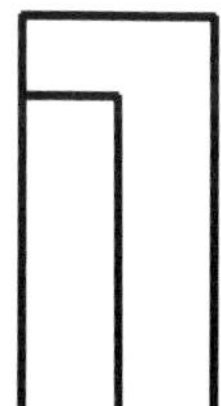

PROBLEM 11–18

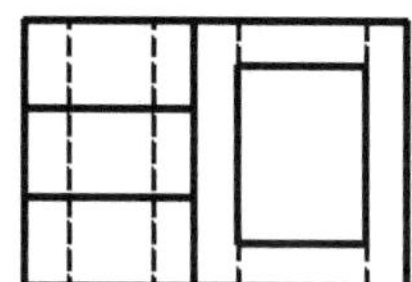

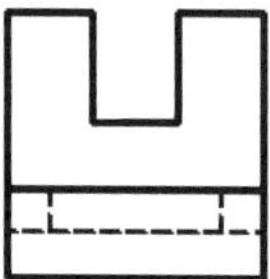

Intermediate Level Problems

Directions for Problems 11–20 and 11–21

Using the provided layout as a guide create the 3 view drawings and fully dimension them.

PROBLEM 11–20 **PROBLEM 11–21**

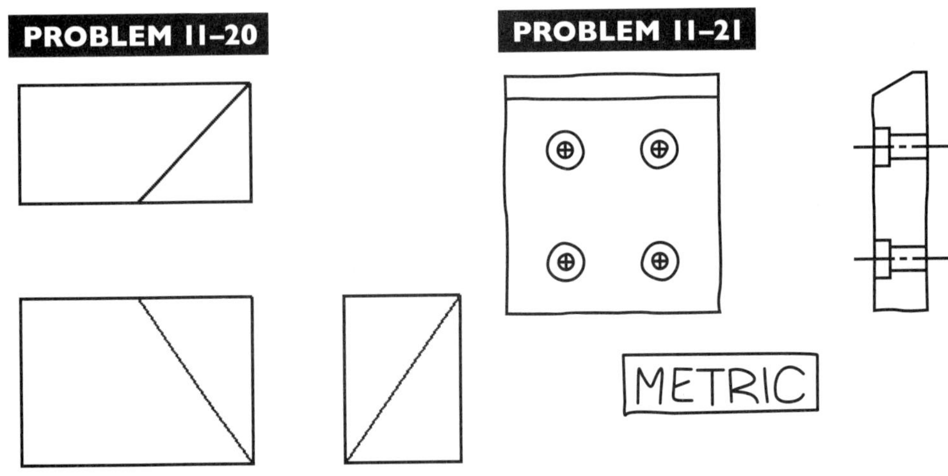

Advanced Level Problems

PROBLEM 11–22

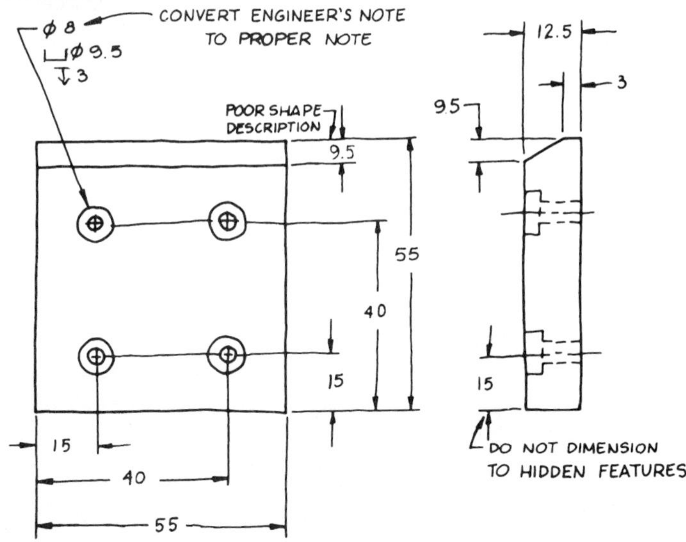

PROBLEM 11–23

Directions for Problem 11–23

Open the drawing *11_plant.dwg*.

Apply the dimension to locate the part bins and the production equipment from the edges of the work cell.

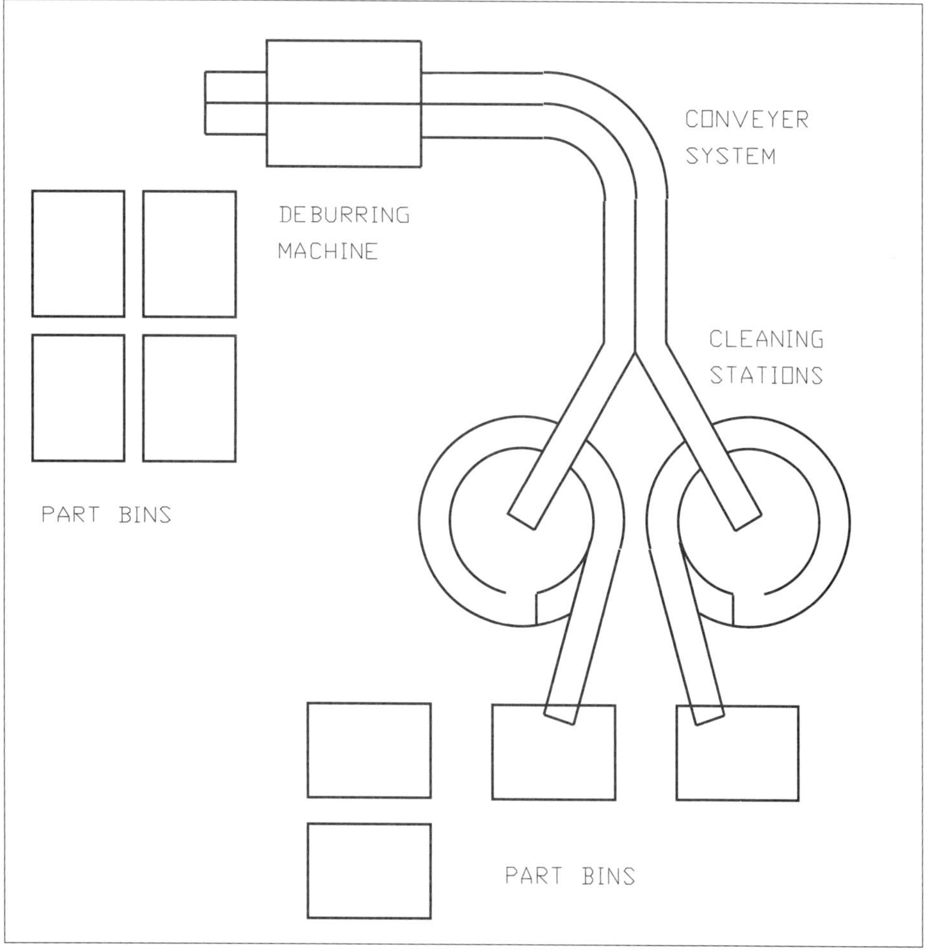

PROBLEM 11–24

Directions for Problem 11–24

Use ordinate dimensioning techniques to dimension this drawing.

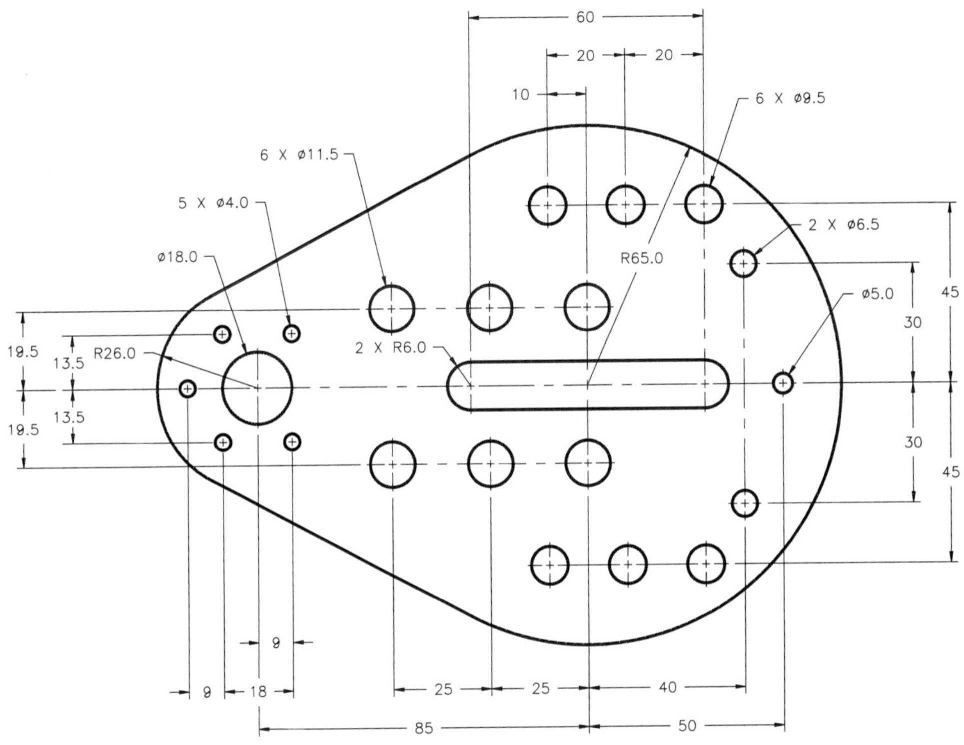

The Dimension Style Manager

The exercises in this chapter introduce:

- Creating dimension styles
- Editing dimension styles
- Scaling dimensions for drawings and layouts
- Tolerancing
- Alternate units and dual dimensioning

TUTORIAL EXERCISE: 12_ARCHITECTURALDIMENSION.DWG

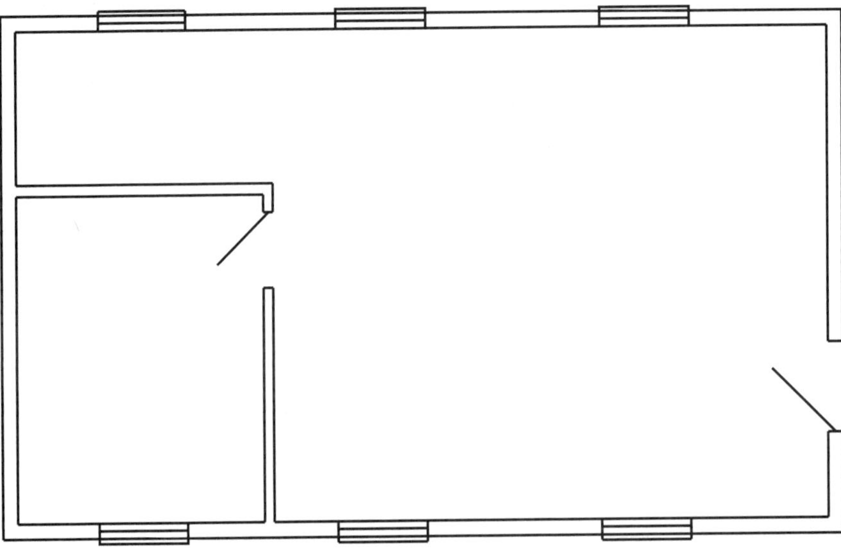

Figure 12–1

Purpose

This tutorial is designed to develop and apply dimension styles to an architectural drawing project. The dimensions will be initially applied to a model space environment, but the final results will be printed from a layout space.

System Settings

No special settings are needed for this project.

Layers

The following layers have already been created:

Name	Color	Linetype
Walls	Blue	Continuous
Windows	Green	Continuous
Dimension	Red	Continuous

Suggested Commands

Dimensioning commands used in the tutorial include:

STEP 1

Open the *12_Architectural dimension.dwg* file. Check to make sure that the dimension layer is the current layer. Select the DIMENSION LINEAR command and place an overall dimension on the drawing by selecting the endpoints of the top wall. Once the dimension is placed, the dimensional text is extremely hard to see. The dimension style tools will help us repair this dimension, so that it can be seen and uses the correct units of measurement.

Command: **dimlin** *(For Dimlinear)*
[Dimension>Linear]

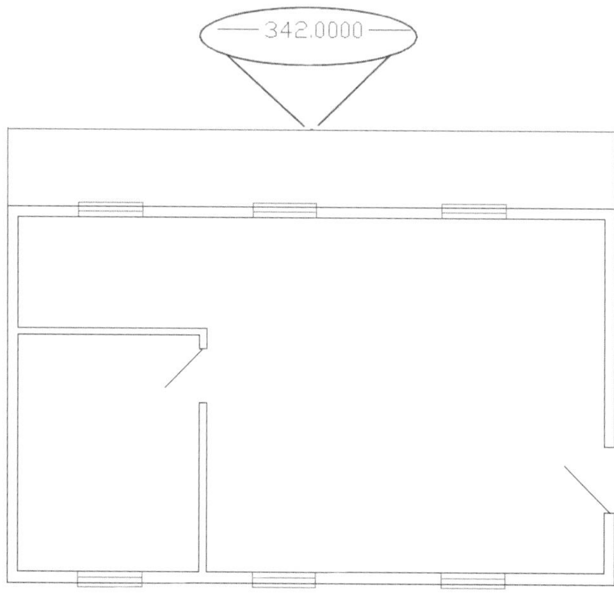

Figure 12–2

STEP 2

Since the dimension cannot be seen without magnification, the Dimension Style Manager will be used to set-up the characteristics of the dimensions so that it can be read. To begin, open the Dimension Style Manager by selecting the Dimstyle button, or the command can be found on either the Format or the Dimension pull down menus.

Command: **Dimstyle**
[Dimension>Style]

STEP 3

Select the New... button and the Create New Dimension Style dialog box appears, as shown in FIgure 12–3. Replace the "Copy of Standard" in the New Style Name edit box with ARCHITECTURAL. This dimension style will initially be based on the current standard dimension style and applied to all dimensions that are placed in the drawing.

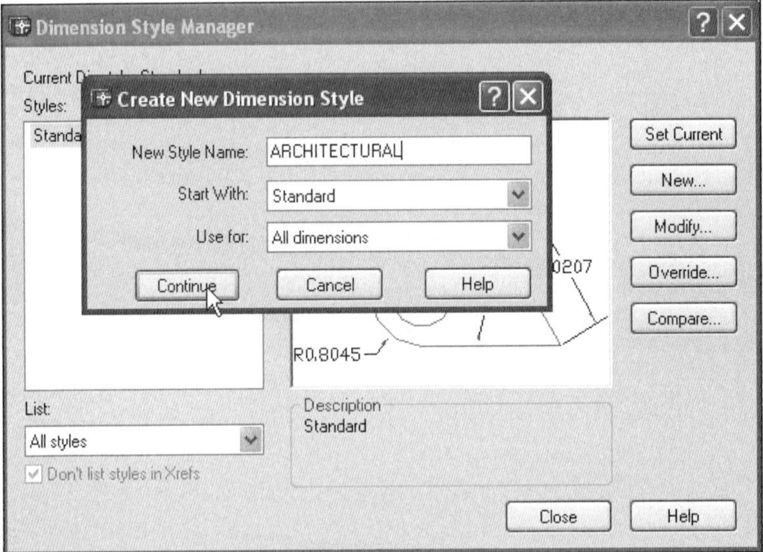

Figure 12–3

STEP 4

In Figure 12–4 note that the name ARCHITECTURAL is now displayed in the title bar of the New Dimension Style dialog box. In the Arroheads section, change the arrowhead style to Architectural tick. When you change the 1st. arrowhead it also changes the 2nd. arrowhead. When using the Architectural ticks arrowhead, the Extend beyond ticks edit box, found in the Dimension Line area, becomes available. This value refers to the linear tail of the dimension line that extends past the tick mark.

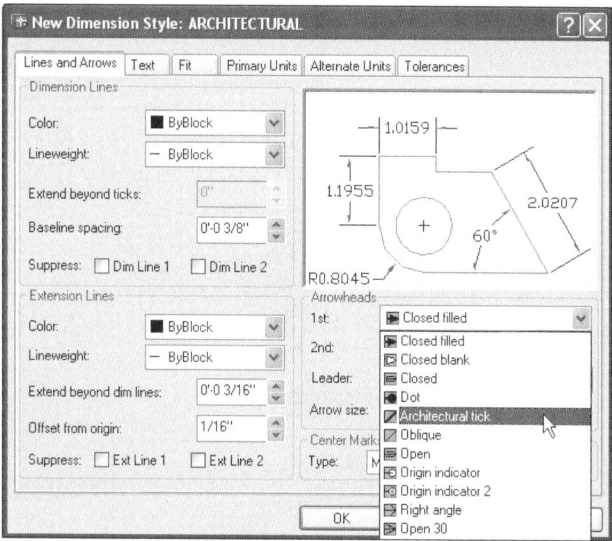

Figure 12–4

STEP 5

Select the Fit tab to change the overall scale of the dimension. The overall scale is based on the plotting scale when plotting is done from the model tab. The anticipated plotting scale for this drawing is ¼"=1'. Because ¼" divides into 1', 48 times, type **48** in the Use overall dimension scale of edit box. (Note: some adjustment to the overall scale may be needed to achieve the plotting look you desire). Place a check mark in the Always draw dim line between extension lines option (see Figure 12–5).

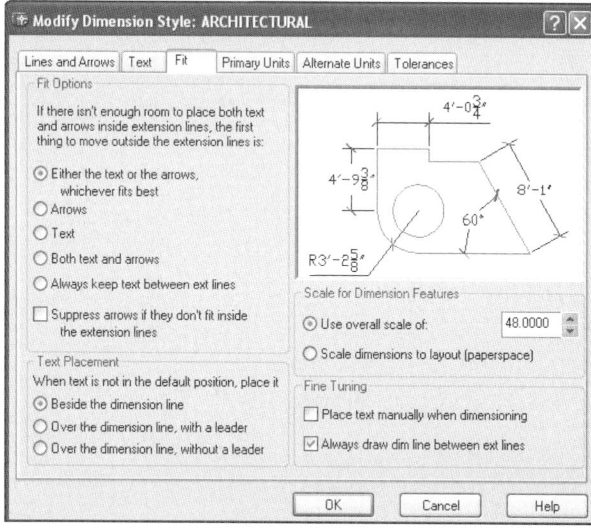

Figure 12–5

STEP 6

Choose OK to exit the New Dimension Style dialog box and click the Close button in the Dimension Style Manager to end the DIMSTYLES command. To change the current dimensional style, use the drop down list on the Styles toolbar shown in Figure 12--6. Choose ARCHITECTURAL from the list.

Figure 12–6

Place a vertical dimension on the right side of the drawing, using the corners of the building as the endpoint locations of the dimension. As shown in Figure 12–7, both the text and the tick marks are now visible on the dimension. Next, you will make adjustments to the units of measurement and text placement attributes of the dimension style.

Command: **dimlin** *(For Dimlinear)* [Dimension>Linear]

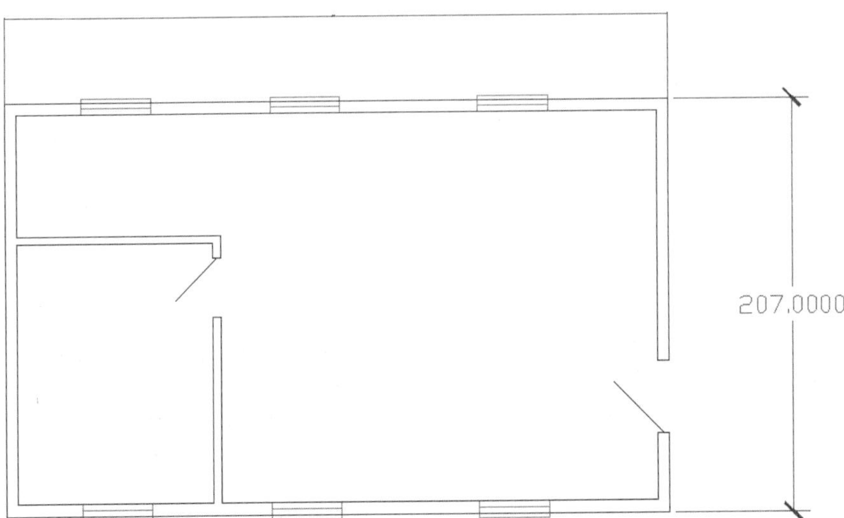

Figure 12–7

STEP 7

To change an existing horizontal dimension to the current style, grip the dimension (just pick the dimension) and right click the mouse. Pick the Dim Style option and a list of available dimension styles appears on the fly-out menu, select ARCHITECTURAL. The horizontal dimension now has the same characteristics as the vertical dimension. Any changes that are done to the ARCHITECTURAL dimension style will now apply to both dimensions.

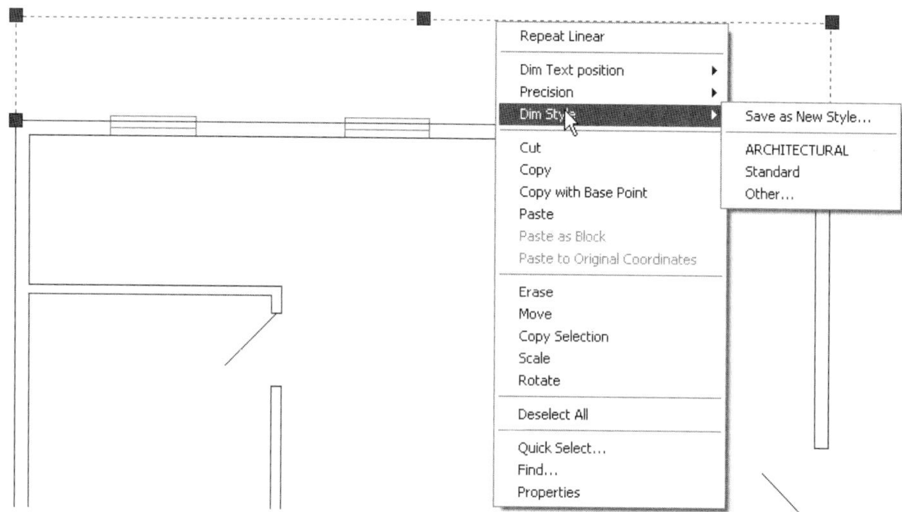

Figure 12–8

STEP 8

Invoke the Dimension Style command. In the Dimension Style Manager, select the ARCHITECTURAL dimension style and choose the Modify... button. This opens the Modify Dimension Style dialog box. Choose the Primary Units tab. In the Linear Dimensions section, change the Unit format option to Architectural. Keep the precision at 0"-0 1/16". Your display will look similar to Figure 12–9.

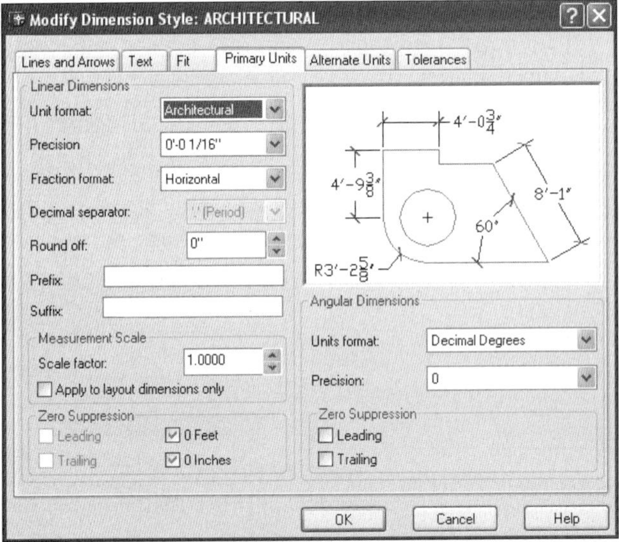

Figure 12–9

STEP 9

Now select the Text tab in the Dimension Style Manager. In the Text Appearance section, next to the Text style drop down list, there is a small button with "..." on it. Select this button to bring up the Text Style dialog box. Create a new text style by picking the New... button and typing a name in the New Text Style dialog box. (see Figure 12-10). Click OK.

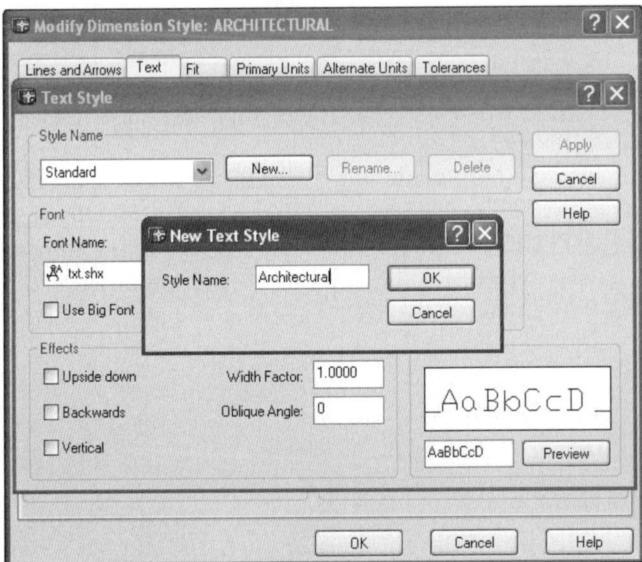

Figure 12–10

Now the current text style is Architectural, the font that will be applied to our dimensions is needed. The font is selected in the drop down list; select Tahoma. A sample of the font is displayed in the lower-right hand corner of the dialog. Leave the text height at 0. Changing the height will override any settings in the dimension style manager. The completed selection is shown in Figure 12-11. Choose the Apply button and then the Close button to return to the Dimension Style Manager. Choose OK again, to close the Dimension Style Manager.

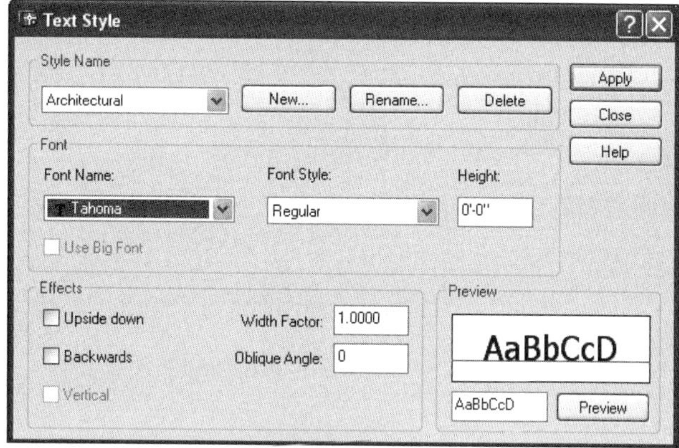

Figure 12–11

STEP 10

Radius dimensions need to use arrowheads. If you scan the Dimension Style Manager dialog, you can see that the architectural tick is currently being used for the radius dimension. Select the New... button, but instead of creating a whole new style, this style will be used for Radius dimensions only. See Figure 12-12.

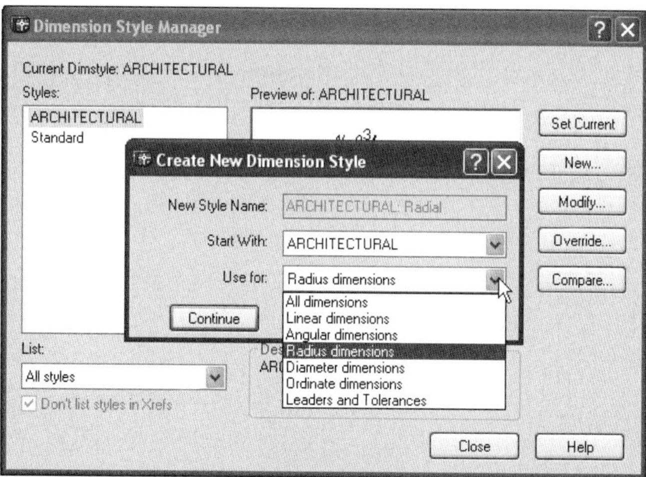

Figure 12–12

STEP 11

Choose Continue, and then select the Lines and Arrows tab. Change the arrowhead selection (only arrowhead 2 is available) from the architectural tick to the Closed filled arrowhead, shown in Figure 12–13 Choose OK to return to the Dimension Style Manager.

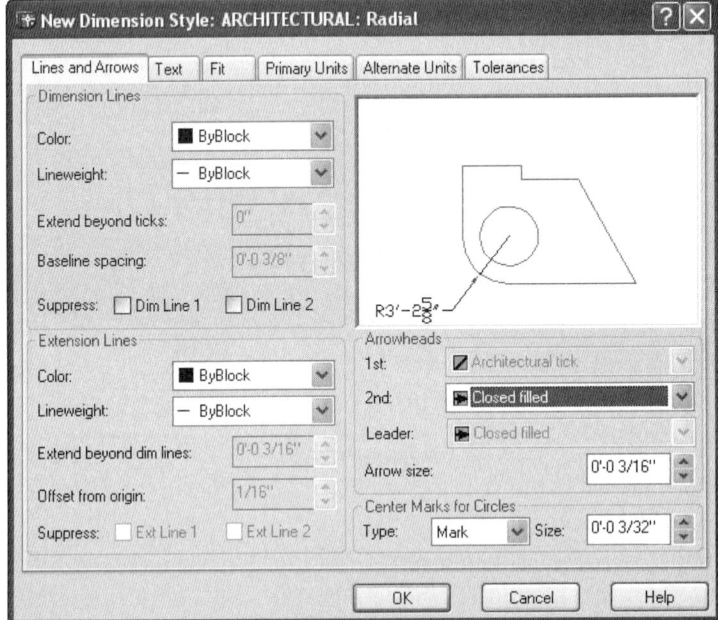

Figure 12–13

STEP 12

A similar task needs to be completed for diameter dimensions. Choose the New... button, and select Diameter dimensions from the Use for drop down menu. Choose Continue and then move to the Lines and Arrows tab to make the change from architectural tick mark (choose 1st arrowhead) to Closed filled arrowheads.

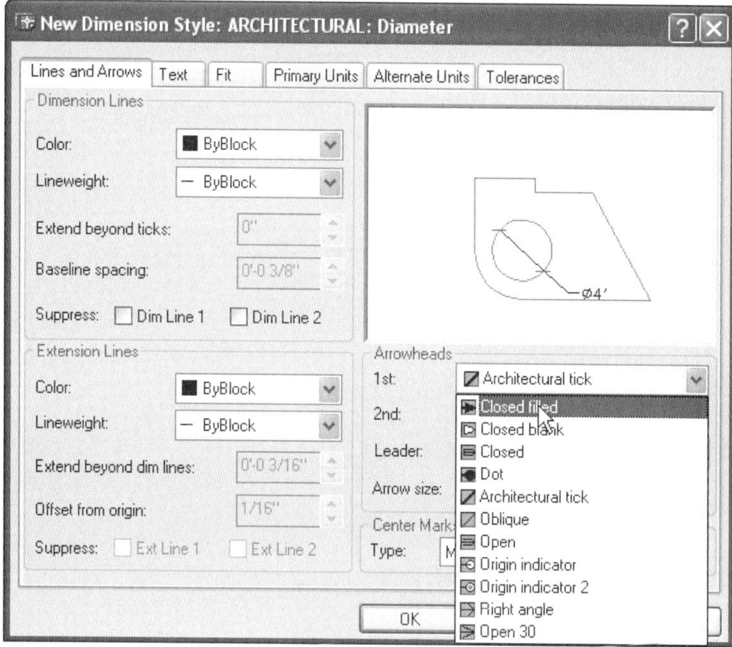

Figure 12–14

STEP 13

Choose OK, and then Close to exit the Dimension Style dialog box. Use Grips to move the existing dimensions further from the building (see Figure 12-15). This will provide the room necessary to dimension the window and door placements.

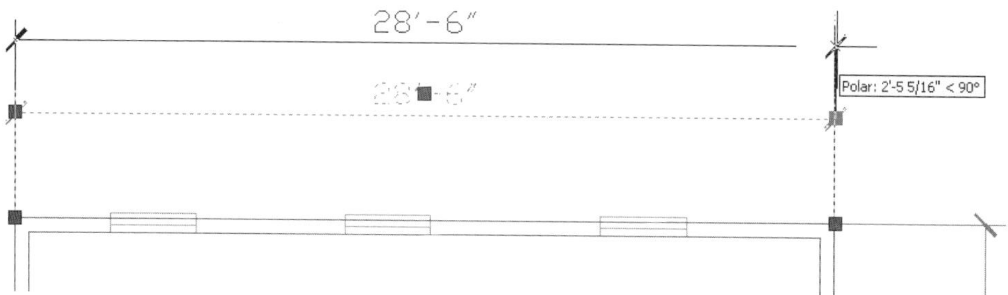

Figure 12–15

STEP 14

Use the Continuous option to apply dimensions succession to dimension the windows. First apply a linear dimension to the midpoint of the first window, from the outside edge of the building, using Object snap.

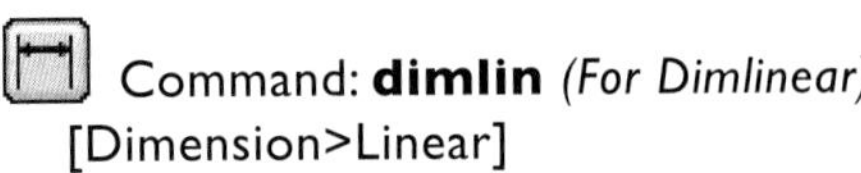
Command: **dimlin** *(For Dimlinear)* [Dimension>Linear]

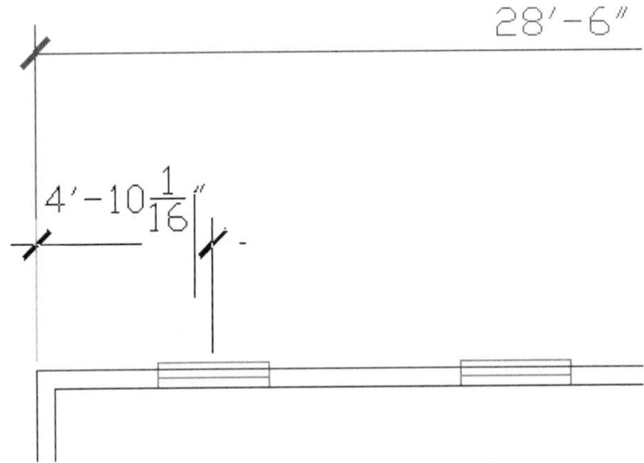

Figure 12–16

The DIMENSION CONTINUE command can be used to apply the next dimension to the midpoint of the next window. Figure 12–17 shows the application of the DIMENSION CONTINUE command.

Command: **dimcont** *(For Dimcontinue)* [Dimension>Continue]

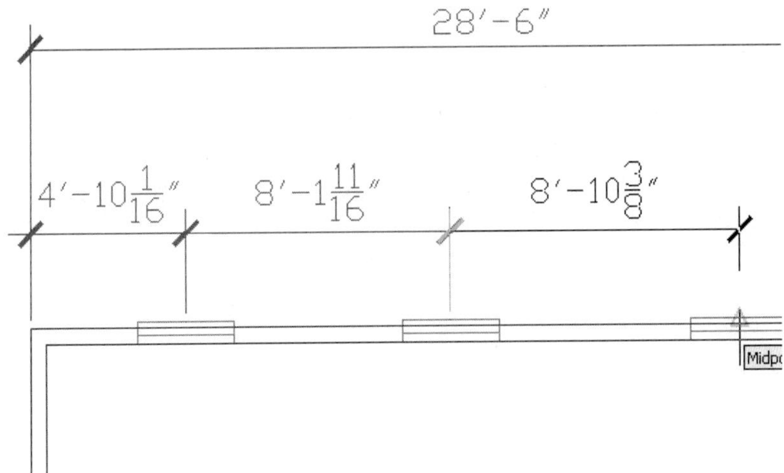

Figure 12–17

STEP 15

Figure 12–18 shows the completed drawing, ready to plot.

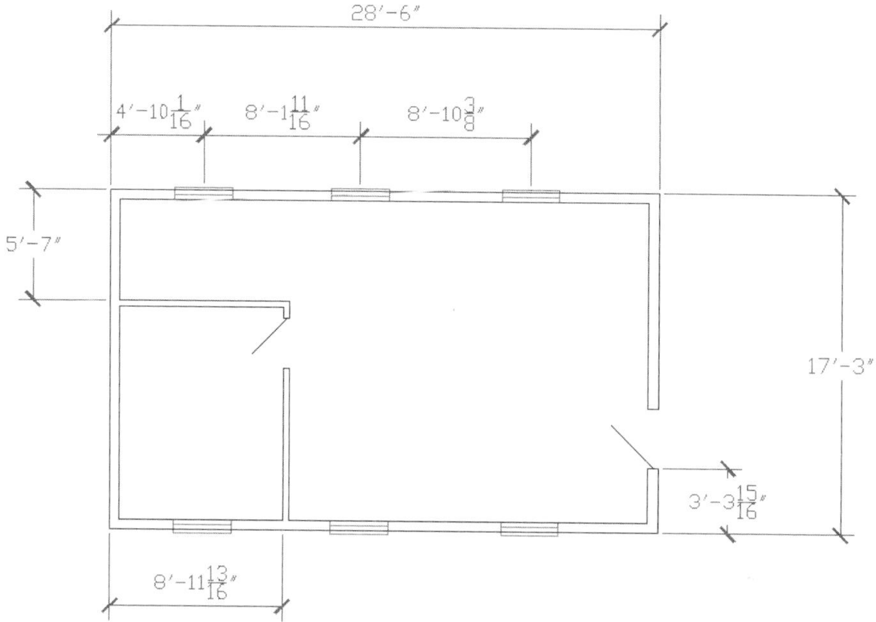

Figure 12–18

STEP 16

To prepare for plotting, select the Layout tab just below the drawing space. This will place the drawing in Layout space, which is specifically designed for plotting. The screen view will change, and a dialogue box will appear for the setup of the layer, as shown in Figure 12–19. Under the Plot Device tab, choose the DWF6 eplot as the plotting device.

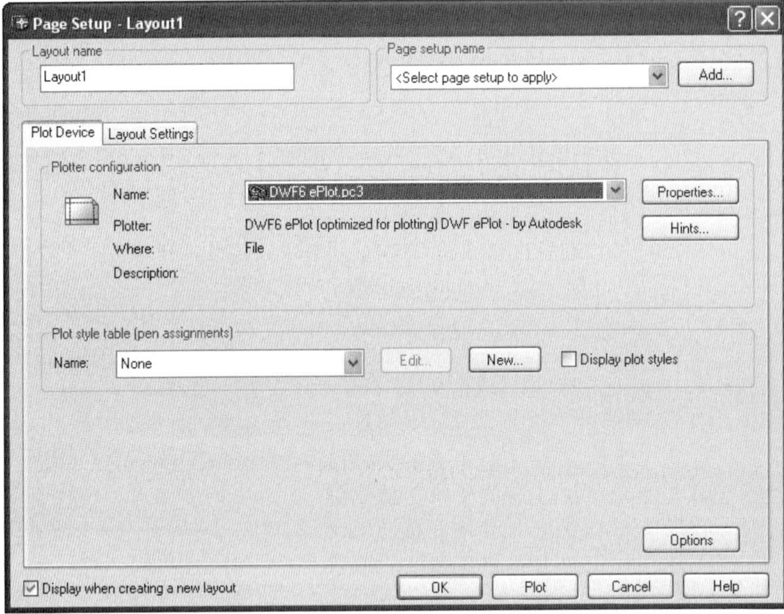

Figure 12–19

Select the Layout Settings tab and use Figure 12–20 to assist in setting up the layout settings. Note that the paper size has been modified to B size (17" x 11"). When you have completed the layout settings choose OK.

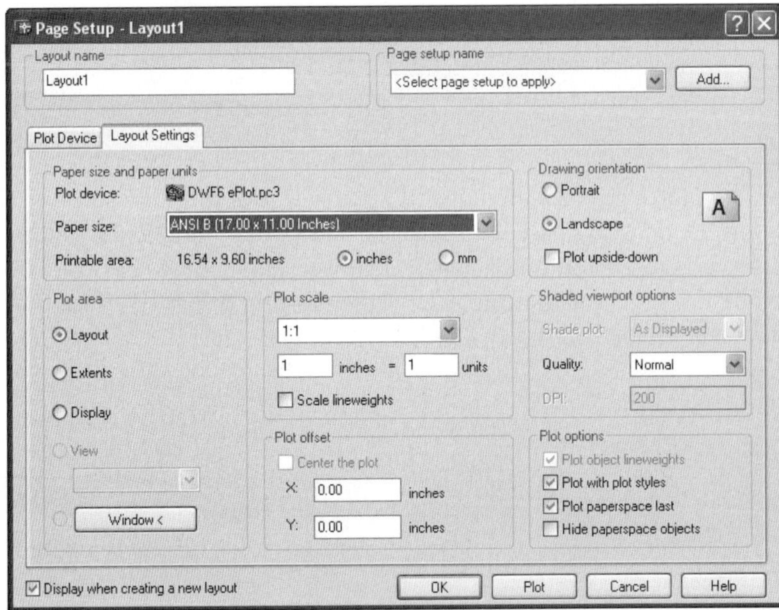

Figure 12–20

STEP 17

Figure 12–21 shows the current paper layout. The thick line represents the viewport. Turn on the viewports toolbar to assist in the plotting process.

Figure 12–21

To maximize the viewport size, double click off the paper, as shown in Figure 12–22. Grip the viewport and drag it to maximize the size to fit inside the dashed lines.

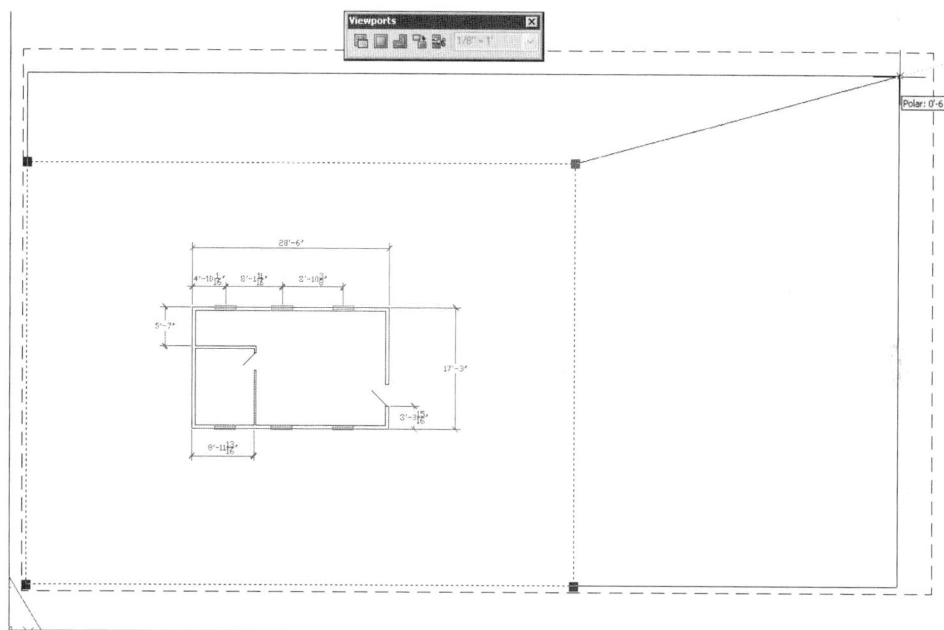

Figure 12–22

To set the scale of the viewport, double click inside the viewport and use the viewports toolbar to apply a scale factor. Set the scale on the drop down to ¼"=1'. Use the PAN command to adjust the image to locate the image on the layout sheet.

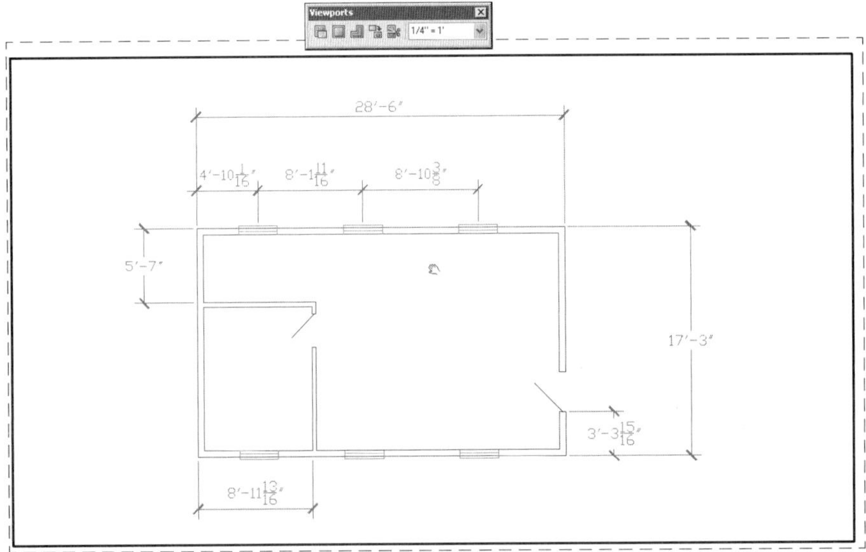

Figure 12–23

STEP 18

The dimensions on your plotting sheet appear too large. To adjust the dimensions to a proper size, the Dimension Style tools will be used. Open up the Dimension Style Manager, select the ARCHITECTURAL style and choose the Modify... button. Select the Fit tab in the Modify Dimension Style dialog box (see Figure 12–24).

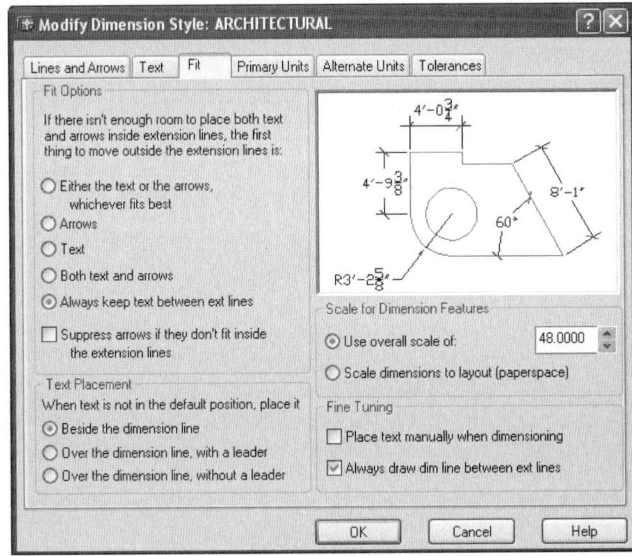

Figure 12–24

STEP 19

Under the Scale for Dimension Features area, select the Scale dimensions to layout space (paper space) radio button. Setting this option insures that the dimensions are scaled to the true height settings that were prescribed in the Dimension Style Manager. For example, the text height that is set in the manager will set the height of the dimension in the Layout space. Choose OK to close the dialog box.

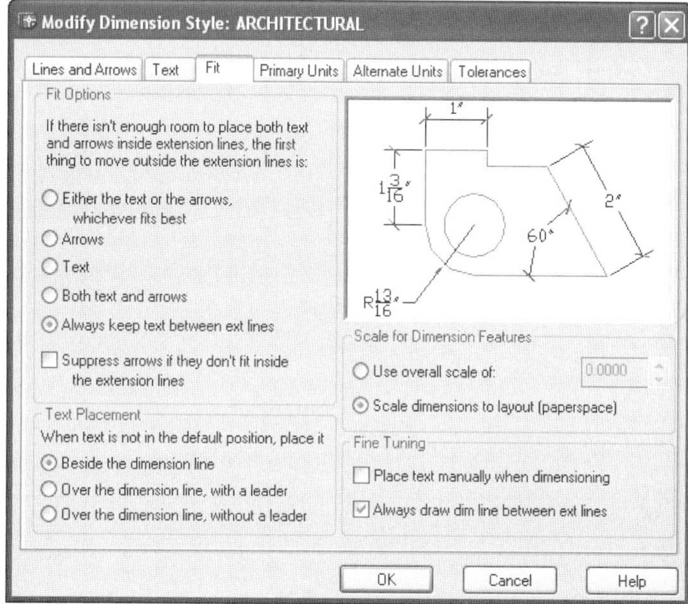

Figure 12–25

STEP 20

Select the Diameter option under the ARCHITECTURAL Style and choose the Modify... button. Select the Fit tab and verify the Layout space radio button is on. Confirm this setting for the Radius option also, using the same process. When finished with the confirmations, choose OK and then Close the Dimension Style Manager.

STEP 21

To apply the new setting, double click inside the viewport to make it active. Choose the Dimension Update command and use a selection window around all the dimensions.

Command: **dimupd** *(For Dimupdate)*
[Dimension>Update]

STEP 22

There will be no change in the dimension text size after this update. To witness a change, modify the viewport scale to 1/8"=1', as shown in Figure 12–26 Then apply the dim update to view the text at true size.

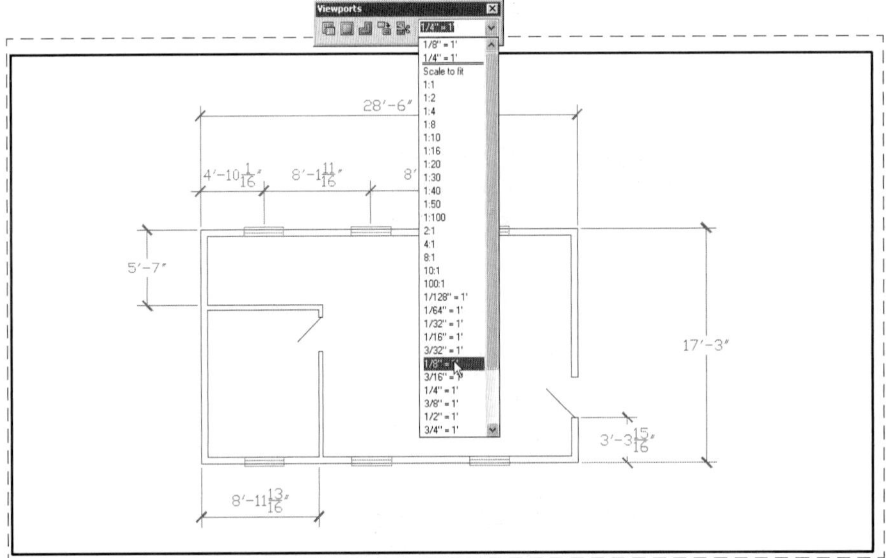

Figure 12–26

STEP 23

The drawing dimensions may need to be adjusted to fit the new drawing scale. Use Grips to accomplish this task.

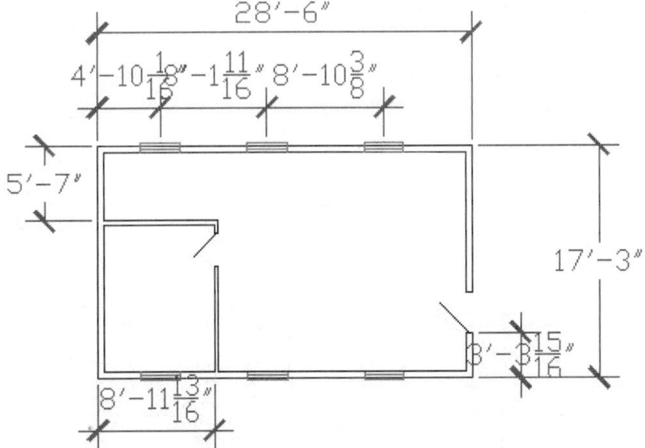

Figure 12–27

Beginning Level Drawings
Problems 12–1 through 12–19

1. Use the grid and spacing of 0.50 units to determine all dimensions.
2. Reproduce the views shown and fully dimension the drawings.
3. The small marks shaped in the lower case letter "v" are called finished marks.These signify surfaces that have been machined.

PROBLEM 12–1

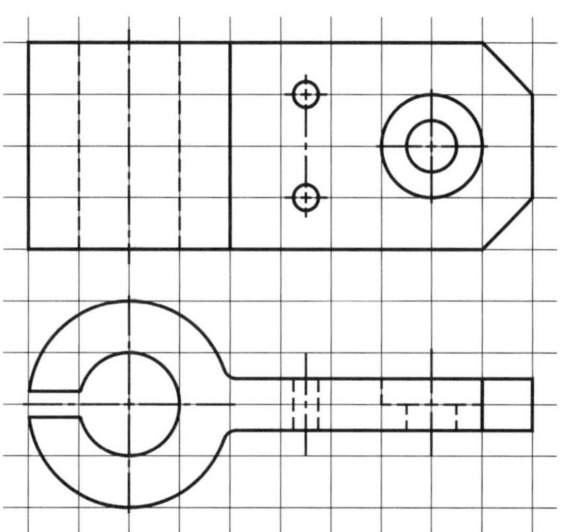

PROBLEM 12–2

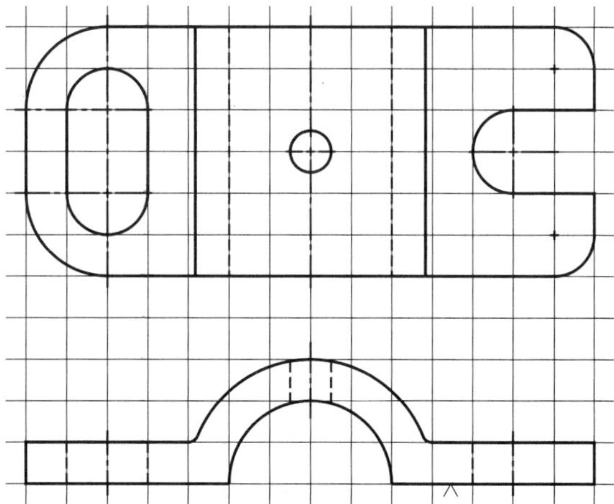

PROBLEM 12–3

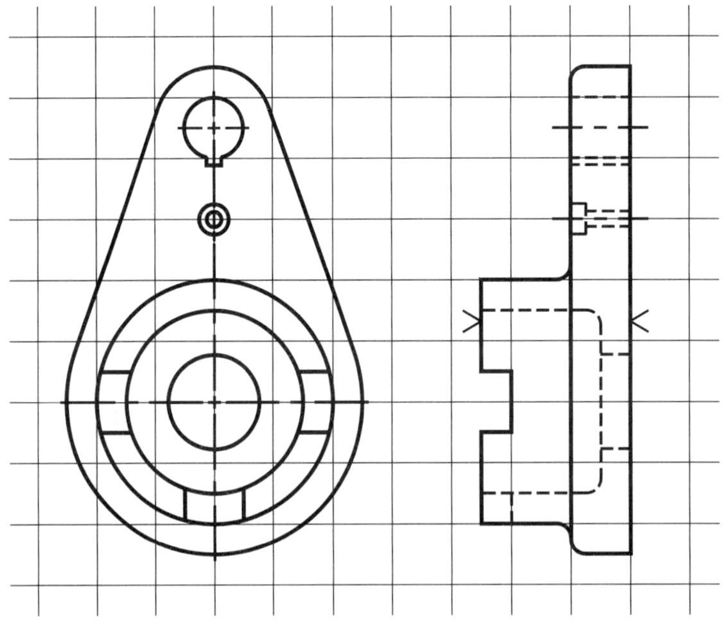

PROBLEM 12–4

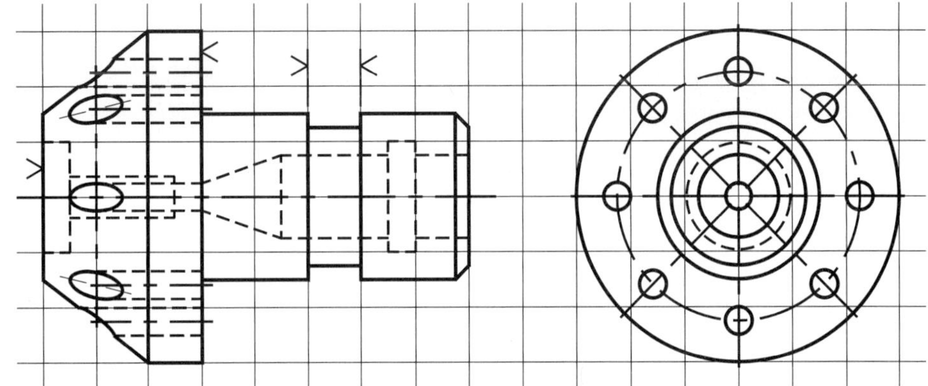

PROBLEM 12–5

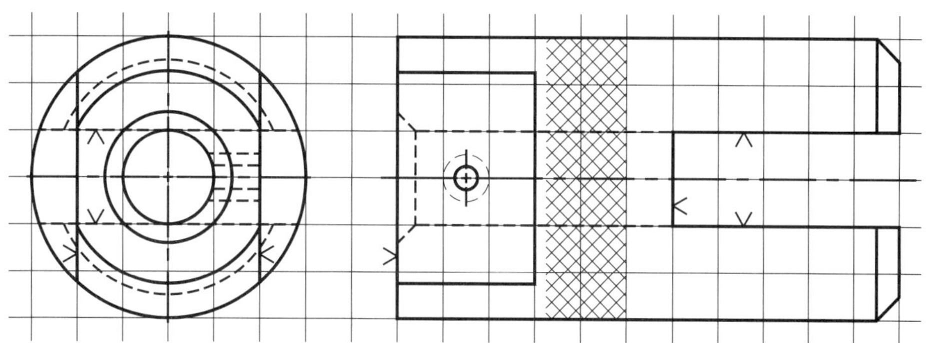

PROBLEM 12–6

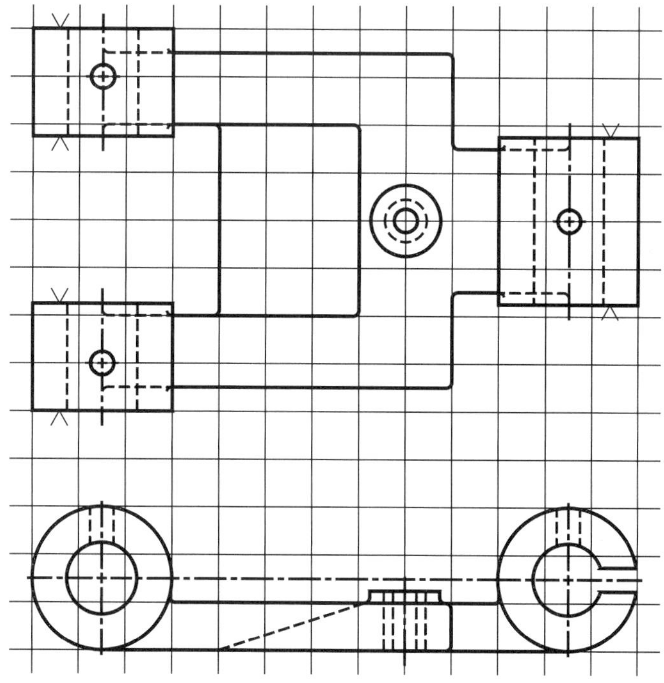

PROBLEM 12–7

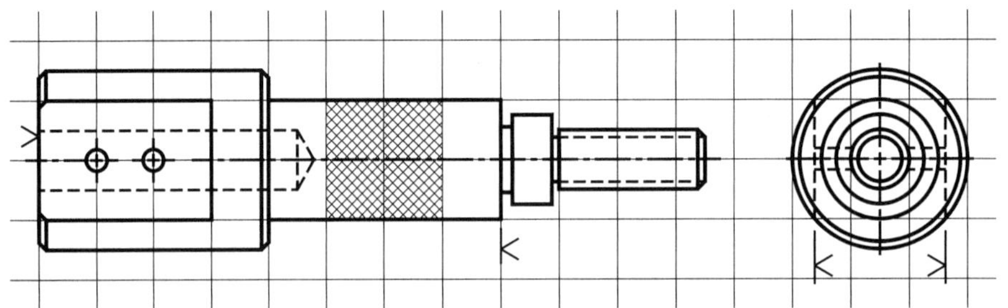

PROBLEM 12–8

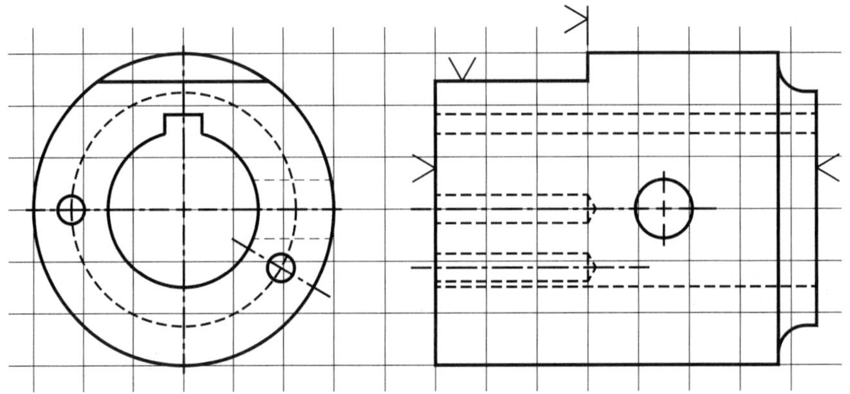

PROBLEM 12–9

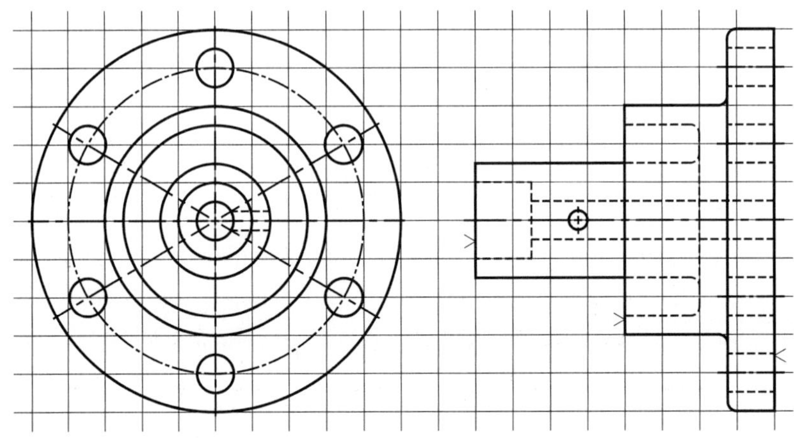

PROBLEM 12–10

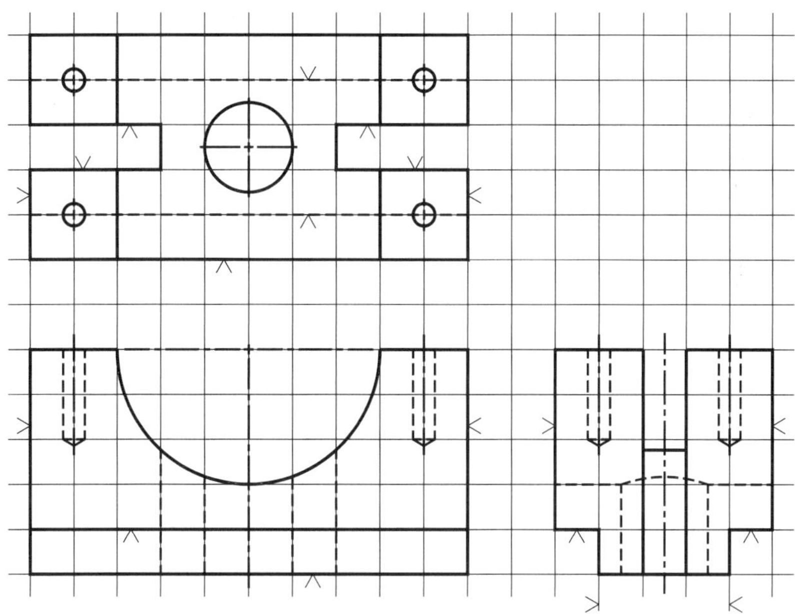

PROBLEM 12–11

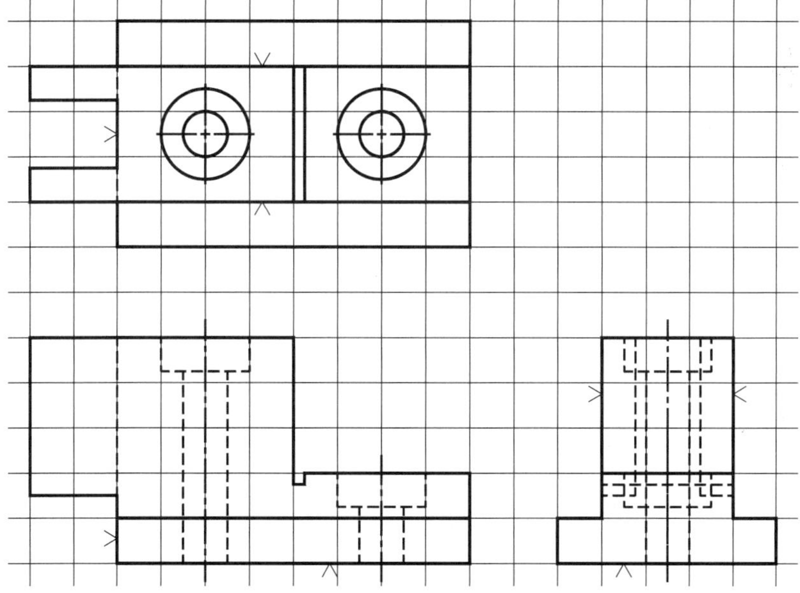

PROBLEM 12–12

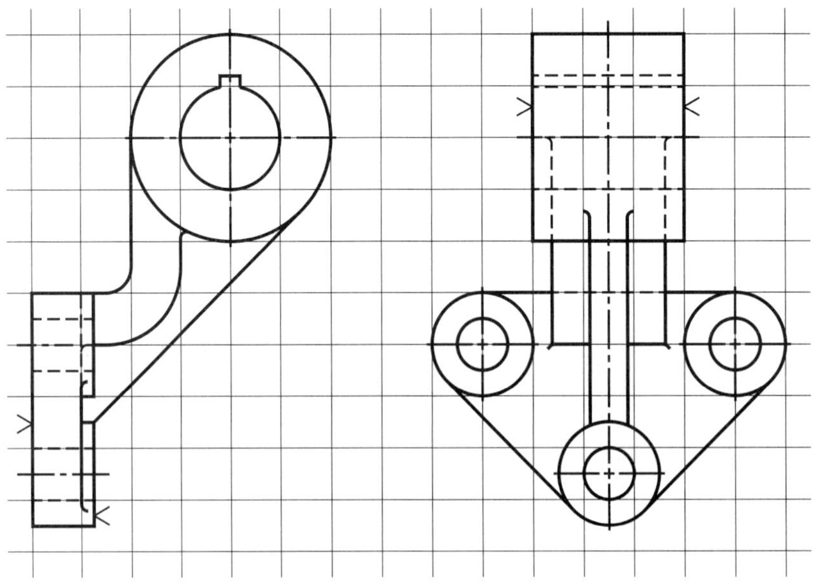

PROBLEM 12–13

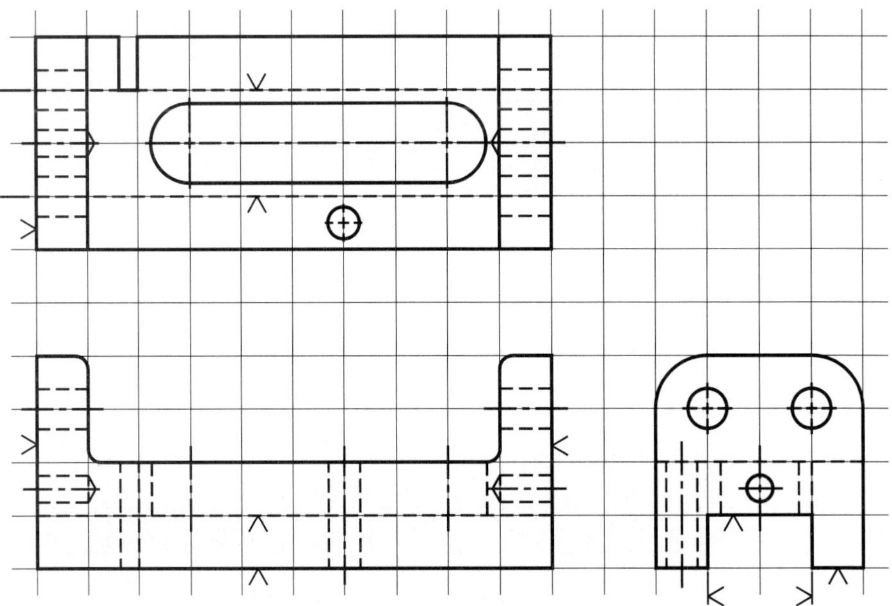

PROBLEM 12–14

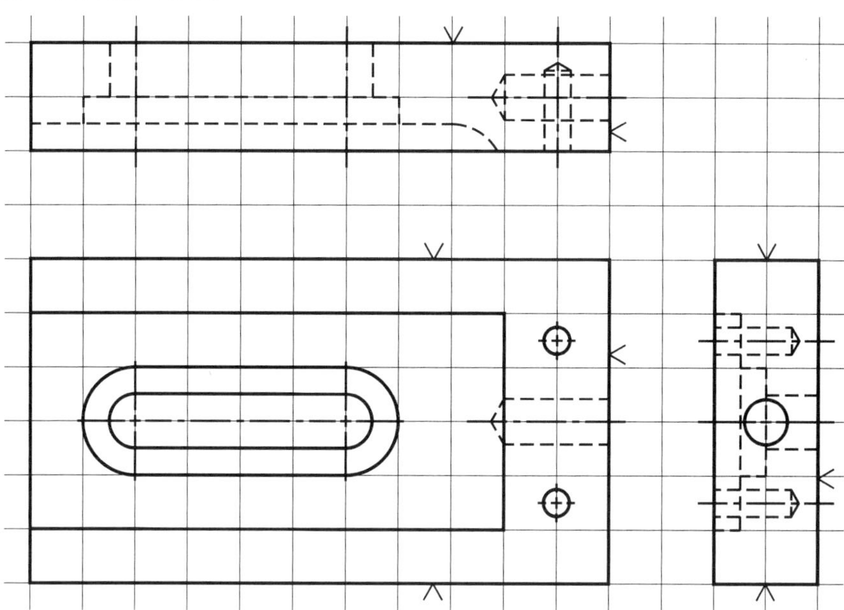

PROBLEM 12–15

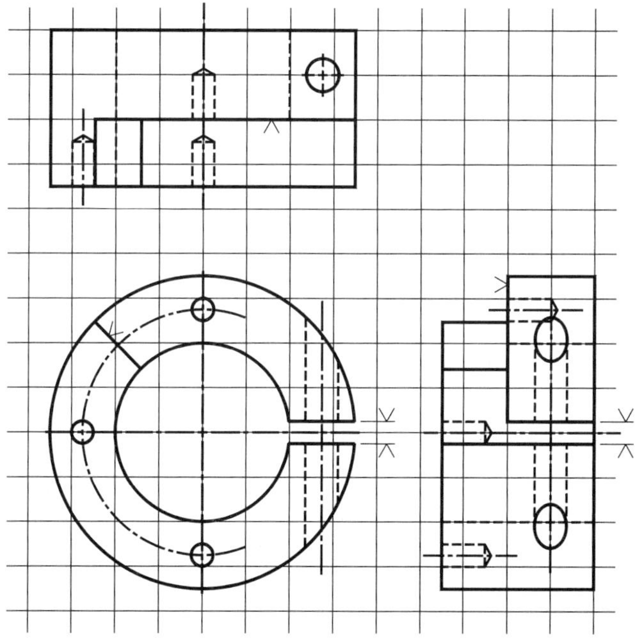

PROBLEM 12–16

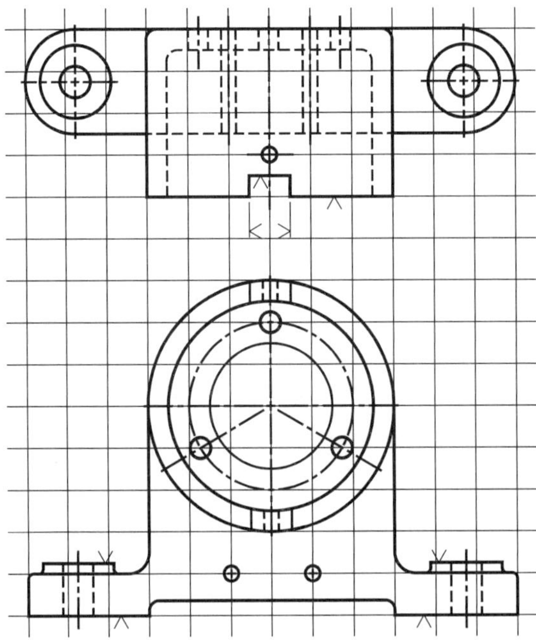

PROBLEM 12–17

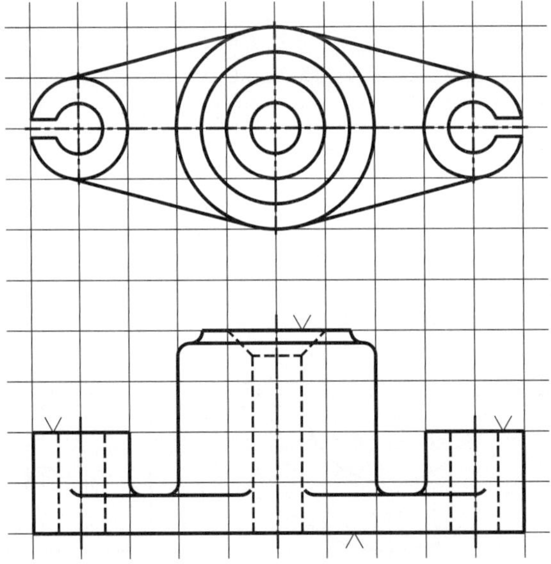

PROBLEM 12–18

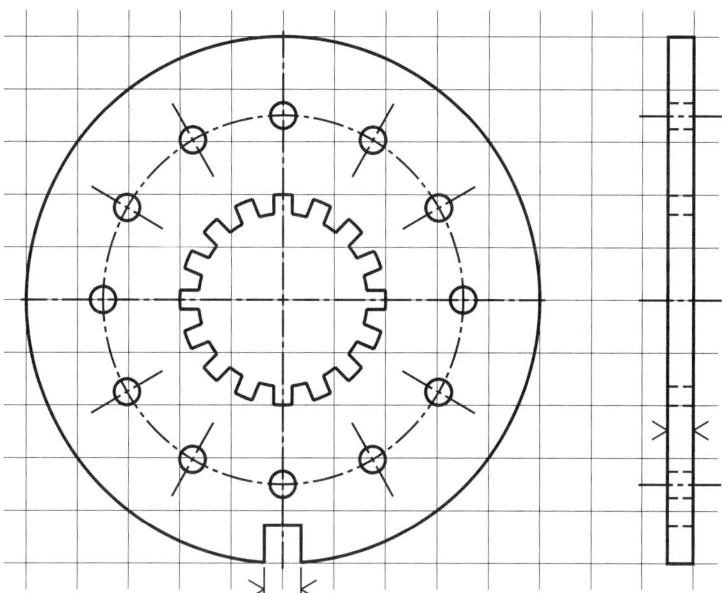

PROBLEM 12–19

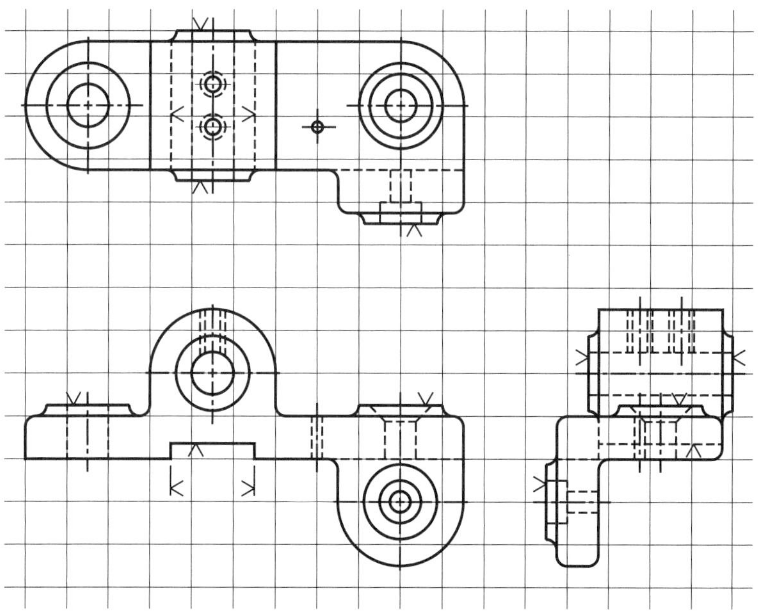

Intermediate Level Drawings
Problems 12–20 through 12–25

1. Convert the isometric drawings provided to orthographic drawings, showing as many views as necessary to communicate the design.
2. Fully dimension your drawings.

PROBLEM 12–20

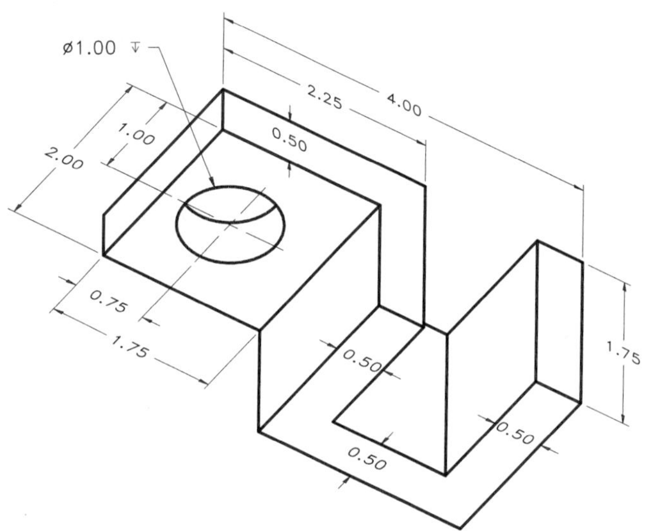

PROBLEM 12–21

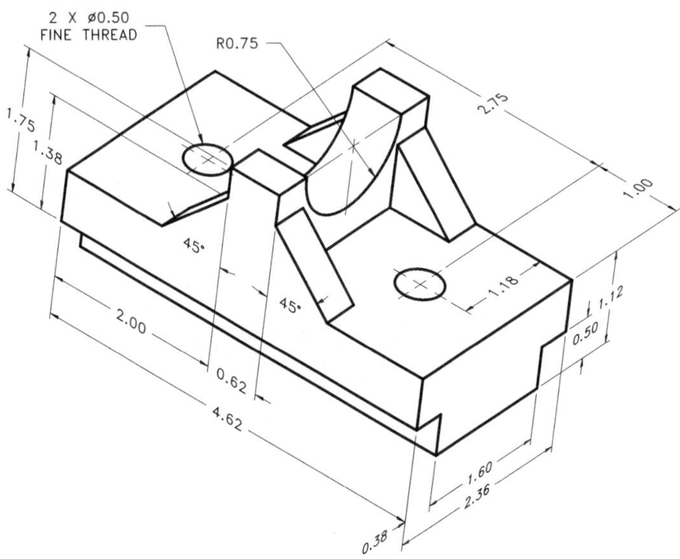

PROBLEM 12–22

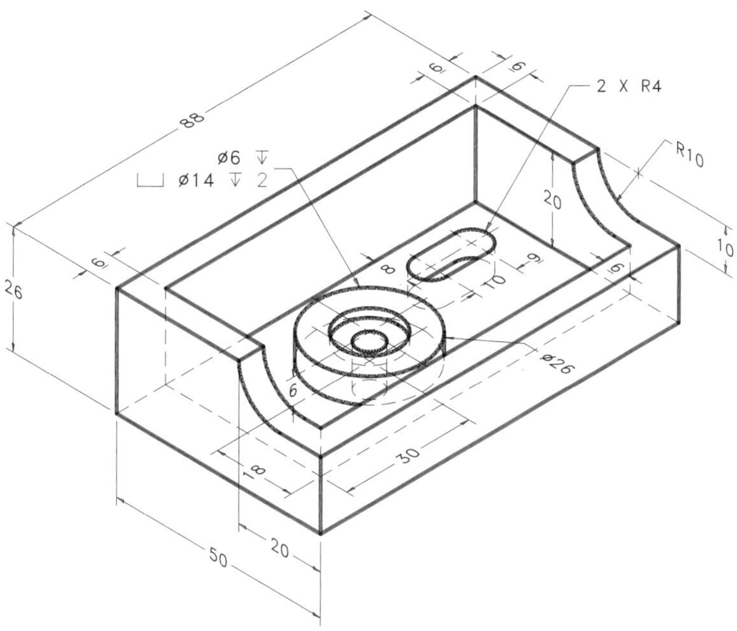

PROBLEM 12–23

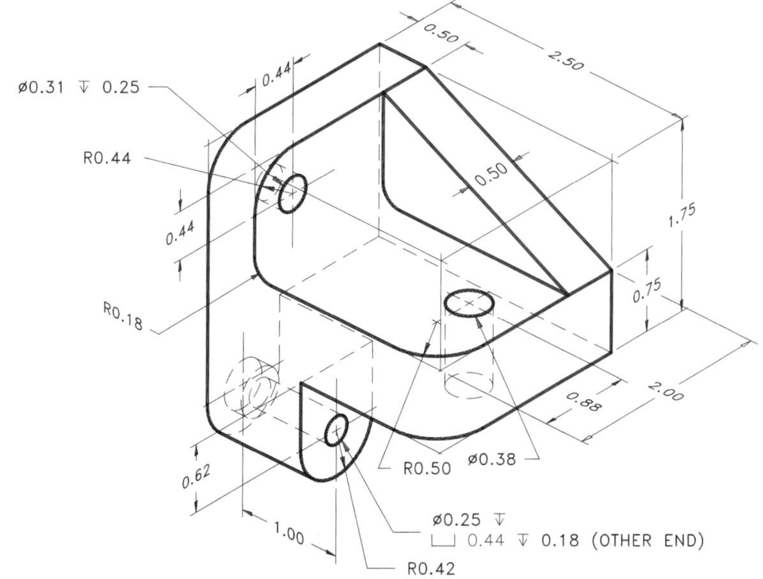

PROBLEM 12–24

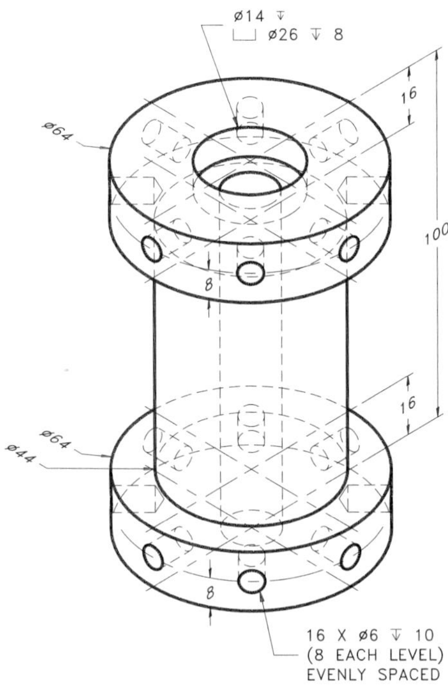

PROBLEM 12–25

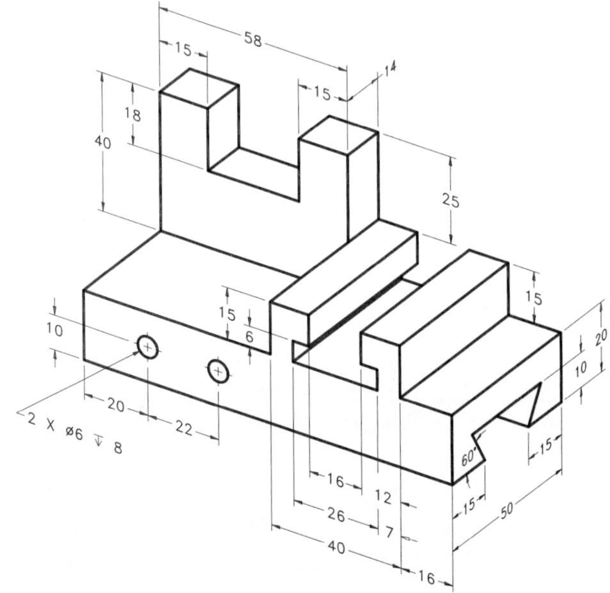

Advanced Level Drawings

PROBLEM 12–26

Directions for Problem 12–26

Create the necessary views of the casting object. Dimension the object fully including any tolerances and notations.

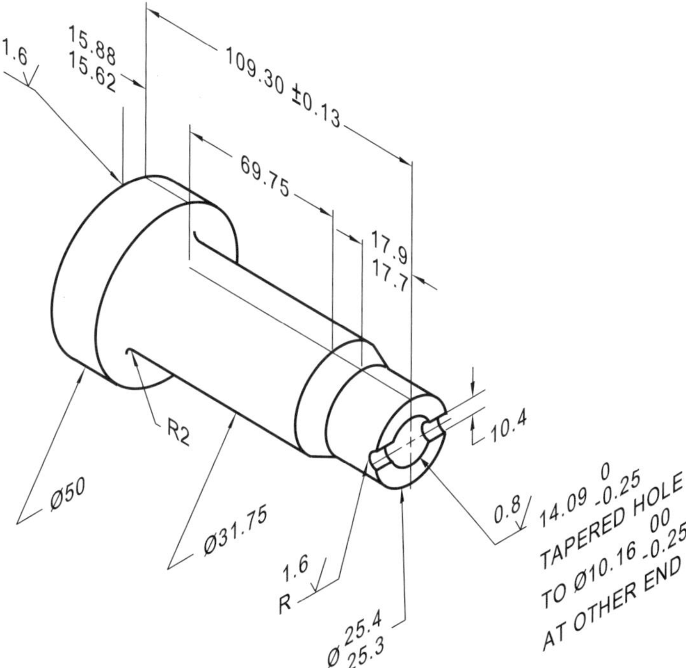

PROBLEM 12–27

Open the drawing *12_footbridge.dwg* from the CD

Create metric (ISO) dimensions for the drawing. Once dimensioned, convert the metric dimensions into English (imperial) dimensions for construction purposes. Alternate solution is to dual dimension the footbridge.

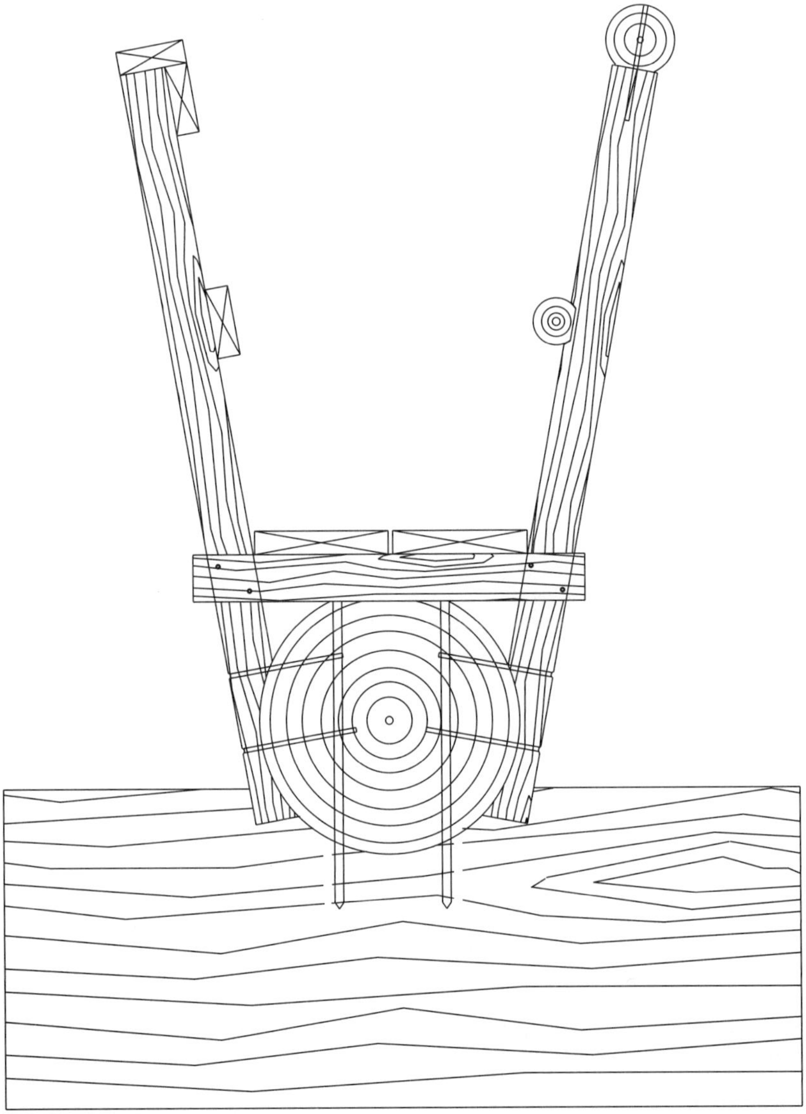

Note: There are 2 different styles of handrail on the drawing.

PROBLEM 12–28

Directions for Problem 12–28.

Create the rim detail using architectural style dimensions.

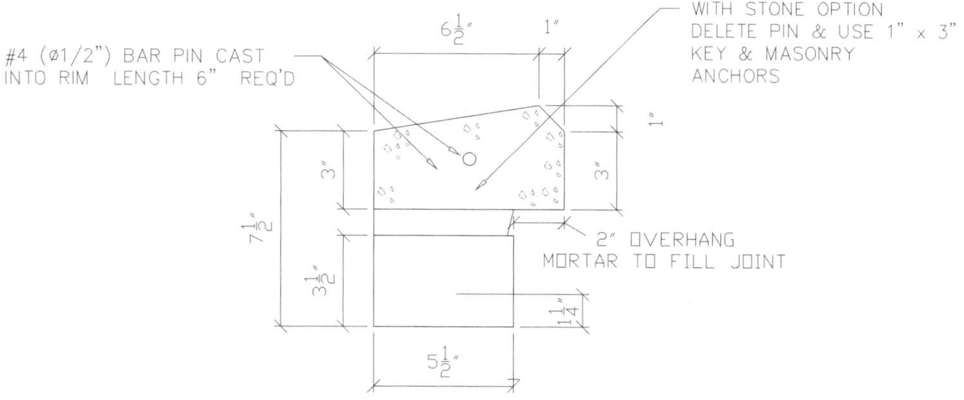

PROBLEM 12–29

Directions for Problem 12–29

Create and dimension the following Architectural drawing.

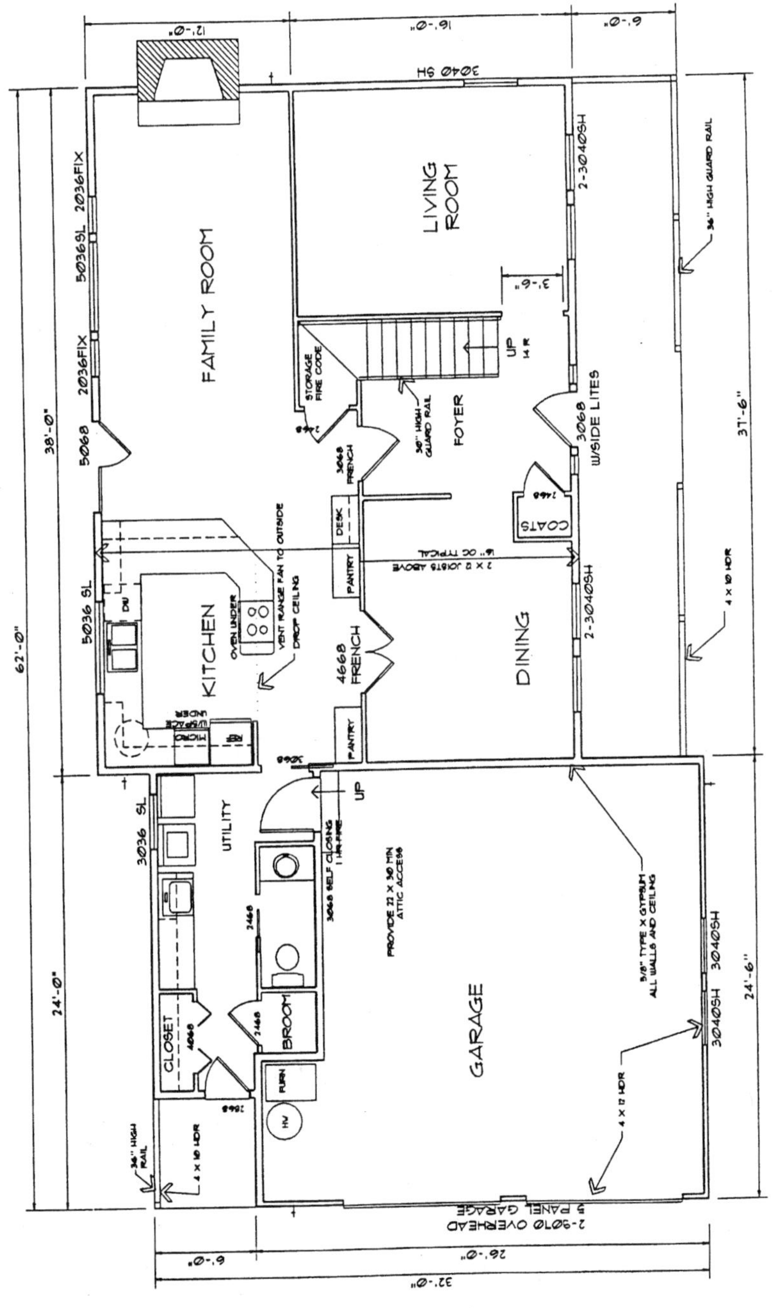

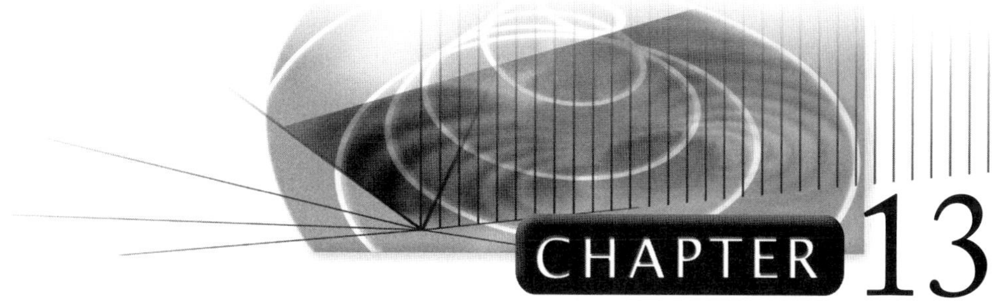

Analyzing 2D Drawings

The exercises in this chapter introduce:

- Extracting information out of an AutoCAD Drawing file
- Using tools like Polyline and Boundary
- Generating the ability to solve problems graphically in AutoCAD
- Using the Inquiry and Properties command tools

TUTORIAL EXERCISE: EXTRUDE.DWG

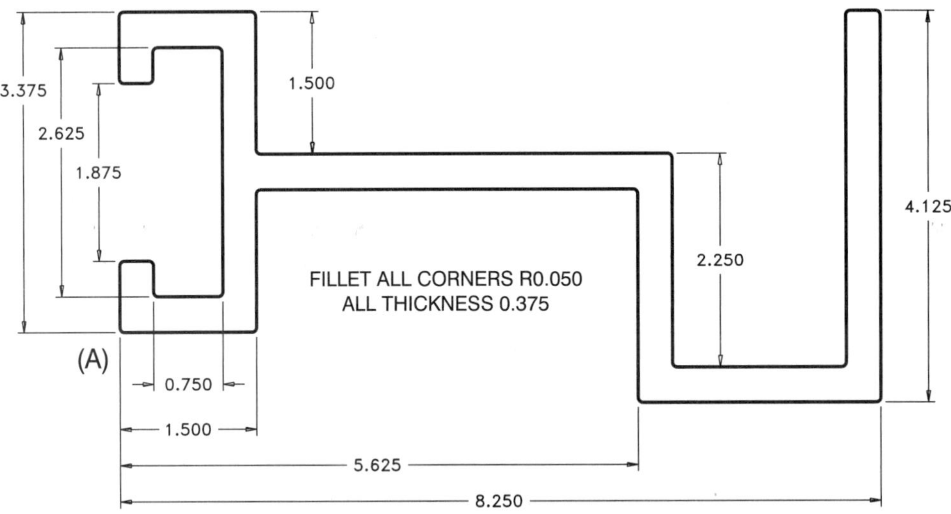

Figure 13–1

Purpose

This tutorial is designed to show you various methods of constructing the extruded pattern in Figure 13–1. The surface area of the extrusion will be found through the AREA command.

System Settings

Use the Drawing Units dialog box and change the Precision setting to three units. Keep the default drawing limits at (0.000,0.000) for the lower-left corner and (12.000,9.000) for the upper-right corner.

Layers

Create the following layers with the format:

Name	Color	Linetype
Boundary	Magenta	Continuous
Object	Yellow	Center

Suggested Commands

Begin drawing the extrusion with point "A," shown in Figure 13–1, at absolute coordinate 2.000,3.000. Use either of the following methods to construct the extrusion:

1. Use a series of absolute, relative, and polar coordinates to construct the profile of the extrusion.
2. Construct a few lines; then use the OFFSET command followed by the TRIM command to construct the extrusion profile.

The FILLET command is used to create the 0.050 radius rounds at all corners of the extrusion.

Before calculating the area of the extrusion, convert and join all objects into one single polyline. This will allow the AREA command to be used in a more productive way. Do not dimension this drawing.

Whenever possible, substitute the appropriate command alias in place of the full AutoCAD command in each tutorial step; for example, use "CP" for the COPY command, "L" for the LINE command, and so on. The complete listing of all command aliases is located in Chapter 1, Table 1–2.

STEP 1

First, make the Object layer current. One method of constructing the extrusion, shown in Figure 13–2, is to use the measurements in, shown Figure 13–1, to calculate a series of polar coordinate system distances. The Direct Distance mode could also be used to accomplish this step.

Command: **L** *(For LINE)* [draw>line]
Specify first point: 2.000,3.000 *(Starting at "A")*
Specify next point or [Undo]: **@1.500<0** *(To "B")*
Specify next point or [Undo]: **@1.500<90** *(To "C")*
Specify next point or [Close/Undo]: **@4.125<0** *(To "D")*
Specify next point or [Close/Undo]: **@2.250<270** *(To "E")*
Specify next point or [Close/Undo]: **@2.625<0** *(To "F")*
Specify next point or [Close/Undo]: **@4.125<90** *(To "G")*
Specify next point or [Close/Undo]: **@0.375<180** *(To "H")*
Specify next point or [Close/Undo]: **@3.750<270** *(To "I")*
Specify next point or [Close/Undo]: **@1.875<180** *(To "J")*
Specify next point or [Close/Undo]: **@2.250<90** *(To "K")*
Specify next point or [Close/Undo]: **@4.500<180** *(To "L")*
Specify next point or [Close/Undo]: **@1.500<90** *(To "M")*
Specify next point or [Close/Undo]: **@1.500<180** *(To "N")*
Specify next point or [Close/Undo]: **@0.750<270** *(To "O")*
Specify next point or [Close/Undo]: **@0.375<0** *(To "P")*
Specify next point or [Close/Undo]: **@0.375<90** *(To "Q")*
Specify next point or [Close/Undo]: **@0.750<0** *(To "R")*
Specify next point or [Close/Undo]: **@2.625<270** *(To "S")*
Specify next point or [Close/Undo]: **@0.750<180** *(To "T")*
Specify next point or [Close/Undo]: **@0.375<90** *(To "U")*
Specify next point or [Close/Undo]: **@0.375<180** *(To "V")*
Specify next point or [Close/Undo]: **C** *(To Close)*

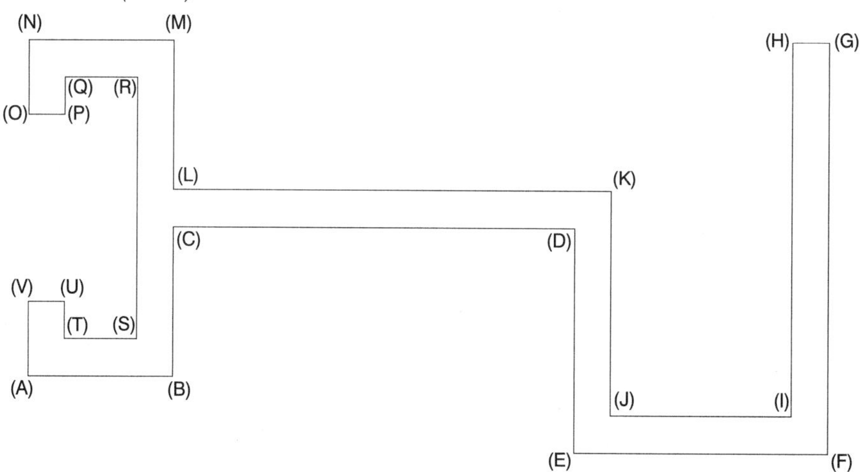

Figure 13–2

STEP 2

Rather than converting individual objects to one polyline using the PEDIT command and Join option, you could use a more efficient means of creating a polyline: the BOUNDARY command. First, make the Boundary layer current. Choosing Boundary… from the Draw pull-down menu, as shown in Figure 13–3, activates the Boundary Creation dialog box, shown in Figure 13–4. Click on the Pick Points button in the upper-right corner of the dialog box.

Then pick an internal point to automatically trace a polyline around a closed shape in the color of the current layer. It must be emphasized that the shape must be completely closed for the BOUNDARY command to function correctly.

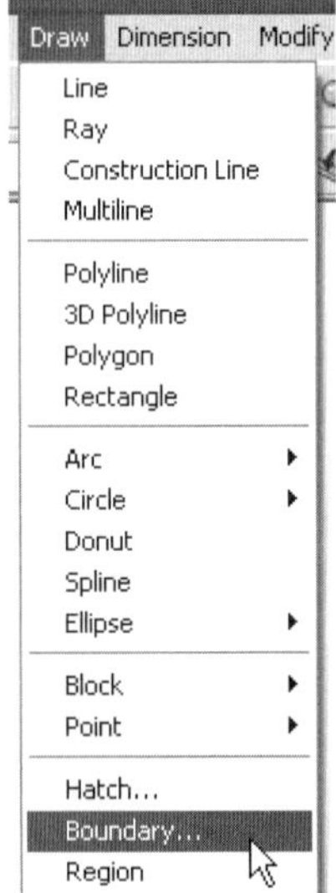

Figure 13–3

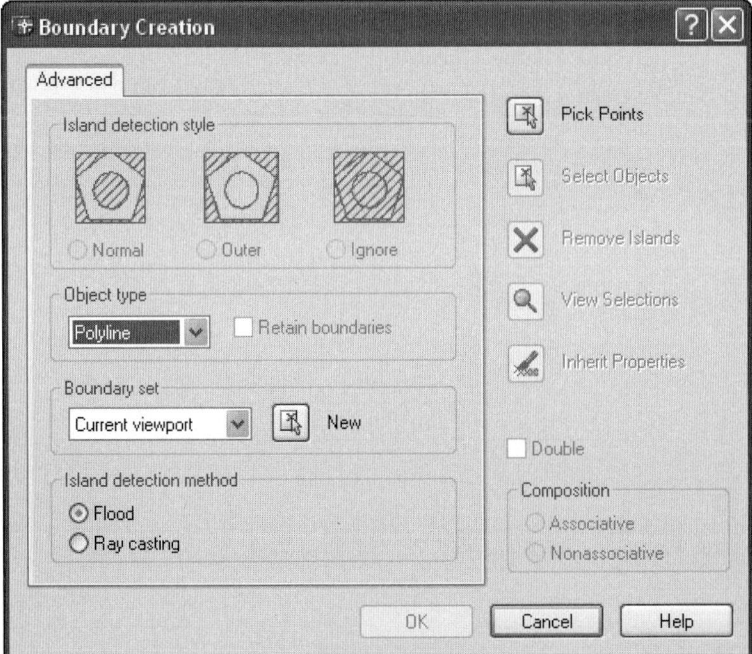

Figure 13–4

Figure 13–5 shows the extrusion. If you open the Boundary Creation dialog box and click on the Pick Points button, AutoCAD prompts you to pick an internal point. Selecting a point inside the extrusion at "A" traces the polyline in the Boundary layer. Turning off the Object layer containing the individual line segments leaves the polyline for performing various calculations.

Command: **BO** *(For BOUNDARY)*
[draw>boundary]
Select internal point: *(Select a point inside of the extrusion at "A")*
Selecting everything...
Selecting everything visible...
Analyzing the selected data...
Analyzing internal islands...
Select internal point: *(Press* ENTER *to create the boundary polyline on the Boundary layer)*
BOUNDARY created 1 polyline

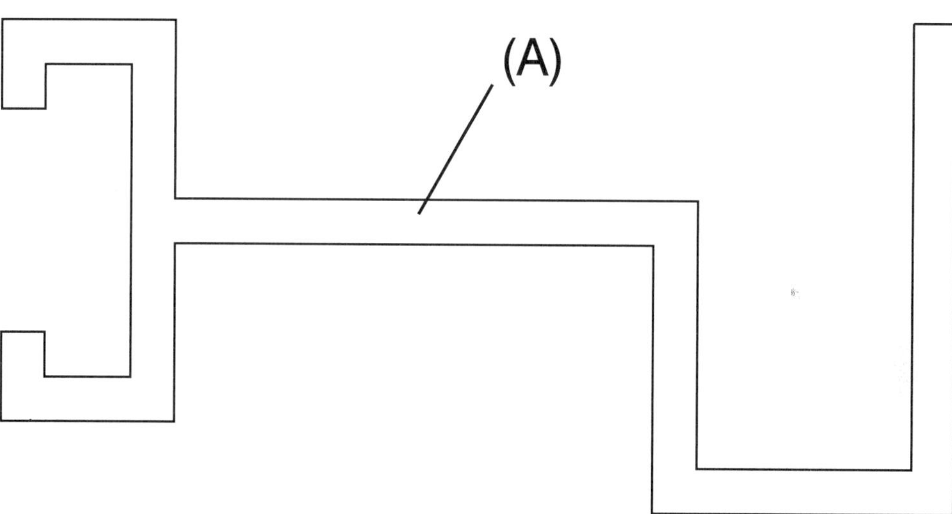

Figure 13–5

STEP 3

With the entire extrusion converted to a polyline, use the FILLET command, set a radius of 0.050, and use the polyline option of the FILLET command to fillet all corners of the extrusion at once. See Figure 13–6.

Command: **F** *(For FILLET)* [modify>fillet]

Current settings: Mode = TRIM, Radius = 0.500 Select first object or [Polyline/Radius/Trim]: **R** (For Radius)

Specify fillet radius <0.500>: **0.050**

Select first object or [Polyline/Radius/Trim]: **P** (For Polyline)

Select 2D polyline: *(Select the polyline in Figure 13–6)*

22 lines were filleted

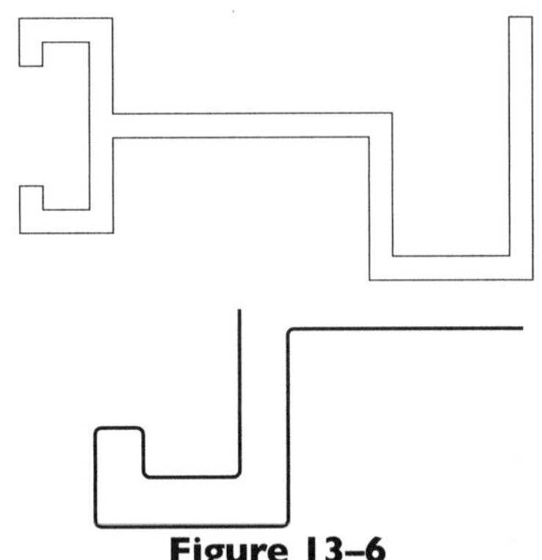

Figure 13–6

CHECKING THE ACCURACY OF EXTRUDE.DWG

Once the extrusion has been constructed, answer the question to determine the accuracy of the drawing in Figure 13–7.

Question 1

What is the total surface area of the extrusion?

(A) 7.020

(B) 7.070

(C) 7.120

(D) 7.170

(E) 7.220

Use the AREA command to calculate the surface area of the extrusion. This is easily accomplished because the extrusion has already been converted to a polyline.

Command: **AA** *(For AREA)*

Specify first corner point or [Object/Add/Subtract]: **O** *(For Object)*

Select objects: *(Select any part of the extrusion in Figure 13–7)*

Area = 7.170, Perimeter = 38.528

The total surface area of the extrusion is "D," 7.170.

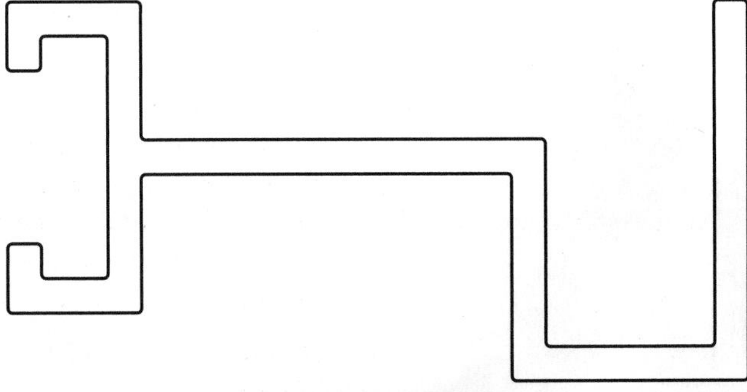

Figure 13–7

Beginning Level Drawings

PROBLEM 13–1 ANGLEBLK.DWG

Directions for Angleblk.Dwg

Use the Drawing Units dialog box to set the units to decimal. Set the precision to two. Be sure the system of angle measure is set to decimal degrees and the number of decimal places for the display of angles is zero.

Keep all remaining default unit values. Keep the default settings for the drawing limits.

Begin the drawing shown in Problem 13–1 by locating the lower-left corner of Angleblk identified by "X" at coordinate (2.35,3.17).

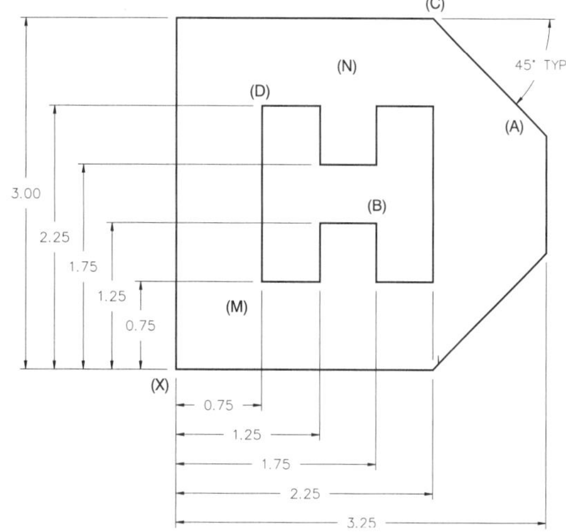

Refer to the drawing of the Angleblk shown in Problem 13–1 to answer the following questions.

1. What is the total surface area of the Angleblk with the inner "H" shape removed?
   (A) 6.66 (D) 7.00
   (B) 6.77 (E) 7.11
   (C) 6.89
2. What is the total area of the inner "H" shape?
   (A) 1.75 (D) 1.93
   (B) 1.81 (E) 1.99
   (C) 1.87
3. What is the total length of line "A"?
   (A) 1.29 (D) 1.47
   (B) 1.35 (E) 1.53
   (C) 1.41
4. What is the absolute coordinate value of the endpoint of the line at "B"?
   (A) 4.04,4.42 (D) 4.22,4.42
   (B) 4.10,4.42 (E) 4.28,4.42
   (C) 4.16,4.42
5. What is the absolute coordinate value of the endpoint of the line at "C"?
   (A) 4.60,6.11 (D) 4.60,6.29
   (B) 4.60,6.17 (E) 4.60,6.35
   (C) 4.60,6.23
6. Use the STRETCH command and extend the inner "H" shape a distance of 0.37 units in the 180° direction. Use "N" as the first corner of the crossing window and "M" as the other corner. Use the endpoint of "D" as the base point of the stretching operation. What is the new surface area of Angleblk with the inner "H" shape removed?
   (A) 6.63 (D) 6.82
   (B) 6.69 (E) 6.87
   (C) 6.75
7. Use the SCALE command with the endpoint of the line at "D" as the base point. Reduce the size of just the inner "H" using a scale factor of 0.77. What is the new surface area of Angleblk with the inner "H" removed?
   (A) 7.48 (D) 7.66
   (B) 7.54 (E) 7.72
   (C) 7.60

problem EXERCISES

PROBLEM 13–2 LEVER1.DWG

DirectionsforLever1.Dwg

Use the Drawing Units dialog box to set the units to decimal. Set the precision to three. Be sure the system of angle measure is set to decimal degrees and the number of decimal places for the display of angles is zero.

Keep the remaining default unit values.

Begin the drawing, shown in Problem 13–2, by locating the center of the 2.000-diameter circle at coordinate (4.500,3.250).

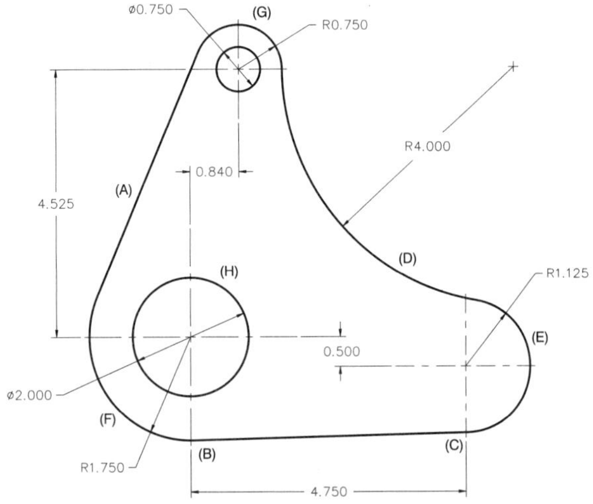

Refer to the drawing of Lever1 in Problem 13–2 to answer the following questions.

1. What is the total length of line segment "A"?
   (A) 4.483 (D) 4.504
   (B) 4.492 (E) 4.515
   (C) 4.499
2. What is the absolute coordinate value of the center of the 4.00-radius arc "D"?
   (A) 10.090,7.773 (D) 10.090,7.782
   (B) 10.090,7.791 (E) 10.090,7.800
   (C) 10.090,7.806
3. What is the length of the 1.125-radius arc segment "E"?
   (A) 3.299 (D) 3.329
   (B) 3.308 (E) 3.337
   (C) 3.319
4. What is the distance from the center of the 1.750-radius arc "F" to the center of the 0.750-radius arc "G"?
   (A) 4.572 (D) 4.602
   (B) 4.584 (E) 4.610
   (C) 4.596
5. What is the total area of Lever1 with the two holes removed?
   (A) 26.271 (D) 26.298
   (B) 26.282 (E) 26.307
   (C) 26.290
6. What is the circumference of the 2.000-diameter circle "H"?
   (A) 6.269 (D) 6.289
   (B) 6.276 (E) 6.294
   (C) 6.283
7. Use the SCALE command to reduce Lever1 in size by a scale factor of 0.333. Use the center of the 2.000-diameter hole as the base point. What is the absolute coordinate value of the center of the 0.750 arc "G"?
   (A) 4.780,4.757 (D) 4.780,4.767
   (B) 4.780,4.777 (E) 4.785,4.757
   (C) 4.793,4.777

PROBLEM 13–3 GASKET1.DWG

DirectionsforGasket1.Dwg

Use the Drawing Units dialog box to set the units to decimal. Set the precision to three. Be sure the system of angle measure is set to decimal degrees and the number of decimal places for the display of angles is zero.

Keep the remaining default unit values.

Begin the drawing, shown in Problem 13–3, by locating the center of the 1.500-radius circle at coordinate (5.750,4.750).

Refer to the drawing of Gasket1, shown in Problem 13–3, to answer the following questions.

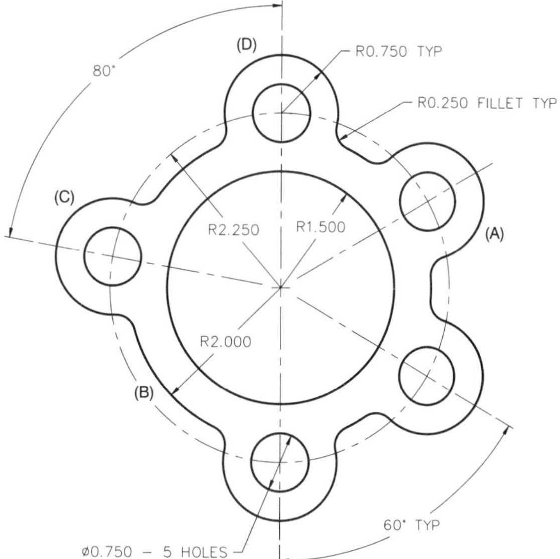

1. What is the total area of Gasket1 with the all holes removed?
   (A) 9.918 (D) 9.927
   (B) 9.921 (E) 9.930
   (C) 9.924
2. What is the absolute coordinate value of the center of the 0.750-radius arc "A"?
   (A) 7.669,5.875 (B) 7.669,5.870
   (B) 7.666,5.875 (D) 7.699,5.875
   (E) 7.699,5.975
3. What is the length of arc segment "B"?
   (A) 1.698 (D) 1.716
   (B) 1.704 (E) 1.722
   (C) 1.710
4. What is the angle formed in the XY plane from the center of the arc "C" to the center of arc "D"?
   (A) 30° (D) 45°
   (B) 35° (E) 50°
   (C) 40°
5. What is the total length of the 0.750-radius arc "C"?
   (A) 2.674 (D) 2.692
   (B) 2.680 (E) 2.698
   (C) 2.686
6. Use the MOVE command to reposition Gasket1 at a distance of 1.832 in the -45° direction. Use the center of the 1.500-radius circle as the base point of the move. What is the new absolute coordinate value of the center of the 1.500-radius circle?
   (A) 7.045,3.437 (C) 7.045,3.443
   (B) 7.045,3.449 (D) 7.045,3.455
   (E) 7.045,3.461
7. What is the new absolute coordinate value of the center of the 0.750-radius arc "C"?
   (A) 4.806,3.845 (C) 4.812,3.845
   (B) 4.818,3.845 (D) 4.824,3.845
   (E) 4.830,3.845

PROBLEM 13–4 LEVER2.DWG

Directions for Lever2.Dwg

Use the Drawing Units dialog box to set the units to decimal. Keep the precision at four places.

Be sure the system of angle measure is set to decimal degrees and the number of decimal places for the display of angles is zero.

Keep the remaining default unit values.

Begin the drawing, shown in Problem 13–4, by locating the center of the 1.0000-diameter circle at coordinate (2.2500,4.0000).

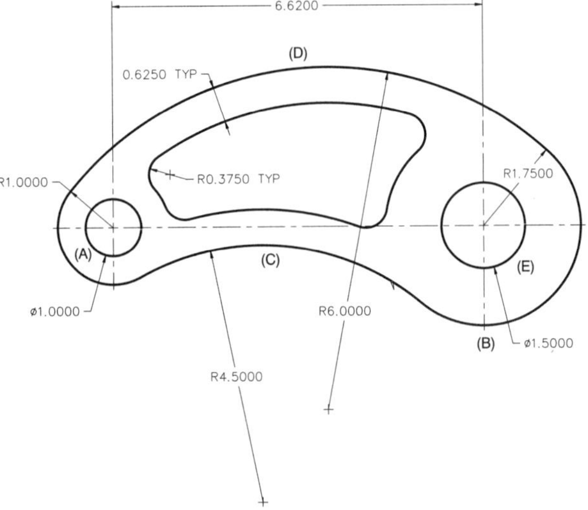

Refer to the drawing of Lever2 in Problem 13–4 to answer the following questions.

1. What is the total area of Lever2 with the inner irregular shape and both holes removed?
   (A) 17.6813
   (B) 17.6819
   (C) 17.6825
   (D) 17.6831
   (E) 17.6837
2. What is the absolute coordinate value of the center of the 4.5000-radius arc "C"?
   (A) 4.8944,-0.0822
   (B) 4.8944,-0.8226
   (C) 4.8950,-0.8232
   (D) 4.8956,-0.8238
   (E) 4.8962,-0.8244
3. What is the absolute coordinate value of the center of the 6.0000-radius arc "D"?
   (A) 6.0828,0.7893
   (B) 6.0834,0.7899
   (C) 6.0840,0.7905
   (D) 6.0846,0.7911
   (E) 6.0852,0.7917
4. What is the total length of arc "C"?
   (A) 5.3583
   (B) 5.3589
   (C) 5.3595
   (D) 5.3601
   (E) 5.3607
5. What is the distance from the center of the 1.0000-diameter circle "A" to the intersection of the circle and centerline at "B"?
   (A) 6.8456
   (B) 6.8462
   (C) 6.8474
   (D) 6.8480
   (E) 6.8486
6. What is the angle formed in the XY plane from the upper quadrant of arc "D" to the center of the 1.5000-circle "E"?
   (A) 313°
   (B) 315°
   (C) 317°
   (D) 319°
   (E) 321°
7. What is the delta X,Y distance from the upper quadrant of arc "C" to the center of the 1.0000-hole "A"?
   (A) -2.6444,0.3220
   (B) -2.6444,0.3226
   (C) -2.6444,0.3232
   (D) -2.6444,0.3238
   (E) -2.6444,0.3244

PROBLEM 13–5 BRACKET1.DWG

DirectionsforBracket1.Dwg

Use the Drawing Units dialog box to set the units to decimal. Set the precision to three. Be sure the system of angle measure is set to decimal degrees and the number of decimal places for the display of angles is zero.

Keep all remaining default unit values. Keep the default values for the limits.

Begin the drawing, shown in Problem 13–5, by locating the center of the 1.500-radius arc "A" at coordinate (4.000,3.500).

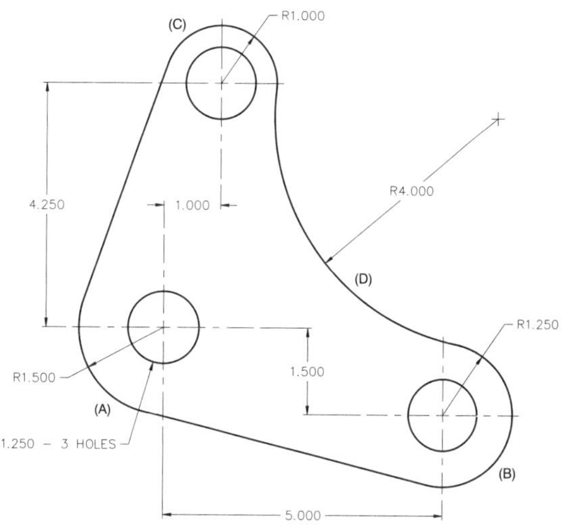

Refer to the drawing of Bracket1, shown in Problem 13–5, to answer the following questions.

1. What is the distance from the center of the 1.500-radius arc "A" to the center of the 1.250-radius arc "B"?
   (A) 5.205
   (B) 5.210
   (C) 5.215
   (D) 5.220
   (E) 5.228
2. What is the distance from the center of the 1.500-radius arc "A" to the center of the 1.000-radius arc "C"?
   (A) 4.366
   (B) 4.370
   (C) 4.374
   (D) 4.378
   (E) 4.382
3. What is the distance from the center of the 1.250-radius arc "B" to the center of the 1.000-radius arc "C"?
   (A) 6.990
   (B) 6.995
   (C) 6.998
   (D) 7.000
   (E) 7.004
4. What is the length of arc "B"?
   (A) 3.994
   (B) 4.000
   (C) 4.006
   (D) 4.012
   (E) 4.018
5. What is the absolute coordinate value of the center of the 4.000-radius arc "D"?
   (A) 9.965,7.112
   (B) 9.965,7.250
   (C) 9.960,7.161
   (D) 9.965,7.161
   (E) 9.995,1.161
6. What is the total area of Bracket1 with all three 1.250-diameter holes removed?
   (A) 27.179
   (B) 27.187
   (C) 27.193
   (D) 27.198
   (E) 28.003
7. What is the angle formed in the XY plane from the center of the 1.250-radius arc "B" to the center of the 1.000-radius arc "C"?
   (A) 121°
   (B) 123°
   (C) 125°
   (D) 127°
   (E) 129°

PROBLEM 13–6 MAIN ROCKER.DWG

Directions for Main Rocker.Dwg

Begin the construction of Main Rocker in Problem 13–6, by keeping the default units set to decimal but changing the precision to two. Be sure the system of angle measure is set to decimal degrees and the number of decimal places for the display of angles is zero. Keep the remaining default unit values.

Begin the drawing in Problem 13–6, by placing the center of the regular hexagon and 4.000-radius arc at coordinate (12.125,10.50).

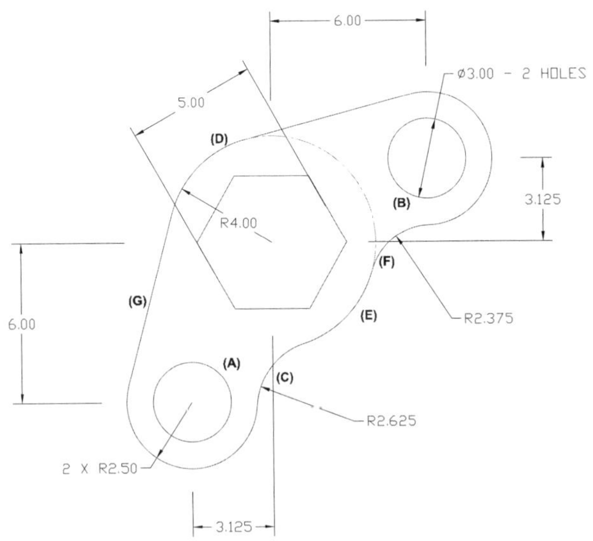

Refer to the drawing of main rocker in Problem 13–6 to answer the following questions.

1. What is the distance from the center of the 2.50-diameter hole "A" to the center of the other 2.50-diameter hole "B"?
   (A) 6.765 (D) 12.905
   (B) 14.779 (E) 11.649
   (C) 13.453
2. What is the absolute coordinate value of the center of the 3.00-radius arc "C"?
   (A) 14.115,4.181 (D) 9.000,4.500
   (B) 18.256,8.752 (E) 13.862,8.291
   (C) 10.597,5.428
3. What is the total length of line "G"?
   (A) 6.748 (D) 6.233
   (B) 6.861 (E) 6.483
   (C) 6.597
4. What is the length of the 4.00-radius arc segment "E"?
   (A) 3.843 (D) 3.931
   (B) 3.850 (E) 3.986
   (C) 3.952
5. What is the angle formed in the XY plane from the center of the 2.50-diameter circle "B" to the center of the 4.00-radius arc "D"?
   (A) 204° (D) 208°
   (B) 206° (E) 212°
   (C) 210°
6. What is the total surface area of main rocker with the hexagon and both 2.25-diameter holes removed?
   (A) 67.365 (D) 68.567
   (B) 67.682 (E) 67.188
   (C) 68.136
7. What is the total length of arc "F"?
   (A) 2.965 (D) 3.159
   (B) 3.007 (E) 3.397
   (C) 3.104

PROBLEM 13–7 PATTERN4.DWG

Directions for Pattern4.Dwg

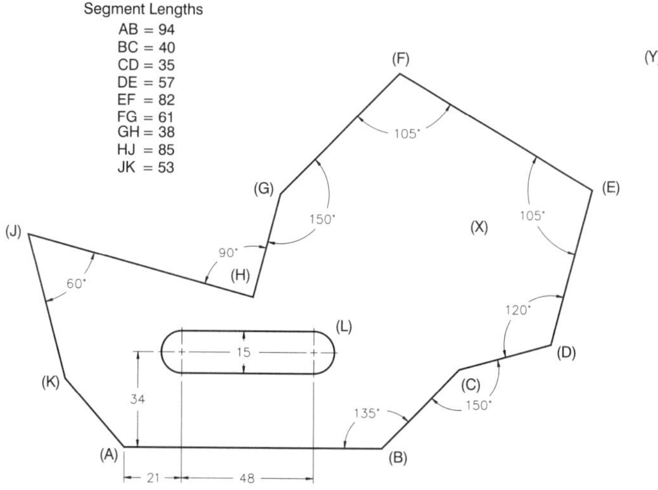

Start a new drawing called Pattern4. Even though this is a metric drawing, no special limits need be set. Keep the default setting of decimal units precision to zero. Be sure the system of angle measure is set to decimal degrees and the number of decimal places for the display of angles is zero. Keep all remaining default unit values.

Begin the drawing, shown in Problem 13–7, by constructing Pattern4 with vertex "A" at absolute coordinate (50,30).

Refer to the drawing of Pattern4, shown in Problem 13–7, to answer the following questions.

1. What is the total distance from the intersection of vertex "K" to the intersection of vertex "A"?
   (A) 33 (B) 34 (C) 35 (D) 36 (E) 37
2. What is the total area of Pattern4 with the slot removed?
   (A) 14493 (B) 14500 (C) 14529 (D) 14539 (E) 14620
3. What is the perimeter of the outline of Pattern4?
   (A) 570 (B) 578 (C) 586 (D) 594 (E) 602
4. What is the distance from the intersection of vertex "A" to the intersection of vertex "E"?
   (A) 186 (B) 190 (C) 194 (D) 198 (E) 202
5. What is the absolute coordinate value of the intersection at vertex "G"?
   (A) 104,117 (B) 105,118 (C) 106,119 (D) 107,120 (E) 108,121
6. What is the total length of arc "L"?
   (A) 20 (B) 21 (C) 22 (D) 23 (E) 24
7. Stretch the portion of Pattern4 around the vicinity of angle "E." Use "Y" as the first corner of the crossing box. Use "X" as the other corner. Use the endpoint of "E" as the base point of the stretching operation. For the new point, enter a polar coordinate value of 26 units in the 40° direction. What is the new degree value of the angle formed at vertex "E"?
   (A) 78° (B) 79° (C) 80° (D) 81° (E) 82°

PROBLEM 13–8 TEMPLATE.DWG

Directions for Template.Dwg

Use the Drawing Units dialog box to set the units to decimal. Set the precision to three. Change the system of degrees from Decimal to Degrees/Minutes/Seconds along with the following angular precision: 0d00'00.00".

Begin the drawing, shown in Problem 13–8, by locating Point "A" at coordinate (4.500,2.750).

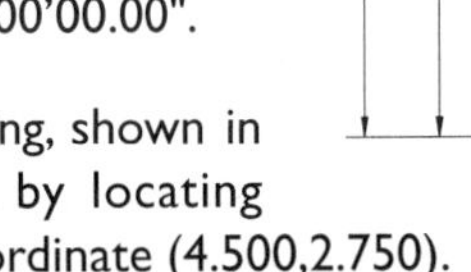

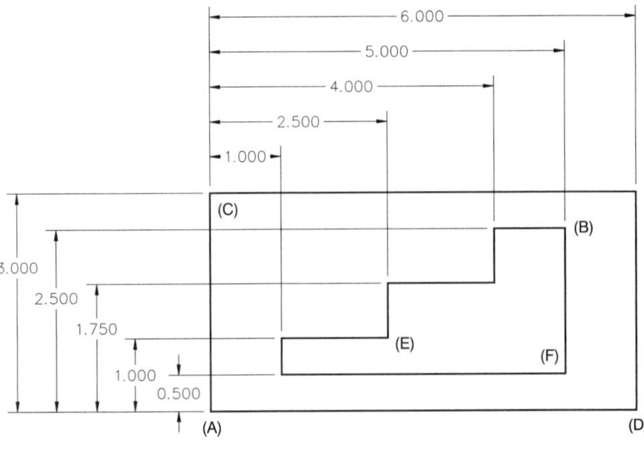

Refer to the drawing of the template, shown in Problem 13–8, to answer the following questions.

1. What is the angle formed in the XY plane from the intersection at "A" to the intersection at "B"?
   (A) 26d25'21" (D) 26d32'37"
   (B) 26d28'3" (E) 26d33'54"
   (C) 26d30'10"
2. What is the distance from the intersection at "C" to the intersection at "D"?
   (A) 6.704 (D) 6.716
   (B) 6.708 (E) 6.720
   (C) 6.712
3. What is the angle formed in the XY plane from the intersection at "E" to the intersection at "F"?
   (A) 347d31'10" (D) 348d20'20"
   (B) 347d47'32" (E) 348d41'24"
   (C) 348d1'53"
4. What is the total area of the template with the step pattern removed?
   (A) 13.367 (D) 13.379
   (B) 13.371 (E) 13.383
   (C) 13.375
5. Using the SCALE command, select all objects to reduce the size of the template. Use Point "A" as the base point and 0.950 as the scale factor. What is the perimeter of the inside step pattern?
   (A) 11.384 (D) 11.396
   (B) 11.388 (E) 11.400
   (C) 11.392
6. What is the new total area of the template with the step pattern removed?
   (A) 12.059 (D) 12.071
   (B) 12.063 (E) 12.075
   (C) 12.067
7. What is the new absolute coordinate value of the endpoint of the line at "B"?
   (A) 9.200,5.100 (D) 9.260,5.135
   (B) 9.250,5.125 (E) 9.260,5.140
   (C) 9.250,5.130

PROBLEM 13–9 ARCHITECTURAL APPLICATION — APARTMENT.DWG

Directions for Apartment.Dwg

Change from decimal units to architectural units. Keep all remaining default values. The thickness of all walls measures 4". Answer the questions on the next page regarding this drawing (see Problem 13–9).

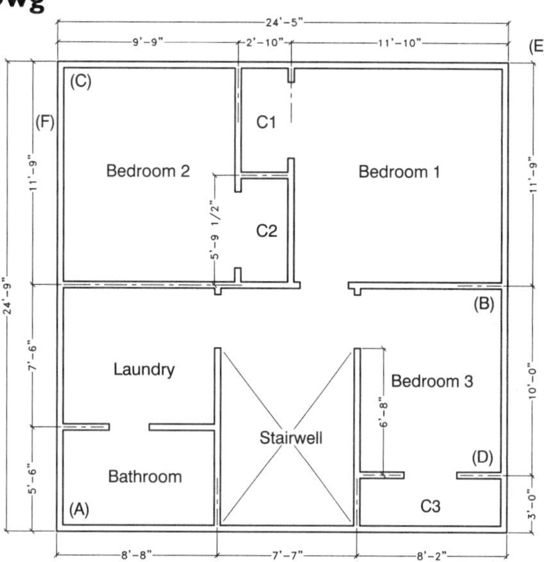

Refer to the drawing of Apartment, shown in Problem 13–9, to answer the following questions.

1. What is the total area in square feet of all bedrooms? (Pick the closest value.)
   (A) 296 sq. ft. (D) 326 sq. ft.
   (B) 306 sq. ft. (E) 336 sq. ft.
   (C) 316 sq. ft.
2. What is the distance from the intersection of the corner at "A" to the intersection of the corner at "B"? (Pick the closest value.)
   (A) 25'-2" (D) 26'-5"
   (B) 25'-7" (E) 26'-10"
   (C) 26'-0"
3. What is the total area in square feet of all closets (C1, C2, and C3)? (Pick the closest value.)
   (A) 46 sq. ft. (D) 52 sq. ft.
   (B) 48 sq. ft. (E) 54 sq. ft.
   (C) 50 sq. ft.
4. What is the total area in square feet of the laundry and bathroom? (Pick the closest value.)
   (A) 91 sq. ft. (D) 97 sq. ft.
   (B) 93 sq. ft. (E) 99 sq. ft.
   (C) 95 sq. ft.
5. What is the distance from the intersection of the corner at "C" to the intersection of the corner at "D"? (Pick the closest value.)
   (A) 31'-0" (D) 32'-3"
   (B) 31'-5" (E) 32'-8"
   (C) 31'-10"
6. What is the angle in the XY plane from the intersection of the wall corner at "A" to the intersection of the wall corner at "D"? (Pick the closest value.)
   (A) 1° (D) 7°
   (B) 3° (E) 9°
   (C) 5°
7. Stretch the floor plan up at a distance of 2'-5". Use "E" and "F" as the locations of the crossing box. What is the new total area of Bedroom #1, Bedroom #2, C1, and C2? (Pick the closest value.)
   (A) 315 sq. ft. (D) 327 sq. ft.
   (B) 319 sq. ft. (E) 331 sq. ft.
   (C) 323 sq. ft.

Intermediate Level Drawings

PROBLEM 13–10 GASKET2.DWG

Directions for Gasket2.Dwg

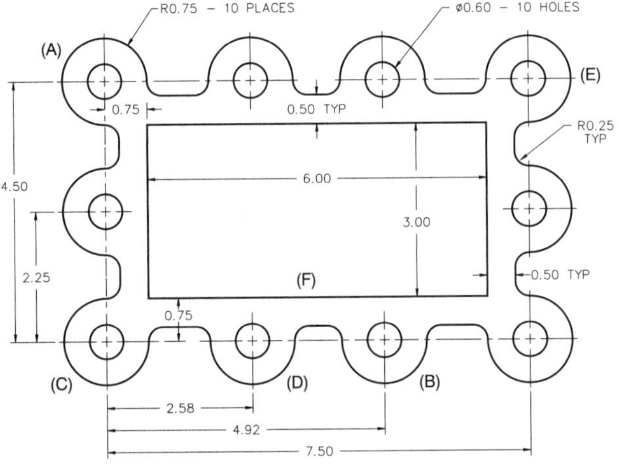

Use the Drawing Units dialog box to set the units to decimal. Set the precision to two. Be sure the system of angle measure is set to decimal degrees and the number of decimal places for the display of angles is zero.

Keep the remaining default unit values.

Begin the drawing in Problem 13–10 by locating the center of the 6.00 × 3.00 rectangle at coordinate (6.00,4.75).

Refer to the drawing of Gasket2 in Problem 13–10 to answer the following questions.

1. What is the total surface area of Gasket2 with the rectangle and all ten holes removed?
   (A) 21.46 (D) 21.52
   (B) 21.48 (E) 21.54
   (C) 21.50
2. What is the distance from the center of arc "A" to the center of arc "B"?
   (A) 6.63 (D) 6.75
   (B) 6.67 (E) 6.79
   (C) 6.71
3. What is the length of arc segment "C"?
   (A) 3.44 (D) 3.53
   (B) 3.47 (E) 3.56
   (C) 3.50
4. What is the absolute coordinate value of the center of the 0.75-radius arc "D"?
   (A) 4.83,2.50 (D) 4.80,2.50
   (B) 4.83,2.47 (E) 4.83,2.56
   (C) 4.83,2.53
5. What is the angle formed in the XY plane from the center of the 0.75-radius arc "D" to the center of the 0.75-radius arc "A"?
   (A) 116° (D) 122°
   (B) 118° (E) 124°
   (C) 120°
6. What is the delta X,Y distance from the intersection at "E" to the midpoint of the line at "F"?
   (A) -4.50,3.65 (D) -4.50,3.75
   (B) -4.50,-3.65 (E) -4.50,-3.75
   (C) -4.50,-3.70
7. Use the SCALE command to reduce the size of the inner rectangle. Use the midpoint of the line at "F" as the base point. Use a scale factor of 0.83 units. What is the new total surface area with the rectangle and all ten holes removed?
   (A) 26.99 (D) 27.14
   (B) 27.04 (E) 27.19
   (C) 27.09

PROBLEM 13–11 PLATE1.DWG

Directions for Plate1.Dwg

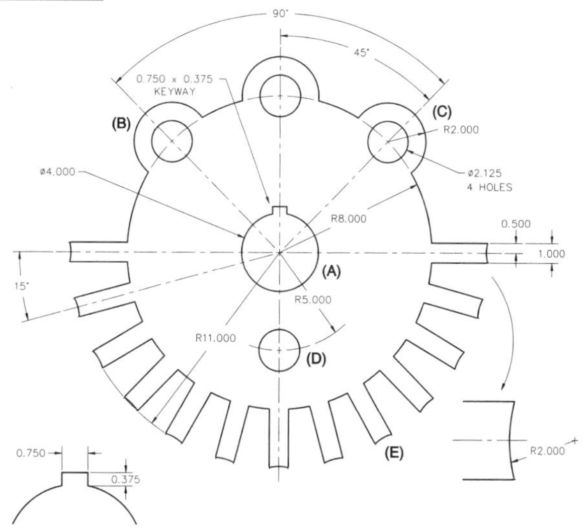

Use the Drawing Units dialog box and set to decimal units. Set the precision to three.

Be sure the system of angle measure is set to decimal degrees and the number of decimal places for the display of angles is zero. Keep all remaining default unit values. Use the LIMITS command and set the upper-right corner of the screen area to a value of (36.000,24.000).

Begin the drawing, shown in Problem 13–3, by placing the center of the 4.000-diameter arc with keyway at coordinate (16.000,13.000).

Refer to the drawing of Plate1, shown in Problem 13–3, to answer the following questions.

1. What is the distance from the center of the 2.000-radius arc "B" to the center of the 2.000-radius arc "C"?
   (A) 10.286 (D) 11.307
   (B) 10.293 (E) 11.314
   (C) 11.300
2. What is the absolute coordinate value of the center of arc "C"?
   (A) 21.657,18.657
   (B) 21.657,18.664
   (C) 21.657,18.671
   (D) 21.657,18.678
   (E) 21.657,18.685
3. What is the angle formed in the XY plane from the center of the 2.000-radius arc "C" to the center of the 2.125-diameter hole "D"?
   (A) 242° (D) 248°
   (B) 244° (E) 250°
   (C) 246°
4. What is the total length of arc "E"?
   (A) 0.999 (D) 1.017
   (B) 1.005 (E) 1.023
   (C) 1.011
5. What is the total length of arc "C"?
   (A) 6.766 (D) 6.784
   (B) 6.772 (E) 6.790
   (C) 6.778
6. What is the total area of Plate1 with all holes including keyway removed?
   (A) 232.259 (D) 232.277
   (B) 232.265 (E) 232.283
   (C) 232.271
7. What is the distance from the center of the 2.000-radius arc "B" to the center of the 2.000-radius arc "E"?
   (A) 20.732 (D) 20.750
   (B) 20.738 (E) 20.756
   (C) 20.744

PROBLEM 13–12 HANGER.DWG

Directions for Hanger.Dwg

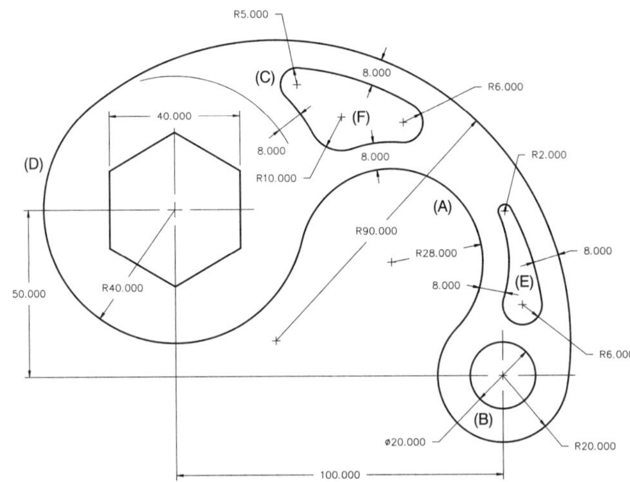

Use the Drawing Units dialog box set units to decimal. Set precision to three. Be sure the system of angle measure is set to decimal degrees and the number of decimal places for the display of angles is zero. Keep all remaining default unit values. Use the LIMITS command and set the upper-right corner of the screen area to a value of (250.000,150.000).

Begin drawing the hanger by locating the center of the 40.000-radius arc at coordinate (55.000,85.000). See Problem 13–12.

Refer to the drawing of the hanger, shown in Problem 13–12, to answer the following questions.

1. What is the area of the outer profile of the hanger?
   (A) 9970.567
   (B) 9965.567
   (C) 9975.567
   (D) 9975.005
   (E) 9980.347
2. What is the area of the hanger with the polygon, circle, and irregular shapes removed?
   (A) 7304.089
   (B) 7305.000
   (C) 7303.890
   (D) 7304.000
   (E) 7306.098
3. What is the absolute coordinate value of the center of the 28.000-radius arc "A"?
   (A) 120.000,69.000
   (B) 121.082,68.964
   (C) 121.520,68.237
   (D) 121.082,69.000
   (E) 121.082,66.937
4. What is the absolute coordinate value of the center of the 20.000-diameter circle "B"?
   (A) 154.000,35.000
   (B) 156.000,36.000
   (C) 156.147,35.256
   (D) 155.000,35.000
   (E) 156.000,37.000
5. What is the angle formed in the XY plane from the center of the 5.000-radius arc "C" to the center of the 40.000-radius arc "D"?
   (A) 219°
   (B) 221°
   (C) 223°
   (D) 225°
   (E) 227°
6. What is the total area of irregular shape "E"?
   (A) 271.613
   (B) 271.723
   (C) 271.784
   (D) 271.801
   (E) 271.822
7. What is the total area of irregular shape "F"?
   (A) 698.511
   (B) 698.621
   (C) 699.817
   (D) 699.856
   (E) 699.891

PROBLEM 13–13 WEDGE.DWG

Directions for Wedge.Dwg

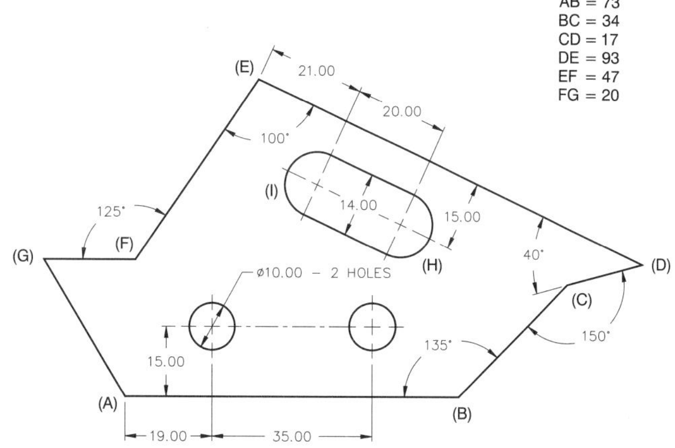

Use the Drawing Units dialog box to set the units to decimal. Set the precision to two. Be sure the system of angle measure is set to decimal degrees and the number of decimal places for the display of angles is zero. Keep all remaining default unit values.

Begin constructing the wedge with vertex "A" located at coordinate (30,30) as shown in Problem 13–13.

Refer to the drawing of the wedge, shown in Problem 13–13, to answer the following questions.

1. What is the total area of the wedge with the two holes and slot removed?

   (A) 4367.97 (B) 4368.54 (C) 4370.12 (D) 4371.83 (E) 4374.91

2. What is the distance from the intersection of vertex "E" to the intersection of vertex "G"?

   (A) 60.72 (B) 60.74 (C) 60.80 (D) 60.85 (E) 60.87

3. What is the distance from the intersection of vertex "D" to the intersection of vertex "G"?

   (A) 131.00 (B) 131.12 (C) 131.25 (D) 131.36 (E) 131.48

4. What is the length of arc "H"?

   (A) 21.00 (B) 21.50 (C) 21.99 (D) 22.50 (E) 22.99

5. What is the overall height of the wedge from the base of line "AB" to the peak at "E"?

   (A) 60.72 (B) 65.87 (C) 67.75 (D) 69.08 (E) 71.98

6. What is the distance from the intersection of vertex "A" to the center of arc "I"?

   (A) 61.09 (B) 61.67 (C) 61.98 (D) 62.93 (E) 63.02

7. What is the length of line "AG"?

   (A) 31.92 (B) 32.47 (C) 33.62 (D) 34.23 (E) 35.33

PROBLEM 13–14 PATTERN1.DWG

Directions for Pattern1.Dwg

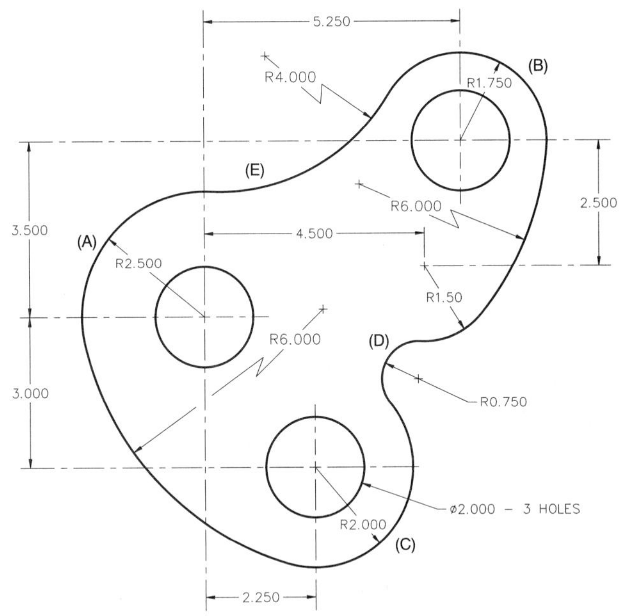

Use the Drawing Units dialog box to set the units to decimal. Set the precision to three. Be sure the system of angle measure is set to decimal degrees and the number of decimal places for the display of angles is zero.

Begin the drawing, shown in Problem 13–14, by locating the center of the 2.500-radius arc at coordinate (4.250,5.750).

Refer to the drawing of Pattern1, shown in Problem 13–14, to answer the following questions.

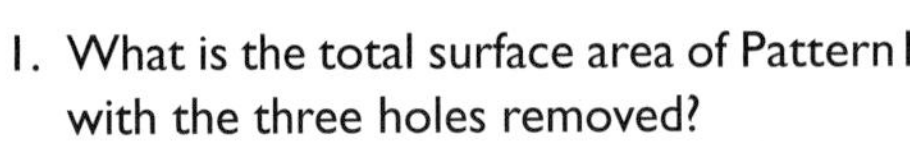

1. What is the total surface area of Pattern1 with the three holes removed?
   (A) 47.340 (D) 47.486
   (B) 47.346 (E) 47.586
   (C) 47.386
2. What is the distance from the center of the 2.500-radius arc "A" to the center of the 1.750-radius arc "B"?
   (A) 6.310 (D) 6.321
   (B) 6.315 (E) 6.305
   (C) 6.210
3. What is the absolute coordinate value of the center of the 4.000-radius arc at "E"?
   (A) 4.580,12.241 (D) 4.480,12.237
   (B) 4.589,12.249 (E) 4.589,12.241
   (C) 4.589,12.237
4. What is the perimeter of the outline of Pattern1?
   (A) 31.741 (D) 31.759
   (B) 31.747 (E) 31.765
   (C) 31.753
5. What is the total length of arc "A"?
   (A) 4.633 (D) 4.651
   (B) 4.639 (E) 4.657
   (C) 4.645
6. What is the angle formed in the XY plane from the center of the 2.500-radius arc "A" to the center of the 2.000-radius arc "C"?
   (A) 301° (D) 307°
   (B) 303° (E) 309°
   (C) 305°
7. Use the MIRROR command to flip but not duplicate Pattern1. Use the center of the 2.000-radius arc "C" as the first point of the mirror line. Use a polar coordinate value of @1.000<90 as the second point. What is the new absolute coordinate value of the center of the 0.750-radius arc "D"?
   (A) 4.378,4.504 (C) 4.386,4.504
   (B) 4.382,4.504 (D) 4.390,4.504
   (E) 4.394,4.504

PROBLEM 13–15 HOUSING1.DWG

Directions for Housing1.Dwg

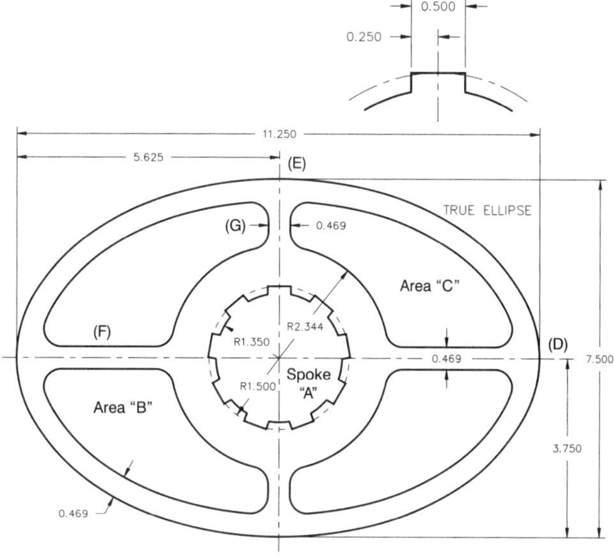

Begin constructing Housing1 in Problem 13–15 by keeping the default units set to decimal but changing the precision to three. Be sure the system of angle measure is set to decimal degrees and the number of decimal places for the display of angles is zero. Place the center of the 1.500-radius circular centerline at coordinate (6.500,5.250).

Before constructing the outer ellipse, set the PELLIPSE command to a value of 1. This will draw the ellipse as a polyline object. Refer to the drawing of Housing1, shown in Problem 13–15, to answer the following questions.

1 What is the perimeter of Spoke "A"?
   (A) 11.564 (D) 11.582
   (B) 11.570 (E) 11.588
   (C) 11.576
2. What is the perimeter of Area "B"?
   (A) 12.513 (D) 12.531
   (B) 12.519 (E) 12.537
   (C) 12.525
3. What is the total area of Area "C"?
   (A) 7.901 (D) 7.919
   (B) 7.907 (E) 7.928
   (C) 7.913
4. What is the absolute coordinate value of the intersection of the ellipse and centerline at "D"?
   (A) 12.125,5.244 (D) 12.125,5.262
   (B) 12.125,5.250 (E) 12.125,5.268
   (C) 12.125,5.256
5. What is the total surface area of Housing1 with the spoke and all slots removed?
   (A) 27.095 (D) 28.113
   (B) 28.101 (E) 28.119
   (C) 28.107
6. What is the distance from the midpoint of the horizontal line segment at "F" to the midpoint of the vertical line segment at "G"?
   (A) 4.235 (D) 4.253
   (B) 4.241 (E) 4.259
   (C) 4.247
7. Increase Spoke "A" in size using the SCALE command. Use the center of the 1.500-radius arc as the base point. Use a scale factor of 1.115 units. What is the new total area of Housing1 with the spoke and all slots removed?
   (A) 26.535 (D) 26.553
   (B) 26.541 (E) 26.559
   (C) 26.547

PROBLEM 13–16 CAM1.DWG

Directions for Cam1.Dwg

Start a new drawing called Cam1 (see Problem 13–16). Keep the default settings of decimal units precision to two. Be sure the system of angle measure is set to decimal degrees and the number of decimal places for the display of angles is zero. Begin the drawing, shown in Problem 13–16, by constructing the center of the 2.00-unit-diameter circle at coordinate (3.00,4.00).

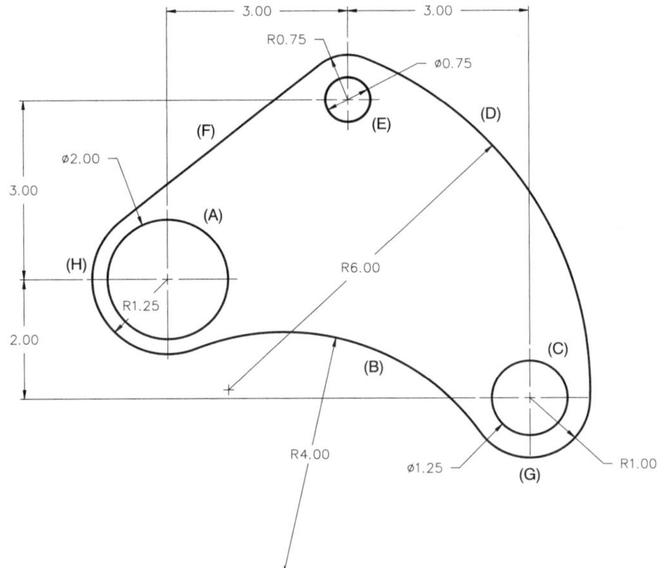

Refer to the drawing of Cam1, shown in Problem 13–16, to answer the following questions.

1. What is the absolute coordinate value of the center of the 4.00-radius arc "B"?
   (A) 4.92,0.89
   (B) 4.92,-0.89
   (C) 4.95,-0.91
   (D) 4.92,-0.85
   (E) -4.95,-0.85
2. What is the angle formed in the XY plane from the center of the 2.00-diameter circle "A" to the center of the 1.25-diameter circle "C"?
   (A) 334°
   (B) 336°
   (C) 338°
   (D) 340°
   (E) 342°
3. What is the length of arc "D"?
   (A) 7.20
   (B) 7.23
   (C) 7.26
   (D) 7.29
   (E) 7.32
4. What is the total area of Cam1 with all three holes removed?
   (A) 26.78
   (B) 26.81
   (C) 26.84
   (D) 26.87
   (E) 26.90
5. What is the total length of line "F"?
   (A) 4.05
   (B) 4.09
   (C) 4.13
   (D) 4.17
   (E) 4.21
6. What is the delta X distance from the quadrant of the 1.00-radius arc at "G" to the quadrant of the 1.25-radius arc at "H"?
   (A) -7.13
   (B) -7.16
   (C) -7.19
   (D) -7.22
   (E) -7.25
7. Use the ROTATE command to realign Cam1. Use the center of the 2.00-diameter circle "A" as the base point of the rotation. Rotate Cam1 from this point at a -10-degree angle. What is the absolute coordinate value of the center of the 0.75-diameter hole "E"?
   (A) 6.48,6.37
   (B) 6.48,6.40
   (C) 6.48,6.43
   (D) 6.51,6.46
   (E) 6.54,6.49

PROBLEM 13–17 ROTOR.DWG

DirectionsforRotor.Dwg

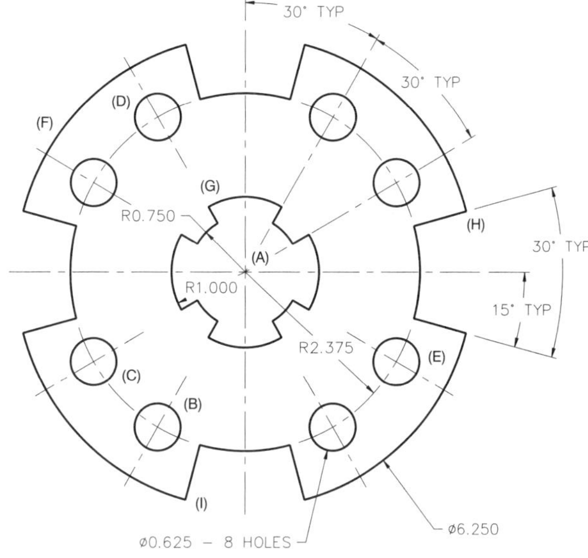

Start a new drawing called Rotor, as shown in Problem 13–17. Keep the default setting of decimal units precision to three.

Be sure the system of angle measure is set to decimal degrees and the number of decimal places for the display of angles is zero. Keep all remaining default unit values. Begin the drawing in Problem 13–17 by constructing the center of the 6.250-unit-diameter circle at "A" at coordinate (5.500,5.000).

Refer to the drawing of the rotor, shown in Problem 13–17, to answer the following questions.

1. What is the absolute coordinate value of the center of the 0.625-diameter circle "B"?
   (A) 4.294,2.943 (D) 4.312,2.943
   (B) 4.300,2.943 (E) 4.318,2.943
   (C) 4.306,2.943
2. What is the total area of the rotor with all eight holes and the center slot removed?
   (A) 21.206 (D) 21.218
   (B) 21.210 (E) 21.222
   (C) 21.214
3. What is the total length of arc "F"?
   (A) 3.260 (D) 3.272
   (B) 3.264 (E) 3.276
   (C) 3.268
4. What is the distance from the center of the 0.625 circle "C" to the center of the 0.625 circle "D"?
   (A) 3.355 (D) 3.367
   (B) 3.359 (E) 3.371
   (C) 3.363
5. What is the angle formed in the XY plane from the center of the 0.625 circle "B" to the center of the 0.625 circle "E"?
   (A) 11° (D) 17°
   (B) 13° (E) 19°
   (C) 15°
6. What is the delta X,Y distance from the intersection at "H" to the intersection at "I"?
   (A) -3.827,-3.827 (D) -3.841,3.813
   (B) -3.827,3.827 (E) -3.848,-3.806
   (C) -3.834,3.820
7. Use the SCALE command to increase the size of just the center slot "G." Use the center of arc "F" as the base point. Use a scale factor of 1.500 units. What is the new surface area of the rotor with the center slot and all eight holes removed?
   (A) 17.861 (D) 17.882
   (B) 17.868 (E) 17.889
   (C) 17.875

PROBLEM 13–18 S-CAM.DWG

Directions for S-Cam.Dwg

Begin the construction of S-Cam, shown in Problem 13–18, by keeping the default units set to decimal but changing the precision to three. Be sure the system of angle measure is set to decimal degrees and the number of decimal places for the display of angles is zero. Keep all remaining default unit values.

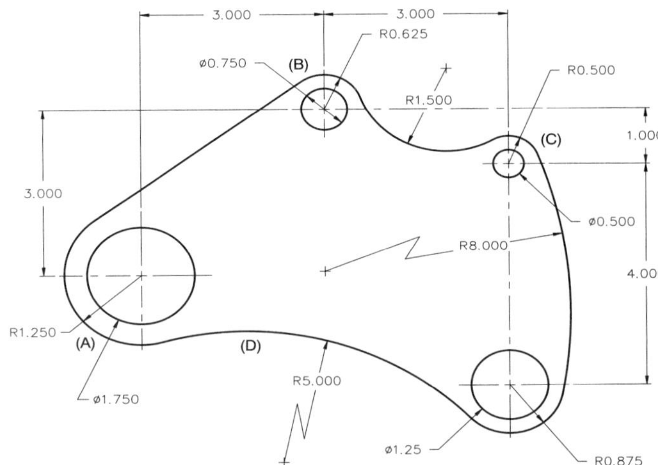

Begin the drawing, shown in Problem 13–18, by placing the center of the 1.750-diameter circle at coordinate (2.500,3.750).

Refer to the drawing of the S-Cam in Problem 13–18 to answer the following questions.

1. What is the total surface area of the S-Cam with all four holes removed?
   (A) 27.654 (D) 27.672
   (B) 27.660 (E) 28.678
   (C) 27.666
2. What is the distance from the center of the 1.250-radius arc "A" to the center of the 0.625-radius arc "B"?
   (A) 4.237 (D) 4.255
   (B) 4.243 (E) 4.261
   (C) 4.249
3. What is the absolute coordinate value of the center of arc "D"?
   (A) 4.208,-2.258 (D) 4.208,-2.270
   (B) 4.208,-2.262 (E) 4.208,-2.274
   (C) 4.208,-2.266
4. What is the total length of arc "D"?
   (A) 5.480 (D) 5.498
   (B) 5.486 (E) 6.004
   (C) 5.492
5. What is the angle formed in the XY plane from the center of the 0.625-arc "B" to the center of the 0.500-arc "C"?
   (A) 334° (D) 340°
   (B) 336° (E) 342°
   (C) 338°
6. What is the delta X,Y distance from the center of the 0.500-arc "C" to the quadrant of the 5.000-arc "D"?
   (A) -4.284,-3.012 (D) -4.296,-3.012
   (B) -4.288,-3.012 (E) -4.300,-3.012
   (C) -4.292,-3.012
7. Use the SCALE command to reduce S-Cam in size. Use the center of the 1.250-radius arc "A" as the base point. Use a scale factor of 0.822 units. What is the new surface area of S-Cam with all four holes removed?
   (A) 18.694 (D) 18.706
   (B) 18.698 (E) 18.710
   (C) 18.702

PROBLEM 13–19 HOUSING2.DWG

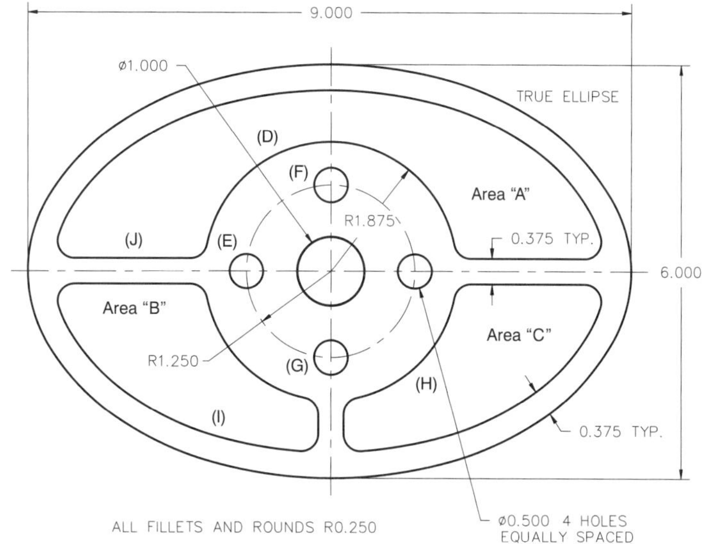

Directions for Housing2.Dwg

Begin the construction of Housing2, shown in Problem 13–19, by keeping the default units set to decimal but changing the precision to three. Be sure the system of angle measure is set to decimal degrees and the number of decimal places for the display of angles is zero. Keep all remaining default unit values.

Before constructing the outer ellipse, set the PELLIPSE command to a value of 1. This will draw the ellipse as a polyline object.

Begin the drawing by placing the center of the ellipse at coordinate (5.500,4.500).

Refer to the drawing of Housing2, shown in Problem 13–19, to answer the following questions.

1. What is the total area of Housing2 with all five holes and Areas "A," "B," and "C" removed?
   (A) 20.082
   (B) 20.088
   (C) 20.094
   (D) 20.100
   (E) 20.106
2. What is the perimeter of Area "A"?
   (A) 19.837
   (B) 19.843
   (C) 19.849
   (D) 19.855
   (E) 19.861
3. What is the length of arc "D"?
   (A) 5.107
   (B) 5.113
   (C) 5.119
   (D) 5.125
   (E) 5.131
4. What is the distance from the center of the 0.500-diameter hole "E" to the center of the 0.500-diameter hole "F"?
   (A) 1.762
   (B) 1.768
   (C) 1.774
   (D) 1.780
   (E) 1.786
5. What is the total length of polyarc "I"?
   (A) 4.374
   (B) 4.380
   (C) 4.386
   (D) 4.392
   (E) 4.398
6. What is the delta X,Y distance from the center of circle "G" to the midpoint of the horizontal line at "J"?
   (A) -2.943,1.438
   (B) 2.943,1.438
   (C) 2.937,-1.438
   (D) -2.931,1.431
   (E) -2.931,-1.425
7. Use the MOVE command to relocate Housing2 at a distance of 0.375 units in a 45° direction. What is the new absolute coordinate value of the center of the 0.500-diameter hole "G"?
   (A) 5.759,3.515
   (B) 5.765,3.503
   (C) 5.759,3.509
   (D) 5.765,3.509
   (E) 5.765,3.515

PROBLEM 13–20 PATTERN5.DWG

Directions for Pattern5.Dwg

Begin the construction of Pattern5, shown in Problem 13–20, by keeping the default units set to decimal but changing the precision to zero. Be sure the system of angle measure is set to decimal degrees and the number of decimal places for the display of angles is zero. Begin the drawing by placing Vertex "A" at absolute coordinate (190,30).

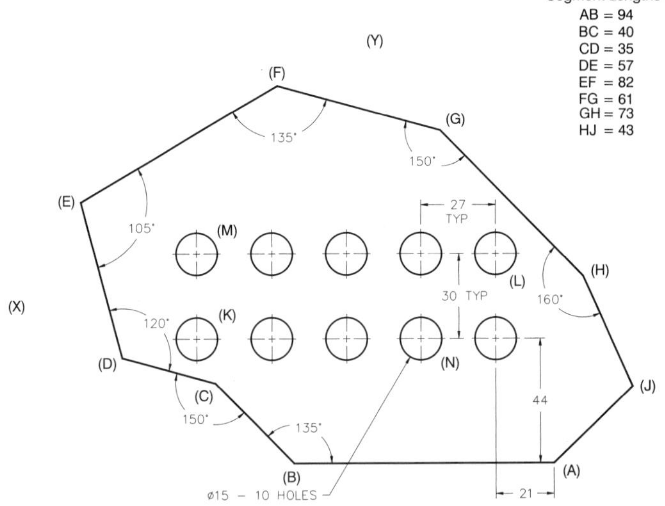

Refer to the drawing of Pattern5, shown in Problem 13–20, to answer the following questions.

1. What is the distance from the intersection of vertex "J" to the intersection of vertex "A"?
   (A) 38 (D) 41
   (B) 39 (E) 42
   (C) 40
2. What is the perimeter of Pattern5?
   (A) 523 (D) 526
   (B) 524 (E) 527
   (C) 525
3. What is the total area of Pattern5 with all 10 holes removed?
   (A) 16369 (D) 16372
   (B) 16370 (E) 16373
   (C) 16371
4. What is the distance from the center of the 15-diameter hole "K" to the center of the 15-diameter hole "L"?
   (A) 109 (D) 112
   (B) 110 (E) 113
   (C) 111
5. What is the angle formed in the XY plane from the center of the 15-diameter hole "M" to the center of the 15-diameter hole "N"?
   (A) 340° (D) 346°
   (B) 342° (E) 348°
   (C) 344°
6. What is the absolute coordinate value of the intersection at "F"?
   (A) 90,163 (D) 96,167
   (B) 92,163 (E) 98,169
   (C) 94,165
7. Use the STRETCH command to lengthen Pattern5. Use "Y" as the first point of the stretch crossing box. Use "X" as the other corner. Pick the intersection at "F" as the base point and stretch Pattern5 a total of 23 units in the 180 direction. What is the new total area of Pattern5 with all ten holes removed?
   (A) 17746 (D) 17767
   (B) 17753 (E) 17774
   (C) 17760

PROBLEM 13–21 BRACKET5.DWG

Directions for Bracket5.Dwg

Begin the construction of Bracket5, shown in Problem 13–21, by keeping the default units set to decimal but changing the precision to two. Be sure the system of angle measure is set to decimal degrees and the number of decimal places for the display of angles is zero Begin the drawing by placing the center of the 1.50-diameter hole at coordinate (7.50,5.75).

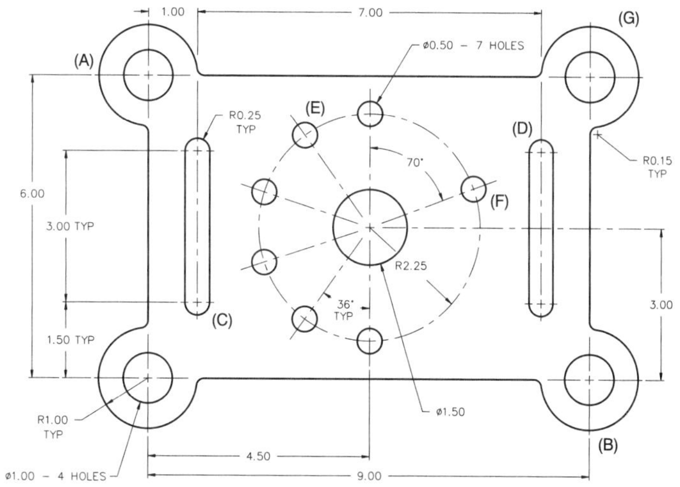

Refer to the drawing of Bracket5, shown in Problem 13–21, to answer the following questions.

1. What is the distance from the quadrant of the 1.00-radius arc at "A" to the quadrant of the 1.00-radius arc at "B"?
   (A) 12.09 (D) 12.18
   (B) 12.12 (E) 12.21
   (C) 12.15
2. What is the distance from the center of the 0.25-radius arc at "C" to the center of the 0.25-radius arc at "D"?
   (A) 7.62 (D) 7.71
   (B) 7.65 (E) 7.74
   (C) 7.68
3. What is the distance from the center of the 0.50-diameter circle at "E" to the center of the 0.50-diameter circle at "F"?
   (A) 3.50 (D) 3.59
   (B) 3.53 (E) 3.62
   (C) 3.56
4. What is the length of arc "G"?
   (A) 4.42 (D) 4.51
   (B) 4.45 (E) 4.54
   (C) 4.48
5. What is the total area of Bracket5 with all holes and slots removed?
   (A) 53.72 (D) 53.81
   (B) 53.75 (E) 53.84
   (C) 53.78
6. What is the delta X,Y distance from the center of the 0.25-radius arc "C" to the center of the 0.50 circle "E"?
   (A) 2.12,3.38 (D) 2.30,3.20
   (B) 2.18,3.32 (E) 2.36,3.14
   (C) 2.24,3.26
7. What is the angle formed in the XY plane from the center of the 0.25-radius arc "D" to the center of the 1.00-radius arc "G"?
   (A) 50° (D) 56°
   (B) 52° (E) 58°
   (C) 54°

PROBLEM 13–22 PLATE2A.DWG

Directions for Plate2A.Dwg

Begin Plate2A by keeping the default units set to decimal but changing the precision to two. Keep the system of angle measure set to decimal degrees, but change the fractional places for display of angles to two.

Keep the remaining default unit values.

Begin constructing Plate2A by starting the lower-left corner "X" at absolute coordinate (2.25,2.25). See Problem 13–22.

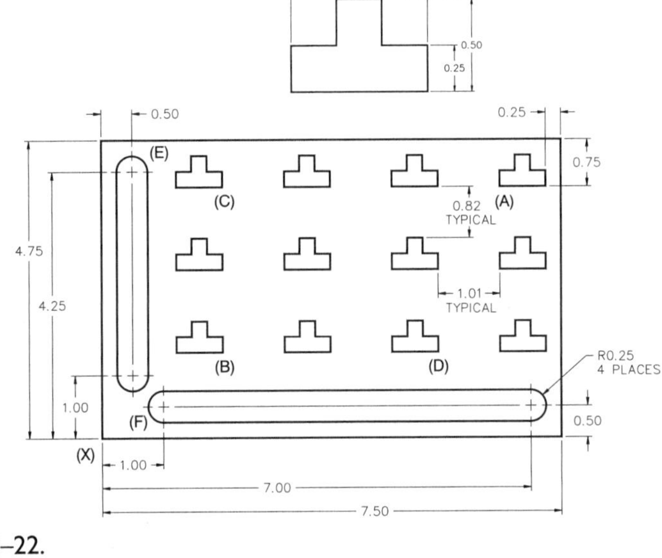

Refer to the drawing of Plate2A, shown in Problem 13–22, to answer the following questions.

1. What is the total perimeter of the horizontal slot?

   (A) 13.42 (B) 13.47 (C) 13.52 (D) 13.57 (E) 13.62

2. What is the total perimeter of the vertical slot?

   (A) 7.98 (B) 8.03 (C) 8.07 (D) 8.14 (E) 8.20

3. What is the total area of the horizontal slot?

   (A) 3.02 (B) 3.08 (C) 3.14 (D) 3.20 (E) 3.26

4. What is the total area of the vertical slot?

   (A) 1.82 (B) 1.88 (C) 1.94 (D) 2.00 (E) 2.06

5. What is the distance from the endpoint of the line at "C" to the endpoint of the line at "D"?

   (A) 4.25 (B) 4.30 (C) 4.35 (D) 4.40 (E) 4.45

6. What is the angle formed in the XY plane formed from the center of arc "E" to the endpoint of the line at "B"?

   (A) 296.90° (B) 296.96° (C) 297.02° (D) 297.08° (E) 297.14°

7. What is the angle formed in the XY plane from the endpoint of the line at "A" to the center of arc "F"?

   (A) 212.47° (B) 212.53° (C) 212.59° (D) 212.65° (E) 212.71°

PROBLEM13–23PLATE2B.DWG

Directions for Plate2B.Dwg

Change the precision to three. Change from decimal degrees to degrees/minutes/seconds. Change the number of places for display of angles to four.

Rotate Plate2B at an angle of 32°, 0', 0" using absolute coordinate (2.250,2.250) at "X" below as the base point of rotation. Perform a ZOOM-EXTENTS. See Problem 13–23.

Refer to the drawing of Plate2B, shown in Problem 13–23, to answer the following questions.

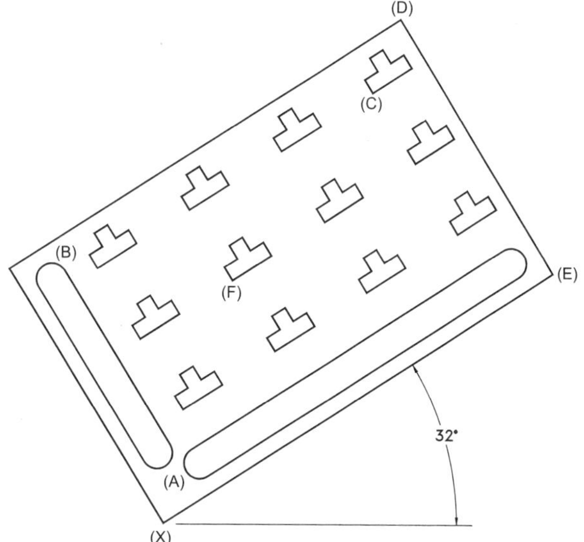

1. What is the angle in the XY plane from the center of arc "A" to the center of arc "B"?
   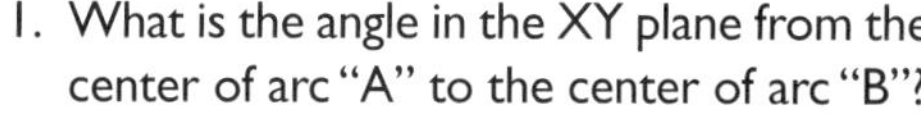
   (A) 122d0'0" (D) 122d40'30'
   (B) 122d20'0" (E) 123d0'0"
   (C) 122d40'0"
2. What is the absolute coordinate value of the endpoint of the line at "C"?
   (A) 5.627,9.071 (D) 5.639,9.083
   (B) 5.631,9.075 (E) 5.643,9.087
   (C) 5.635,9.079
3. What is the absolute coordinate value of the endpoint of the line at "D"?
   (A) 6.077,10.237 (D) 6.089,10.249
   (B) 6.081,10.241 (E) 6.093,10.253
   (C) 6.085,10.245
4. What is the absolute coordinate value of the endpoint of the line at "E"?
   (A) 8.606,6.220 (D) 8.618,6.232
   (B) 8.610,6.224 (E) 8.622,6.236
   (C) 8.614,6.228
5. What is the angle formed in the XY plane from the endpoint of the line at "F" to the center of arc "B"?
   (A) 178d1'2" (D) 179d39'49"
   (B) 178d27'37" (E) 180d34'29"
   (C) 179d15'12"
6. What is the delta X,Y distance from the intersection at "D" to the center of the 0.250-radius arc "A"?
   (A) -3.937,-6.878 (D) -3.949,-6.890
   (B) -3.941,-6.882 (E) -3.953,-6.894
   (C) -3.945,-6.886
7. Use the SCALE command to reduce Plate2B in size. Use absolute coordinate 2.250,2.250 as the base point. Use 0.777 as the scale factor. What is the area of Plate2B with both slots and the 12 T-shaped objects removed?
   (A) 16.659 (D) 16.671
   (B) 16.663 (E) 16.675
   (C) 16.667

PROBLEM 13–24 RATCHET.DWG

Directions for Ratchet.Dwg

Use the Drawing Units dialog box to change the precision to two. Be sure the system of angle measure is set to decimal degrees and the number of decimal places for the display of angles is zero. Keep the remaining default unit values. Begin by drawing the center of the 1.00-radius arc of the ratchet at absolute coordinate (6.00,4.50), as shown in Problem 13–24.

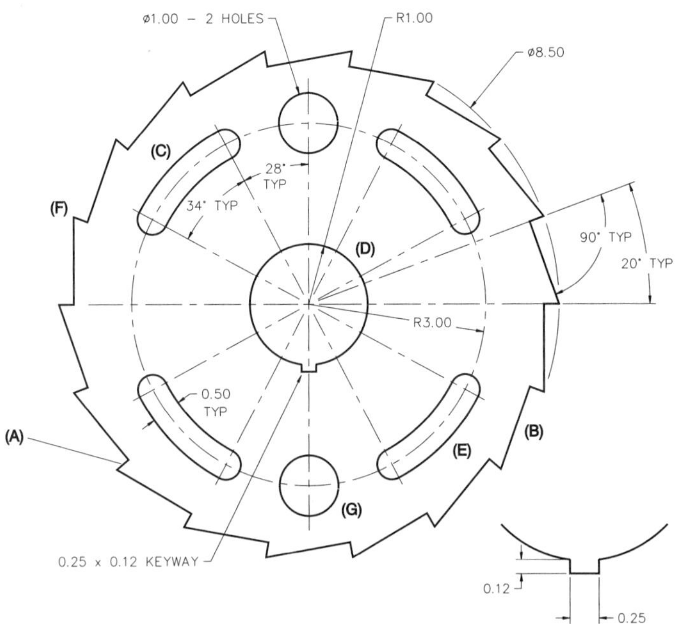

Refer to the drawing of the ratchet, shown in Problem 13–24, to answer the following questions.

1. What is the total length of the short line segment "A"?
   (A) 0.22 (B) 0.24 (C) 0.26 (D) 0.28 (E) 0.30
2. What is the total length of line "B"?
   (A) 1.06 (B) 1.19 (C) 1.33 (D) 1.45 (E) 1.57
3. What is the total length of arc "C"?
   (A) 1.81 (B) 1.93 (C) 2.08 (D) 2.19 (E) 2.31
4. What is the perimeter of the 1.00-radius arc "D" with the 0.25 × 0.12 keyway?
   (A) 6.52 (B) 6.63 (C) 6.77 (D) 6.89 (E) 7.02
5. What is the total surface area of the ratchet with all 4 slots, the two 1.00-diameter holes, and the 1.00-radius arc with the keyway removed?
   (A) 42.98 (B) 43.04 (C) 43.10 (D) 43.16 (E) 43.22
6. What is the absolute coordinate value of the endpoint at "F"?
   (A) 2.01,5.10 (B) 2.01,5.63 (C) 2.01,5.95 (D) 2.20,5.95 (E) 2.37,5.95
7. What is the angle formed in the XY plane from the endpoint of the line at "F" to the center of the 1.00-diameter hole "G"?
   (A) 304° (B) 306° (C) 308° (D) 310° (E) 312°

PROBLEM 13–25 SLIDE.DWG

DirectionsforSlide.Dwg

Use the current unit settings and limits settings for this drawing (see Problem 13–25). The precision should already be set to four. Be sure the system of angle measure is set to decimal degrees and the number of decimal places for the display of angles is zero.

Begin by drawing the center of the 0.5000-diameter circle of the slide at absolute coordinate (2.0000,2.2500).

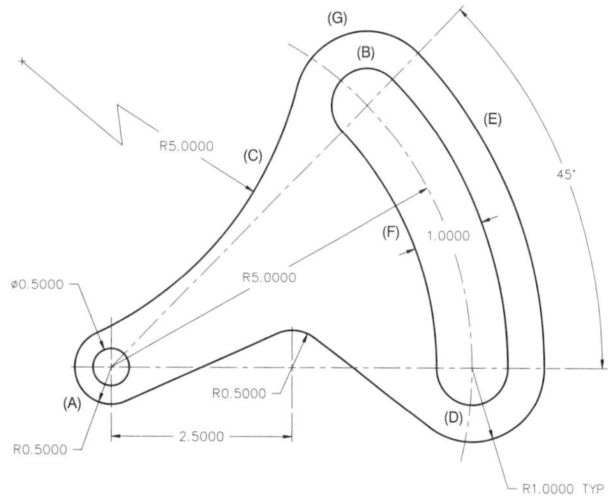

Refer to the drawing of the slide, shown in Problem 13–25, to answer the following questions.

1. What is the absolute coordinate value of the center of the 5.0000-radius arc "C"?
   (A) -0.2800,6.5603
   (B) 0.2887,-5.4393
   (C) -0.3953,7.2503
   (D) -0.2819,7.2543
   (E) -0.3953,6.2543
2. What is the total length of arc "E"?
   (A) 4.7264
   (B) 4.7124
   (C) 5.4302
   (D) 4.8710
   (E) 4.6711
3. What is the distance from the center of the 0.5000-diameter circle "A" to the center of the 5.0000-radius arc "C"?
   (A) 5.5000
   (B) 5.0043
   (C) 4.8768
   (D) 4.6691
   (E) 4.0001
4. What is the total area of the of the 1.0000-diameter slot?
   (A) 4.7124
   (B) 4.7863
   (C) 6.8370
   (D) 4.7625
   (E) 5.9102
5. What is the angle formed in the XY plane from the upper quadrant of the 1.0000-diameter slot at "B" to the lower quadrant of the slot at "D"?
   (A) 282°
   (B) 284°
   (C) 286°
   (D) 288°
   (E) 290°
6. What is the total area of the slide with the 0.5000-diameter hole and slot "F" removed?
   (A) 8.9246
   (B) 13.6370
   (C) 13.8750
   (D) 13.8333
   (E) 14.0297
7. What is the absolute coordinate value of the center of the 1.0000-radius arc "G"?
   (A) 5.5120,5.5621
   (B) 5.5237,5.5551
   (C) 5.5355,5.5590
   (D) 5.5355,5.6123
   (E) 5.5355,5.7855

PROBLEM 13–26 GENEVA.DWG

Directions for Geneva.Dwg

Start a new drawing called Geneva (see Problem 13–26). Keep the default settings of decimal units, and precision set to two. Be sure the system of angle measure is set to decimal degrees and the number of decimal places for the display of angles is zero. Begin the drawing by constructing the 1.50-diameter arc at absolute coordinate (7.50,5.50).

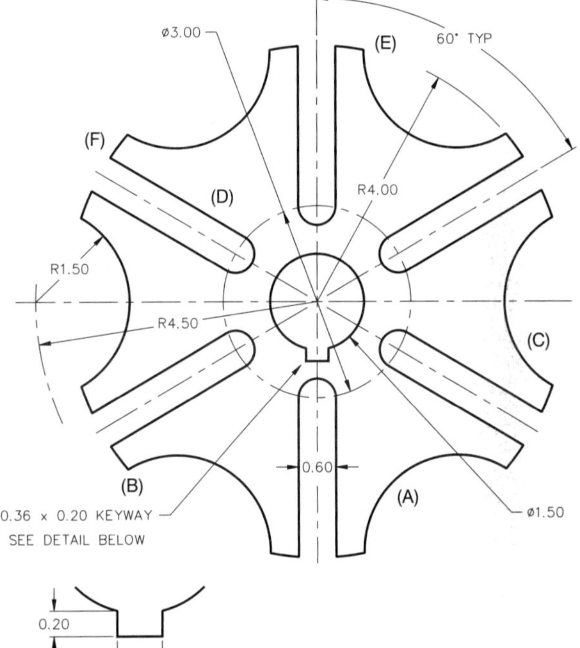

Refer to the drawing of the Geneva, shown in Problem 13–26, to answer the following questions.

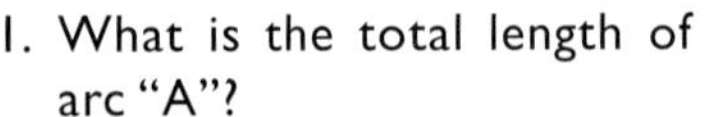

1. What is the total length of arc "A"?
   (A) 3.00 (D) 3.30
   (B) 3.10 (E) 3.40
   (C) 3.21
2. What is the angle formed in the XY plane from the intersection at "B" to the center of arc "C"?
   (A) 11° (D) 17° (B) 13° (E) 19° (C) 15°
3. What is the absolute coordinate value of the midpoint of line "D"?
   (A) 5.27,7.13 (D) 5.31,7.13
   (B) 5.27,7.17 (E) 5.31,7.09
   (C) 5.23,7.13
4. What is the total area of the Geneva with the 1.50-diameter hole and keyway removed?
   (A) 27.40 (D) 27.52
   (B) 27.44 (E) 27.66
   (C) 27.48
5. What is the total distance from the midpoint of arc "F" to the center of arc "A"?
   (A) 8.24 (D) 8.39
   (B) 8.29 (E) 8.44
   (C) 8.34
6. What is the delta X,Y distance from the intersection at "E" to the center of arc "C"?
   (A) 3.93,-3.75 (D) 3.75,3.93
   (B) 3.93,3.75 (E) 3.75,-3.93
   (C) 3.75,3.80
7. Use the SCALE command to reduce the Geneva in size. Use 7.50,5.50 as the base point; use a scale factor of 0.83 units. What is the absolute coordinate value of the intersection at "E"?
   (A) 8.12,8.71 (D) 8.12,8.86
   (B) 8.12,8.76 (E) 8.12,8.91
   (C) 8.12,8.81

PROBLEM 13–27 ANGLE_GUIDE.DWG

Directions for Angle_Guide.Dwg

Change the precision to two decimal places. Change the number of decinal places for display of angles to zero. Line "AB" is represented as a horizontal line. The centerline of the slot is parallel to vertex DE. Begin this drawing by constructing vertex "A" at absolute coordinate 4.24,2.81 (see Problem 13–27).

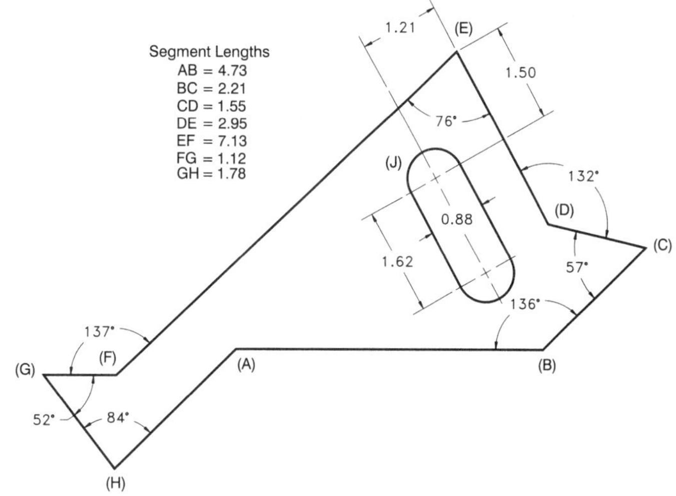

Refer to the drawing of the Angle_Guide, shown in Problem 13–27, to answer the following questions.

1. What is the distance of the line from the intersection of vertex "H" to the intersection vertex "A" is?

   (A) 2.49 (D) 2.64
   (B) 2.54 (E) 2.69
   (C) 2.59

2. What angle does Vertex "A" measure?

   (A) 128° (D) 134°
   (B) 130° (E) 136°
   (C) 132°

3. What is the angle formed in the X-Y plane from the center of arc "J" to the intersection of vertex "B"?

   (A) 297° (D) 303°
   (B) 299° (E) 305°
   (C) 301°

4. What is the total surface area of the Angle Plate with the slot removed?

   (A) 16.90 (D) 17.05
   (B) 16.95 (E) 17.10
   (C) 17.00

5. What is the perimeter of the Angle Plate?

   (A) 23.96 (D) 24.11
   (B) 24.01 (E) 24.16
   (C) 24.06

6. What is the absolute coordinate value of the center of arc "J"?

   (A) 7.24,5.38 (D) 7.34,5.38
   (B) 7.29,5.38 (E) 7.39,5.43
   (C) 7.34,5.33

7. Stretch vertex "E" 1.62 units to the left and 1.47 units straight up. Also, round off all vertex corners of the outline using a radius value of 0.25 units. What is the new surface area of the Angle Plate with the slot removed?

   (A) 23.33 (D) 23.48
   (B) 23.38 (E) 23.53
   (C) 23.43

PROBLEM 13–28 CONTROL_BEARING.DWG

Directions for Control_Bearing.Dwg

Change the precision to two decimals. Keep the remaining default values. Begin this drawing by locating the center of the 9.00-diameter circle at (11.50,11.50). See Problem 13–28

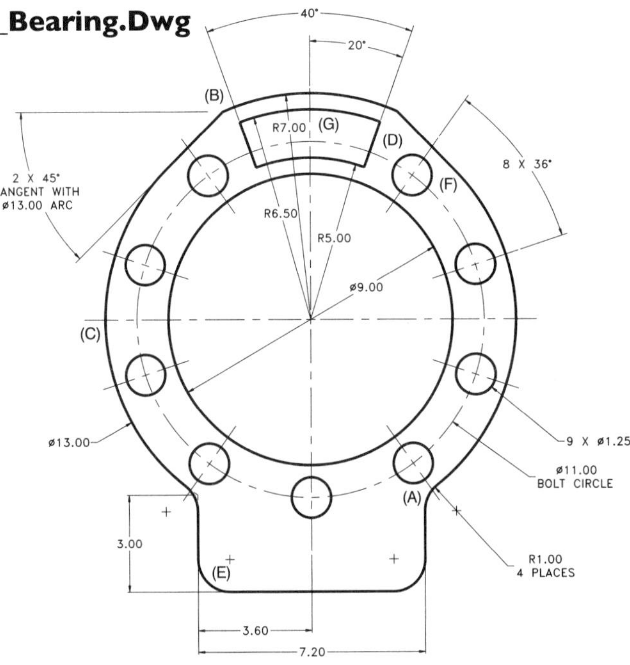

Refer to the drawing of the Control_Bearing, shown in Problem 13–28, to answer the following questions.

1. What is the distance from the center of arc "A" to the intersection at "B"?
   (A) 14.26
   (B) 14.30
   (C) 14.34
   (D) 14.38
   (E) 14.42
2. What is the total length of arc "C"?
   (A) 11.02
   (B) 11.06
   (C) 11.10
   (D) 11.14
   (E) 11.18
3. What is the perimeter of inner slot "G"?
   (A) 10.87
   (B) 10.91
   (C) 10.95
   (D) 10.99
   (E) 11.03
4. What is the absolute coordinate value of the center of the 0.44-diameter hole "F"?
   (A) 14.61,15.91
   (B) 14.65,15.91
   (C) 14.69,15.95
   (D) 14.73,15.95
   (E) 14.73,15.99
5. What is the perimeter of the outermost shape of the Control_Bearing?
   (A) 46.01
   (B) 46.05
   (C) 46.09
   (D) 46.13
   (E) 46.17
6. What is the angle formed in the XY plane from the intersection at "D" to the center of the arc "E"?
   (A) 250°
   (B) 252°
   (C) 254°
   (D) 256°
   (E) 258°
7. What is the area of the Control_Bearing with all holes and inner slot removed?
   (A) 71.45
   (B) 71.49
   (C) 71.53
   (D) 71.57
   (E) 71.61

PROBLEM 13–29 FACILITIES APPLICATION — BUILDING.DWG

Directions for Building.Dwg

Change from decimal units to architectural units. Keep all remaining default values. All block wall thicknesses identified by crosshatching measure 12". All other interior wall thicknesses measure 6". Do not add any dimensions to this drawing (see Problem 13–29).

Refer to the drawing of the building, shown in Problem 13–29, to answer the questions on the following page.

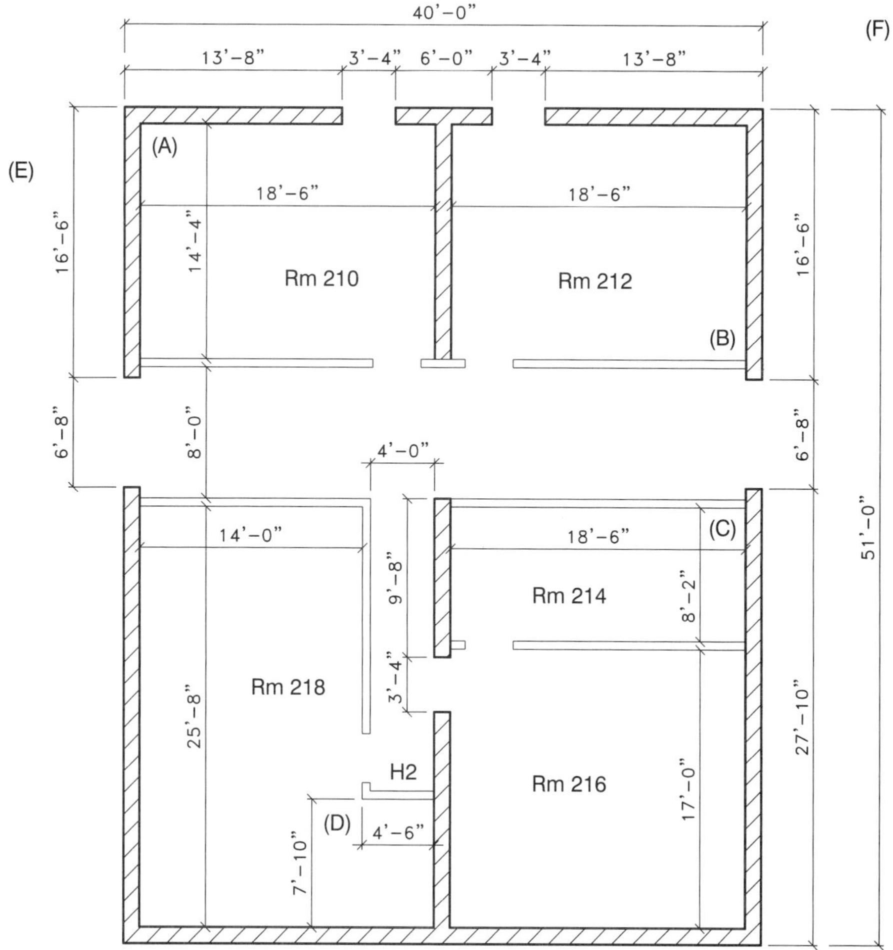

1. What is the total area of all concrete block walls identified by the crosshatching pattern? (Pick the closest value.)
   (A) 183 sq. ft. (D) 192 sq. ft.
   (B) 186 sq. ft. (E) 195 sq. ft.
   (C) 189 sq. ft.
2. What is the total area of rooms 210, 216, and 218? (Pick the closest value.)
   (A) 971 sq. ft. (D) 980 sq. ft.
   (B) 974 sq. ft. (E) 983 sq. ft.
   (C) 977 sq. ft.
3. What is the angle formed in the XY plane from the inside corner intersection at "A" to the inside corner intersection at "B"? (Pick the closest value.)
   (A) 333° (D) 339°
   (B) 335° (E) 341°
   (C) 337°
4. What is the delta X,Y distance from the inside corner intersection at "C" to the outside intersection of the hallway corner at "D"? (Pick the closest value.)
   (A) -24'-0",-17'-10" (D) -24'-0",-18'-7"
   (B) -24'-0",-18'-1" (E) -24'-0",-18'-10"
   (C) -24'-0",-18'-4"
5. What is the total area of rooms 212, 214, and 218? (Pick the closest value.)
   (A) 802 sq. ft. (D) 811 sq. ft.
   (B) 805 sq. ft. (E) 814 sq. ft.
   (C) 808 sq. ft.
6. What is the total distance from the intersection of the wall corner at "A" to the intersection of the interior wall corner at "D"? (Pick the closest value.)
   (A) 43'-2" (D) 43'-11"
   (B) 43'-5" (E) 44'-2"
   (C) 43'-8"
7. Stretch the building straight up using a crossing box from "E" to "F" and at a distance of 5'-4". What is the total area of rooms 212, 214, and 218? (Pick the closest value.)
   (A) 900 sq. ft. (D) 909 sq. ft.
   (B) 903 sq. ft. (E) 912 sq. ft.
   (C) 906 sq. ft.

PROBLEM 13–30 STRUCTURAL APPLICATION — GUSSET.DWG

Directions for Gusset.Dwg

Begin the construction of the gusset, illustrated in Problem 13–30, by keeping the default units set to decimal, but changing the precision to two.

Refer to the drawing of the gusset in Problem 13–30 to answer the following questions.

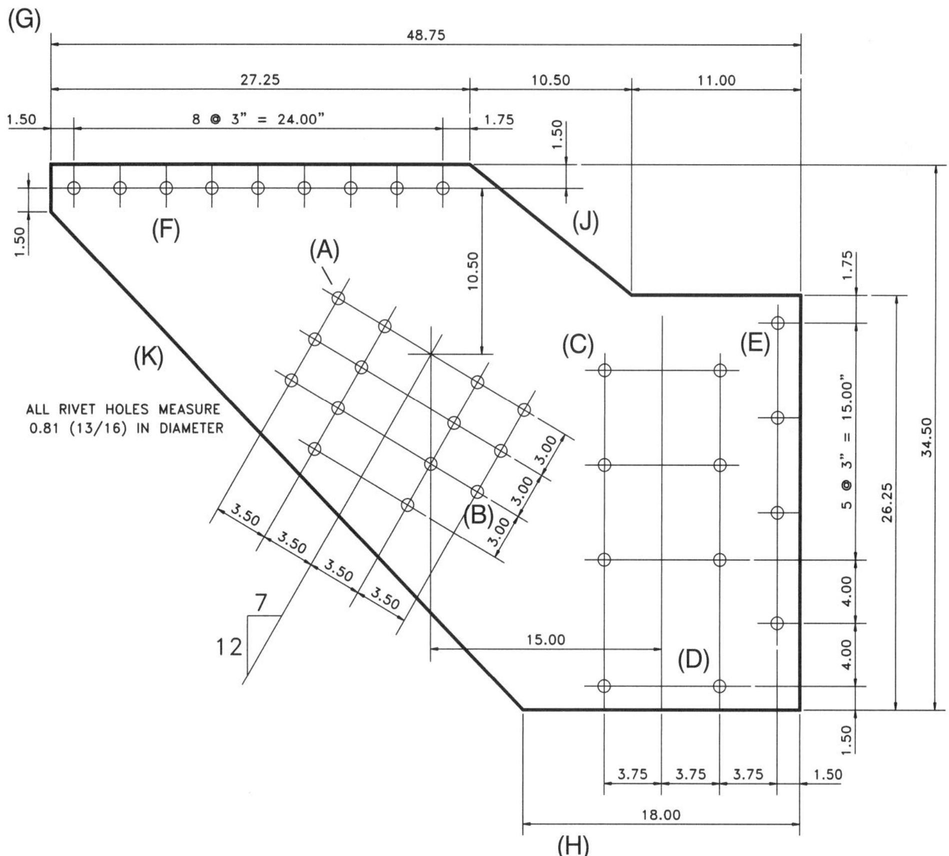

1. What is the total area of the gusset plate with all 35-rivet holes removed?
   (A) 1045.46
   (B) 1050.67
   (C) 1055.21
   (D) 1060.40
   (E) 1065.88
2. What is the delta X,Y distance from the center of hole "A" to the center of hole "B"?
   (A) 9.07,-12.19
   (B) 9.07,-12.24
   (C) 9.07,-12.29
   (D) 9.07,-12.34
   (E) 9.07,-12.39
3. What is the angle formed in the XY plane from the center of hole "C" to the center of hole "D"?
   (A) 285°
   (B) 288°
   (C) 291°
   (D) 294°
   (E) 297°
4. What is the angle formed in the XY plane from the center of hole "E" to the center of hole "F"?
   (A) 165°
   (B) 168°
   (C) 171°
   (D) 174°
   (E) 177°
5. What is the total length of line "J"?
   (A) 13.15
   (B) 13.20
   (C) 13.25
   (D) 13.30
   (E) 13.35
6. What is the total length of line "K"?
   (A) 44.02
   (B) 44.07
   (C) 44.12
   (D) 44.17
   (E) 44.22
7. Stretch the gusset plate directly to the left using a crossing box from "H" to "G" and at a distance of 3.00 units. What is the new area of the gusset plate with all 35-rivet holes removed?
   (A) 1121.98
   (B) 1126.72
   (C) 1131.11
   (D) 1136.59
   (E) 1141.34

PROBLEM 13–31 CIVIL APPLICATION — PLATPLAN.DWG

Directions for Platplan.Dwg

Use the Units Control dialog box to change the system of units from decimal to engineering; change the precision to two. Change the system of angle measurement to Surveyors units. Keep the remaining Units Control dialog box default values. Do not add any dimensions to this drawing.

Use the detail measurements to construct the house and all decks outlined by the hatching. The house is aligned parallel with line AB of the plat. Answer the questions on the next page regarding this drawing (see Problem 13–31).

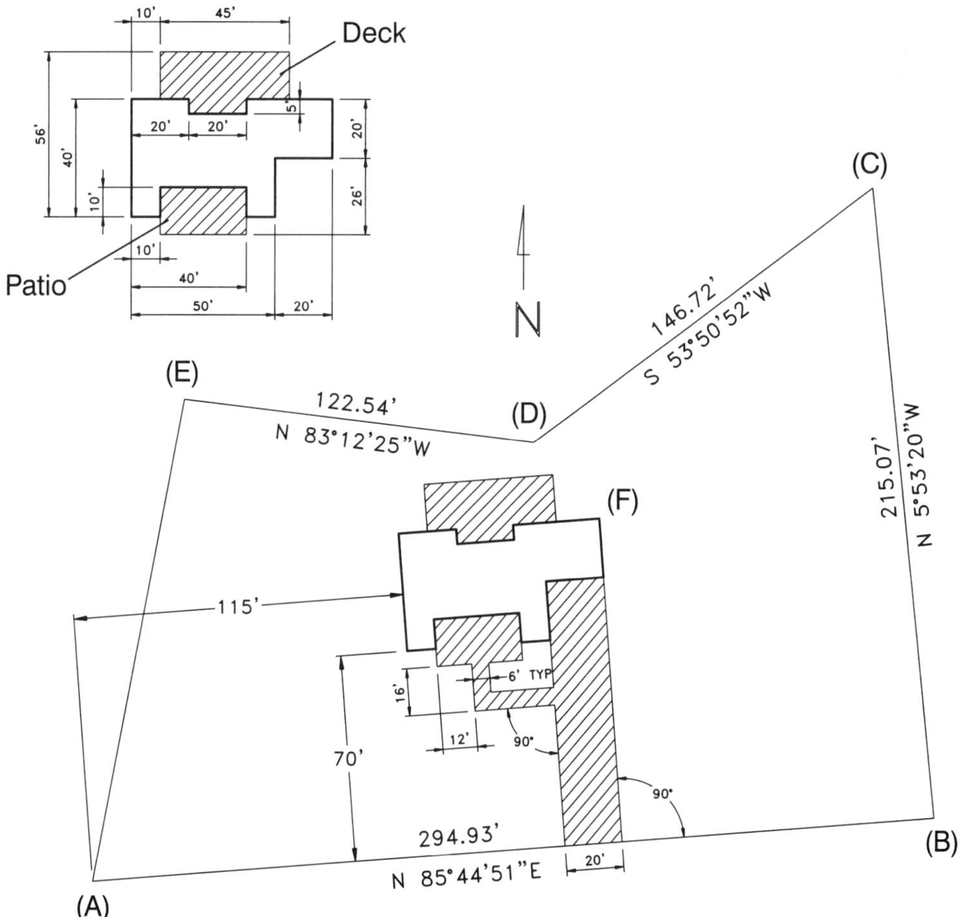

1. What is the correct direction of the line segment from the endpoint at "E" to the endpoint at "A"? (Pick the closest value.)
   (A) S 10d30'20" W
   (B) S 11d1'30" W
   (C) S 12d52'10" W
   (D) S 13d23'17" W
   (E) S 14d9'24" W
2. What is the total area of the plat with only the house removed? (Pick the closest value.)
   (A) 42082.00 sq. ft.
   (B) 42086.00 sq. ft.
   (C) 42090.00 sq. ft
   (D) 42094.00 sq. ft.
   (E) 42098.00 sq. ft.
3. What is the total area of the deck, patio, sidewalks, and driveway? (Pick the closest value.)
   (A) 3328.00 sq. ft.
   (B) 3332.00 sq. ft.
   (C) 3336.00 sq. ft.
   (D) 3340.00 sq. ft.
   (E) 3344.00 sq. ft.
4. What is the distance from the intersection of the corner of the house at "F" to the intersection of vertex "B"? (Pick the closest value.)
   (A) 154'-6.00"
   (B) 154'-9.00"
   (C) 155'-0.00"
   (D) 155'-3.00"
   (E) 155'-6.00"
5. What is the distance from the intersection of vertex "A" to the intersection of vertex "C"? (Pick the closest value.)
   (A) 352'-0.00
   (B) 356'-0.00
   (C) 360'-0.00
   (D) 364'-0.00
   (E) 368'-0.00
6. What is the angle formed in the XY plane from the intersection of vertex "E" to the intersection of vertex "B"? (Pick the closest value.)
   (A) S 53d55'9" E
   (B) S 55d49'7" E
   (C) S 57d27'5" E
   (D) S 59d10'4" E
   (E) S 61d35'2" E
7. An error has been discovered in the original plat. Vertex "C" needs to be stretched straight down a distance of 35.00'. What is the total area of the plat with the house, deck, patio, sidewalks, and driveway removed? (Pick the closest value.)
   (A) 36302.00 sq. ft.
   (B) 36306.00 sq. ft.
   (C) 36310.00 sq. ft.
   (D) 36314.00 sq. ft.
   (E) 36318.00 sq. ft.

PROBLEM 13–32 ARCHITECTURAL APPLICATION — FOUNDATION.DWG

Directions for Foundation.Dwg

Start a new drawing called Foundation. Change the default units from decimal to architectural. Do not add dimensions to this drawing. Unless otherwise noted, the thickness of all concrete block walls is 8" and the thickness of all footings is 16" (see Problem 13–32).

Refer to the drawing of Foundation, shown in Problem 13–32, to answer the questions on the following page.

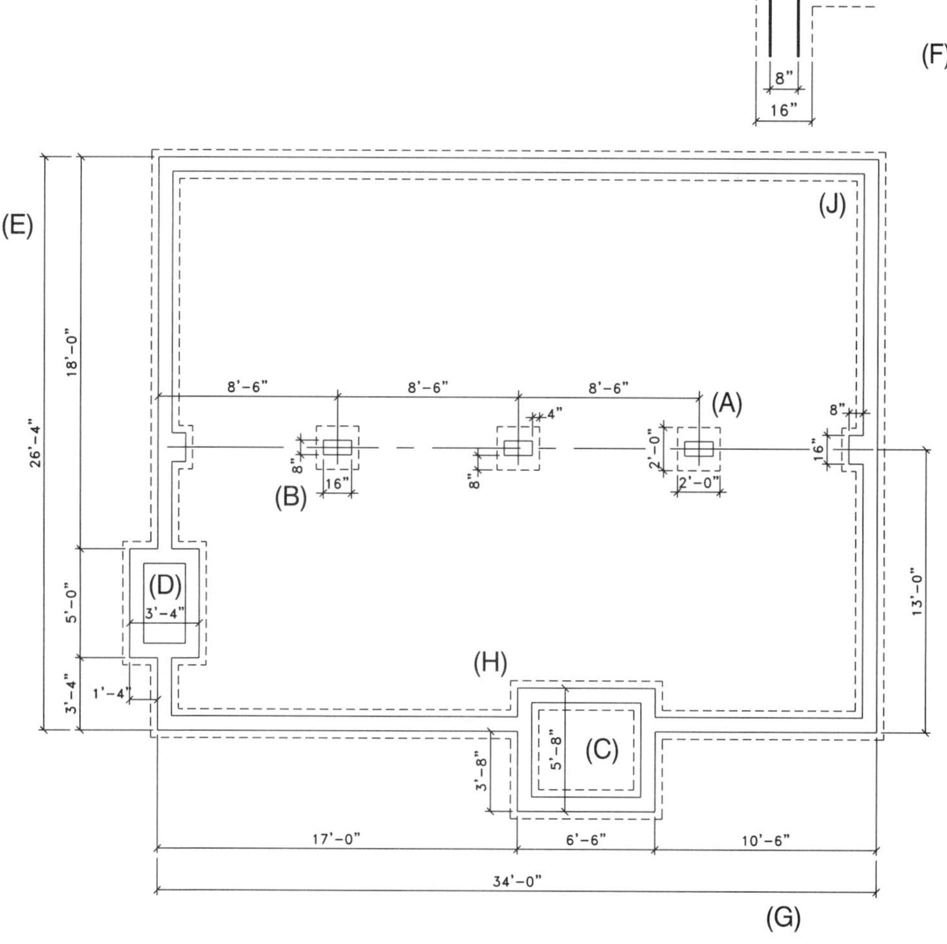

1. What is the total area of all concrete block regions including the three piers located in the center of the foundation plan? (Pick the closest value.)
   (A) 91 sq. ft.
   (B) 93 sq. ft.
   (C) 95 sq. ft.
   (D) 97 sq. ft.
   (E) 99 sq. ft.
2. What is the total distance from the intersection of the footing at "A" to the intersection of the footing at "B"? (Pick the closest value.)
   (B) 15 ft.
   (C) 17 ft.
   (D) 19 ft.
   (E) 21 ft.
3. What is the total area of all footing regions including the three footings that support the piers located in the center of the foundation plan? (Pick the closest value.)
   (A) 206 sq. ft.
   (B) 208 sq. ft.
   (C) 210 sq. ft.
   (D) 212 sq. ft.
   (E) 214 sq. ft.
4. What is the total distance from the middle of the concrete slab area at "C" to the middle of the fireplace area at "D"? (Pick the closest value.)
   (A) 15 ft.
   (B) 17 ft.
   (C) 19 ft.
   (D) 21 ft.
   (E) 23 ft.
5. What is the angle formed in the XY plane from the intersection of the footing at "A" to the intersection of the footing at "H"? (Pick the closest value.)
   (A) 228°
   (B) 230°
   (C) 232°
   (D) 234°
   (E) 236°
6. What is the delta Y distance from the intersection of the footing at "B" to the intersection of the footing at "J"? (Pick the closest value.)
   (A) 11'-4"
   (B) 11'-10"
   (C) 12'-4"
   (D) 12'-10"
   (E) 13'-4"
7. Increase the size of the foundation plan using the STRETCH command according to the following specifications:
   (A) Use a crossing box from "F" to "E" to stretch the foundation plan straight up at a distance of 2'-6".
   (B) Use a crossing box from "F" to "G" to stretch the foundation plan directly to the left at a distance of 18".
8. What is the new area of the concrete block region not including the three piers located at the middle of the foundation plan? (Pick the closest value.)
   (A) 94 sq. ft.
   (B) 96 sq. ft.
   (C) 98 sq. ft.
   (D) 100 sq. ft.
   (E) 102 sq. ft.

Advanced Level Drawings

PROBLEM 13–33 FLANGE1.DWG

Directions for Flange1.Dwg

Use the Drawing Units dialog box to set the units to decimal. Set the precision to two. Be sure the system of angle measure is set to decimal degrees and the number of decimal places for the display of angles is zero.

Keep the remaining default unit values.

Begin the drawing, shown in Problem 13–33, by locating the center of the 2.00-diameter circle at coordinate (6.00,5.50).

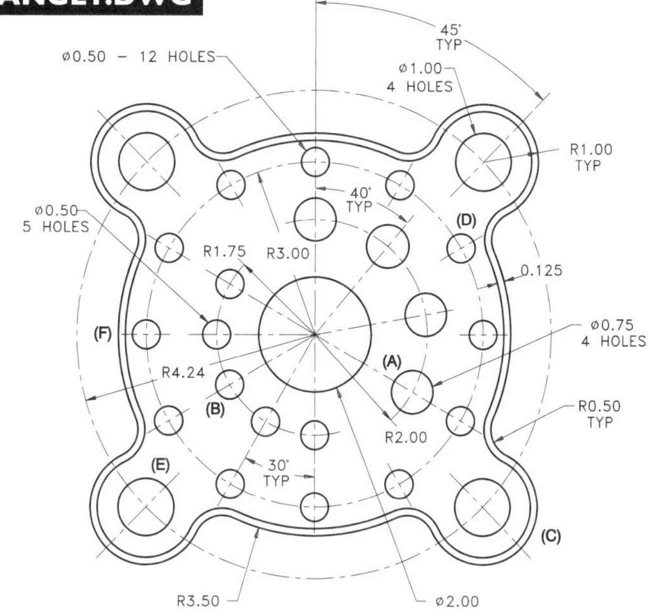

Refer to the drawing of Flange1, shown in Problem 13–33, to answer the following questions.

1. What is the total area of the 0.125 strip around the perimeter of Flange1?
   (A) 4.10 (D) 4.16
   (B) 4.12 (E) 4.18
   (C) 4.14
2. What is the absolute coordinate value of the center of the 0.75-diameter circle "A"?
   (A) 7.73,4.45 (D) 7.75,4.55
   (B) 7.73,4.50 (E) 7.75,4.60
   (C) 7.73,4.55
3. What is the absolute coordinate value of the center of the 0.50-diameter circle "B"?
   (A) 4.45,4.55 (D) 4.48,4.69
   (B) 4.48,4.55 (E) 4.50,4.56
   (C) 4.48,4.63
4. What is the length of the 1.00-radius arc "C"?
   (A) 3.82 (D) 3.91
   (B) 3.85 (E) 3.97
   (C) 3.88
5. What is the total surface area of the inner part of Flange1 with all holes removed?
   (A) 35.12 (D) 35.30
   (B) 35.18 (E) 35.36
   (C) 35.24
6. What is the total length of outer arc "F"?
   (A) 2.91 (D) 3.09
   (B) 2.97 (E) 3.15
   (C) 3.03
7. What is the angle formed in the XY plane from the center of the 0.50 hole "D" to the center of the 0.50 hole "B"?
   (A) 204° (D) 210°
   (B) 206° (E) 212°
   (C) 208°

PROBLEM 13–34 MECHANICAL APPLICATION — PULLEYS.DWG

Directions for Pulleys.Dwg

Change the decimal precision to two. Construct the image of the pulley, shown in Figure A. Use the UCS command and the Origin option to locate a new User Coordinate System at (5.00,5.00). Create a block of this object and call it Pulley. Insert Pulley "A" in the current drawing at insertion point (0.00,0.00) and at a scale of 0.50. Use Figure B as a guide for inserting the remainder of the pulleys at the following scales: Pulley "B" has a scale of 1.00; Pulley "C" has a scale of 0.75; Pulley "D" has a scale of 0.50; and Pulley "E" has a scale of 1.00. The dimensions in Figure B are given as ordinate or datum dimensions, which are all calculated from the center of Pulley "A" located at 0,0. Construct a continuous belt of thickness 0.30 around all pulleys, as shown in Figure B.

Refer to the drawing of Pulleys, shown in Figure B of Problem 13–34, to answer the questions on the following page.

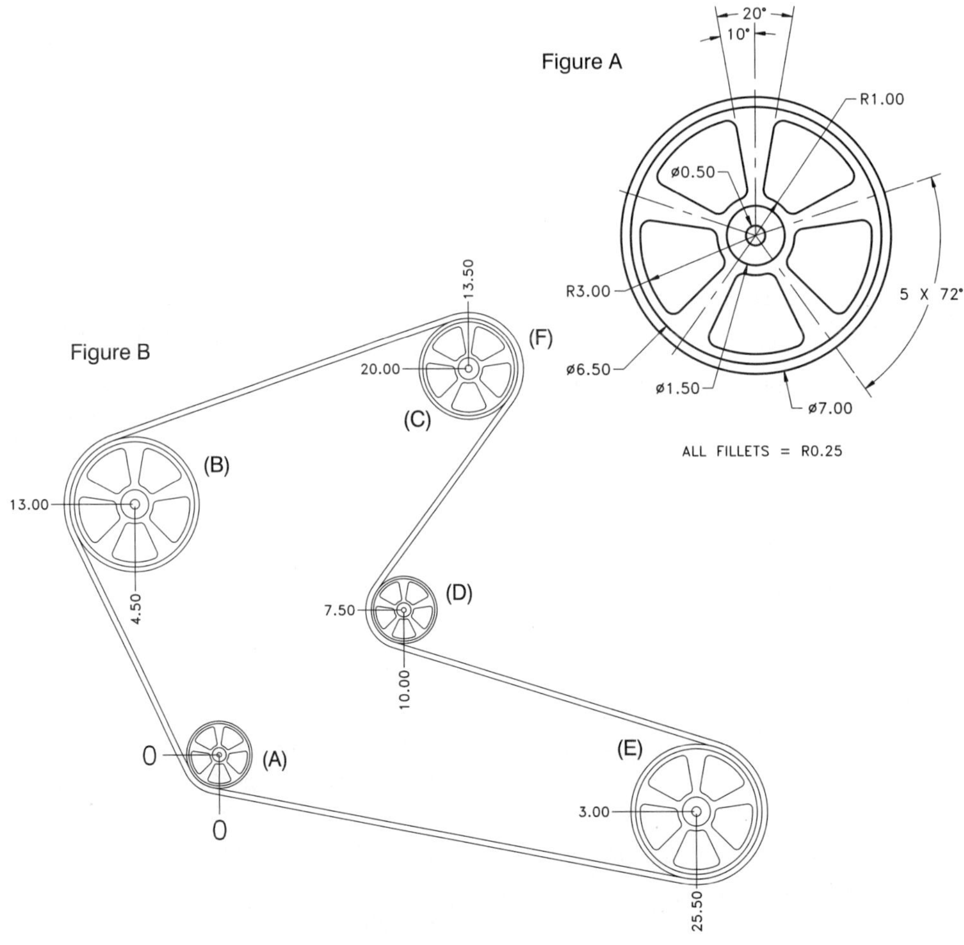

1. What is the angle in the XY plane from the center of Pulley "B" to the center of Pulley "E"?
   (A) 320°
   (B) 323°
   (C) 326°
   (D) 329°
   (E) 332°
2. What is the distance from the quadrant of the belt at "F" to the center of Pulley "D"?
   (A) 14.00
   (B) 14.05
   (C) 14.10
   (D) 14.15
   (E) 14.20
3. What is the total area of the belt?
   (A) 35.38
   (B) 35.43
   (C) 35.48
   (D) 35.53
   (E) 35.58
4. What is the total area of the shape inside of Pulley "B" at "G"?
   (A) 3.38
   (B) 3.43
   (C) 3.48
   (D) 3.53
   (E) 3.58
5. What is the angle in the XY plane from the center of Pulley "E" to the center of Pulley "A"?
   (A) 167°
   (B) 170°
   (C) 173°
   (D) 176°
   (E) 179°
6. What is the distance from the center of Pulley "A" to the center of Pulley "D"?
   (A) 12.45
   (B) 12.50
   (C) 12.55
   (D) 12.60
   (E) 12.65
7. Move Pulley "E" 5.25 units to the right and reconstruct the belt. What is the new area of the belt? (Pick the closest value.)
   (A) 38.56
   (B) 38.61
   (C) 38.66
   (D) 38.71
   (E) 38.76
8. What is the length of the belt? (Use the midpoint of the belt as the basis for the length) _________

PROBLEM 13–35 ROTOR2.DWG

Directions for Rotor2.Dwg

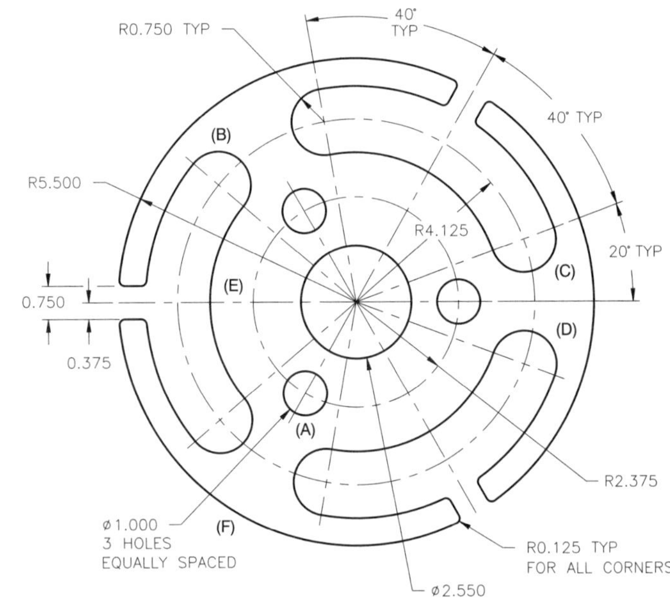

Start a new drawing called Rotor2 (See Problem 13–35). Keep the default settings of decimal units, precision to three. Be sure the system of angle measure is set to decimal degrees and the number of decimal places for the display of angles is zero. Keep all remaining default unit values.

Begin by constructing the 2.550-diameter circle at absolute coordinate (11.125,9.225).

Refer to the drawing of Rotor2, shown in Problem 13–27, to answer the following questions.

1. What is the absolute coordinate value of the center of hole "A"?
   (A) 9.937,7.158 (D) 9.942,7.168
   (B) 9.937,7.163 (E) 9.947,7.173
   (C) 9.937,7.168
2. What is the perimeter of Rotor2?
   (A) 81.850 (D) 81.865
   (B) 81.855 (E) 81.870
   (C) 81.860
3. What is the distance from the center of arc "B" to the center of arc "C"?
   (A) 7.125 (D) 7.140
   (B) 7.130 (E) 7.145
   (C) 7.135
4. What is the absolute coordinate value of the center of arc "D"?
   (A) 15.001,7.814 (D) 15.006,7.829
   (B) 15.001,7.819 (E) 15.011,7.834
   (C) 15.001,7.824
5. What is the total length of arc "F"?
   (A) 10.474 (D) 10.489
   (B) 10.479 (E) 10.494
   (C) 10.484
6. What is the total area of Rotor2 with all four holes removed?
   (A) 54.902 (D) 54.917
   (B) 54.907 (E) 54.922
   (C) 54.912
7. Change the diameter of all three 1.000-diameter holes to 0.700 diameter. Change the diameter of the 2.550 hole to a new diameter of 1.625. What is the new area of Rotor2, with all four holes removed?
   (A) 59.122 (D) 59.137
   (B) 59.127 (E) 59.142
   (C) 59.132

PROBLEM 13-36 LAND_DEVELOPMENT.DWG

Directions for Land_development

Create the drawing in Land development (See Problem 13-36). Set units to engineering and angular units to surveyors. Do not change the angle direction or location for this drawing. Typical line input will follow this style in the line command, @544.75'<N16d48'13"E This input reads; the line will be 544.75' at a direction of North 16 degrees 48 minutes and 13 seconds East.

Refer to the drawing of Land development in Problem 13–36, to answer the questions on the next page.

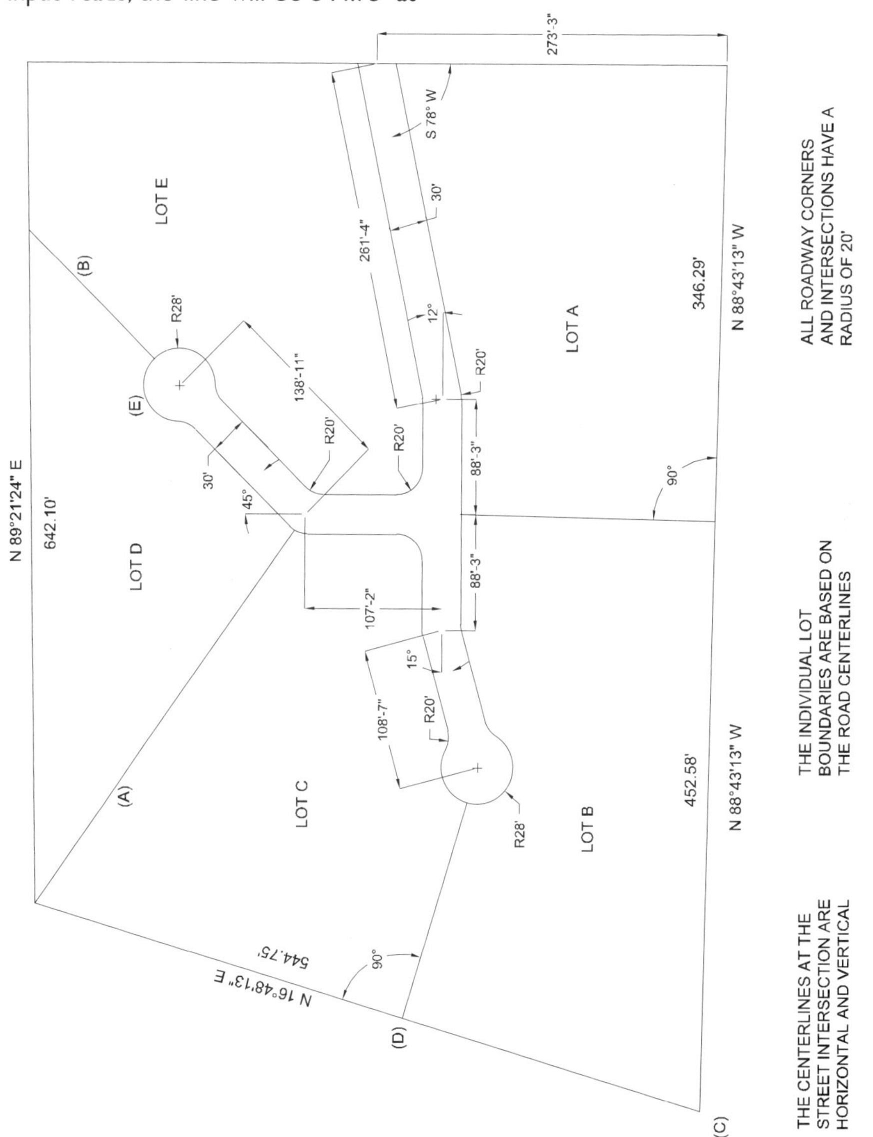

1. Which Lot has the most acreage?
   (A) Lot A
   (B) Lot B
   (C) Lot C
   (D) Lot D
   (E) Lot E
2. What is the Distance between the Endpoint of "C" and the Endpoint of "D"?
   (A) 239'1
   (B) 227'9
   (C) 240'6
   (D) 253'6
   (E) 261'7
3. What is the length of lot line "A"?
   (A) 352'2"
   (B) 351'6"
   (C) 357'9"
   (D) 356'8"
   (E) 359'4"
4. If 1 acre = 43560 square feet then how many acres is lot "C"?
   (A) 1.88 acres
   (B) 2.34 acres
   (C) 1.52 acres
   (D) 1.73 acres
   (E) 1.02 acres
5. What is the total length of arc "E"?
   (A)130'2"
   (B) 134'8"
   (C) 28'
   (D) 144'3"
   (E) 127'5"
6. What is the angle from Line "B" and the top line of the parcel?
   (A) 41
   (B) 43
   (C) 40
   (D) 42
   (E) 44
7. What is the percentage of land that the road takes up in this 5 parcel development?
   (A) 7.0%
   (B) 9.7%
   (C) 6.5%
   (D) 3.1%
   (E) 5.4%
8. What is the total square footage of the lots (this does not include the streets)?
   (A) 358377
   (B) 358513
   (C) 358446
   (D) 358239
   (E) 358733
9. How many lineal feet of curbs will need to be constructed in this plat (select closest distance)?
   (A) 1685
   (B) 1692
   (C) 1700
   (D) 1712
   (E) 1722

Section Views

The exercises in this chapter introduce:

- Using the various section views:
  - Full
  - Half
  - Offset
  - Revolved
  - Removed
  - Broken out
  - Aligned
- Using the hatch tools as section lines

TUTORIAL EXERCISE: 14_ASSEMBLY.DWG

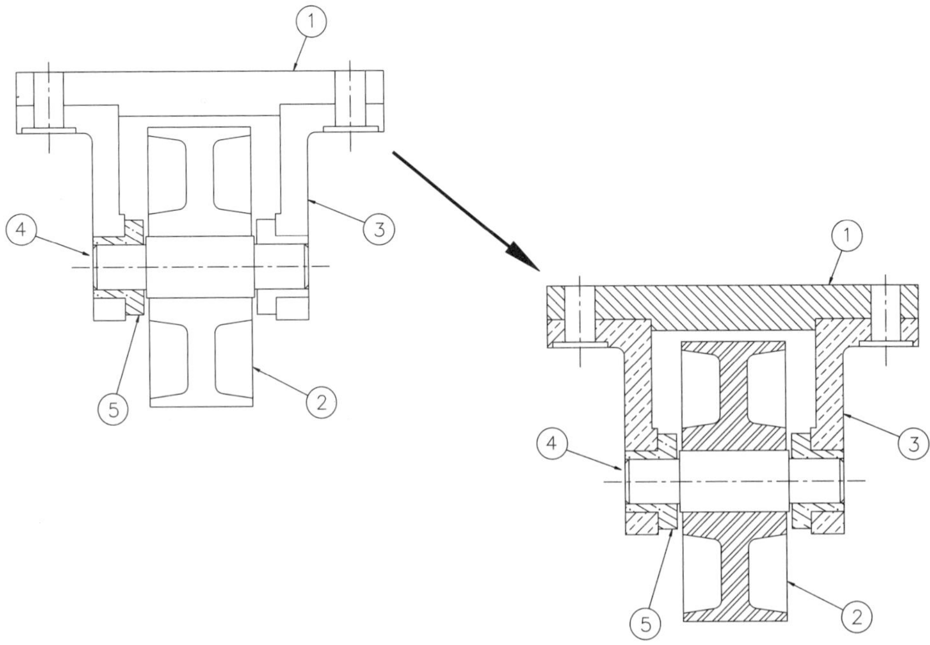

Figure 14–1

Purpose

This tutorial is designed to use the BHATCH command to hatch an assembly drawing.

System Settings

This drawing is provided on the CD, you will edit the existing drawing named *14_Assembly.dwg*. Follow the steps in this tutorial for converting the object to an assembly section. All Units, Limits, Grid, and Snap values have been previously set.

Layers

Layers have already been created for this tutorial exercise.

Name	Color	Linetype
Center	Yellow	Center
Leader	Cyan	Continuous
Object	White	Continuous
Section	Magenta	Continuous

Suggested Commands

The BHATCH command will be used exclusively during this tutorial exercise.

Whenever possible, substitute the appropriate command alias in place of the full AutoCAD command in each tutorial step. For example, use "CP" for the COPY command, "L" for the LINE command, and so on. The complete listing of all command aliases is located in Chapter 1, Table 1–2.

STEP 1

Before beginning the hatching operations, use the Layer Control box, shown in Figure 14–2, to ensure that the SECTION layer is the current layer, and then turn off the layers called LEADER and CENTER. The LEADER layer holds all bubbles identifying numbers and leaders pointing to the specific parts. The CENTER layer will turn off all centerlines in the drawing.

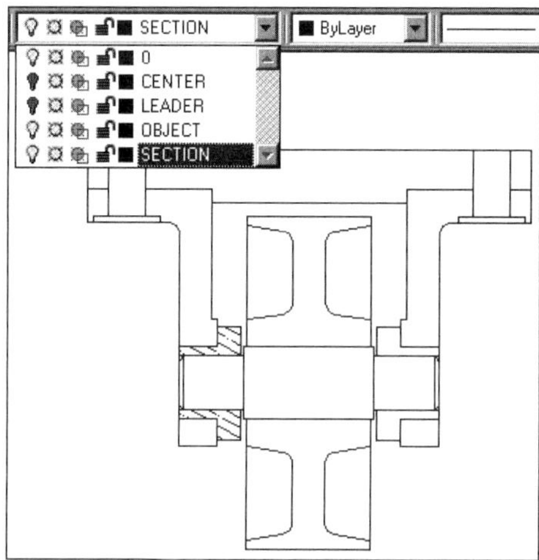

Figure 14–2

STEP 2

Issue the BHATCH command and begin crosshatching the assembly; first crosshatch Part 1, the Plate. Use the default hatch pattern of ANSI31 in addition to a scale factor of 1.0000. Change the angle to 90°, as shown in Figure 14–3. Pick three internal points "A," "B," and "C" to identify the areas to hatch (see Figure 14–4).

Command: **BH** *(For BHATCH)* [draw>boundary]

(The Boundary Hatch and Fill dialog box appears.)

(Make changes to match Figure 14–3. When finished, click on the Pick Points button.)

Select internal point: *(Pick a point at "A")*

Selecting everything... Selecting everything visible... Analyzing the selected data... Analyzing internal islands... Select internal point: *(Pick a point at "B")*

Analyzing internal islands... Select internal point: *(Pick a point at "C")*

Analyzing internal islands... Select internal point: *(Press* ENTER *to exit this area and return to the Boundary Hatch and Fill dialog box)*

When the Boundary Hatch and Fill dialog box reappears, click on the Preview button to view the results. If the preview does not appear correct, press ESC to return to the Boundary Hatch and Fill dialog box to make changes. When the hatch results are acceptable, click on the OK button to place the hatch pattern.

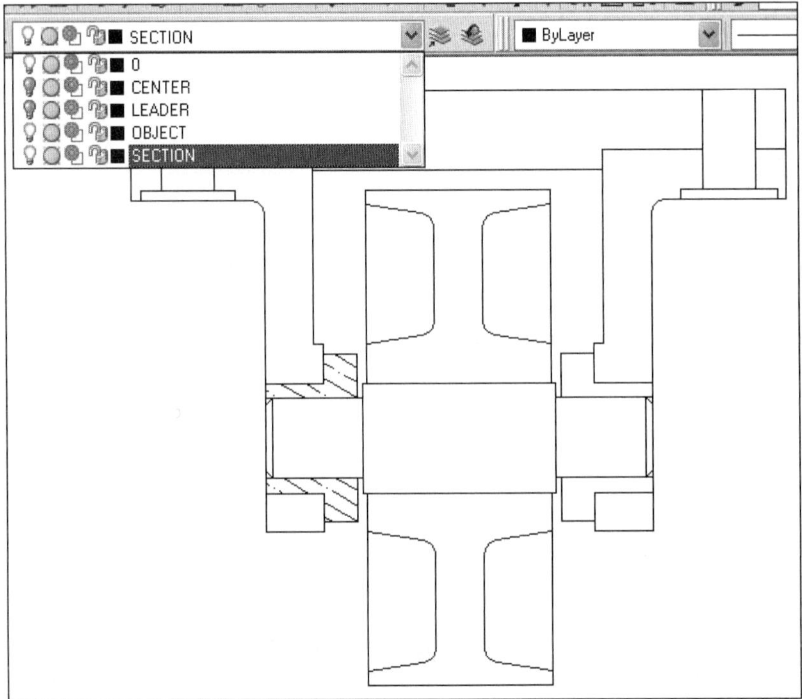

Figure 14–3

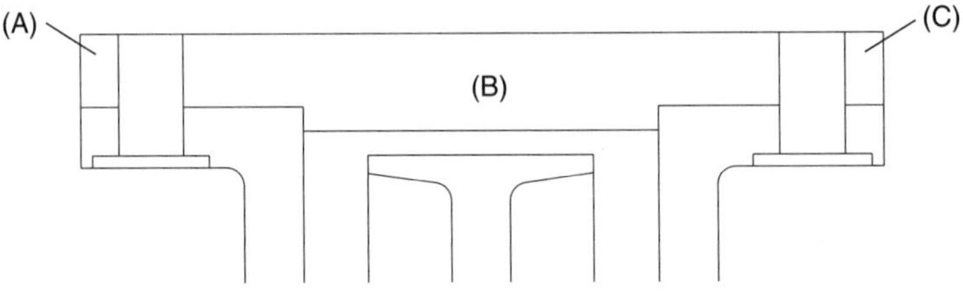

Figure 14–4

STEP 3

Next, crosshatch the pulley in the assembly drawing. Execute the BHATCH command; when the Boundary Hatch and Fill dialog box appears, choose the pattern "ANSI32" and change the angle to 0°, as shown in Figure 14–5. Click on two internal points inside the pulley at "A" and "B," shown in Figure 14–6. If the proper boundaries highlight, press ENTER to return to the Boundary Hatch and Fill dialog box.

Command: **BH** *(For BHATCH)*[draw>boundary]

(The Boundary Hatch and Fil dialog box appears. Make changes to match Figure 14–5. When finished, click on the Pick Points button.)

Select internal point: *(Pick a point at "A")*

Selecting everything...

Selecting everything visible...

Analyzing the selected data...

Analyzing internal islands... Select internal point: *(Pick a point at "B")*

Analyzing internal islands... Select internal point: *(Press* ENTER *to exit this area and return to the Boundary Hatch and Fill dialog box)*

Click on the Preview button to preview the hatch pattern; if the results are desirable, press ESC to return to the Boundary Hatch and Fill dialog box and click the OK button to place the hatch pattern.

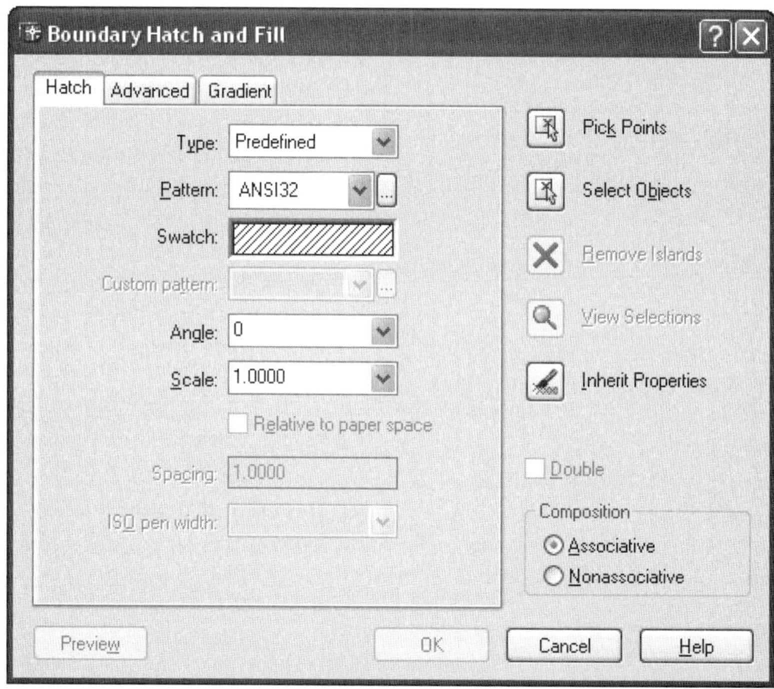

Figure 14–5

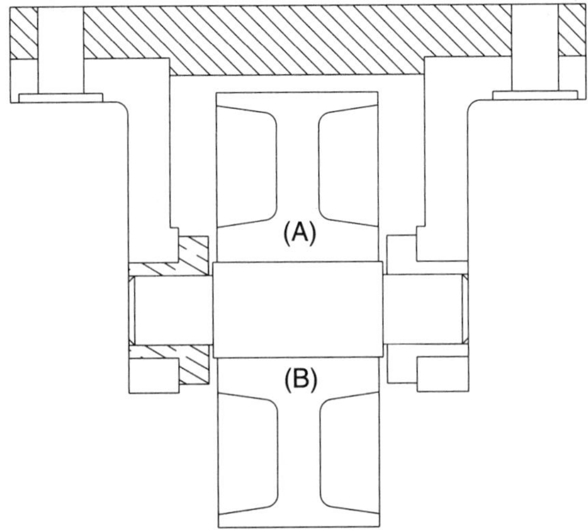

Figure 14–6

STEP 4

Next, hatch the two brackets that support the pulley axle and are connected to the plate. Execute the BHATCH command and select the "ANSI33" hatch pattern; keep all other defaults, as shown in Figure 14–7. When prompted to select internal points, click inside the areas identified by "A" through "F," as shown in Figure 14–8.

Command: **BH** *(For BHATCH)*
[draw>boundary]

(The Boundary Hatch and Fill dialog box appears. Make changes to match Figure 14–7. When finished, click on the Pick Points button.)

Select internal point: *(Pick a point at "A")*
Selecting everything... Selecting everything visible... Analyzing the selected data... Analyzing internal islands... Select internal point: *(Pick a point at "B")*
Analyzing internal islands... Select internal point: *(Pick a point at "C")*
Analyzing internal islands... Select internal point: *(Pick a point at "D")*
Analyzing internal islands... Select internal point: *(Pick a point at "E")*
Analyzing internal islands... Select internal point: *(Pick a point at "F")*
Analyzing internal islands... Select internal point: *(Press ENTER to exit this area and return to the Boundary Hatch and Fill dialog box)*

Click on the Preview button to preview the hatch pattern; if the results are desirable, press ESC to return to the Boundary Hatch and Fill dialog box and click the OK button to place the hatch pattern.

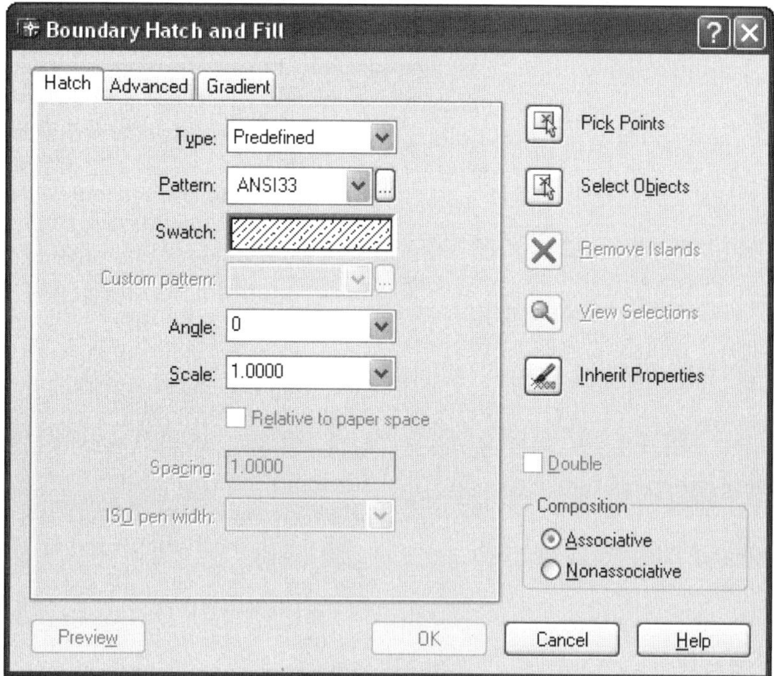

Figure 14–7

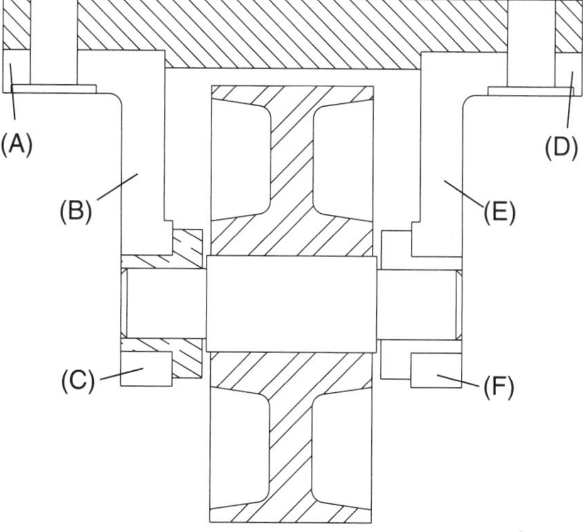

Figure 14–8

STEP 5

Notice that, although one of the bushings has already been hatched, it is difficult to determine the hatch pattern used along with scale and angle. Issue the BHATCH command, click on the Inherit Properties button, shown in Figure 14–9, and select the existing hatch pattern at "A", as shown in Figure 14–10, using the Property paintbrush. This will transfer the hatch properties of the bushing to the Boundary Hatch and Fill dialog box. Pick internal points at "A" and "B," and the object will be hatched to these new current parameters.

Command: **BH** *(For BHATCH)* [draw>boundary]

(When the Boundary Hatch and Fill dialog box appears, click on the Inherit Properties button.)

Select associative hatch object: *(Select the hatch pattern already visible in the bushing at "A," shown in Figure 14–10. Press* ENTER *when finished. When the Boundary Hatch and Fill dialog box reappears, click on the Pick Points < button.)*

Inherited Properties: Name <ANSI35>, Scale <1.0000>, Angle <105> Select internal point: *(Pick a point at "B")*

Select internal point: Selecting everything...

Selecting everything visible...

Analyzing the selected data...

Analyzing internal islands...

Select internal point: *(Pick a point at "C")*

Analyzing internal islands...

Select internal point: *(Press* ENTER *to return to the Boundary Hatch and Fill dialog box)*

Click on the Preview button to preview the hatch pattern; if the results are desirable, press ESC to return to the Boundary Hatch and Fill dialog box and click the OK button to place the hatch pattern.

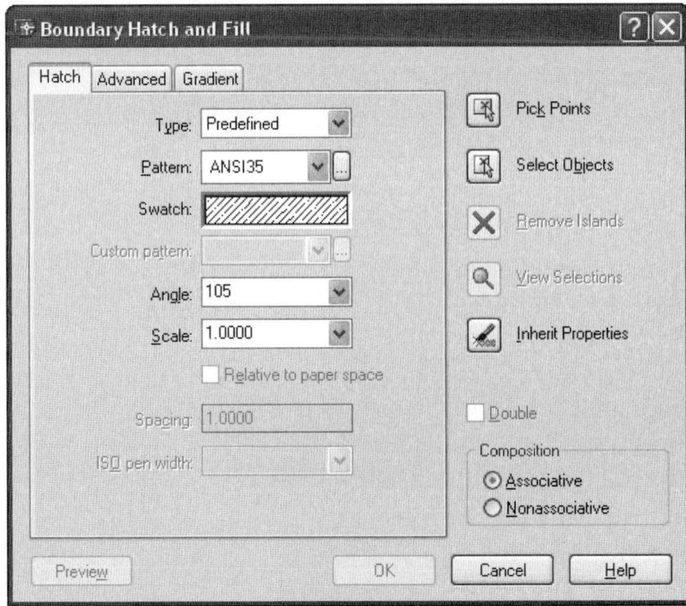

Figure 14–9

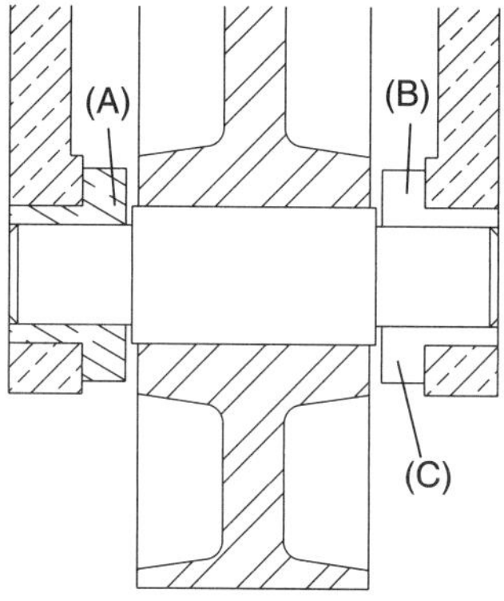

Figure 14–10

STEP 6

The hatch pattern currently displayed in the pulley is too large for the area it occupies; the spacing needs to be scaled to half. Issue the HATCHEDIT command and select the hatch pattern anywhere in the pulley. When the Boundary Hatch amd Fill dialog box appears, change the scale from a value of 1.0000 to 0.5000 units, as shown in Figure 14–11. Click the OK button to update the hatch pattern. The results are illustrated in Figure 14–12.

Command: **HE** *(For HATCHEDIT)* [modify>object>hatch] Select hatch object: *(Select the hatch pattern inside of the pulley)*

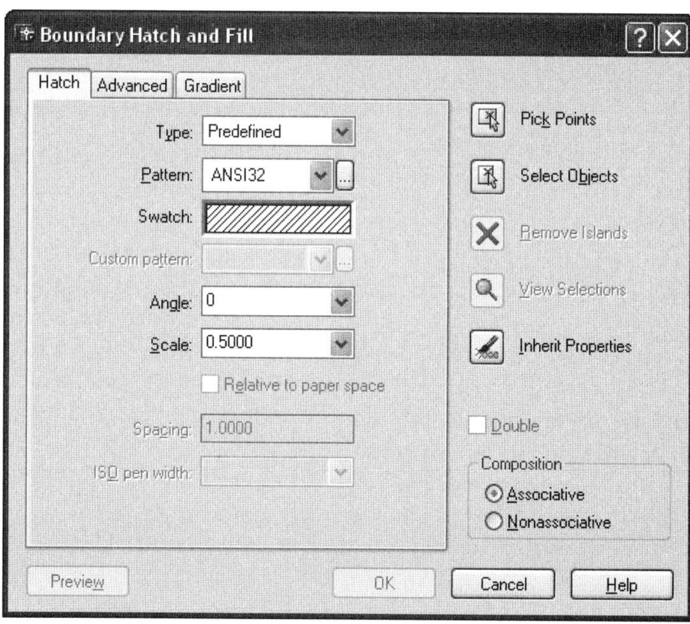

Figure 14–11

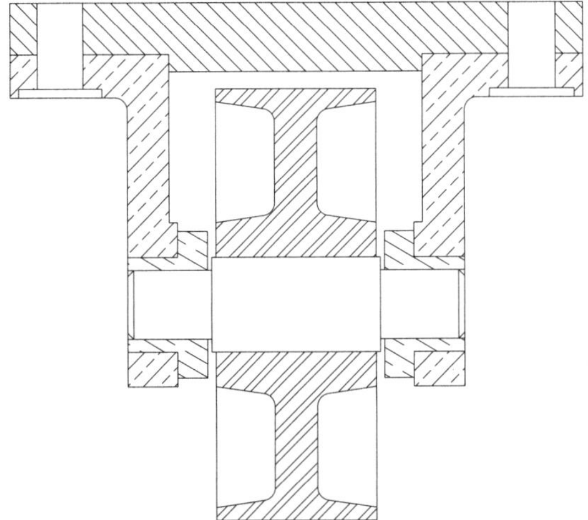

Figure 14–12

STEP 7

Stretch the base to make it 0.25 units longer on both sides. With Ortho mode turned on, use the Direct Distance mode to accomplish this task. When selecting the objects to stretch, take care not to accidentally select the edge of the counterbore hole. Use the ZOOM command to magnify this part of the screen. See Figure 14–13.

 Command: **S** *(For STRETCH)* [modify>stretch]

Select objects to stretch by crossing window or crossing-polygon...
Select objects: *(Pick a point at "A")*
Specify opposite corner: *(Pick a point at "B")*
Select objects: *(Press* ENTER *to continue)*
Specify base point or displacement: *(Pick a blank part of the display screen at "C")*
Specify second point of displacement: *(With Ortho on, move your cursor directly to the left and type a value of* **0.25***)*

 Command: **S** *(For STRETCH)* [modify>stretch]

Select objects to stretch by crossing window or crossing-polygon...
Select objects: *(Pick a point at "D")*
Specify opposite corner: *(Pick a point at "E")*
Select objects: *(Press* ENTER *to continue)*
Specify base point or displacement: *(Pick a blank part of the display screen at "F")*
Specify second point of displacement: *(With Ortho on, move your cursor directly to the right and enter a value of* **0.25***)*

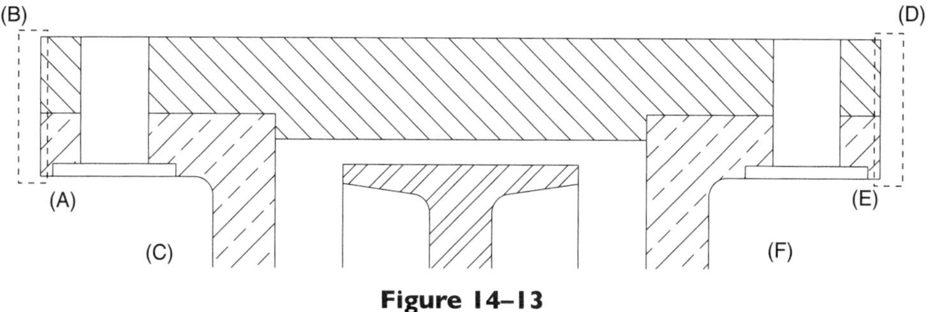

Figure 14–13

STEP 8

Turn all layers back on to complete this hatching exercise (see Figure 14–14).

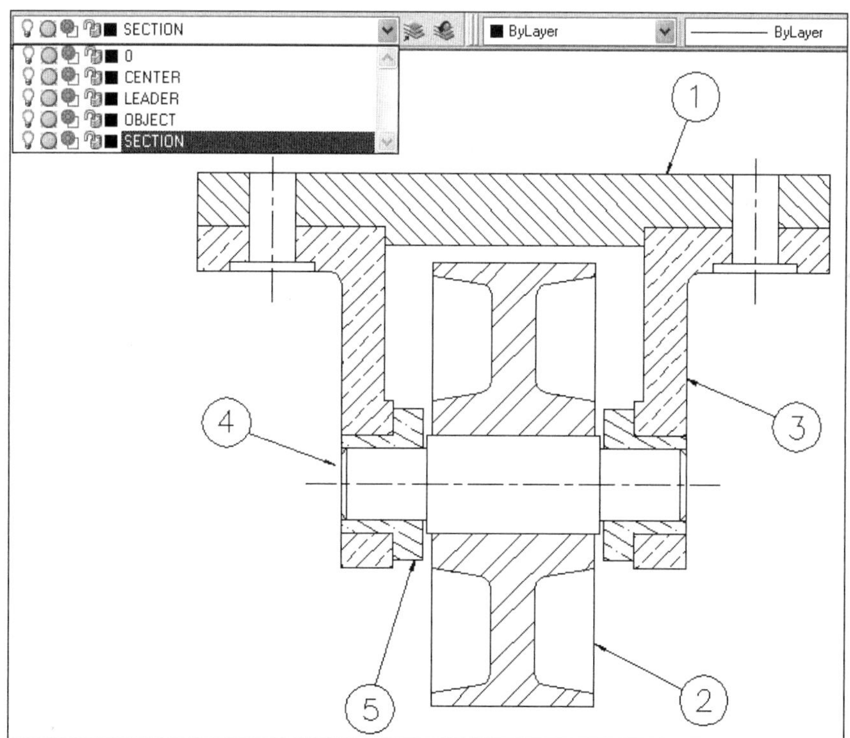

Figure 14–14

Beginning Level Drawings

PROBLEM 14–1

Center three-view drawing and make the Front view a full section.

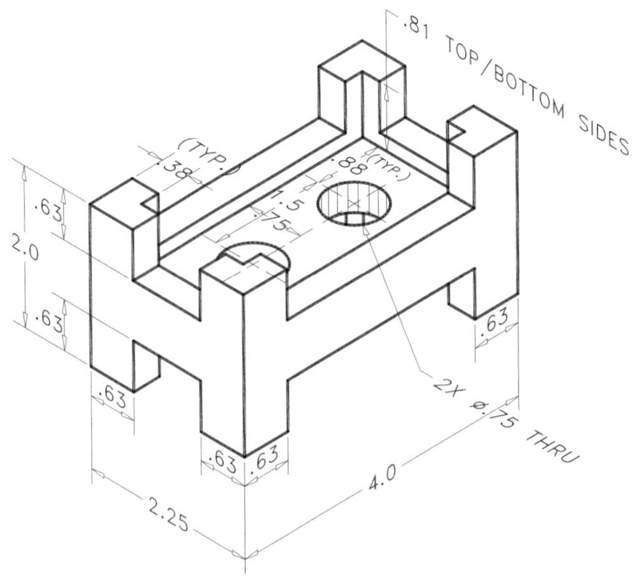

PROBLEM 14–2

Center two views within the work area, and make one view a full section. Use correct drafting practices for the ribs.

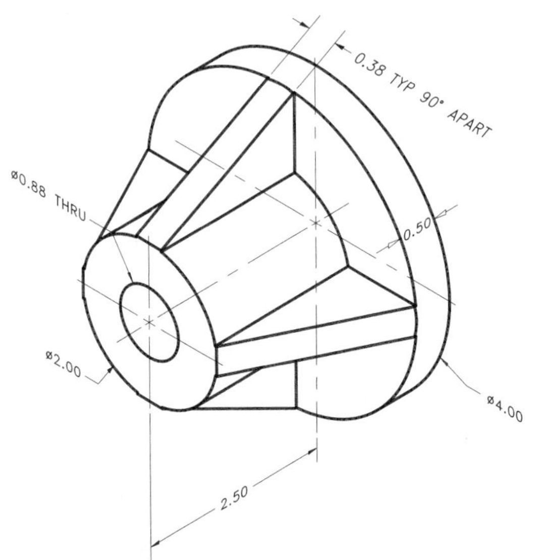

PROBLEM 14–3

Center two views within the work area, and make one view a full section.

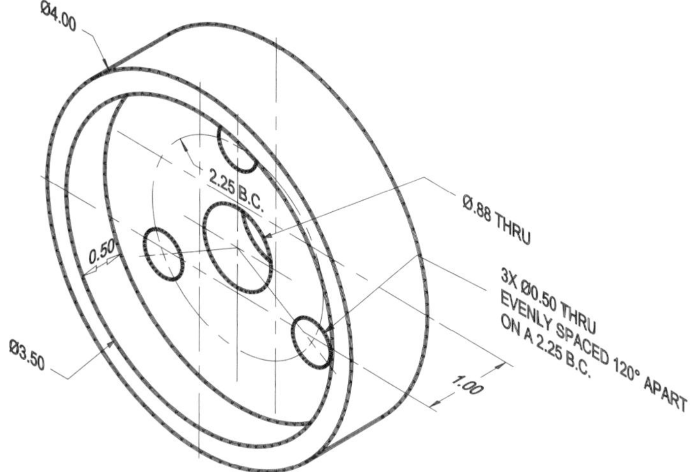

PROBLEM 14–4

Center the Front view and Top views within the work area. Make one view a full section.

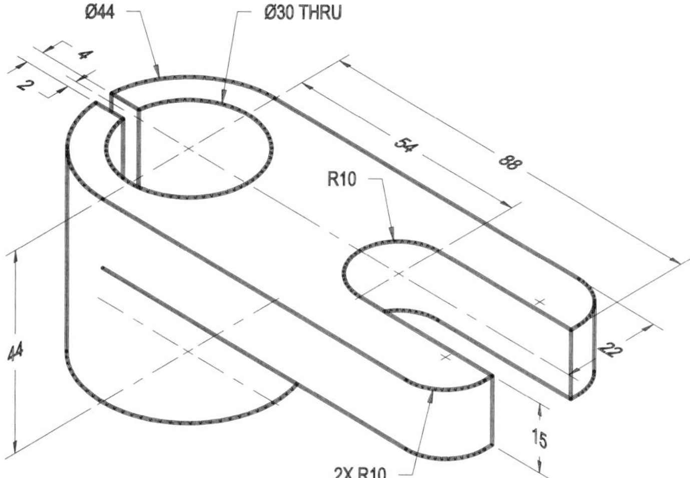

PROBLEM 14–5

Center the Front view and Top view within the work area. Make one view a half section.

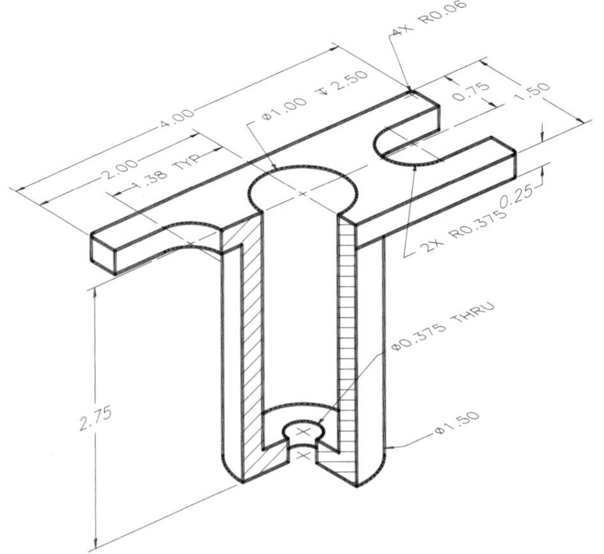

PROBLEM 14–6

Center two views within the work area, and make one view a half section.

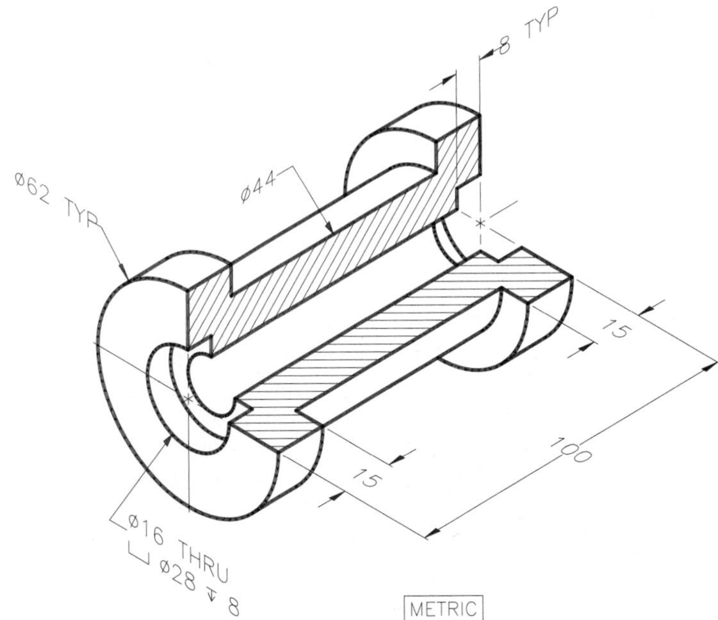

PROBLEM 14–7

Center the two views within the work area, and make one view a half section.

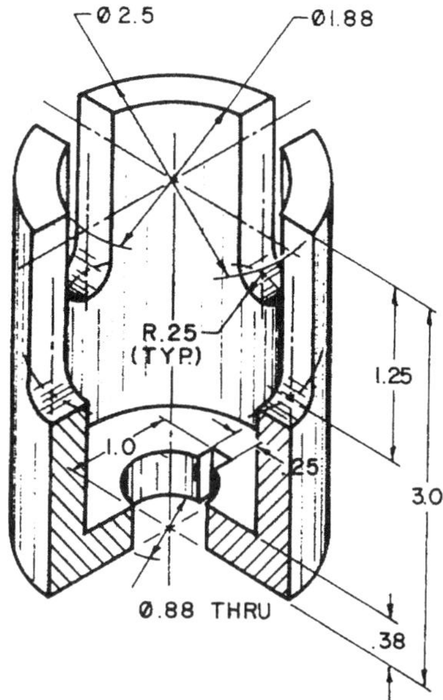

PROBLEM 14–8

Center two views within the work area, and make one view a half section.

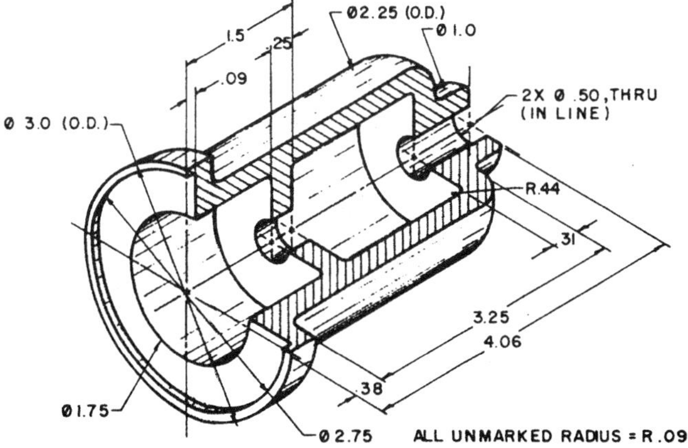

PROBLEM 14–9

Create the grill placement in a single view as shown in the drawing.

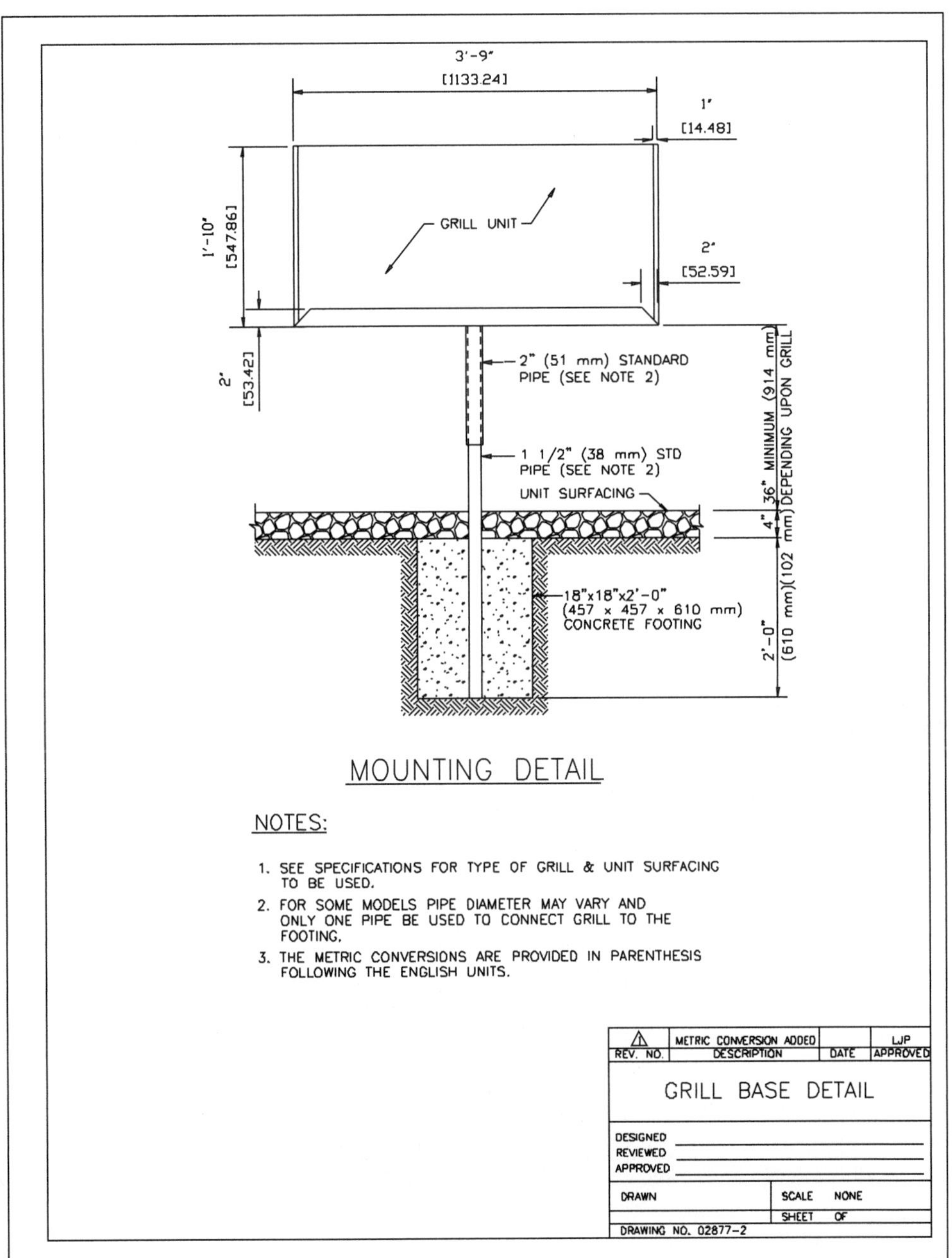

PROBLEM 14–10

Create the trench in a single view as shown in the drawing; apply the hatch patterns as necessary.

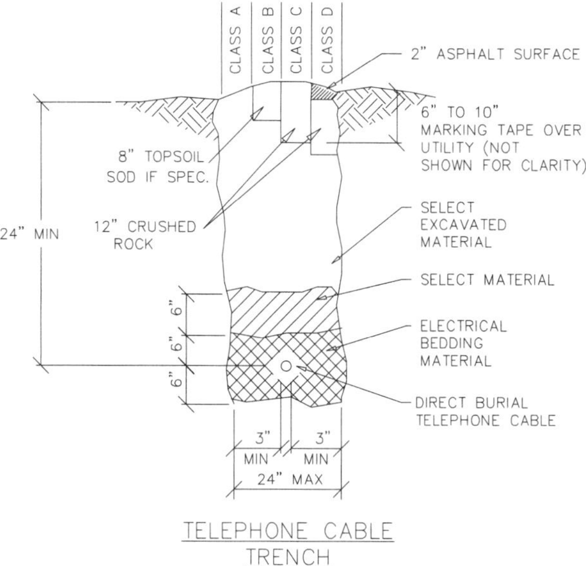

Intermediate Level Drawings

PROBLEM 14–11

Center the required views within the work area, and add removed section A-A.

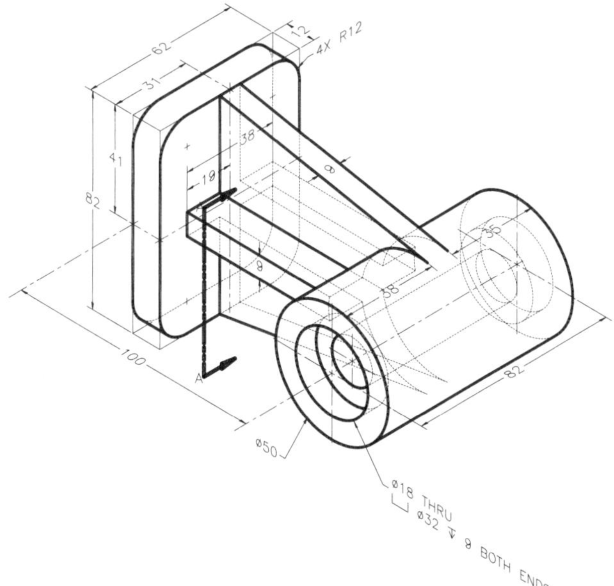

PROBLEM 14–12

Center three views within the work area, and make one view an offset section.

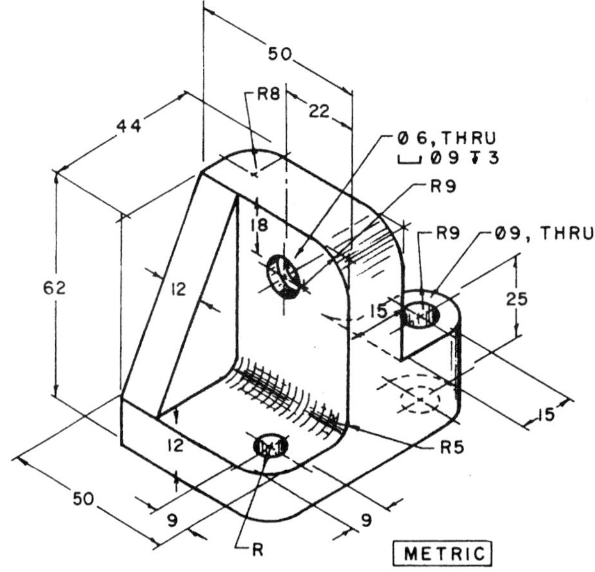

PROBLEM 14–13

Center three views within the work area, and make one view an offset section.

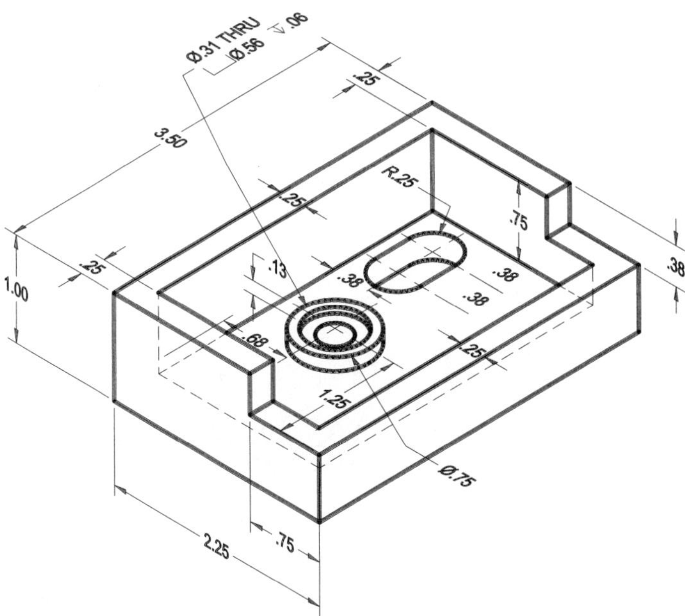

PROBLEM 14–14

Center three views within the work area, and make one view an offset section.

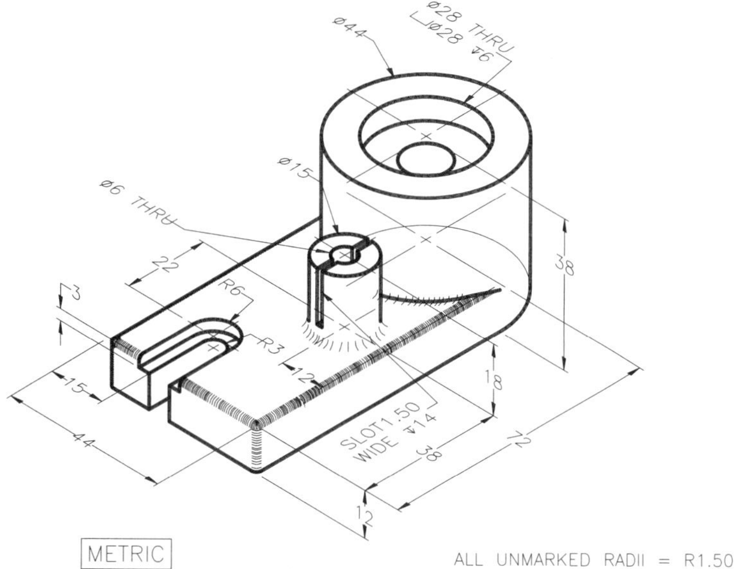

PROBLEM 14–15

Center two views within the work area, and make one view a half section.

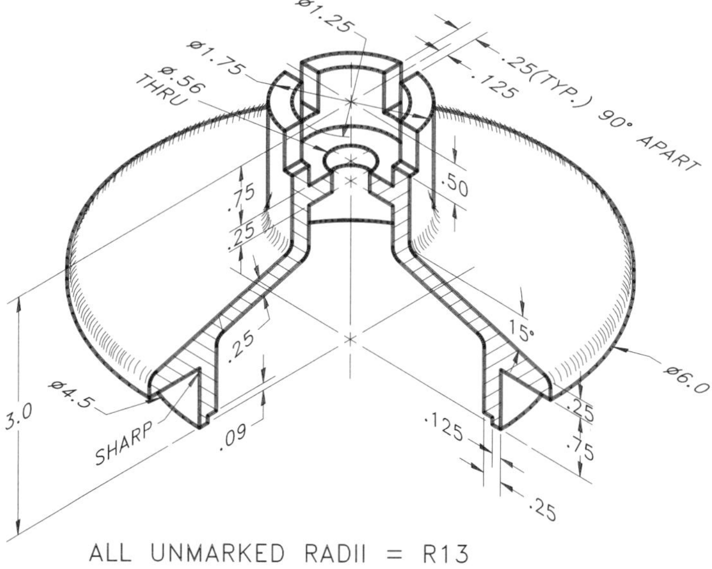

PROBLEM 14–16

Center the required views within the work area, and make one view a broken-out section to illustrate the complicated interior area.

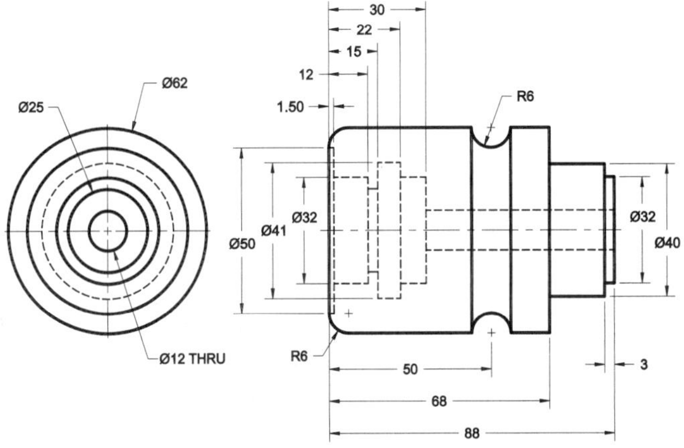

PROBLEM 14–17

Center the required views within the work area, and add a full removed section A-A.

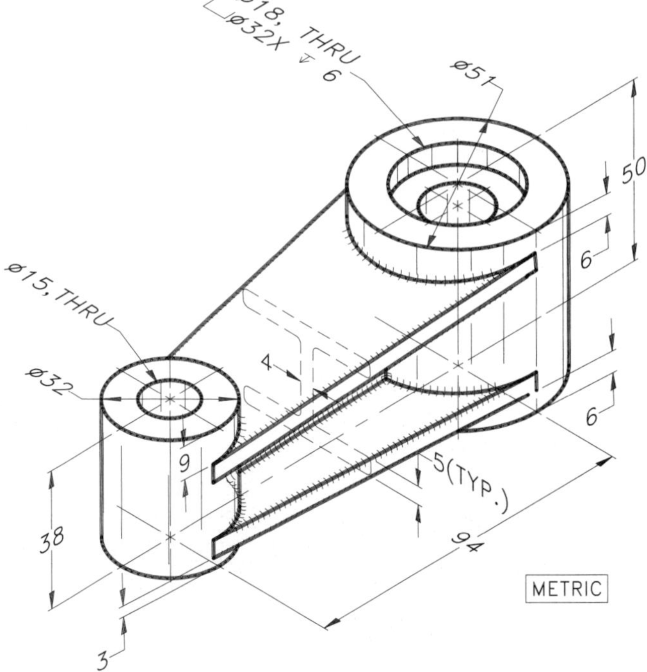

Advanced Level Drawings

PROBLEM 14–18

Center the required views within the work area, and add a full removed section A-A.

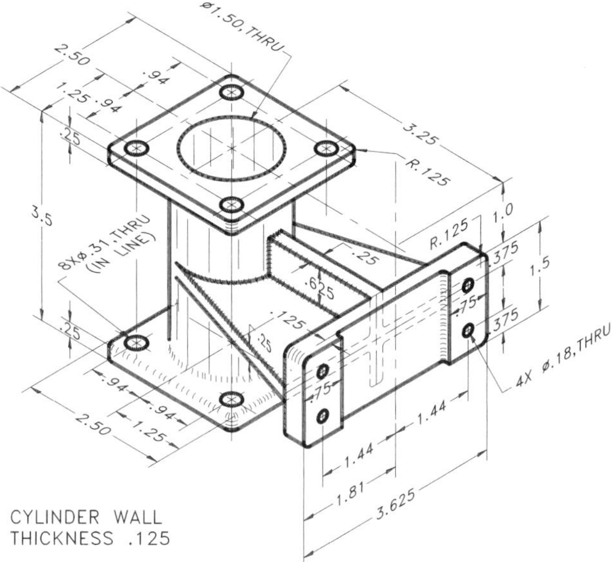

ALL UNMARKED RADII = R.06

PROBLEM 14–19

Center three views within the work area, and make one view an offset section.

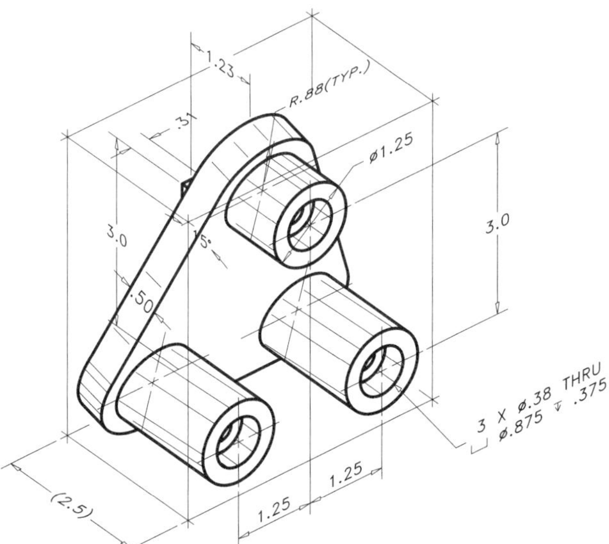

Directions for Problems 14–20 through 14–22:
Center required views within the work area. Leave a 1-inch or 25 mm space between views. Make one view a section view to fully illustrate the object. Use a full half, offset, broken-out, revolved, or removed section. Consult your instructor if you need to add dimensions.

PROBLEM 14–20

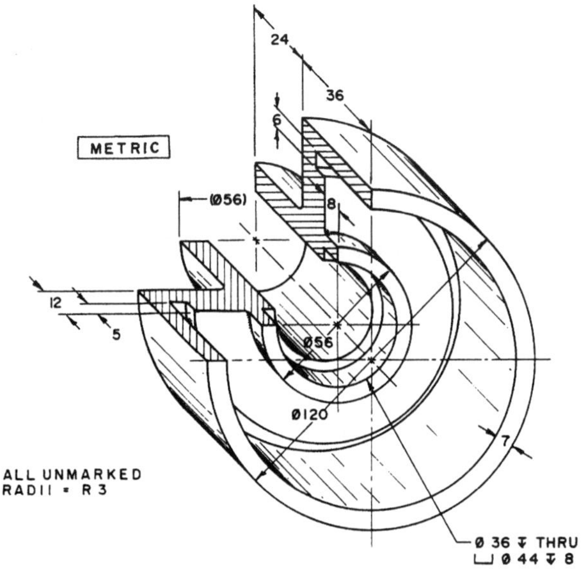

PROBLEM 14–21

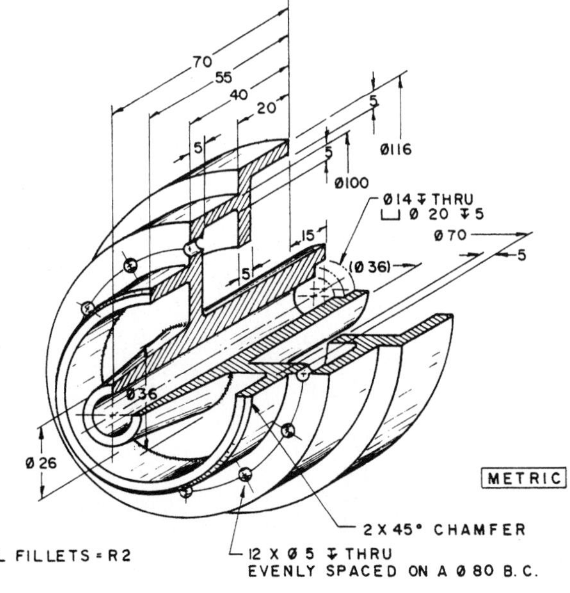

PROBLEM 14–22

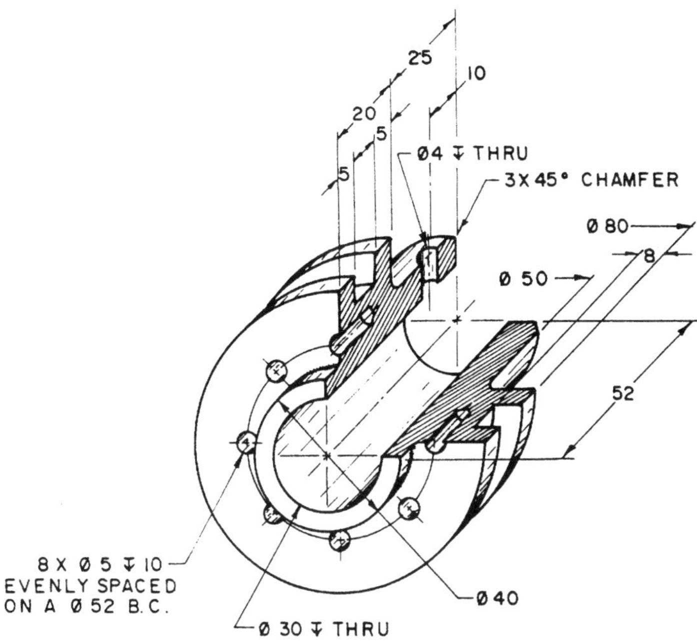

PROBLEM 14–23

Create the manhole section using the cover detail and the section, use appropriate hatch patterns in the drawing.

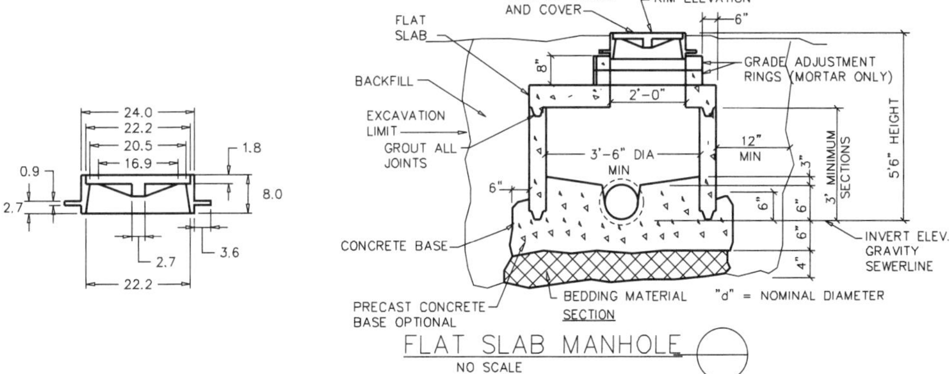

PROBLEM 14–24

Create the typical residential wall section using the figure shown. Create the drawing from measuring the drawing on the page. Scale the drawing at ¾"=1'

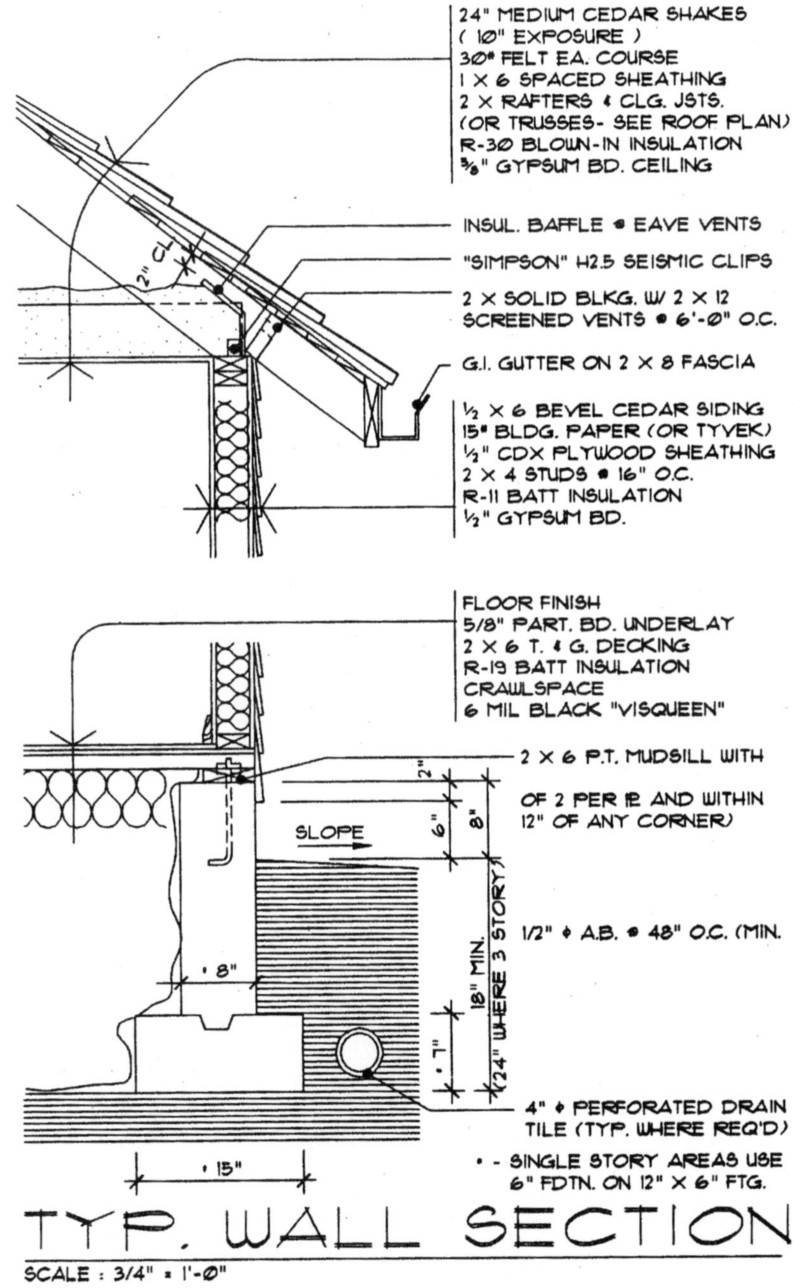

Auxiliary Views

The exercises in this chapter introduce:

- Auxiliary views and the creation of true size views
- Using the Otrack projection tools in AutoCAD
- Determining the necessary layout for an auxiliary view drawing
- Applying projection techniques

TUTORIAL EXERCISE: FIXTURE.DWG

Purpose

This tutorial is designed to show the development of an auxiliary view. The auxiliary view projections and interactivity between each of the views are the key to creating a "true size" drawing of the angled surface.

System Settings

No special systems settings are needed.

Layers

The following layers will need to be created:

Name	Color	Linetype
Object	Black	Continuous
Center	Green	Center
Hidden	Blue	Hidden
Projection	Magenta	Continuous
Dimension	Black	Continuous

Suggested Commands

Offset, Otrack, Grips, Line, Extend, Trim, Osnap

The tutorial that follows will show different techniques used in the development of auxiliary views. These techniques are similar to the textbook's tutorial but some unique commands will be used in the development of the views.

The Otrack tool will be used in the projection of locations between the views. The application of dimensions will also be looked at.

The key to success with auxiliary views is to work and project the details between all 3 (or 4) views concurrently. The object being created is a wedge block with a bolthole pattern on the face (see Figures 15–1 through 15–3).

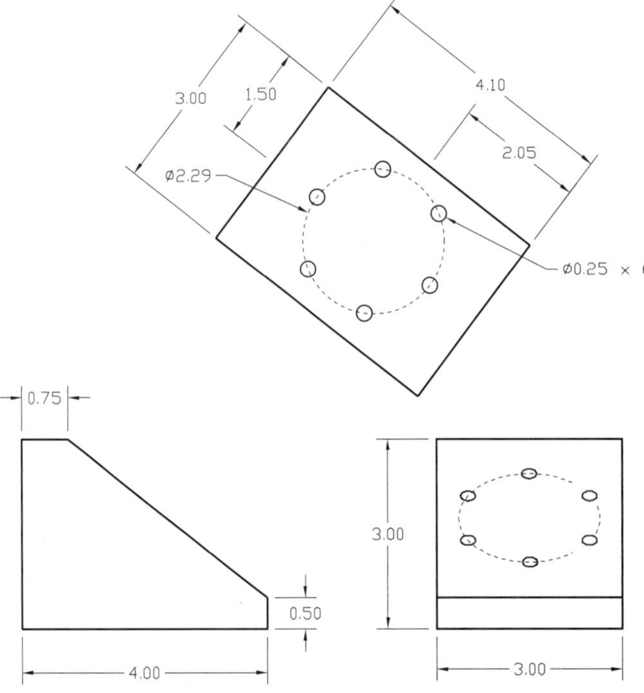

Figure 15–1

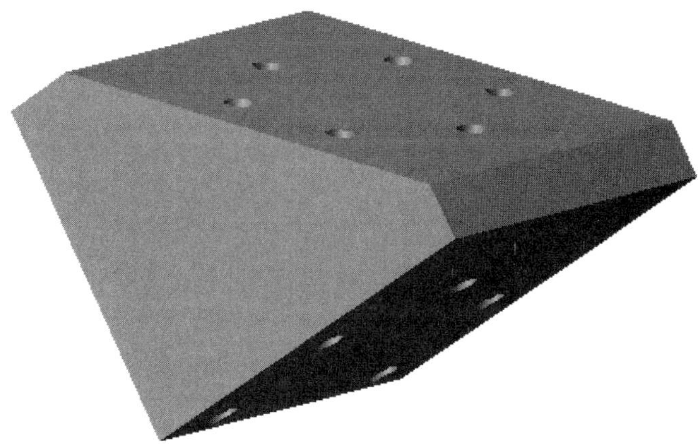

Figure 15–2

STEP 1

We will begin the process by starting with an English based blank drawing file. The first step is the creation of the Front and Right Side orthographic views, without any details. Use the following drawing and dimensions. as a guide. Do not add demensions to the drawing at this step.

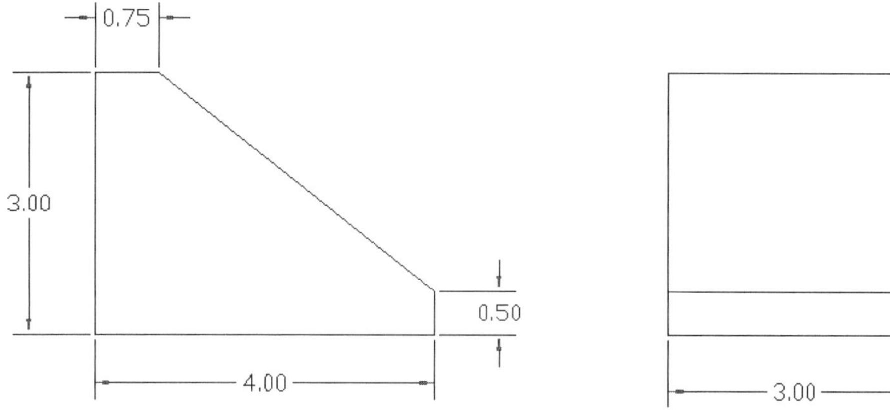

Figure 15–3

As part of the initial setup process, make sure that the Osnap selections: endpoint, intersection, center, extension, midpoint and perpendicular are set, as shown in Figure 15–4.

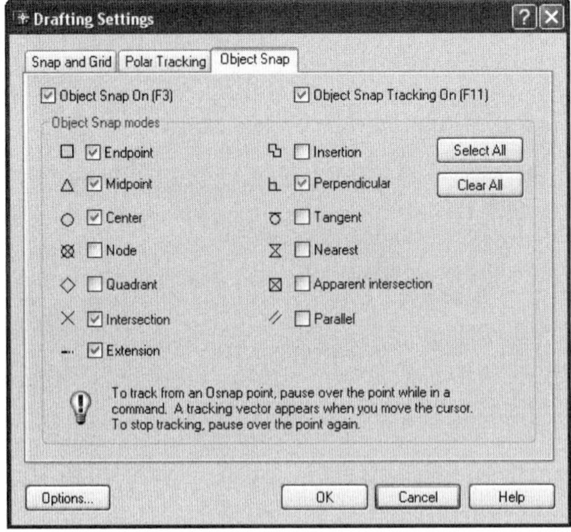

Figure 15–4

STEP 2

Once the Front and Right Side views have been created, the Auxiliary view will be projected from the angled surface on the Front view. Before any detailing occurs on the orthographic views, the Auxiliary view needs to be created, since it will display the bolt hole pattern in true size.

Start the process by offsetting the angled line 4 inches from the angled surface in the front view.

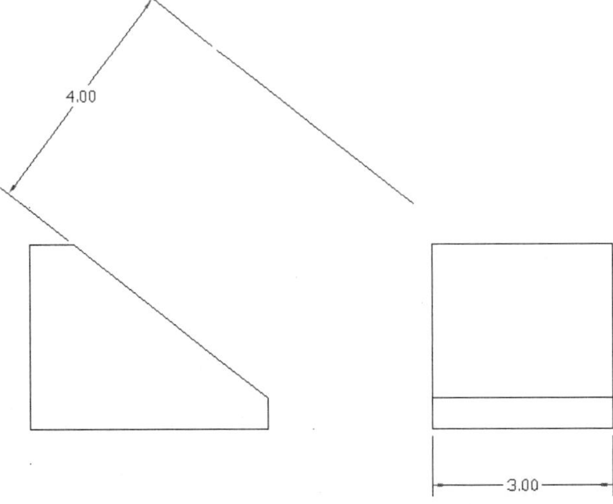

Figure 15–5

STEP 3

The depth of the object in the Auxiliary view is created by offsetting the angle line (previously drawn in the Auxiliary view). This offset will be 3.00 inches.

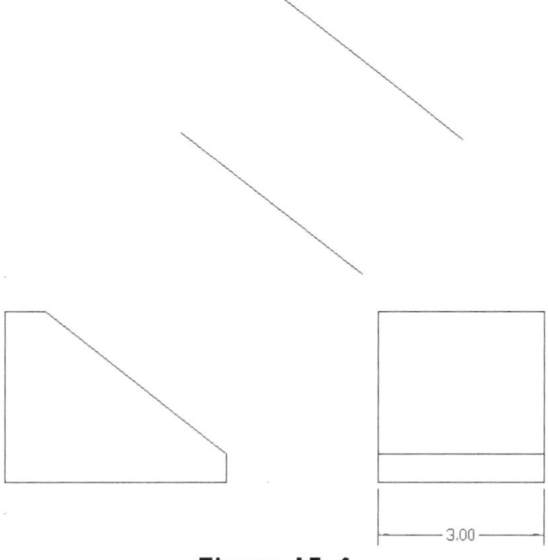

Figure 15–6

STEP 4

Using the existing surface as the projection (offset) media allows for easy creation of the perpendicular projection lines since everything now lines up with the Front view.

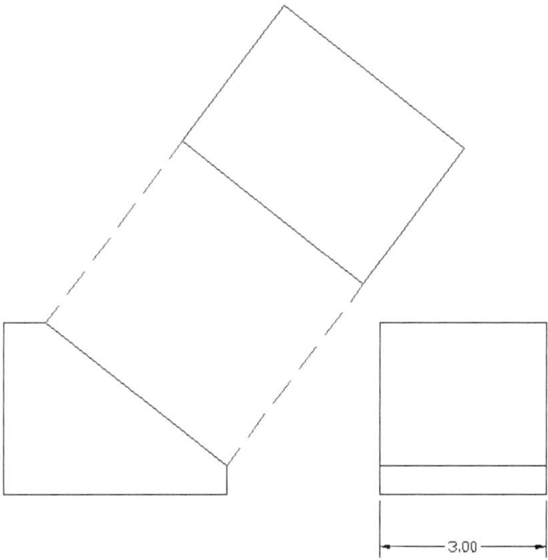

Figure 15–7

STEP 5

Setting up the location of the bolthole pattern is the next step. This is done completely on the Auxiliary view. Draw the offset lines that represent the placement of centerlines in the view.

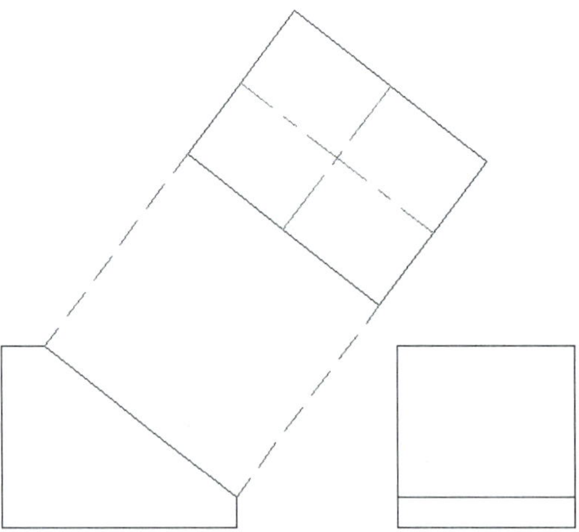

Figure 15–8

STEP 6

Using the center of the Auxiliary view as a guide, create a circle with a diameter of 2.29 inches. Change the circle to the centerline layer if it isn't already placed on it. This circle becomes the centerline for the boltholes. The boltholes have a diameter of .25 inches. Create a circle at the intersection of the bolthole circle and the linear centerlines. See Figure 15–8 for guidance.

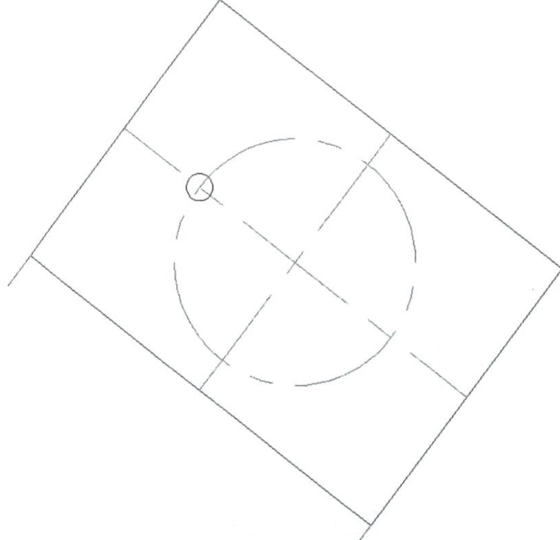

Figure 15–9

STEP 7

To copy the bolthole around the circle, the ARRAY command is the appropriate tool. Set up the ARRAY command similar to Figure 15–9. Select the center point of rotation in the center of the bolthole circle. (Note: This value will be different than what is represented in the figure since the starting point of the drawing will have differed.)

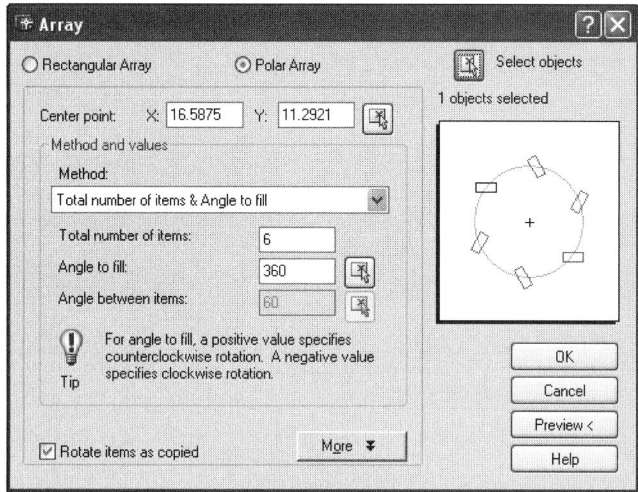

Figure 15–10

The results of the array are in Figure 15–11

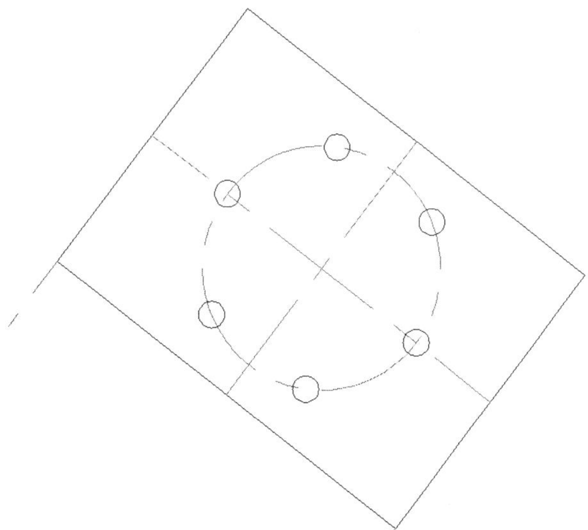

Figure 15–11

STEP 8

The projection process is now ready to begin. The projection process will use the Otrack tool and the drawing of temporary construction lines. First, the centerlines need to be projected to the orthographic views. The second items projected are the center points of the holes. In Figure 15–11 the Osnap selections are shown, check to make sure that the endpoint, intersection midpoint and perpendicular are set.

To use the Otrack, *do not pick a point.* The Otrack tool sets temporary projection points (up to 3) on object snap locations on the drawing. Begin by starting the LINE command; Otrack points are placed when the crosshair is held on the location (look for the osnap icon) for 1-2 seconds. (See Figure 15–12) As the crosshair is pulled away, the construction line is shown. Use this line to project a line perpendicular to the Front view. When the perpendicular Osnap is reached, left mouse click to set the starting point of the line. Extend the line to the next perpendicular location and set the second point. See Figure 15–13.

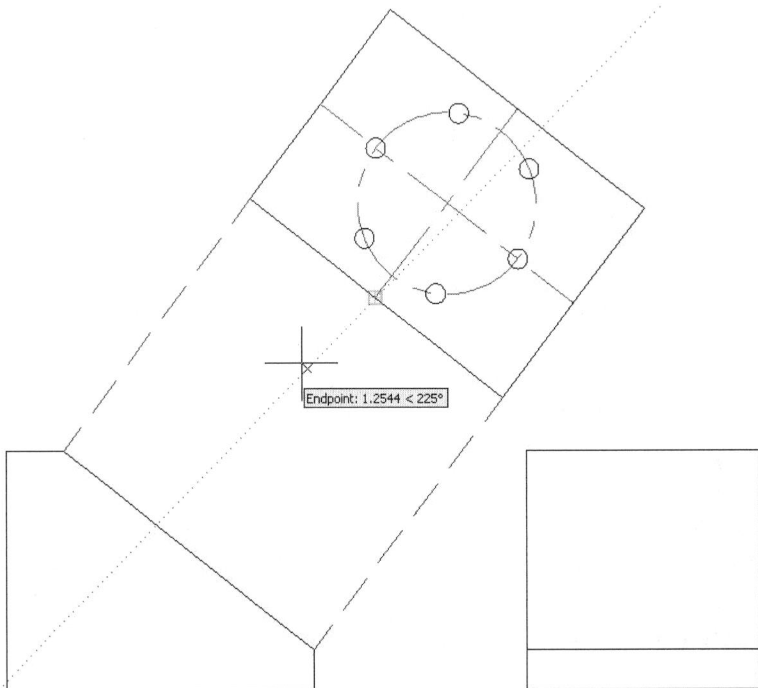

Figure 15–12

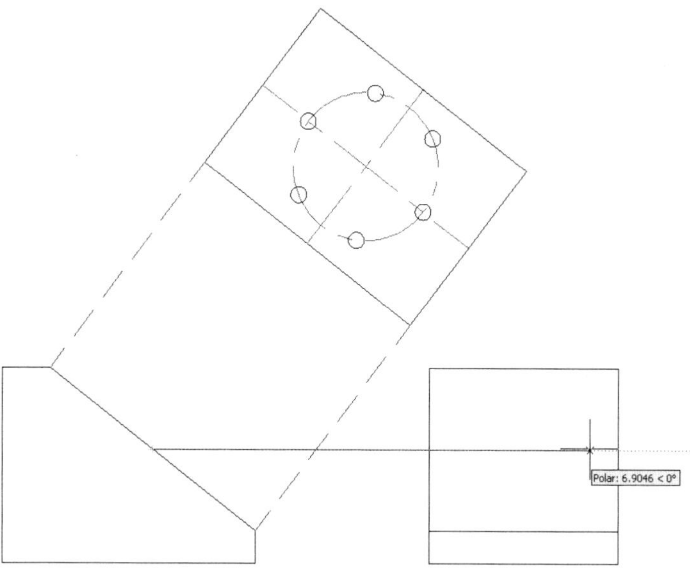

Figure 15–13

STEP 9

Complete the projection process for each of the hole center points. (There are 4 total, since some will overlap others). If the Otrack becomes too cumbersome it is okay to draw construction lines for the projections points as in Figure 15–14.

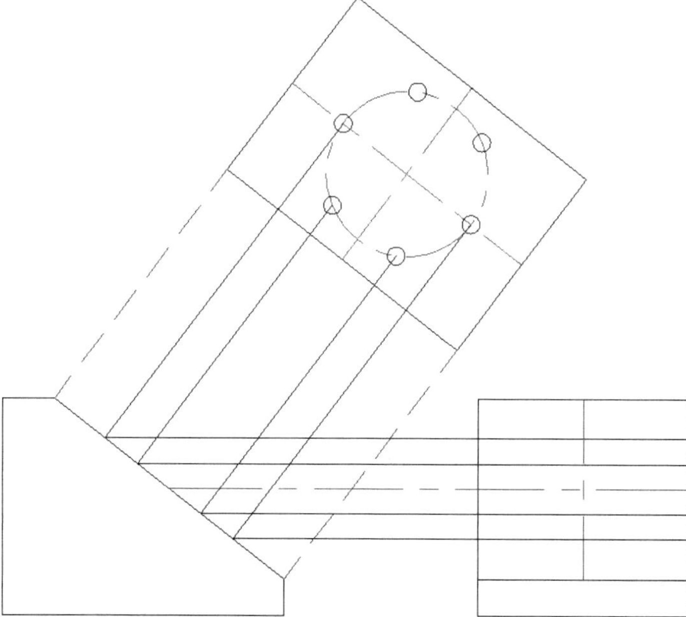

Figure 15–14

STEP 10

Next, we need to make a measurement of the bolthole circle on the Auxiliary view to the front edge of the part, as shown in Figure 15–15. This will provide the offset distance needed in the Right Side view to create the bolthole circle's elliptical shape. Use the aligned dimension along the centerline on the Auxiliary view. Transfer this dimensional value to the OFFSET command and offset the vertical lines on the Right Side view.

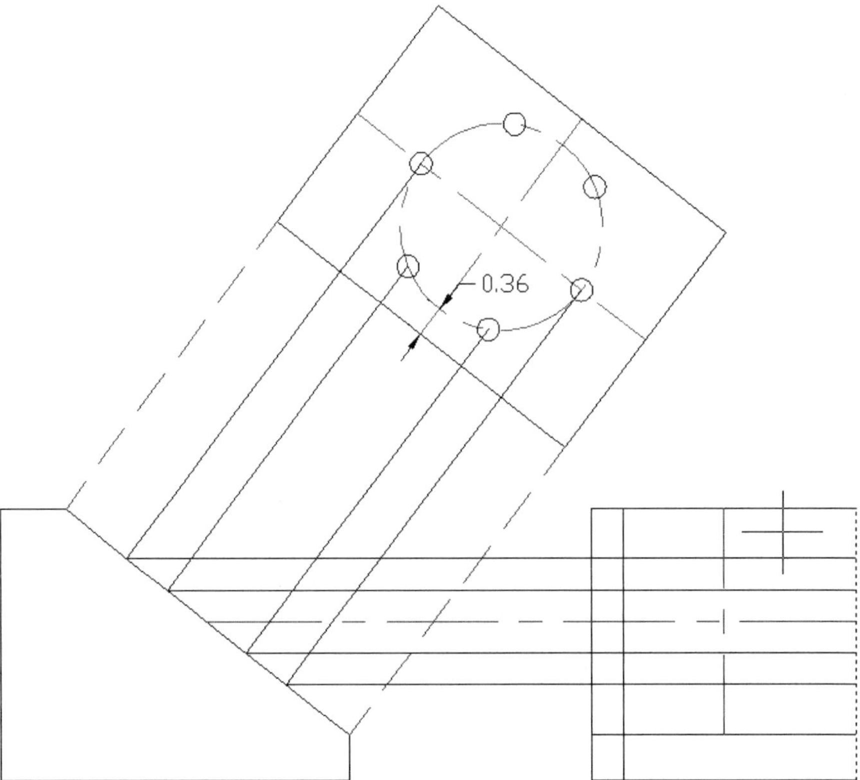

Figure 15–15

STEP 11

To create the elliptical circle, use the Center option on the ellipse. Select the intersection of the centerlines and use the offset vertical lines as a guide on the major axis. On the minor axis use, the intersection between the top projection line and the centerline. See Figure 15–16. Change the ellipse to a centerline when completed.

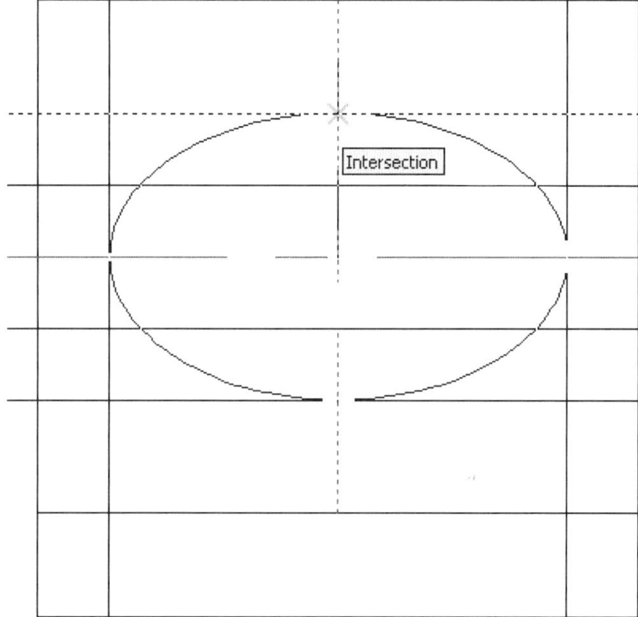

Figure 15–16

STEP 12

The bolt holes will be created on the intersections of the horizontal projection lines and the ellipse. These holes will also need to be elliptical. Additional projections will be needed. With that in mind, move the projection lines onto a separate layer and turn off most of them. Select one hole that needs to be projected. Use the same process of projecting the top and bottom location of the hole (see Figure 15–17). Measure the hole from the center point to the circumference, as shown in Figure 15–18.

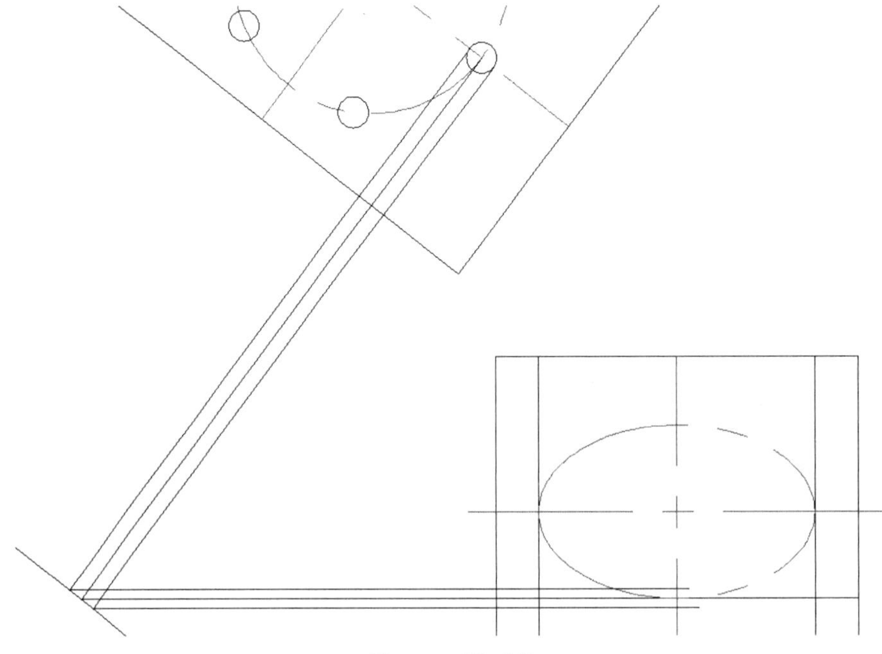

Figure 15–17

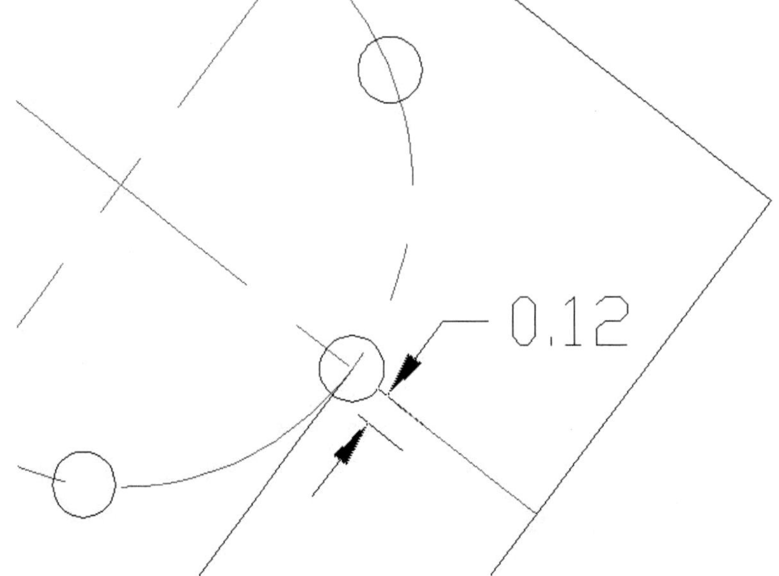

Figure 15–18

The ellipse is created inside the rectangle that is outlined from the offset and projected lines (see Figure 15–19). Use the ELLIPSE command, with the Center option, to create the ellipse shown in Figure 15–20.

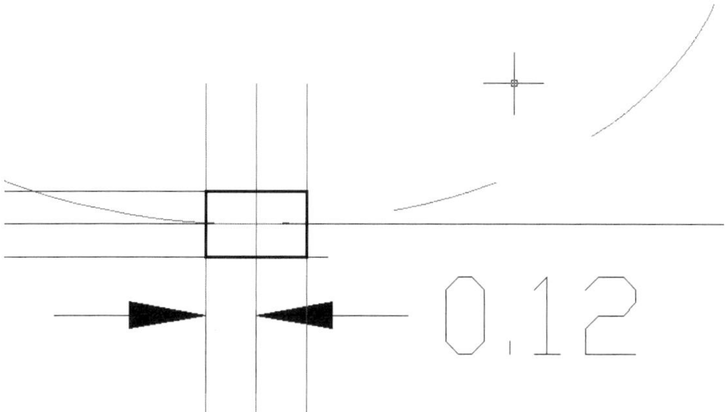

Figure 15–19

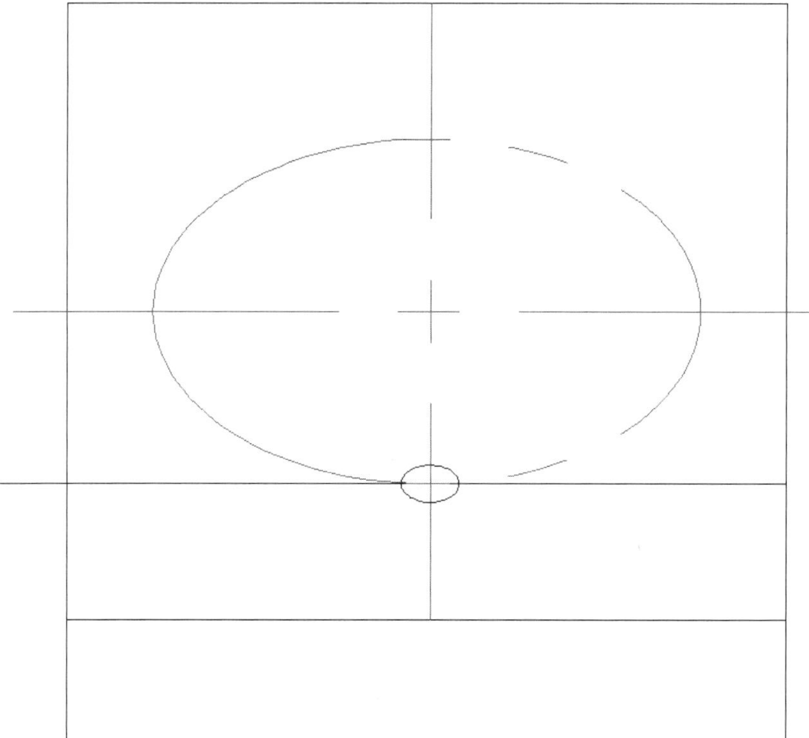

Figure 15–20

Turn the projection lines back on and copy the ellipse around the bolthole circle. This will complete the pattern on the Right Side view, as shown in Figure 15–21.

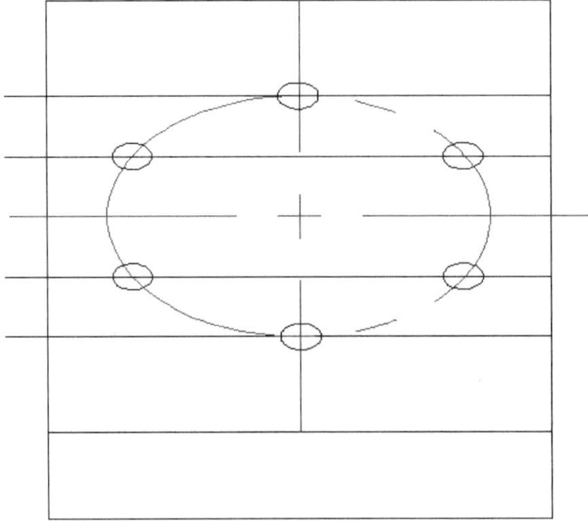

Figure 15–21

STEP 13

Putting in the hidden lines in the Right Side view is also a snap. Set the current working layer to Hidden, and use the projection lines as a guide. At the intersection of the projection line and the small ellipse, draw a line vertically. Only draw lines from the 3 ellipses above the horizontal line.

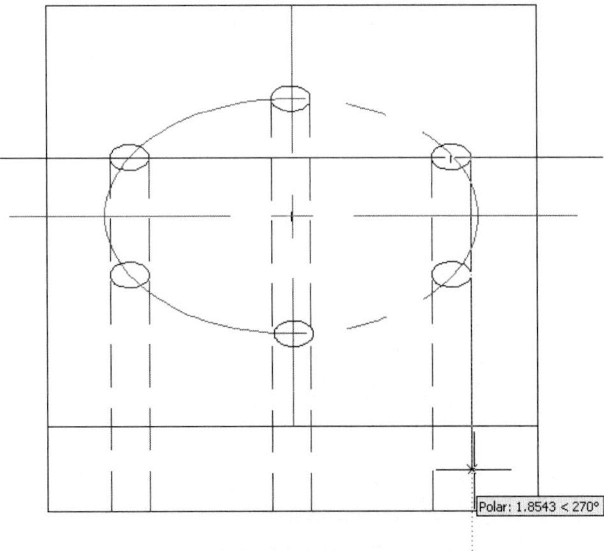

Figure 15–22

STEP 14

Applying the center and hidden lines to the Front view is next. Extend the projection lines into the Front view. Then use the OFFSET command and offset the lines .125 distance on each side. See Figure 15–23. Use EXTEND and TRIM to bring the lines into the front view. Change the lines to the appropriate layers, when you complete the offset. Apply grips to extend the centerlines past the edges of the front view. See Figure 15–24. Hidden lines will not be placed in the Auxiliary view since it will make it difficult to visualize. The designer will need to make the decision regarding whether to place hidden lines in an Auxiliary view.

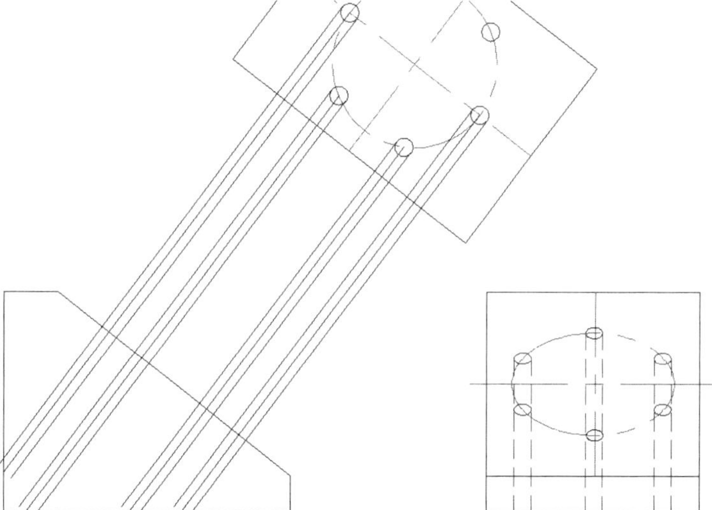

Figure 15–23

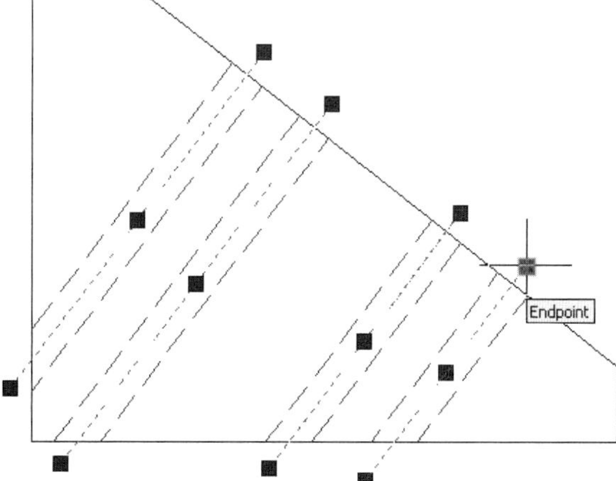

Figure 15–24

STEP 15

Dimension the part. Remember that Auxiliary views will only have the true size dimensions in that view, in this case the hole pattern and face size. The Auxiliary view will use an aligned dimension for any "linear style" dimension, due to the angle. Dimension the overall shape of the part on the Front and Right Side views. See Figure 15–25.

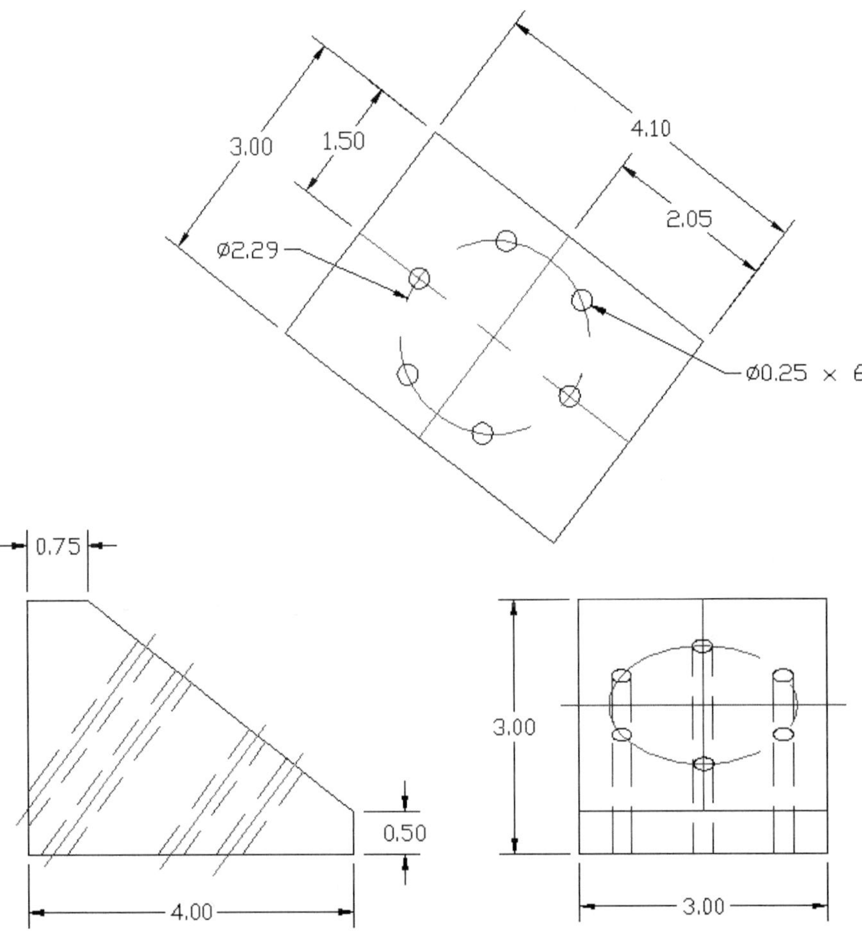

Figure 15–25

Beginning Level Drawings
Directions for Problems 15–1 through 15–12

Draw the required views to fully illustrate each object. Be sure to include an Auxiliary view.

PROBLEM 15–1

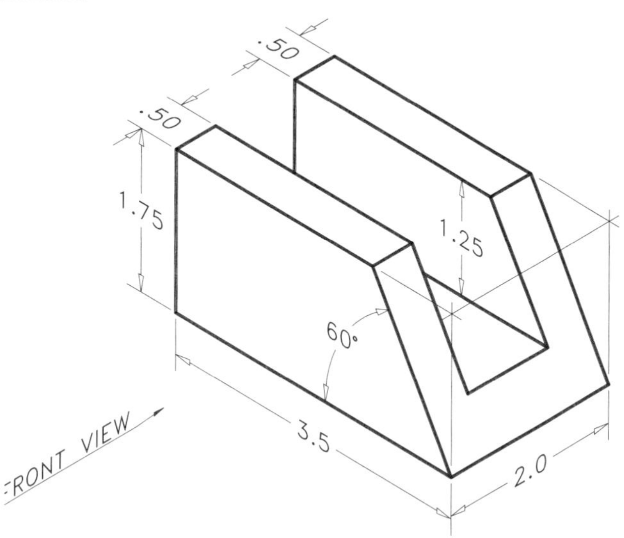

PROBLEM 15–2

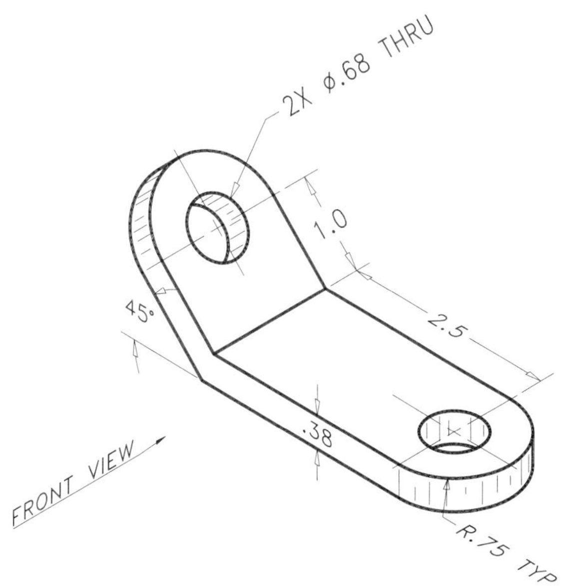

PROBLEM 15–3

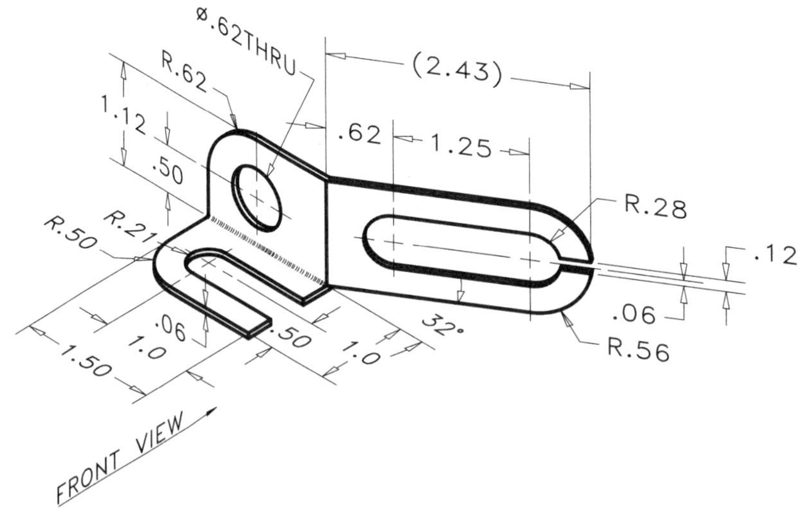

PROBLEM 15–4

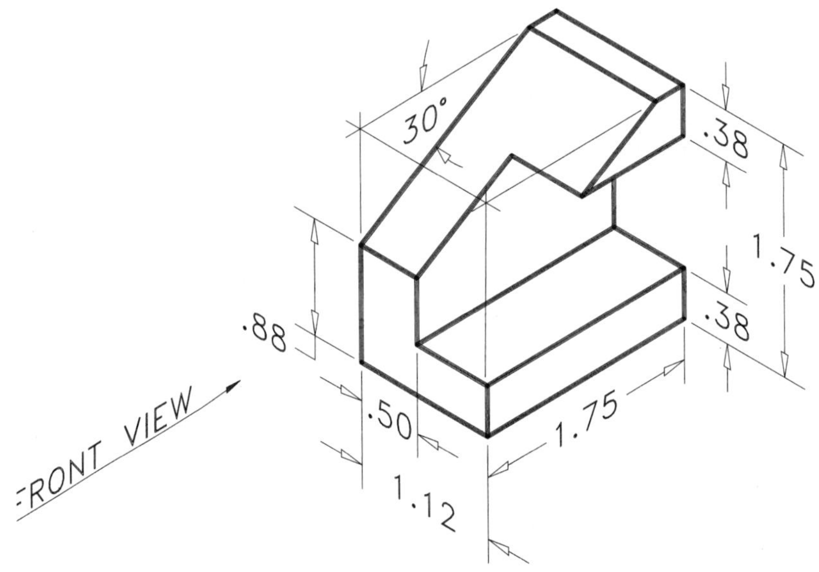

PROBLEM 15–5

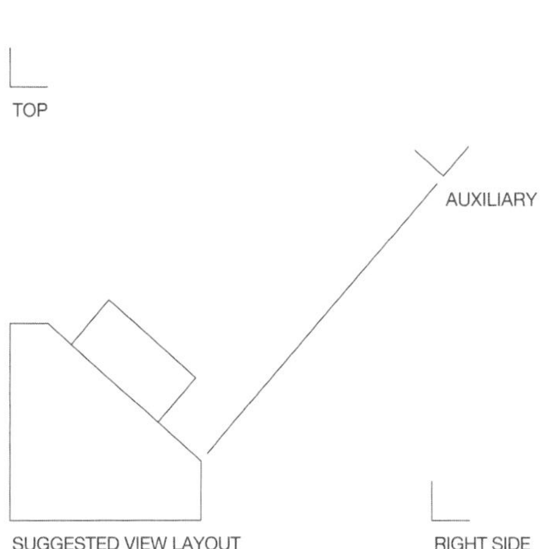

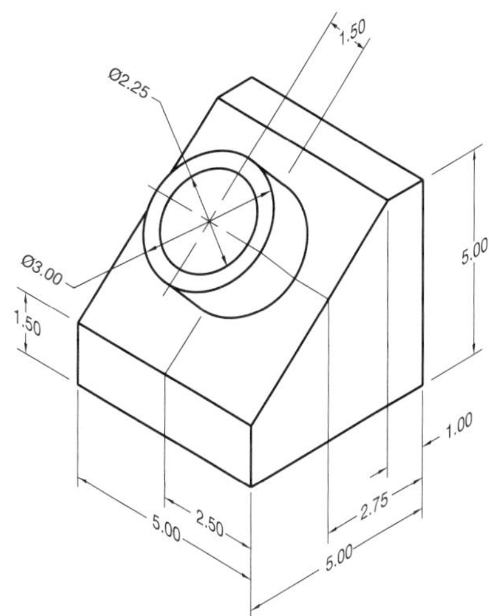

PROBLEM 15–6

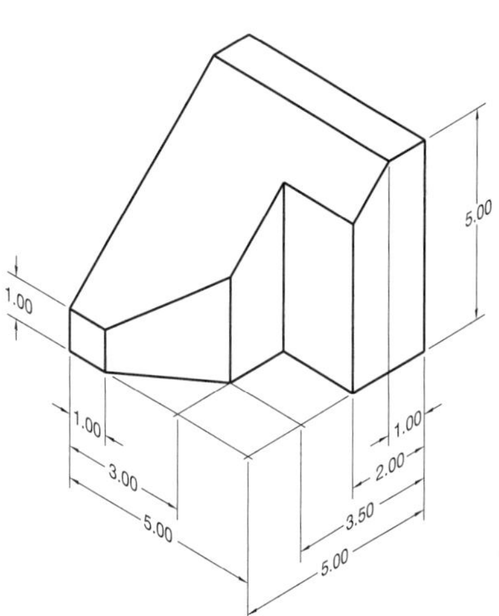

PROBLEM 15–7

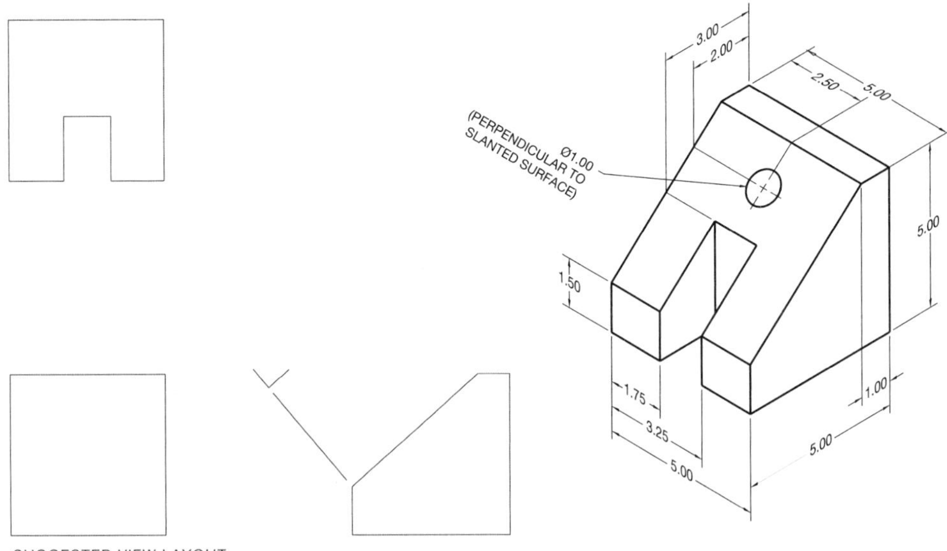

Intermediate Level Drawings

PROBLEM 15–8

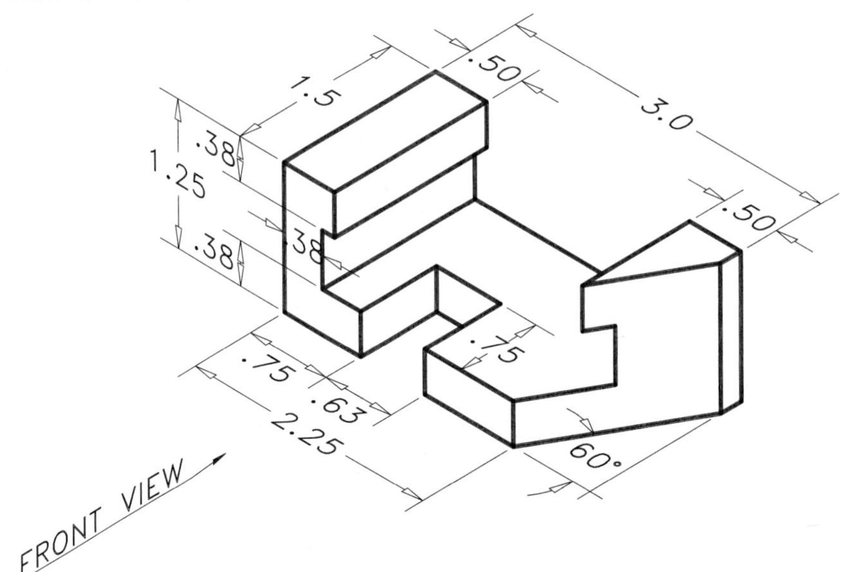

PROBLEM 15–9

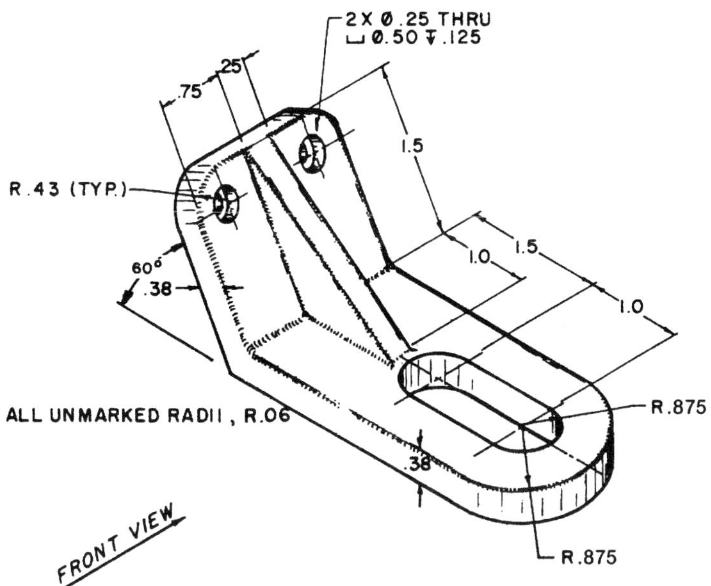

PROBLEM 15–10

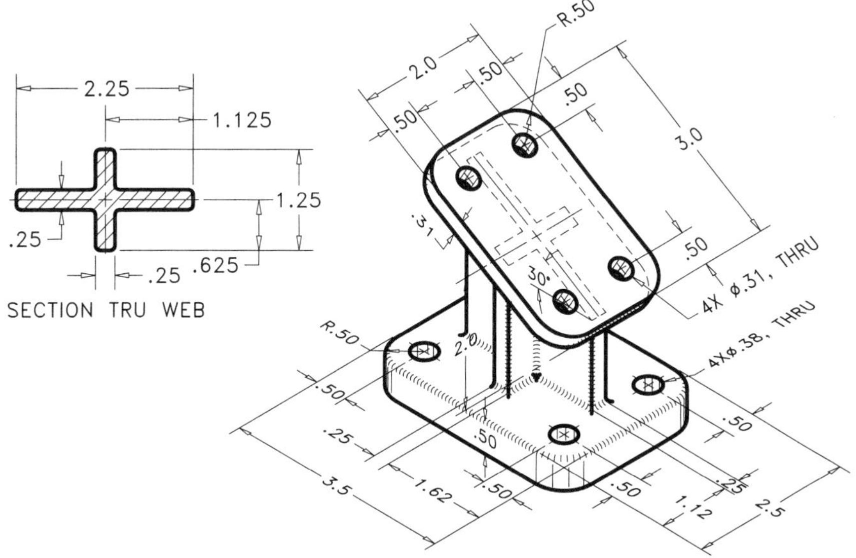

PROBLEM 15–11

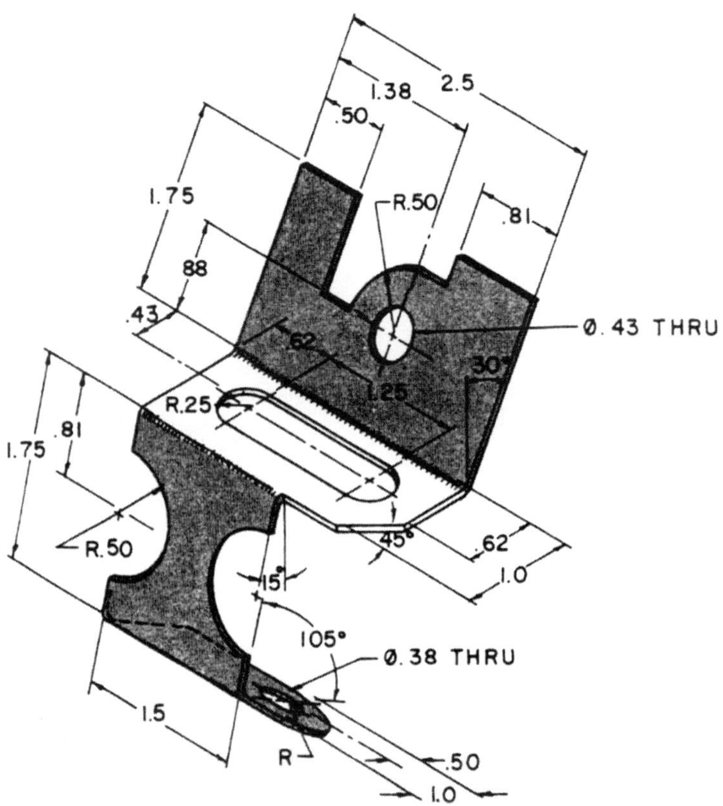

Thickness - 0.0625

PROBLEM 15–12

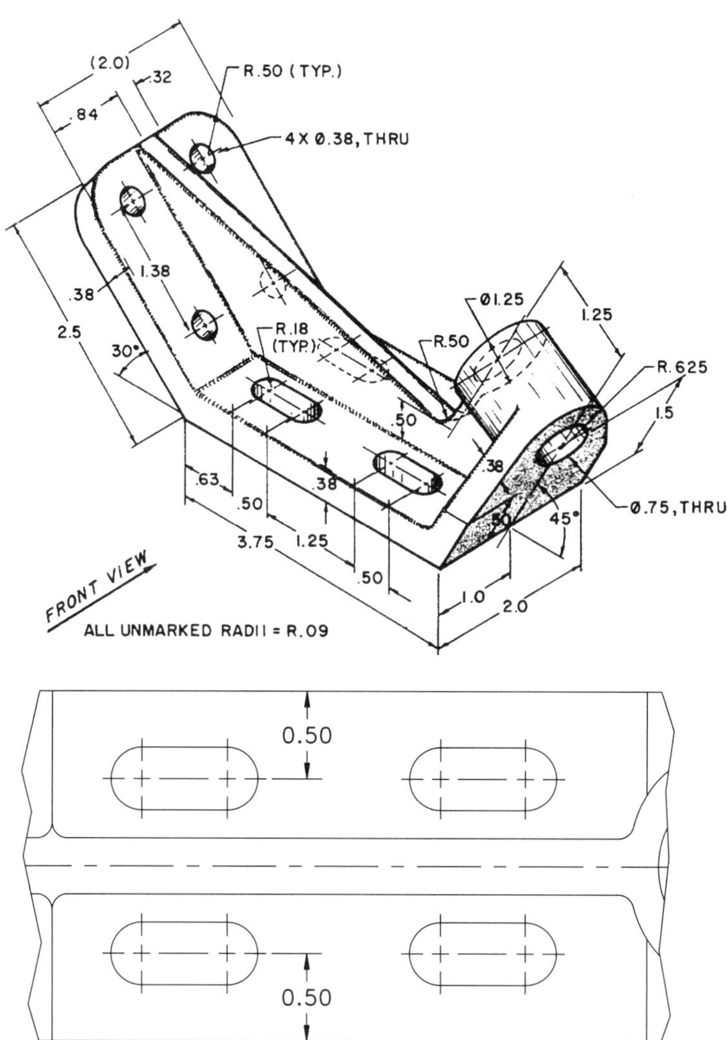

Directions for Problems 15–13 through 15–15

Create the missing Auxiliary view through projection methods.

PROBLEM 15–13

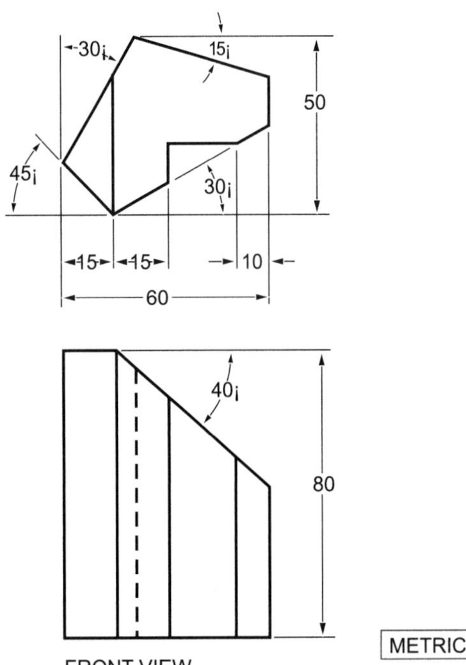

PROBLEM 15–14

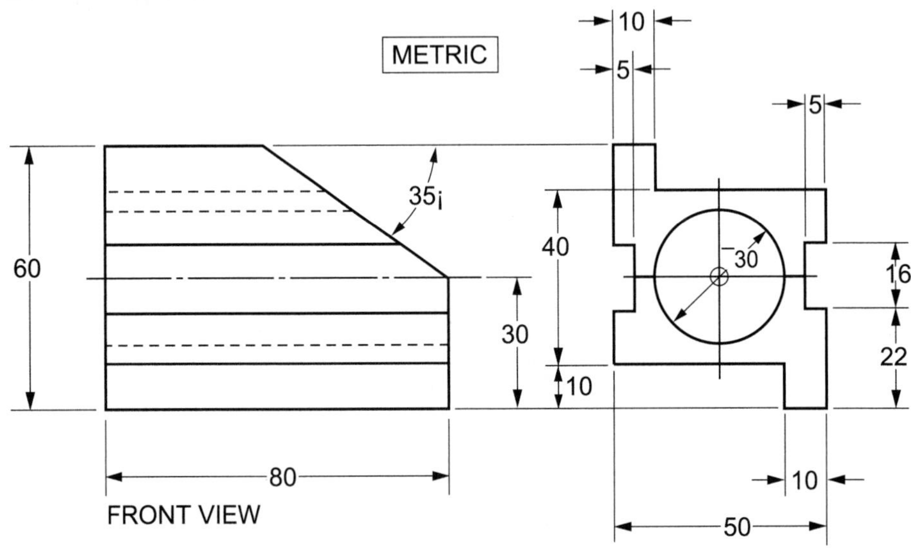

PROBLEM 15–15

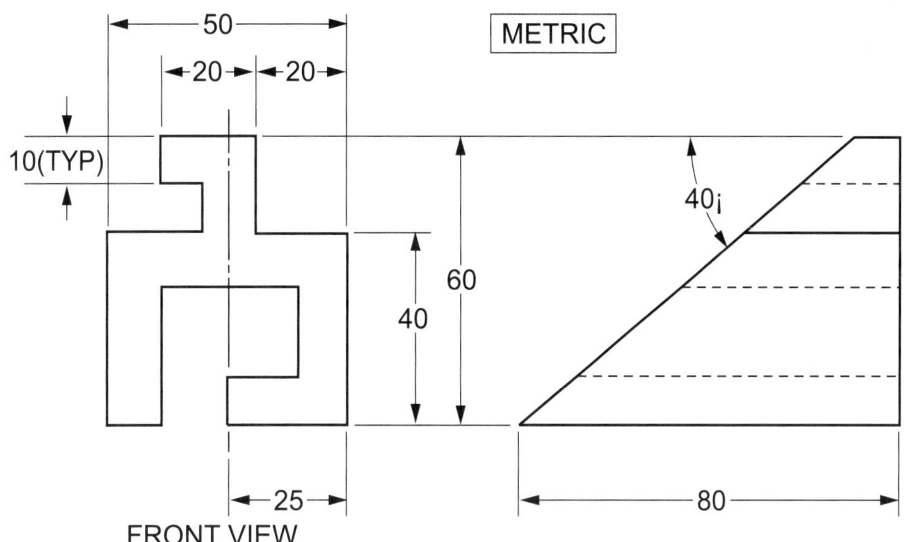

Block Creation, AutoCAD DesignCenter

The exercises of this chapter introduce:

- Developing a Wblock
- Inserting blocks on a drawing
- Using DesignCenter to manage and insert blocks

TUTORIAL EXERCISE: SUBDIVISION.DWG

Purpose

This tutorial is designed to show the application of blocks in a drawing environment. The scope of the tutorial will include block creation and the insertion of the blocks into a drawing file.

System Settings

No special systems settings are needed.

Layers

The following layer has already been created:

Name	Color	Linetype
Streets	Black	Continuous

Suggested Commands

WBLOCK (Write Block)

DESIGNCENTER

RECTANGLE

ZOOM

PAN

STEP 1

Open the drawing file *subdivision block.dwg*. This drawing shows a sample street layout for a subdivision. Our goal is to create and place sewers on the subdivision streets.

Survey the layout of the subdivision. The guidelines for sewer placement are:

- Sewers must be placed on both sides of the streets
- Sewers cannot be farther than 300' apart

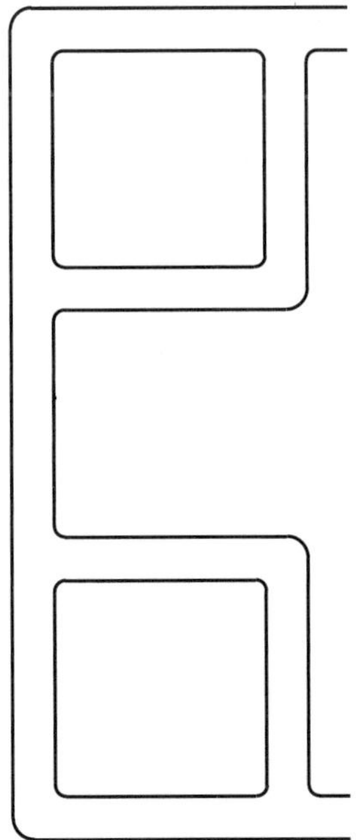

Figure 16–1

STEP 2

The next step is to develop a drawing for a sewer grate that can be used throughout the subdivision. Using the ZOOM command magnify the upper-left hand corner of the drawing, as shown in Figure 16–2.

Figure 16–2

STEP 3

The sewer will be initially drawn off to the left of the subdivision drawing. Once completed it will be made into a block. To begin select the RECTANGLE command from the Draw toolbar or type RECTANGLE at the AutoCAD command prompt. Once you have selected the first point, right mouse click and select Dimensions from the menu (see Figure 16–3).

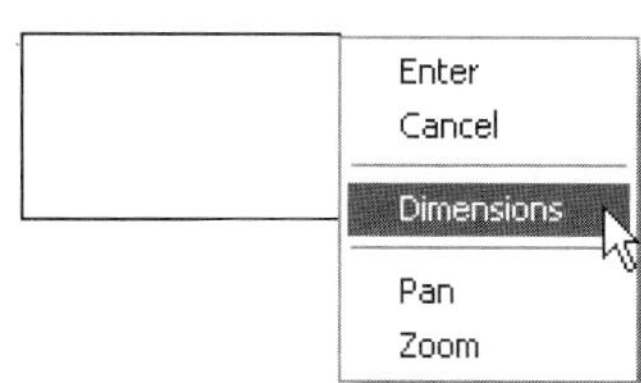

Figure 16–3

The length dimension (which is always the X axis) is 4' long. The width dimension (Y axis) is 2'-6" high. Don't forget to left mouse click a point on the drawing surface to place the rectangle.

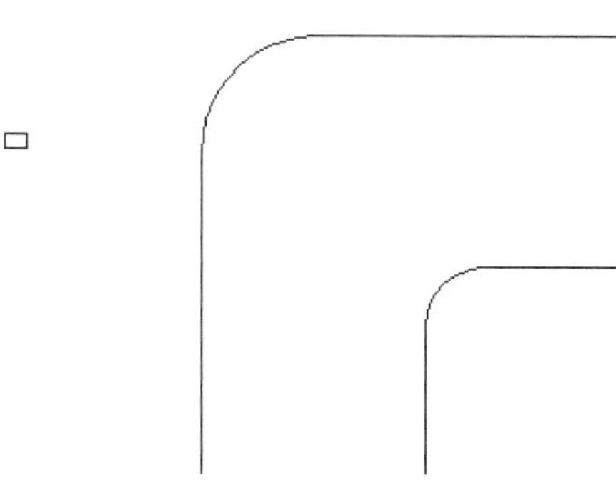

Figure 16–4

STEP 4

The rectangle of the sewer is very small compared to the 40-foot wide streets. The next step is to create the block of the sewer, so that it can be used in this drawing and future drawings. Type the letter **W** at the command prompt to start the WBLOCK command and open the Write Block dialog box shown in Figure 16–5.

Command: **W**

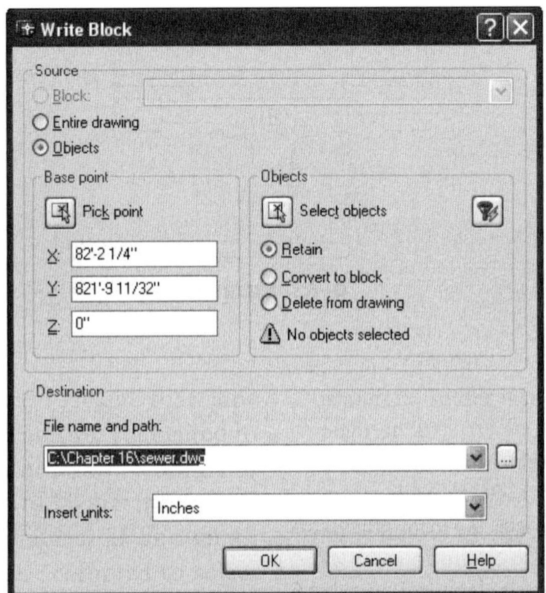

Figure 16–5

STEP 5

For this block, select the Objects option under the Source section, (we don't want the Entire drawing). Next, the base point of the sewer is selected. Base point locations need to be a repeatable selections typically on the object. For the base point selection, click the Pick point button in the Base Point section of the dialog box (see Figure 16–6). Set the location of the base point selection on the sewer, as shown in Figure 16–7.

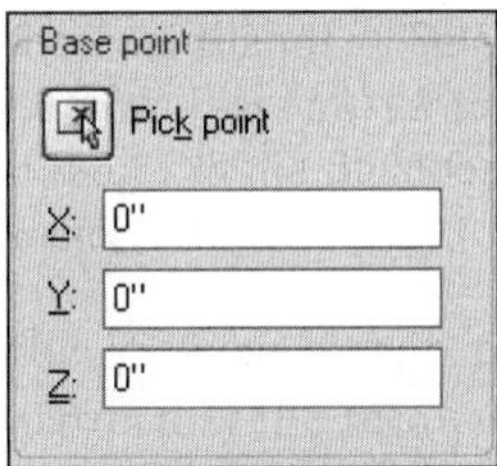

Figure 16–6

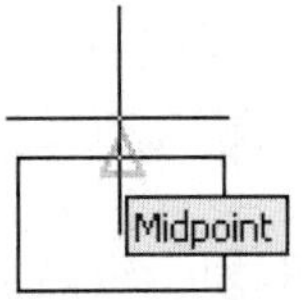

Figure 16–7

STEP 6

Press the Select objects button shown in Figure 16–8. Then click anywhere on the rectangle. After the rectangle is selected, right mouse click to return to the dialog box. (The rectangle should be deleted once it is used to create the block. After Step 7, select it and delete it from drawing.)

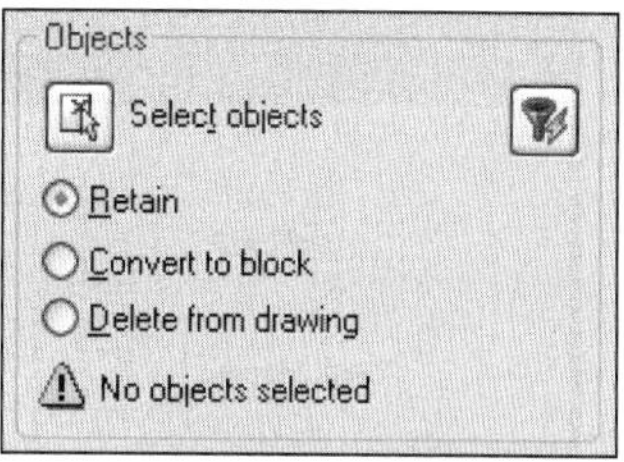

Figure 16–8

STEP 7

The Destination area of the Write Block dialog box is use to name and specify a location for the file that the selected objects will be saved into. (See Figure 16–9) Save the drawing into a folder called *Chapter 16*. Use the Create New Folder button in the Browse for Drawing File dialog box to set it up (see Figure 16–10). Name the drawing *sewer*. The Insertion units setting is utilized when inserting drawings using the Design-Center. It will scale objects so they will match the current drawing scale.

Choose the OK button to finish the block creation process.

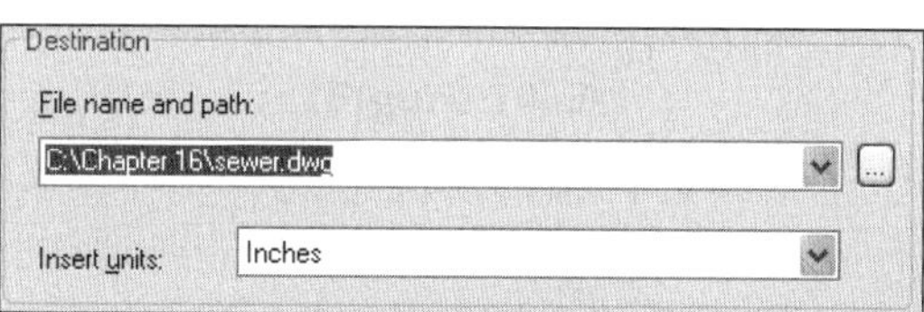

Figure 16–9

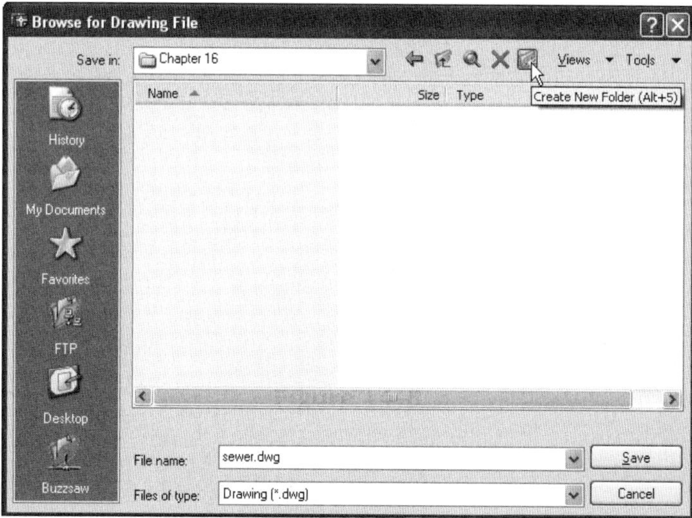

Figure 16–10

STEP 8

The DesignCenter is the block manager that allows you to drag and drop blocks into a drawing. Start the DesignCenter by selecting it from the Tools pull-down menu (see Figure 16–11) or from the Standard toolbar (see Figure 16–12).

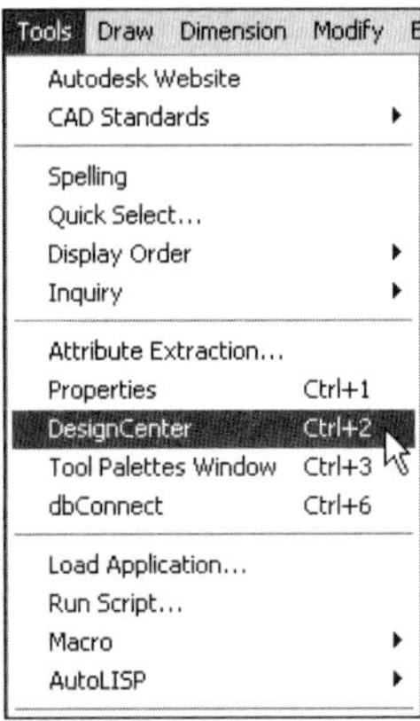

Figure 16–11

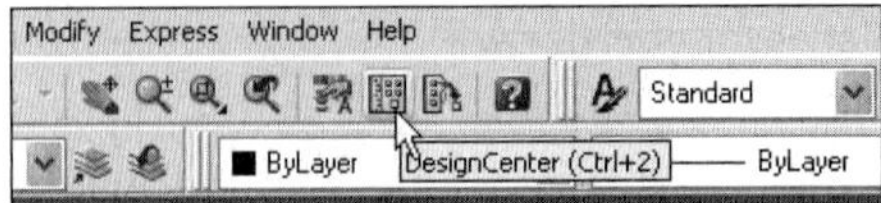

Figure 16–12

STEP 9

The DesignCenter's dialog box is split into two areas. The left side is similar to a Windows Explorer browser window (used to display folder information) and the right side is where the drawing components may be selected and placed via the drag and drop method (see Figure 16–13). Begin by locating the *Chapter 16* folder where you saved the sewer drawing block.

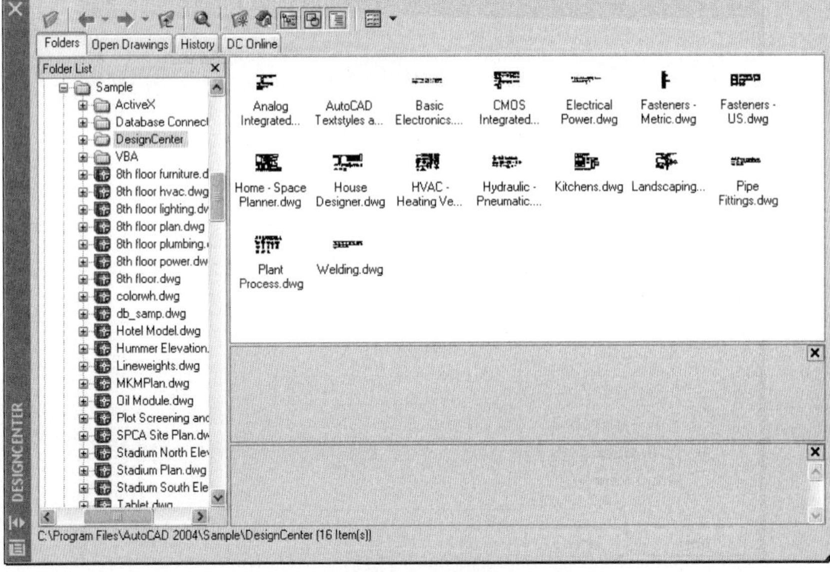

Figure 16–13

Select only the folder level item for *Chapter 16*, as shown in Figure 16–14. When you select a folder, any drawing files it contains are displayed on right side of the DesignCenter dialog box. If you select a file, then only items located inside that file (such as blocks, layers, and text styles) are available for drag and drop placement.

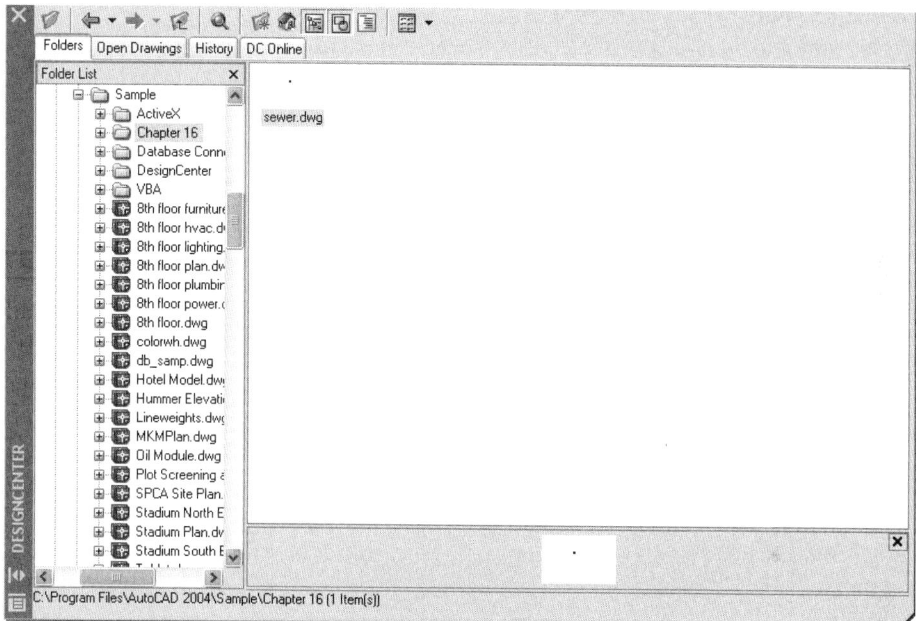

Figure 16–14

STEP 10

Left mouse click on the picture of the sewer drawing on the right side of the DesignCenter dialog box. While holding the left mouse button down, drag the picture outside of the dialog box and then release the left mouse button. This places the sewer drawing on the street drawing (see Figure 16–15). Now, the sewer is attached to the crosshairs at the base point. Locate the sewer near one of the curved parts of the street by left mouse clicking, again (see Figure 16–16). Then finish the block placement, as shown in Figure 16–17 by entering the following command prompt.

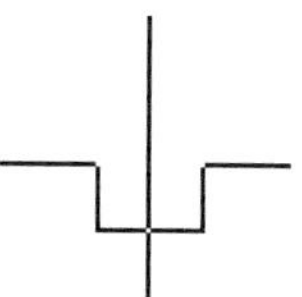

Figure 16–15

Command: _**-INSERT** [Insert>block]
Enter block name or [?]: "C: \sewer.dwg"
Specify insertion point or [Scale/X/Y/Z/Rotate/PScale/PX/PY/PZ/PRotate]: *(Select the end of the curve)*
Enter X scale factor, specify opposite corner, or [Corner/XYZ] <1>: **1**
Enter Y scale factor <use X scale factor>: *(Press* ENTER *or right mouse button)*
Specify rotation angle <0>: **90** *(No rotation may be necessary depending on your placement)*

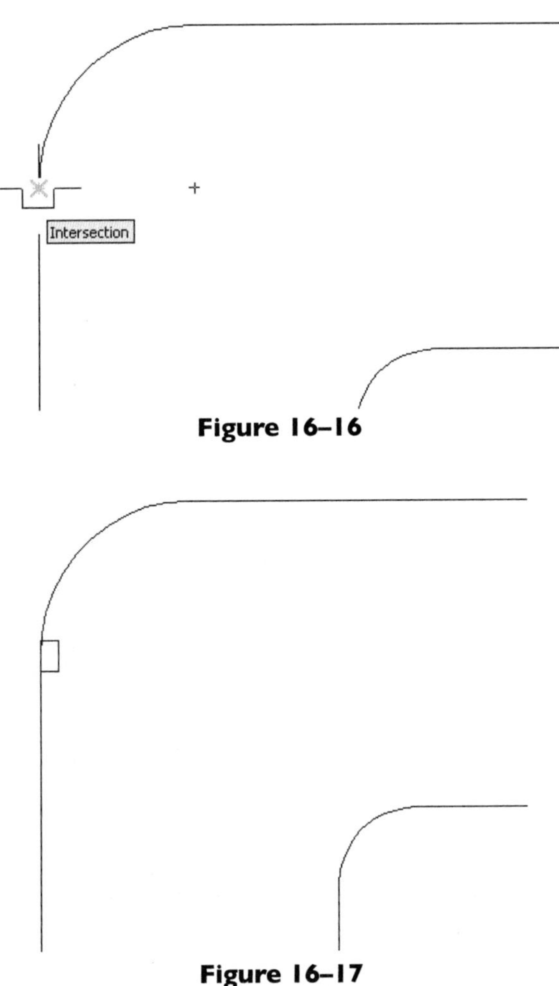

Figure 16–16

Figure 16–17

STEP 11

Before more sewers are inserted, the sewer drawing needs to be modified. The sewer model we are putting in is no longer manufactured. To update to the current model the Edit Reference In-Place and Xref and Block Editing features can be used. Figure 16–18 shows the construction of the new sewer grate.

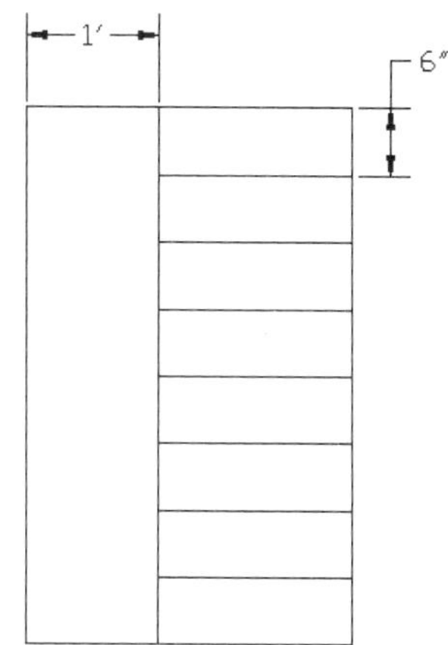

Figure 16–18

To start making the changes, select the Modify pull-down menu, then choose Xref and Block Editing and Edit Reference In-Place from the fly-out menu. (The Refedit toolbar, shown in Figure 16–20, can also be used to start the editing process.) Select the sewer block. The Refedit dialog box appears. Select the sewer drawing from the Reference name selection box, as shown in Figure 16–21. Click OK at the bottom of the dialog box.

Command: **REFEDIT**

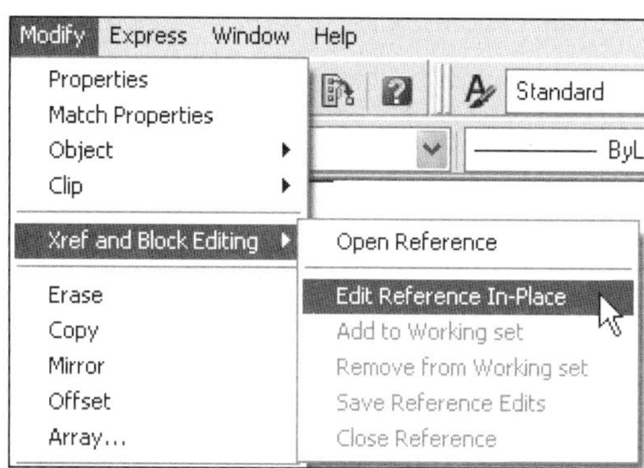

Figure 16–19

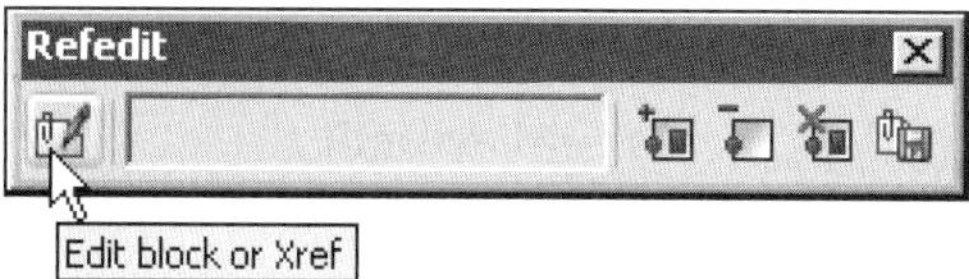

Figure 16–20

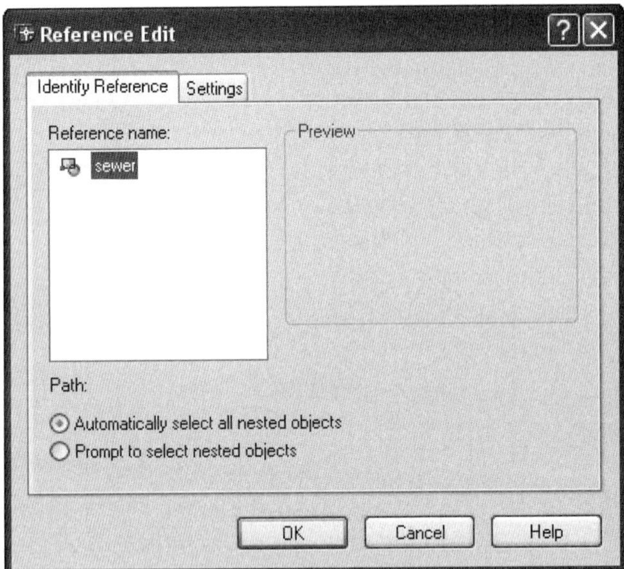

Figure 16–21

STEP 12

Magnify the sewer drawing and start to make the modifications. The easiest way is to start the LINE command and highlight the upper-left corner of the sewer. Move the cursor to the right, as shown in Figure 16–22 and then type **1'**. This will start the line 1' from the corner. Now use the perpendicular object snap to locate the line to the other side, as shown in Figure 16–23.

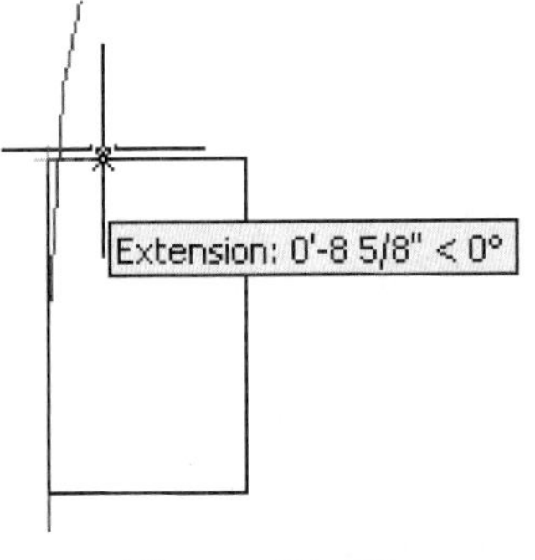

Figure 16–22

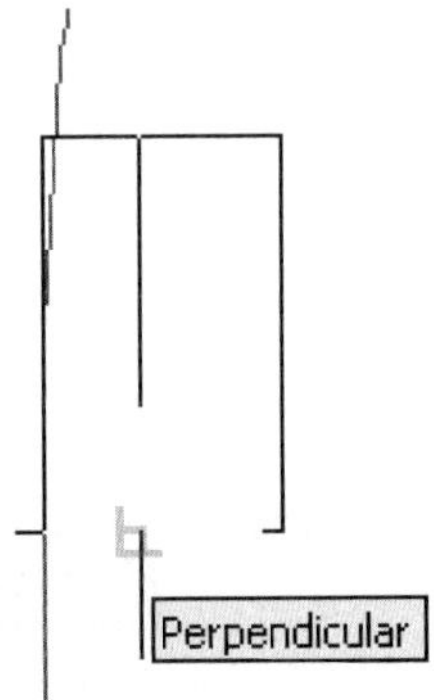

Figure 16–23

STEP 13

A similar process can be used to create the smaller lines that are 6" apart. Use the point and distance method again to locate the first 6" distance line. Then use the OFFSET command to place the rest of the lines on the sewer drawing. The changes being made will be saved into the original *sewer.dwg*, so all future sewer placements will use the updated model.

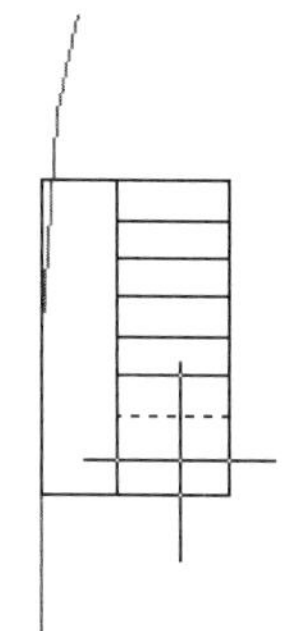

Figure 16–24

STEP 14

To save all the changes, choose the Save back changes to reference button on the Refedit toolbar or type REFCLOSE at the command prompt. As shown Figure 16–26, a warning message appears confirming the overwite of the original block file. Select the OK button.

Figure 16–25

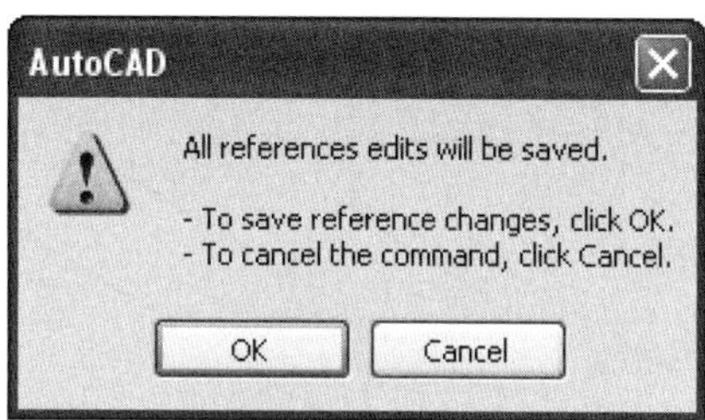

Figure 16–26

A command prompt, similar to the following, wil appear:

Command: **REFCLOSE**
Enter option [Save/Discard reference changes] <Save>: **Enter**
Regenerating model.
8 objects added to sewer
2 block instances updated
sewer redefined.

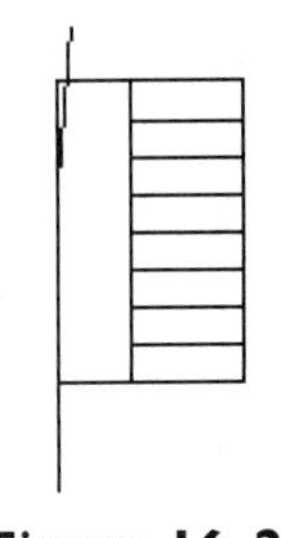

Figure 16–27

Figure 16–27 shows the result of updating the sewer block.

Continue the tutorial by using either the DesignCenter or the INSERT command to insert more sewer blocks into the drawing. Figure 16-28 shows a section of the completed sewer layout for the subdivision.

Figure 16–28

Beginning Level Drawings

PROBLEM 16–1

Directions for Problem 16–1

Using the DesignCenter construct the Welding drawing using the stock AutoCAD symbols found in *Welding.dwg,* located in the *AutoCAD/Sample/DesignCenter* directory.

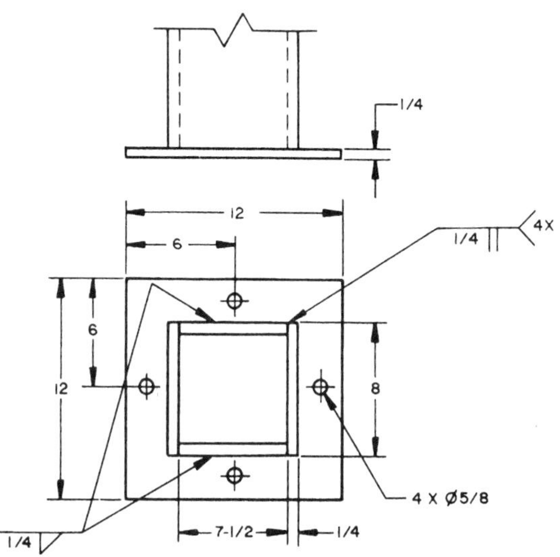

PROBLEM 16–2

Directions for Problem 16–2

Create 2 views and insert the necessary welding symbols.

Dimensions for Members: W10 X 39: depth = 9 7/8", width = 8", web thickness= 5/16", flange thickness = ½", web fillet radius = 1 ½"

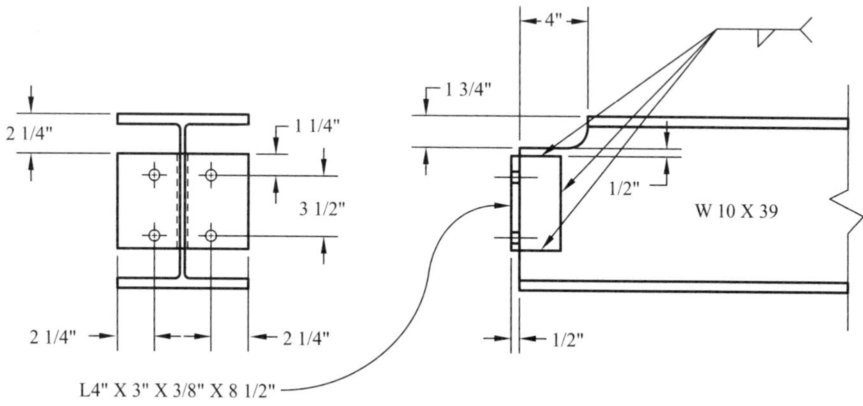

PROBLEM 16–3

Directions for Problem 16–3

Construct this battery pack using the block library supplied with AutoCAD: *Basic Electronics.dwg*. This drawing is not to scale.

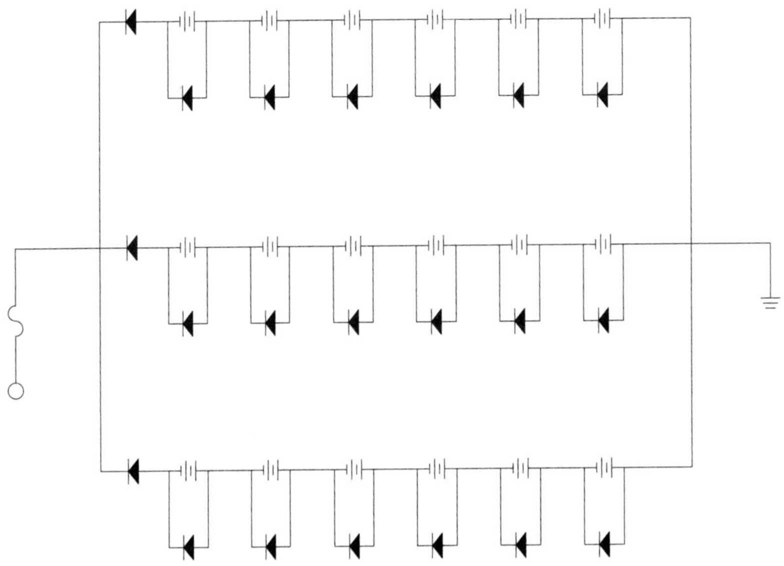

PROBLEM 16–4

Directions for Problem 16–4

Construct this electrical schematic using the block library supplied with AutoCAD: *Basic Electronics.dwg*. This drawing is not to scale.

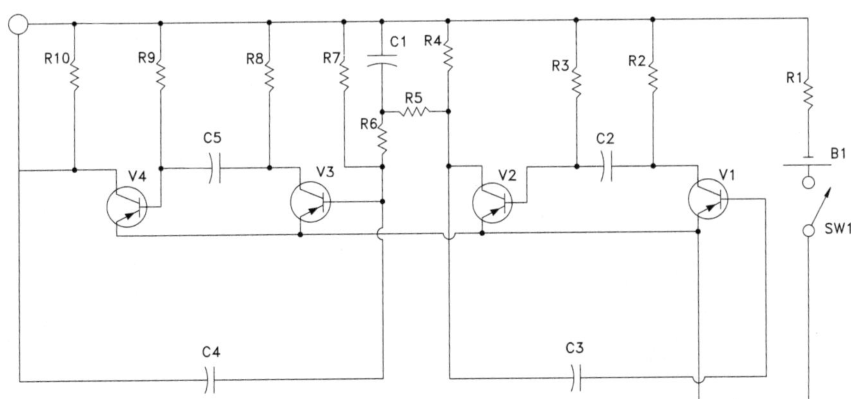

PROBLEM 16–5

Directions for Problem 16–5

Construct this electrical schematic using the block library supplied with AutoCAD: *Basic Electronics.dwg.* This drawing is not to scale.

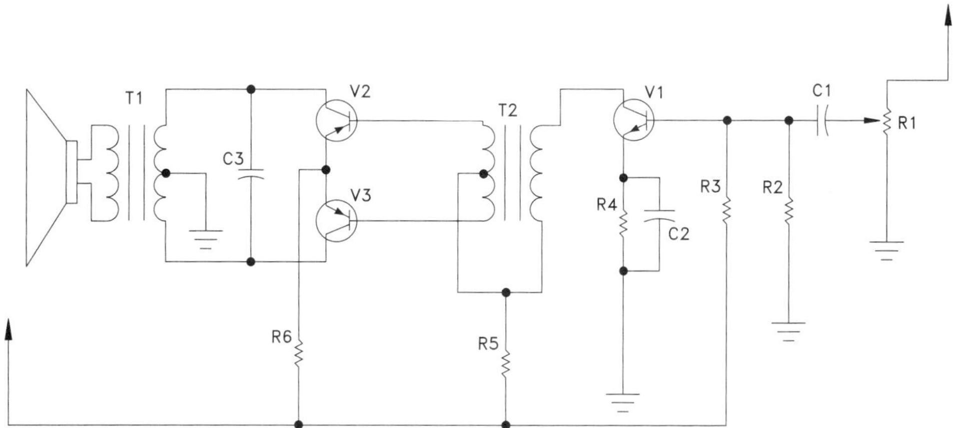

PROBLEM 16–6

Directions for Problem 16–6

Construct this logic gate schematic using the following block libraries supplied with AutoCAD: *CMOS Integrated Circuits.dwg* and *Basic Electronics.dwg.* This drawing is not to scale.

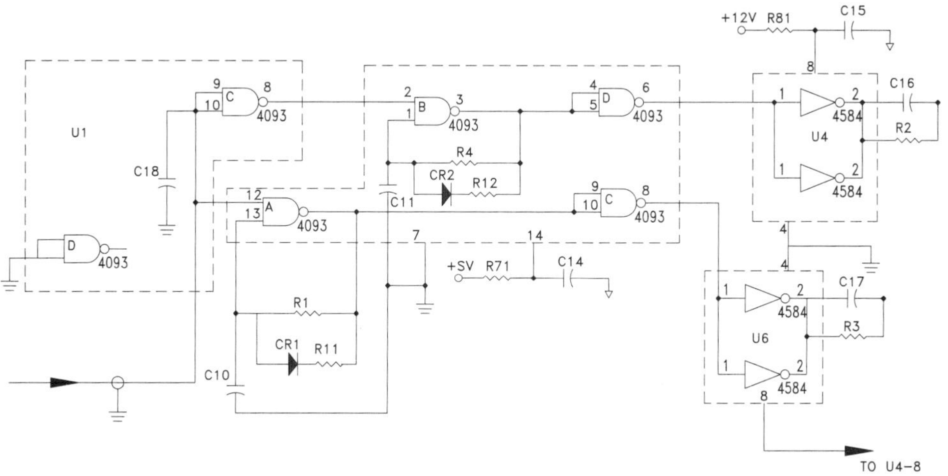

Intermediate Level Drawings

PROBLEM 16–7

Directions for Problem 16–7

Construct this instrumentation pipe diagram using the following block library supplied with AutoCAD: *Pipe Fittings.dwg.* This drawing is not to scale.

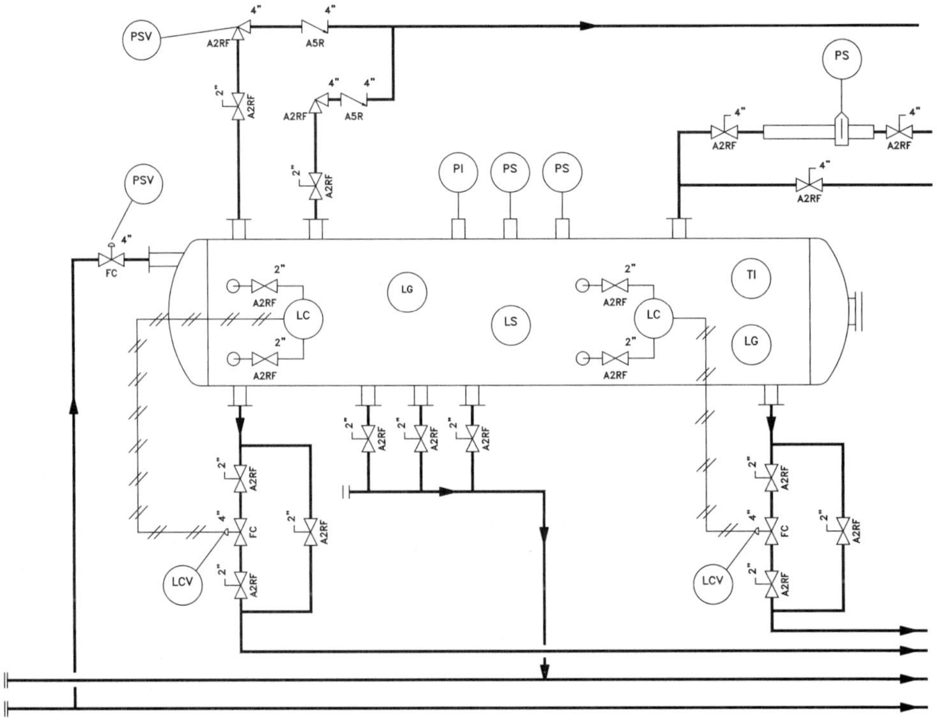

PROBLEM 16–8

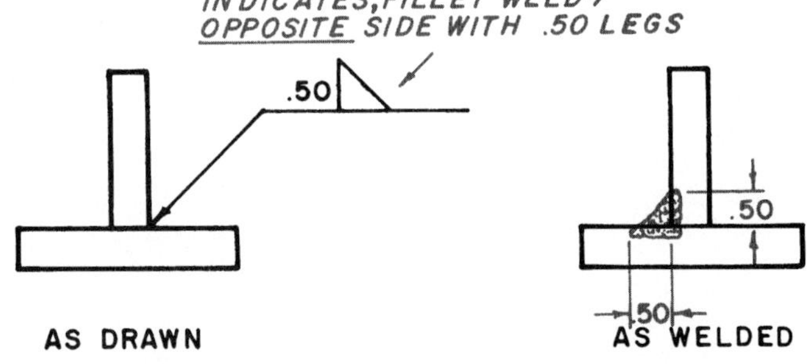

PROBLEM 16–9

Using the Home and Kitchen symbols develop a 2 room house, one room must have a kitchenette and the other room can be flexible in design. The maximum size of the 2 rooms cannot exceed 24' X 24'. Use separate layers for the structure and for the blocks.

Advanced Level Drawings

PROBLEM 16–10

Directions for Problem 16–10

Create the logic diagram using the AutoCAD symbols from the sketch. This drawing is not to scale.

NAME	SYMBOL	PICTORIAL REPRESENTATION
ANTENNA		
BATTERY	+ –	+ –
CAPACITOR		
DIODE		DIODE
GROUND		
INDUCTOR		
RESISTOR		
SWITCH		OR
TRANSFORMER		N/A
TRANSISTORS	NPN PNP	

NAME	SYMBOL	PICTORIAL REPRESENTATION
AMPLIFIER		N/A
BELL		N/A
BUZZER		N/A
CIRCUIT BREAKER		N/A
CONNECTORS–POWER SUPPLIES	FEMALE MALE	N/A
CRYSTAL		N/A
DELAY FUNCTION		N/A
ENVELOPES		N/A
FUSE	OR	N/A
HEADSET		N/A

PROBLEM 16–11

Directions for Problem 16–11

Construct the floor plan of the house using the dimensions provided. Create the appropriate layers to separate the walls from the dimensions, etc. Once the floor plan is completed, create an interior plan consisting of furniture and appliances. Place the blocks in their appropriate spaces using the designated room titles. Use the following block libraries supplied with AutoCAD: *House Designer.dwg*, *Home – Space Planner.dwg*, and *Kitchens.dwg*.

Use the following suggestions for the creation of doors and windows:

All windows measure 2'-8" wide
All bedroom doors measure 2'-6" wide
The bathroom door measures 2'-0" wide
The main entrance door measures 3'-0" wide
The kitchen and laundry doors measures 2'-8" wide
All closet openings measure 4'-0" wide
All interior walls measure 4"
The exterior walls with brick veneer measures 5"

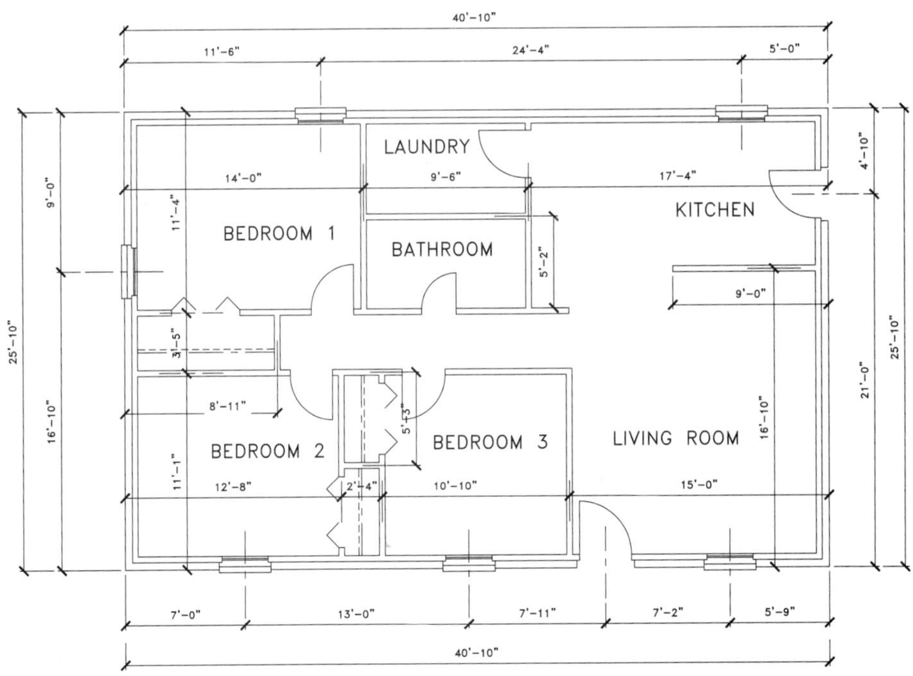

FLOOR PLAN
1/4" = 1'-0"

PROBLEM 16–12

Directions for Problem 16–12

Use blocks to help layout the service garage and adjoining parking facilities. Provided are three sketches, in meters. The first is the proposed layout. Sketch two and three are details of the service work area and parking spaces. One issue that needs resolution with the proposed layout is the access to the service work areas.

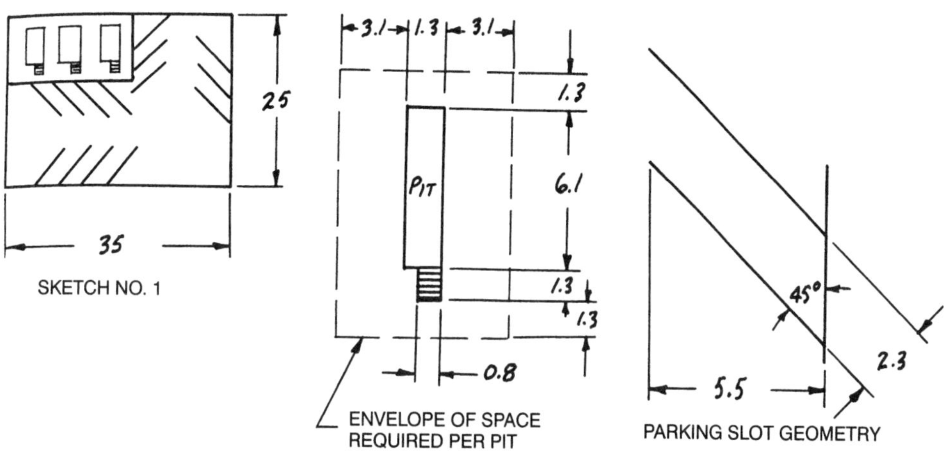

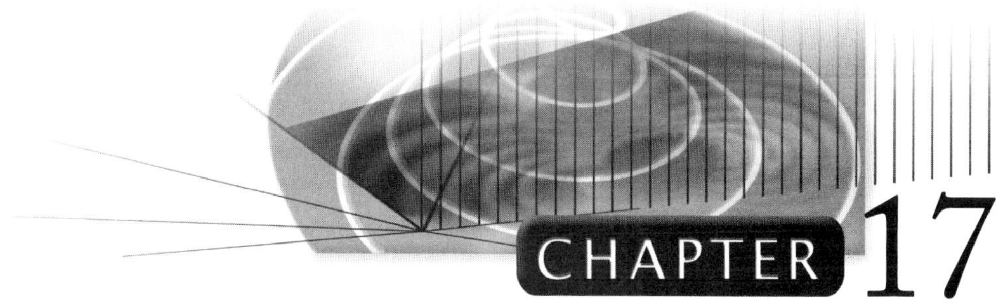

Using Attributes

The exercises in this chapter introduce:

- Creating attributes
- Attribute application within blocks
- Editing attributes
- Extracting attributes into a spreadsheet

TUTORIAL EXERCISE: 17_TAPERED BEARING.DWG

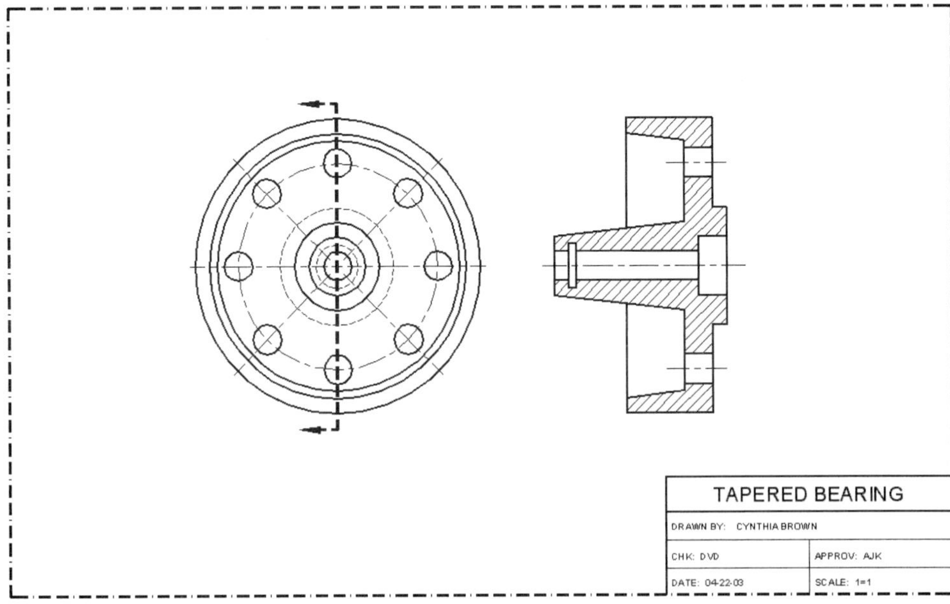

Figure 17–1

Purpose

This tutorial is designed to edit attribute values through the Edit Attributes dialog box using the drawing illustrated in Figure 17–1.

System Settings

Open the drawing *17_Tapered Bearing*. All units and drawing limits have already been set.

Layers

All layers have already been created for this drawing.

Suggested Commands

Activate the Edit Attributes dialog box by entering the ATTEDIT command and clicking anywhere on the title block or border. Make changes to the attribute values from inside the Edit Attributes dialog box. Click the OK button to update the attribute values.

Whenever possible, substitute the appropriate command alias in place of the full AutoCAD command in each tutorial step. For example, use "CP" for the COPY command, "L" for the LINE command, and so on. The complete listing of all command aliases is located in Chapter 1, Table 1–2.

STEP 1

Open the drawing *17_Tapered Bearing.dwg* shown in Figure 17–2. Use the ATTEDIT command and click the title block.

Command: **ATTEDIT** [modify>object>attribute>single]

Select block reference: *(Select the edge of the title block)*

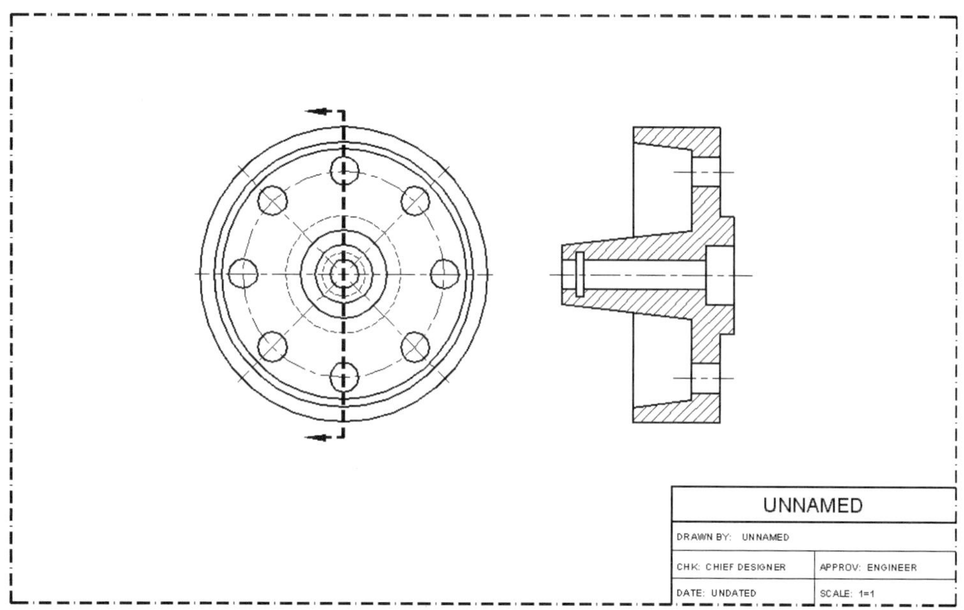

Figure 17–2

STEP 2

When the Edit Attributes dialog box appears, make changes to the various boxes, as illustrated in Figure 17–3.

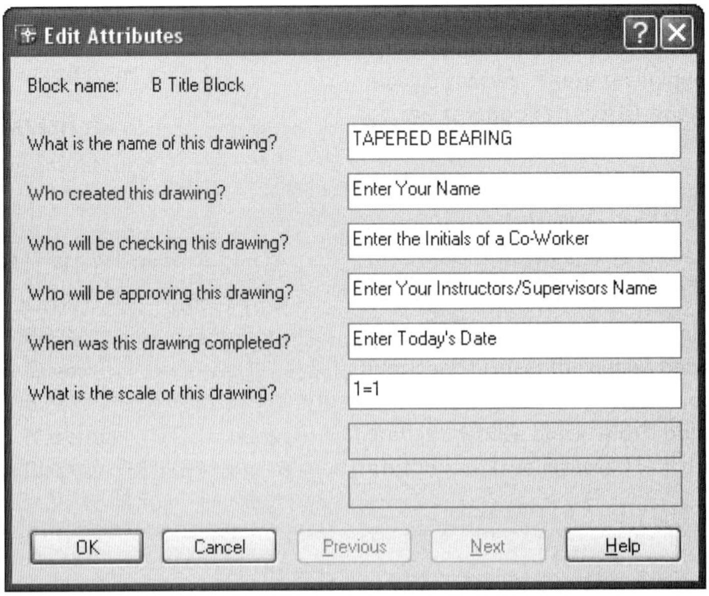

Edit Attributes
Block name: B Title Block
What is the name of this drawing?
TAPERED BEARING
Who created this drawing?
Enter Your Name
Who will be checking this drawing?
Enter the Initials of a Co-Worker
Who will be approving this drawing?
Enter Your Instructors/Supervisors Name
When was this drawing completed?
Enter Today's Date
What is the scale of this drawing?
1=1
OK
Cancel
Previous
Next
Help

Figure 17–3

STEP 3

Clicking the OK button of the Edit Attributes dialog box displays a modified title block similar to the one shown in Figure 17-4. This completes this tutorial exercise.

TAPERED BEARING
DRAWN BY: CYNTHIA BROWN
CHK: DVD
APPROV: AJK
DATE: 04-22-03
SCALE: 1=1

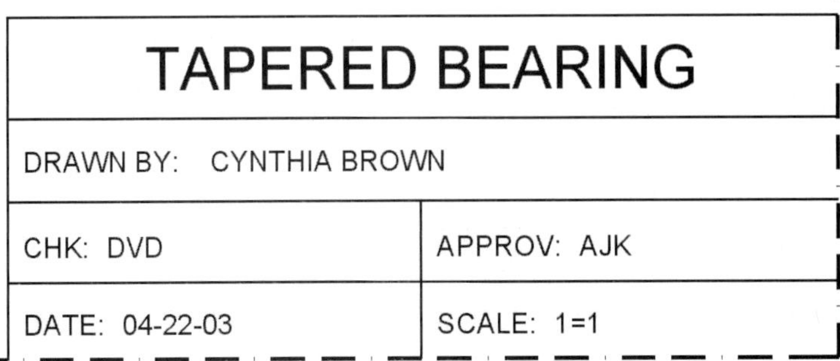

Figure 17–4

TUTORIAL EXERCISE: 17_CAD DEPARTMENT.DWG

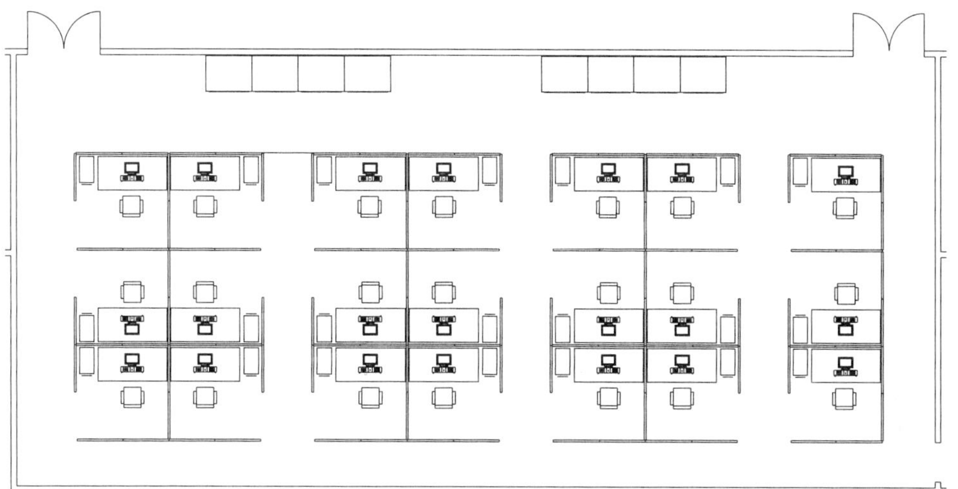

Figure 17–5

Purpose

This tutorial is designed to extract attributes in Comma Delimited file (CDF) and Space Delimited file (SDF) formats.

System Settings

Open the drawing *17_CAD Department* in Figure 17–5. All units and drawing limits have already been set. Attributes have also been assigned to computer components.

Layers

All layers have already been created for this drawing.

Suggested Commands

Before extracting attributes, you must first create a template file consisting of the following tags, character classifications, and column widths:

PRODUCT_NAME	C015000
MANUFACTURER	C020000
COST	N010002

Save this *.txt* file by as *CAD Department.txt*. Be sure to keep track of where this file is stored.

Activate the Attribute Extraction dialog box through the ATTEXT command and extract two text files. These extractions will illustrate the differences between the CDF and SDF files.

Whenever possible, substitute the appropriate command alias in place of the full AutoCAD command in each tutorial step. For example, use "CP" for the COPY command, "L" for the LINE command, and so on. The complete listing of all command aliases is located in Chapter 1, Table 1–2.

tutorial EXERCISE

STEP 1

Use the Notepad application shown in Figure 17–6 to create a template file consisting of the following tags, character classifications, and column widths:

PRODUCT_NAME	C015000
MANUFACTURER	C020000
COST	N010002

Save this template file as *CAD Department.txt.*

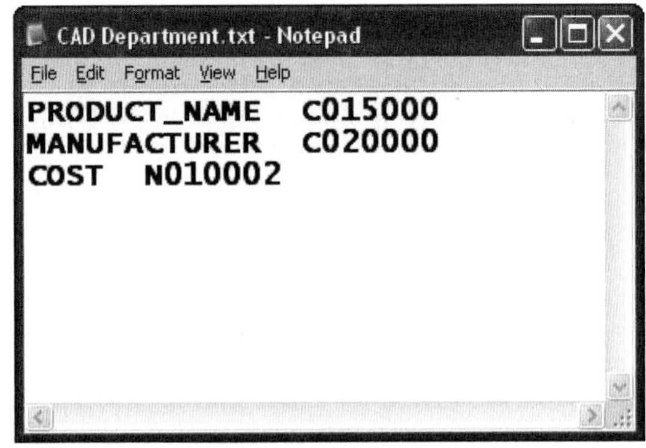

Figure 17–6

STEP 2

Use the ATTEXT command to display the Attribute Extraction dialog box. Be sure the Comma Delimited File (CDF) radio button is selected. In the Template File edit box, be sure the template file you just created is listed there (*CAD Department.txt*). In the Output File edit box, type the name **CAD Department CDF.txt**. This file will hold all of the extracted attribute records in comma delimited format. The dialog box should appear similar to Figure 17–7. Click the Select Objects< button and select all attributes using the Window option. When finished selecting the attribute objects, click the OK button. Notice, in your command line, a statement that reports you have "42 records in extract file." These 42 records can be found in the file *CAD Department CDF.txt.*

Note When creating the output file, keep track of which folder it is stored in. If you had errors during your extraction, open the template file and verify its contents. You can use uppercase or lowercase letters in a template file (it is not case sensitive) and the spacing between the tag and character classification/width format does not matter (they must be on the same line). However, misspelling a tag or an incorrect format specification will cause extraction errors.

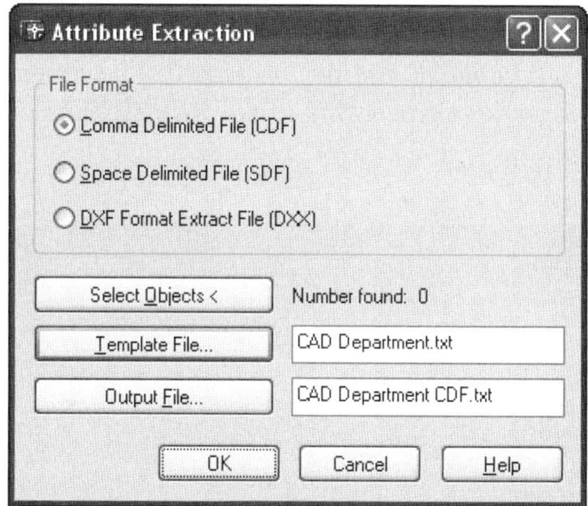

Figure 17–7

STEP 3

Activate the ATTEXT command again to display the Attribute Extraction dialog box. This time, click the Space Delimited File (SDF) radio button. In the Template File edit box, be sure the template file is *CAD Department.txt*. In the Output File edit box, type the name **CAD Department SDF.txt.** This file will hold all of the extracted attribute records in space delimited format. The dialog box should appear similar to Figure 17–8. Click the Select Objects< button and select all attributes again using the Window option. When finished, click the OK button. Notice, in your command line, a statement that reports you have "42 records in extract file." These 42 records can be found in the file *CAD Department SDF.tx*t.

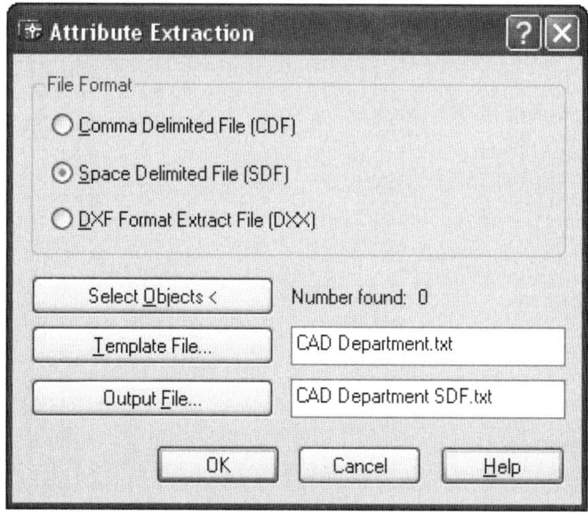

Figure 17–8

STEP 4

View the contents of the CDF file by opening *CAD Department CDF.txt* in Notepad. The contents of this file should be similar to Figure 17–9.

CAD Department CDF.txt - Notepad

File Edit Format View Help

```
'CPU','DELL', 2400.00
'CPU','DELL', 2400.00
'CPU','DELL', 2400.00
'MONITOR','DELL', 400.00
'MONITOR','DELL', 400.00
'CPU','DELL', 2400.00
'MONITOR','DELL', 400.00
'MONITOR','DELL', 400.00
'CPU','DELL', 2400.00
'MONITOR','DELL', 400.00
'CPU','DELL', 2400.00
'MONITOR','DELL', 400.00
'MONITOR','DELL', 400.00
'MONITOR','DELL', 400.00
'CPU','DELL', 2400.00
'CPU','DELL', 2400.00
'CPU','DELL', 2400.00
'CPU','DELL', 2400.00
'MONITOR','DELL', 400.00
'MONITOR','DELL', 400.00
```

Figure 17–9

STEP 5

View the contents of the SDF file by opening *CAD Department SDF.txt* in Notepad. The contents of this file should be similar to Figure 17–10.

CAD Department SDF.txt - Notepad

File Edit Format View Help

```
CPU            DELL                     2400.00
CPU            DELL                     2400.00
CPU            DELL                     2400.00
MONITOR        DELL                      400.00
MONITOR        DELL                      400.00
CPU            DELL                     2400.00
MONITOR        DELL                      400.00
MONITOR        DELL                      400.00
CPU            DELL                     2400.00
MONITOR        DELL                      400.00
CPU            DELL                     2400.00
MONITOR        DELL                      400.00
MONITOR        DELL                      400.00
MONITOR        DELL                      400.00
CPU            DELL                     2400.00
CPU            DELL                     2400.00
CPU            DELL                     2400.00
CPU            DELL                     2400.00
MONITOR        DELL                      400.00
MONITOR        DELL                      400.00
```

Figure 17–10

TUTORIAL EXERCISE: 17_CAD DEPT MODULES.DWG

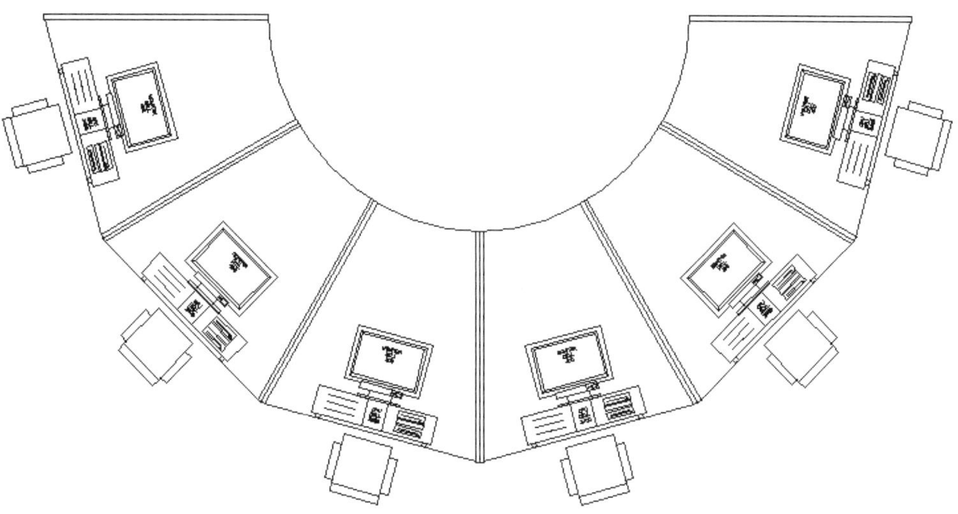

Figure 17–11

Purpose

This tutorial is designed to extract attribute data through the use of the Attribute Extraction Wizard for the drawing *17_CAD Dept Modules,* illustrated in Figure 17–11.

System Settings

All units and drawing limits have already been set. Attributes have also been assigned to computer components.

Layers

All layers have already been created for this drawing.

Suggested Commands

Follow the steps provided by the Attribute Extraction Wizard to create an XLS file and view the results using Microsoft Excel.

Whenever possible, substitute the appropriate command alias in place of the full AutoCAD command in each tutorial step. For example, use "CP" for the COPY command, "L" for the LINE command, and so on. The complete listing of all command aliases is located in Chapter 1, Table 1–2.

STEP 1

Open the drawing file *17_CAD Dept Modules.dwg* in Figure 17–11. Activate the Attribute Extraction Wizard, as shown in Figure 17–12 using the EATTEXT command. Although you will be extracting attribute information from the current drawing, notice that the wizard allows you to select any drawing to perform the extraction process. With the radio button set to Current Drawing, click the Next> button to continue.

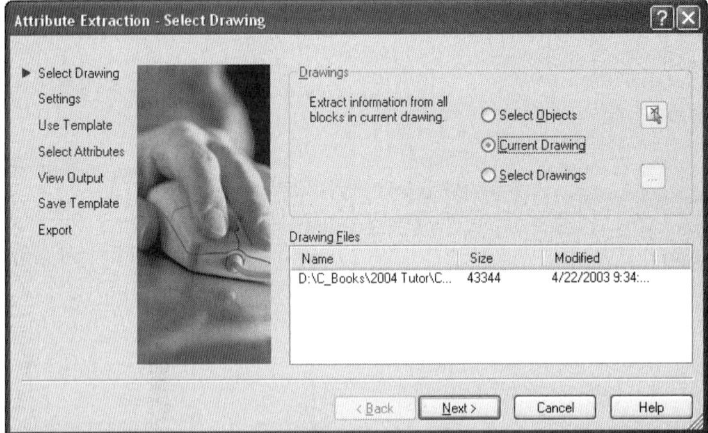

Figure 17–12

STEP 2

With the Attribute Extraction - Settings dialog box displayed in Figure 17–13, you have the option of extracting attributes that are part of external references or nested blocks. Click the Next> button to continue.

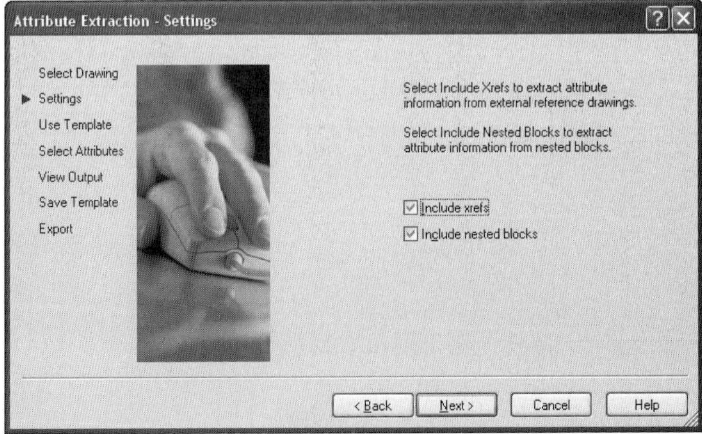

Figure 17–13

STEP 3

In the Attribute Extraction - Use Template dialog box (see Figure 17–14), you can choose whether or not to use a template.

Using block templates is very different from the past attribute extraction operations. Instead of the standard TXT format, the Attribute Extraction Wizard requires a block template with the BLK extension. Don't worry, you can still extract attributes through the wizard without a template. If you will be performing extraction operations numerous times to gain the same information, it would be useful to create and save your own black template. However, you cannot create the black template in such applications as Notepad. You will create this type of template with the help of the wizard in a later step. For now, verify that the No template option is selected and click the Next> button to continue.

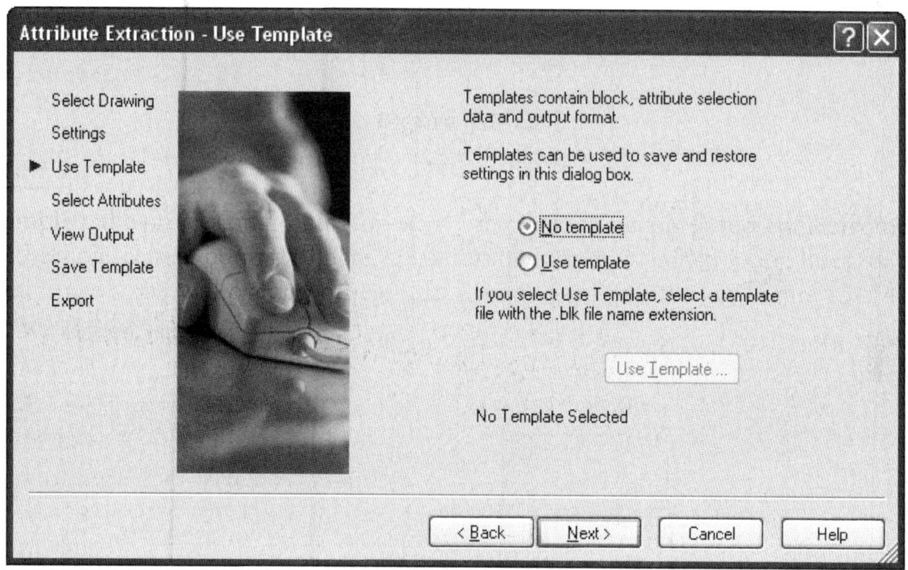

Figure 17–14

STEP 4

The Attribute Extraction - Select Attributes dialog box is shown in Figure 17–15. Here is where you specify the attribute information to be extracted. The left side of the dialog box displays all blocks in the drawing. Attribute information, as it relates to each block is displayed on the right. First, select the block names Monitor and CPU; remove the check marks from the other block names. Next, use the right side of the dialog box to determine what information to extract. With the Monitor block highlighted, click the Uncheck All button and then individually select the following attributes: Layer, COST, MANUFACTURER, and PRODUCT_NAME. Then, click the CPU block name and perform the same operations and select the same attributes as with the Monitor block. Click the Next> button to continue.

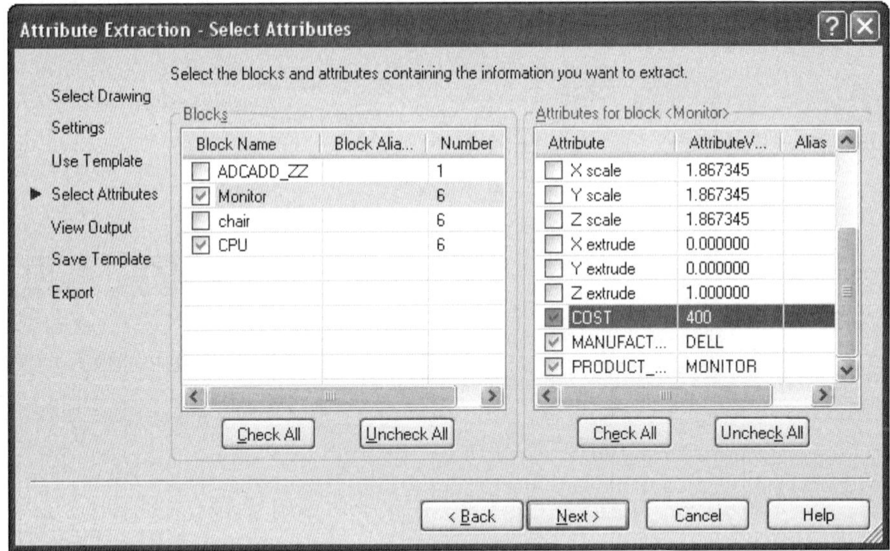

Figure 17–15

STEP 5

Observe the results when the Attribute Extraction - View Output dialog box shown in Figure 17–16, is displayed. This dialog box provides an excellent means of previewing your results. If you accidentally left out an attribute category, you could click the <Back button and place more checks in the appropriate attribute information boxes. You could also copy these results to the Windows Clipboard by clicking the Copy to Clipboard button. This would enable you to paste the contents of the clipboard into such applications as Notepad, as shown in Figure 17–17. Click the Next> button to continue.

Note The results displayed in Figure 17–16 were achieved by clicking on the Alternate View button located at the bottom of the View Output dialog box.

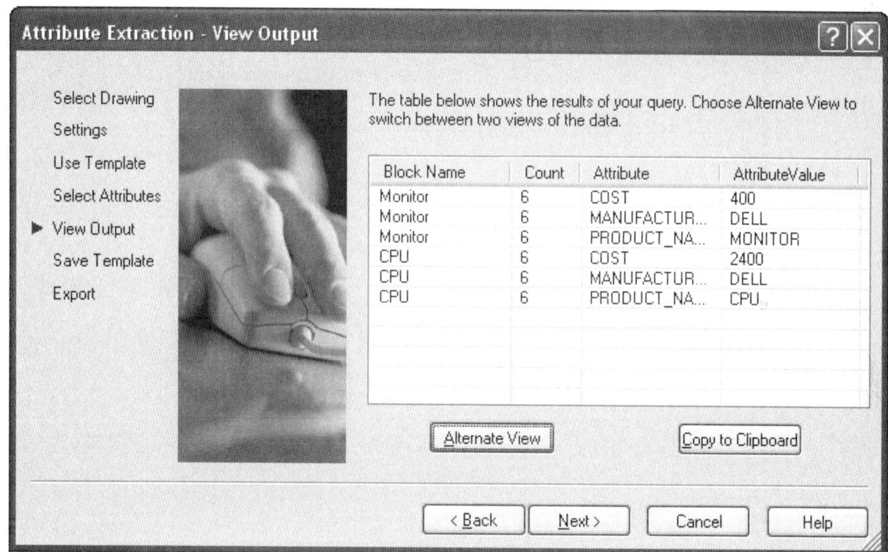

Figure 17–16

Untitled - Notepad

File Edit Format View Help

Block Name	Count	Attribute	AttributeValue
Monitor	6	COST	400
Monitor	6	MANUFACTURER	DELL
Monitor	6	PRODUCT_NAME	MONITOR
CPU	6	COST	2400
CPU	6	MANUFACTURER	DELL
CPU	6	PRODUCT_NAME	CPU

Figure 17–17

STEP 6

This next dialog box, illustrated in Figure 17–18, allows you to save your block template. The information saved in this template will match the attribute specifications you selected in Step 5. The purpose of saving a template is to use it again, in the future. Click the Save Template button to display the Save As dialog box, as shown in Figure 17–19. Give the template a name such as *CAD Dept Modules* and notice that the BLK extension will be used as the file extension. After clicking the Save button, you will be returned to the Attribute Extraction - Save Template dialog box; click the Next> button to continue.

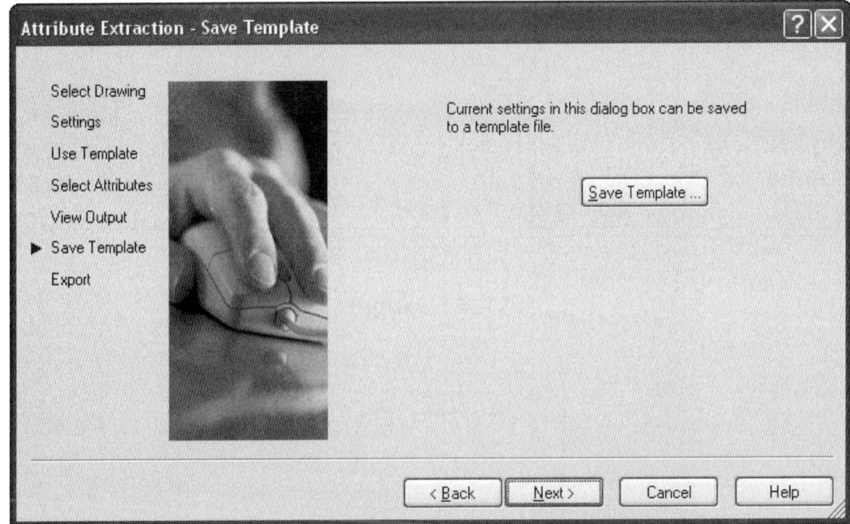

Figure 17–18

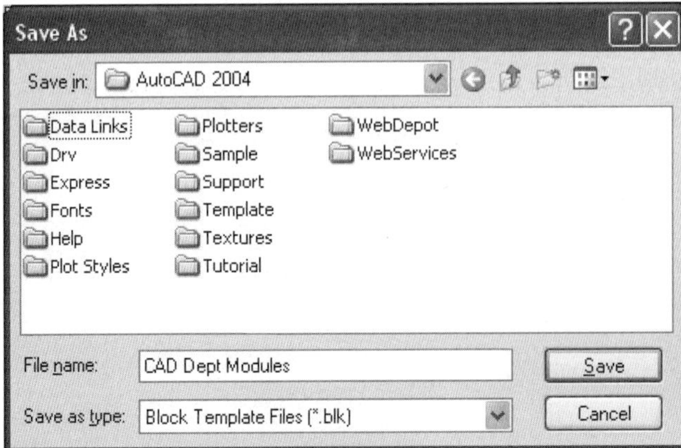

Figure 17–19

STEP 7

The last step in the extraction process is to select the file type. As shown in Figure 17–20, three file types are available through the wizard. They are CVS, TXT, and XLS. Enter **CAD Dept Modules** as the File Name and click the Microsoft Excel (*.xls) file type. When you click the Finish button, the Enhanced Attribute Extract Alert dialog box appears, as shown in Figure 17–21. Click the Yes button to write the XLS file to the specified folder.

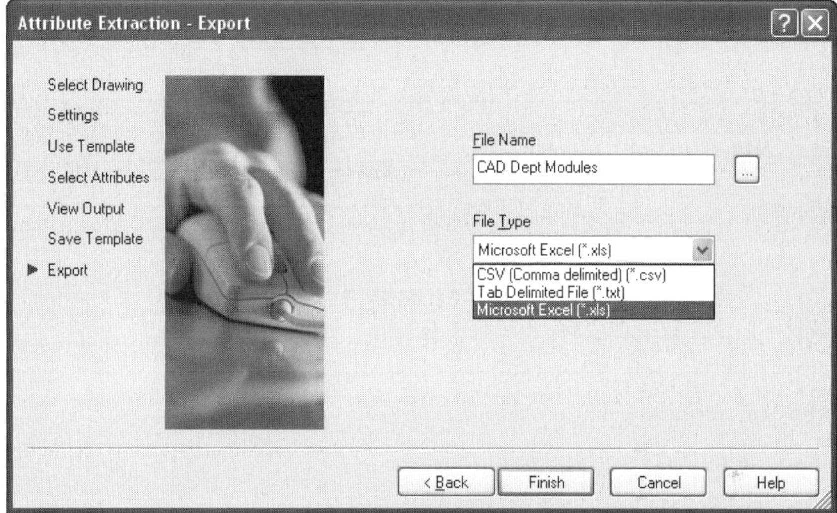

Figure 17–20

Figure 17–21

STEP 8

To verify the results, open Microsoft Excel and open the XLS file that you just created using the Attribute Extraction Wizard. The result should be similar to Figure 17–22.

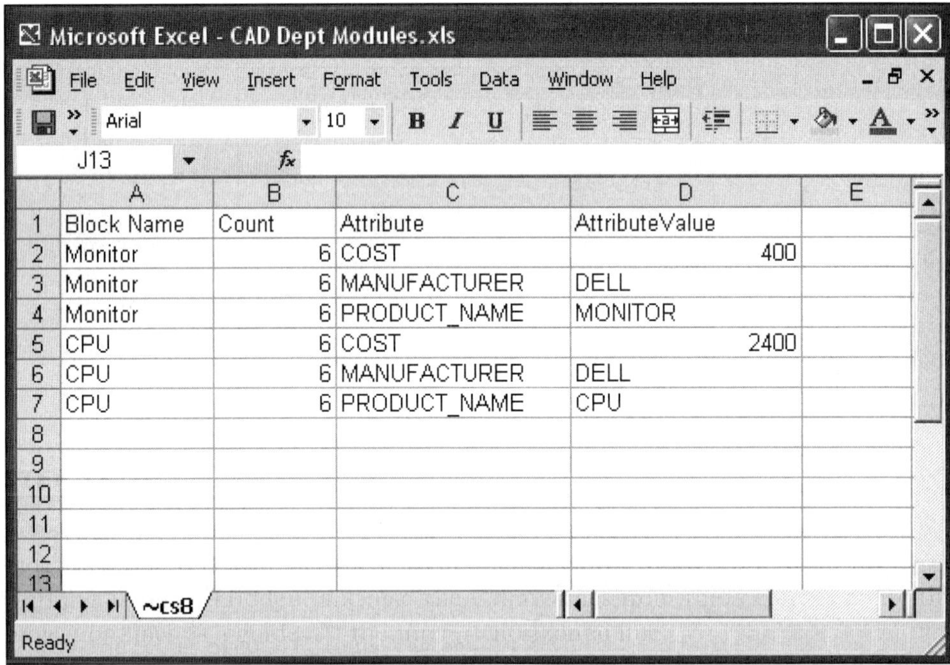

Figure 17–22

Beginning Level Drawings

PROBLEM 17–1

Directions for Problem 17–1

This is a two-part project. Part 1 is the development of the equipment found in an office suite with attributes applied in blocks. Part 2 is the extraction of the attributes into a spreadsheet to tally the equipment costs.

Part 1 – Drawings and Attribute Information

Create the following items as full size, apply two attributes to each item (Name and Cost). Use the WBLOCK command to create each of the equipment items.

Object	Size	Attribute 1	Value 1	Attribute 2	Value 2
Desk	3' X 5'	Mfg	Hon	Cost	$265
Chair	2' X 2'	Mfg	Ikea	Cost	$138
Computer	1' X 2'	Mfg	Dell	Cost	$739
File cabinet	3' X 4'	Mfg	Hon	Cost	$372
Trash can	1' X 1'	Mfg	Rubbermaid	Cost	$4

Part 2 – The Office complex

Create an office complex that has 12 offices in it.

The office will be 10' x 10' in size, door/openings into the office will be 3'-0".

Each office will have the required equipment (created in part one of this exercise) through the block insertion process.

Once the drawing is complete, extract the attributes into a spreadsheet file and total the costs for the company.

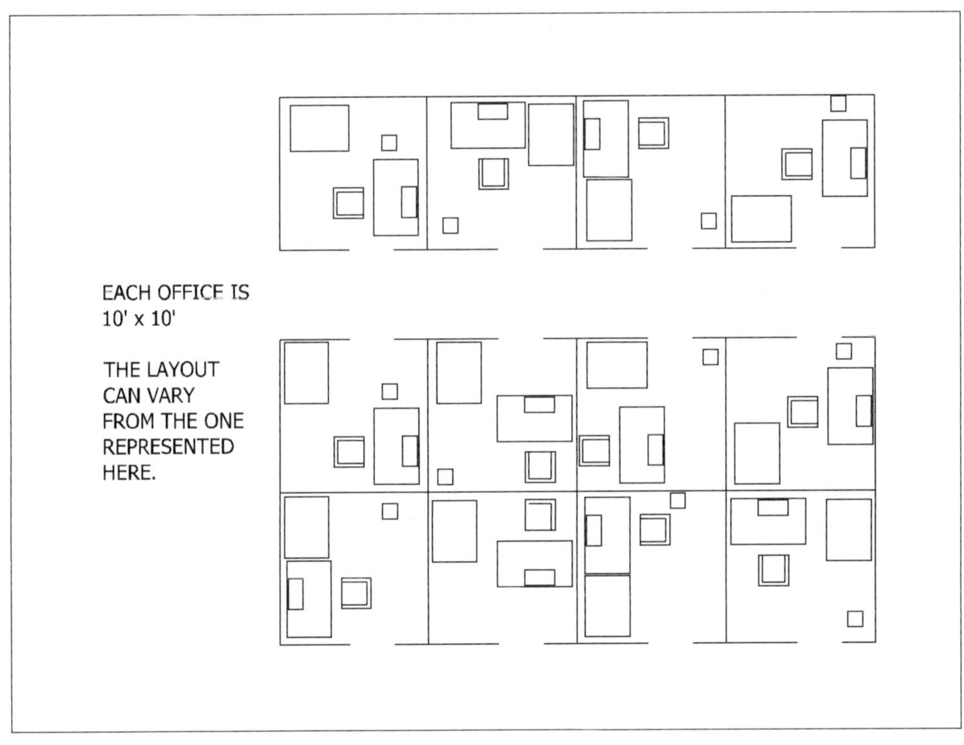

External References

The exercises in this chapter introduce:

- Attaching External references
- Editing External references in a drawing
- Managing External references

TUTORIAL EXERCISE: SUBDIVISION-XREF.DWG

Purpose

This tutorial is designed to show the application of external references in a drawing environment. The external references will be applied as an attachment. The ability to place edit an external reference will be accomplished to show the ease of modifying an existing external reference.

Suggested Commands

XREF

REFEDIT

ZOOM

PAN

STEP 1

Open the drawing file *subdivision.dwg*

Survey the layout of the portion of the subdivision, shown in Figure 18–1. In this tutorial, placement of the fire hydrants will be needed.

The general rules on the placement of fire hydrants in our subdivision, are that they cannot be farther than 400' apart. Another guideline is that there must be a visible line of site between hydrants. From one hydrant, there must be at least one other visible. These guidlines are dependant on local building codes.

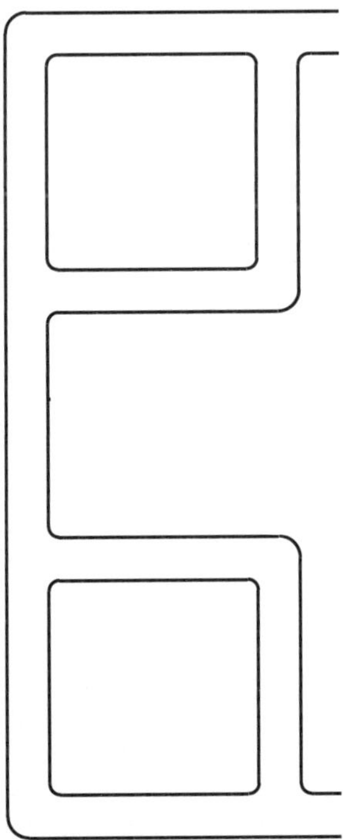

Figure 18–1

STEP 2

Attaching an external reference is much like inserting a block. Select External Reference from the Insert pull-down menu or click the External Reference button on the Insert toolbar and then press the Attach button in the Xref Manager dialog box (see Figure 18–2). Both methods will bring you to the Select Reference File dialog box, shown in Figure 18–3.) Select the fire hydrant symbol drawing from the *Chapter 18* folder on the accompanying CD.

Command: **XREF** [insert>external reference]

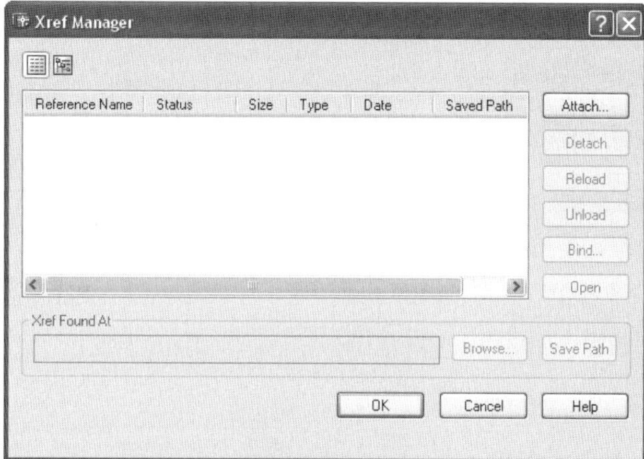

Figure 18–2

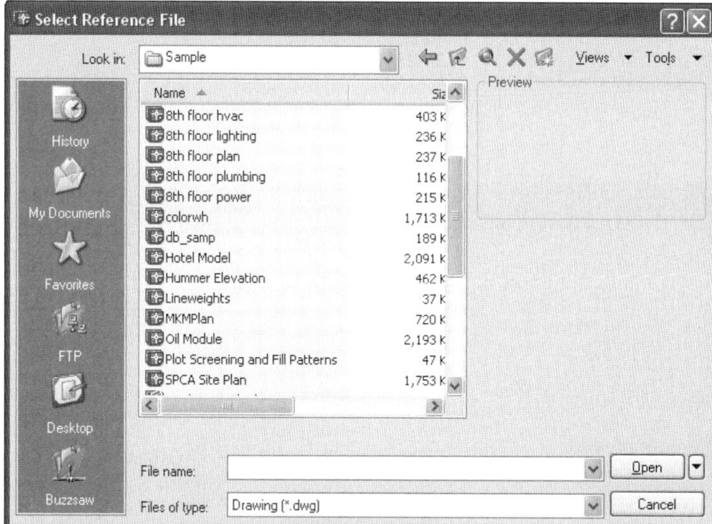

Figure 18–3

STEP 3

Once the file is selected, the External Reference dialog box appears in order to make the final insertion adjustments. This dialog box is very similar to the Insert dialog box used when adding blocks. Make sure you select Attachment for the Reference Type. The Path type will select the linking path information stored in the external reference.

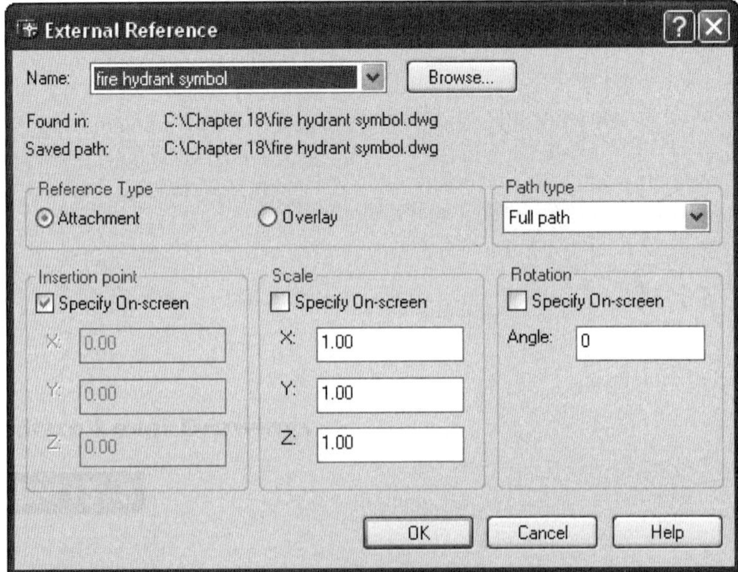

Figure 18–4

STEP 4

The fire hydrant symbol is very small compared to the drawing size, so the scale must be modified in order for it to be seen. Set the scale value to 5 for the X, Y, and Z scale factors. Attach the fire hydrant near the corner of the upper left street, as shown in Figure 18–5.

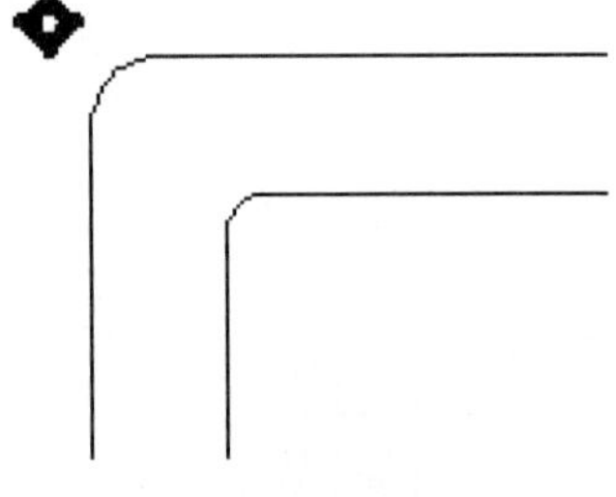

Figure 18–5

STEP 5

The next step is to add more fire hydrant symbols to the drawing. See Figure 18–6 for a finished example. Regardless of whether the symbols are attached using external reference, or are copied into new positions, the symbol retains the original characteristics and path information. This allows all the symbols that have been placed to be changed quickly throughout a drawing.

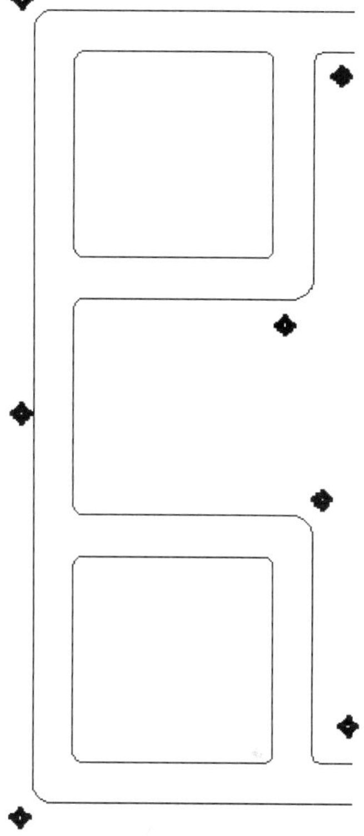

Figure 18–6

STEP 6

To change an external reference you can use In-place reference editing. This modification tool allows you to change the original external reference file or block file without exiting the current drawing. To start the command, choose Xref and Block editing from the modify-pull down menu or select the Edit block or Xref button from the Refedit toolbar or type the REFEDIT command at the AutoCAD command prompt.

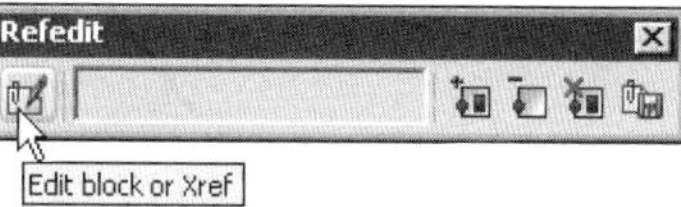

Command: **REFEDIT** [modify>xref and block editing>Edit reference in-place]
Select reference: *(Pick any of the fire hydrants on the drawing)*

To end the reference editing session use REFCLOSE or the Save back changes to reference button on the Refedit toolbar.

STEP 7

If using the pull-down menu or toolbar, select the fire hydrant drawing from the Reference name list in the left column (see Figure 18–7) and click OK. The single hydrant that is on the drawing can now be modified. This modification will be saved over the original drawing of the object. Add a small circle in the center of the hydrant, as shown Figure 18–8.

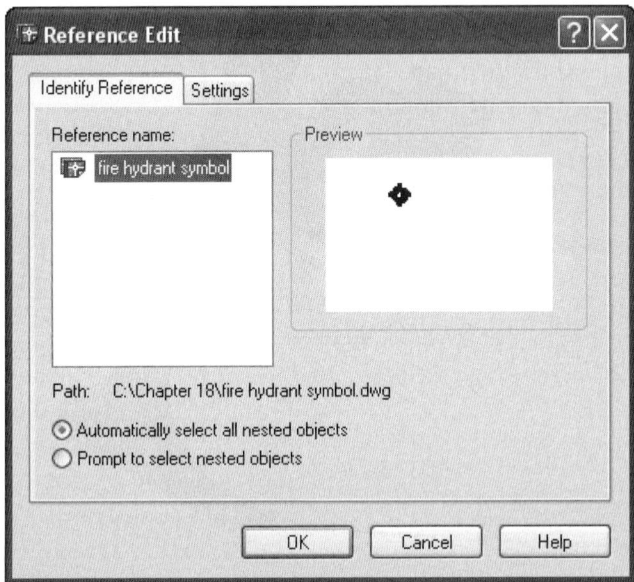

Figure 18–7

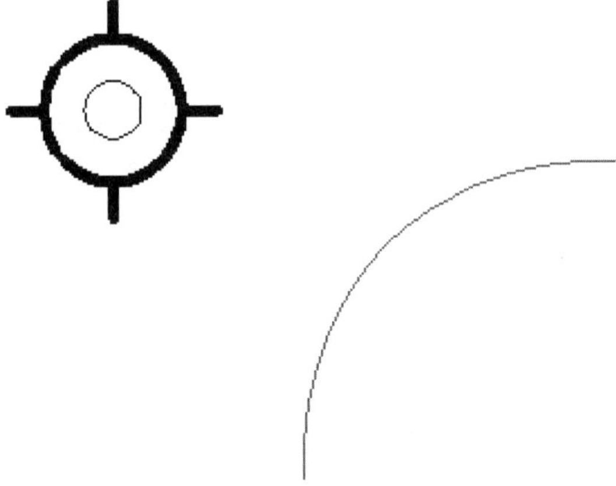

Figure 18–8

STEP 8

Choose the Save back changes to reference button on the right end of the Refedit toolbar or type REFCLOSE at the command prompt.

Command: **REFCLOSE**
Enter option [Save/Discard reference changes] <Save>: **Enter**

An AutoCAD alert dialog box, shown in Figure 18–9, displays to confirm that you want to save these changes to the reference. Choose the OK button. The command prompt confirms that the reference was updated.

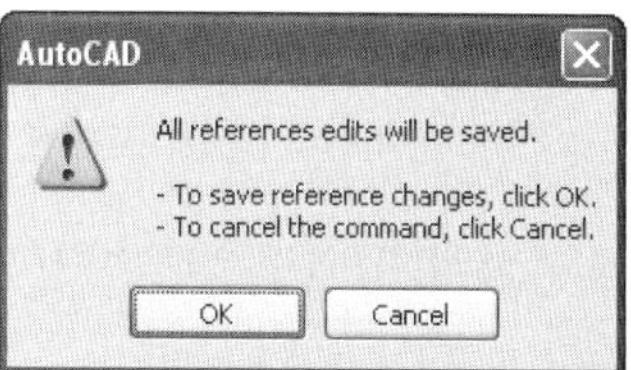

Figure 18–9

Regenerating model.
1 object added to fire hydrant symbol 7 xref instances updated fire hydrant symbol redefined and reloaded.

Your drawing should now look similar to Figure 18–10.

Figure 18–10

STEP 9

The external reference has been modified throughout the drawing. To confirm the change has been completed, open the Xref Manager, select the reference in the list, and click the Reload button (see Figure 18–11). This will cause AutoCAD to access the original drawing and reload the image, to make sure that it has been completely updated.

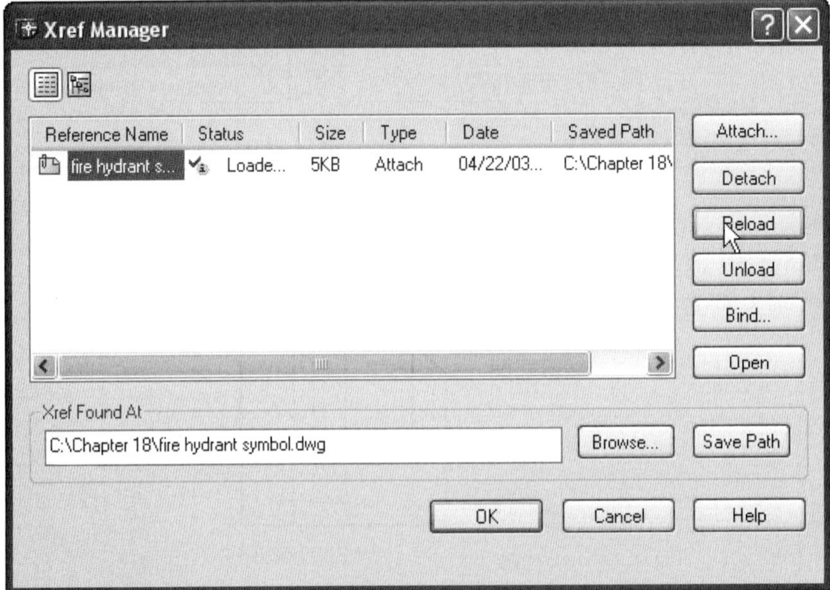

Figure 18–11

Beginning Level Drawing

PROBLEM 18–1

Directions for Problem 18–1

Open up the drawing *18_track layout.dwg*. This 4' X 8' area will be home to an HO scale train layout by using external references. Two styles of track are provided; *18_track HO straight.dwg* & *18_track HO curve.dwg*. To create a more complex layout, develop some switches.

Note: The basic angle between the straight track and switch is approx 10 degrees. To create more accurate switches use Internet references on model railroading or model railroading handbooks.

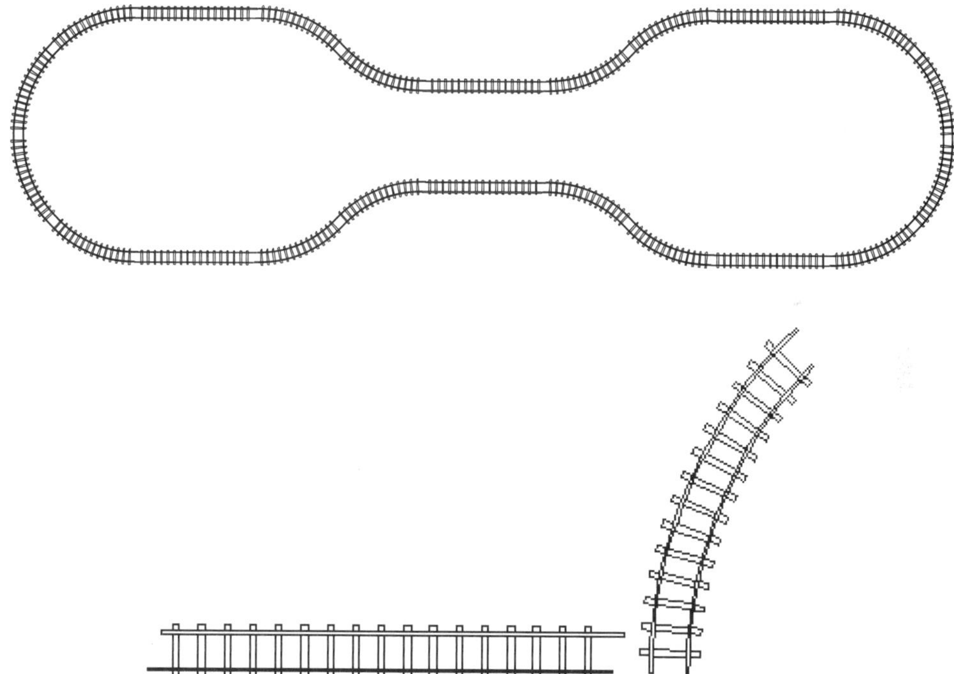

problem EXERCISE

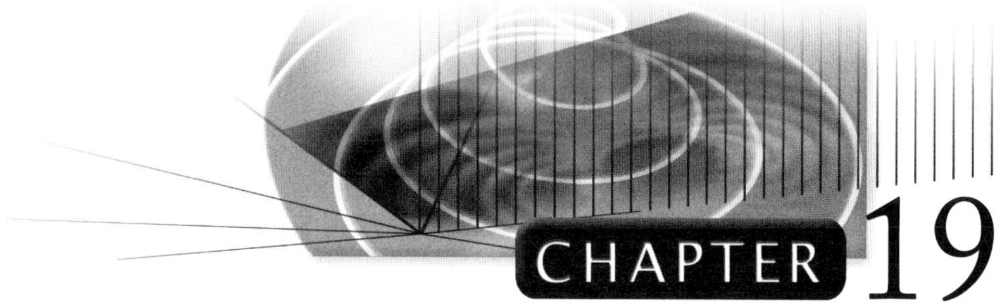

Multiple Viewport Drawing Layouts

The exercises in this chapter introduce:

- Apply viewports to layouts
- Scale viewports within layouts
- Controlling layer properties within a viewport
- Developing detail views
- Viewport visibility controls for printing

TUTORIAL EXERCISE: 19_HOUSING2.DWG

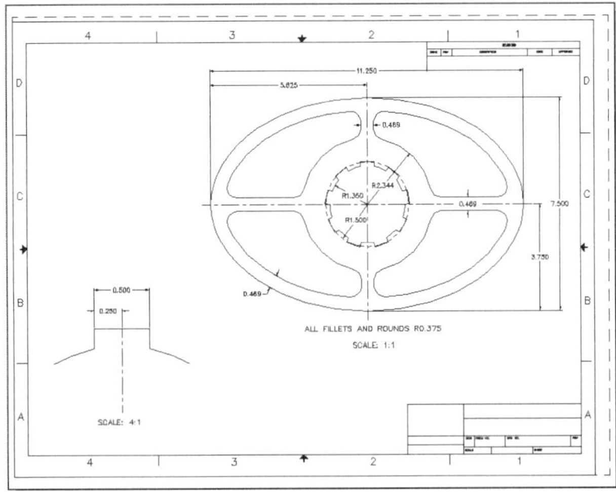

Figure 19–1

Purpose

This tutorial is designed to lay out two views of the same drawing at different scales on the same sheet, using the layout mode (Paper Space environment) in Figure 19–1. This drawing is provided on the CD under the name of *19_Housing2.dwg*.

System Settings

Keep the default drawing limits at (0.0000,0.0000) for the lower-left corner and (12.0000,9.0000) for the upper-right corner.

Use the Drawing Units Dialog box (activated through the UNITS command) to change the number of decimal places past the zero from four to three.

Layers

In addition to the current layers of the drawing, create the following layers:

NAME	COLOR	LINETYPE
Border	Blue	Continuous
Detail-Cen	Magenta	Center
Detail-Dim	Magenta	Continuous
Vports	Gray (8)	Continuous

Suggested Commands

Open the drawing file *19_Housing2*. Enter the layout mode (Paper Space environment), insert a "C" size title block, and create two viewports: one large and the other small. As the image of 19_Housing2 appears in both, use the ZOOM command to scale the first image to Paper Space units at a factor of 1XP or full size. Then scale the detail using a scale factor of 4XP, which is four times its original size. Freeze the DIM layer in the detail viewport while leaving it intact in the large viewport.

Create a new dimension style called DETAIL and change the dimension scale to reflect the current scale of the detail image inside the viewport. Place two dimensions inside this viewport. Then Freeze the Detail-Dim layer in the large viewport. Plot the drawing out at a scale of 1=1.

Whenever possible, substitute the appropriate command alias in place of the full AutoCAD command in each tutorial step; for example, use "CP" for the COPY command, "L" for the line command, and so on. The complete listing of all command aliases is located in Table 1–2.

STEP 1

Before beginning the layout of 19_Housing2 and detail, create the following new layers, as illustrated in Figure 19–2, in the Layer Properties Manager dialog box: Border, Detail-Cen, Detail-Dim, and Vports. Here is an explanation of the purpose of these layers:

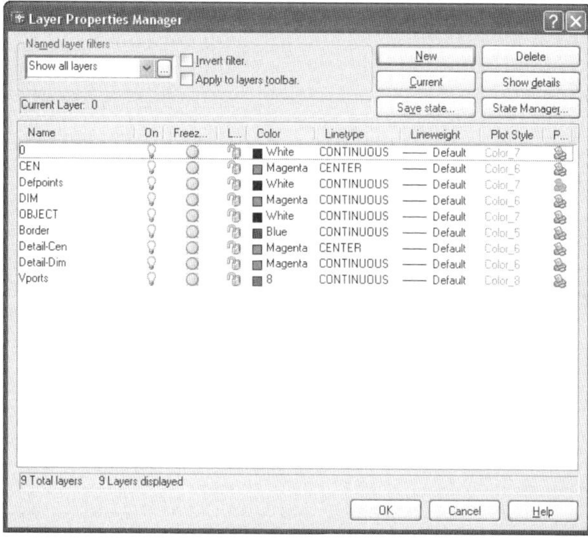

Figure 19–2

BORDER

This layer will hold the title block information that is inserted in Paper Space.

DETAIL-CEN

This layer is designed specifically for the detail that will accompany the main image of 19_Housing2. A special centerline will need to be drawn in the detail. However, a centerline will already be visible in the detail on the CEN layer. Paper Space allows for layers to be visible in certain viewports and frozen in others. Once the CEN layer is frozen in the detail viewport, a new centerline will be drawn in the detail viewport on the Detail-Cen layer.

DETAIL-DIM

This layer is specifically designed to hold any dimension drawn in the detail viewport. Again, as new dimensions are drawn in the detail viewport, they must be frozen in the main 19_Housing2 viewport. This is easy to accomplish while you are in the Paper Space environment.

VPORTS

This layer is designed to hold all viewport information as each is drawn. Viewports need to be visible for sizing them with grips, if the image inside the viewport is too large or small in relation to the image. Once the image has been properly scaled in the viewport and all information is visible in the viewport, the Layer Properties Manager dialog box is used to turn off, freeze, or set to no-plot the Vports layer. In this way, only the images and title block will plot.

STEP 2

Enter layout mode by selecting a layout tab at the bottom of the drawing screen or by clicking on the MODEL/PAPER toggle button located in the bottom Status bar area in Figure 19–3.

Figure 19–3

The Page Setup dialog box, shown in Figure 19–4 is displayed. Change the layout name in the edit box, to **Housing2 and Detail**. With the Layout Setting tab selected, choose a "ANSI ecpand C (22.00 x 17.00 Inches) size sheet of paper from the drop down list box. Be sure the current plotter is the DesignJet 750C Plus. Verify that the plotting scale is 1:1 and click OK to accept the settings.

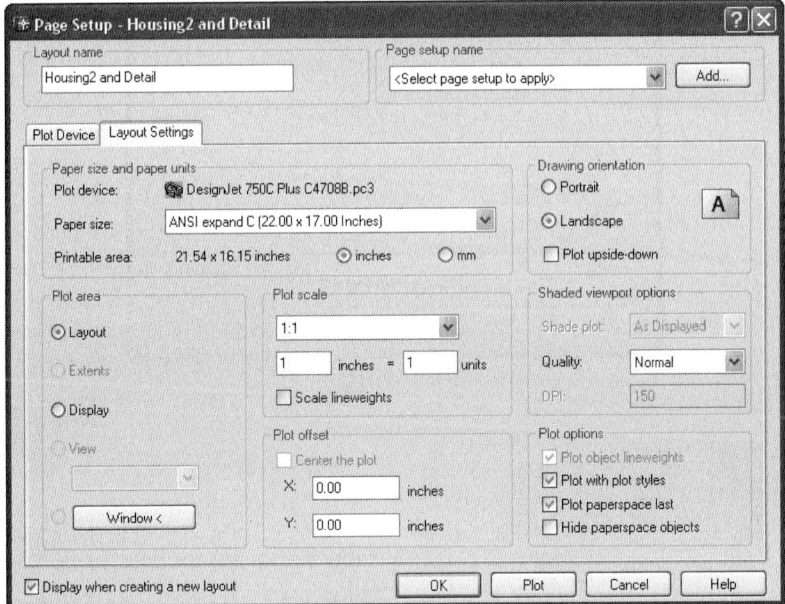

Figure 19–4

You will be returned to the screen in Paper Space and provided a view of your new drawing sheet layout (see Figure 19–5). The "C" size sheet is shown with a dashed line representing the printable area. You have also been provided with a floating viewport, which allows you to see the housing.

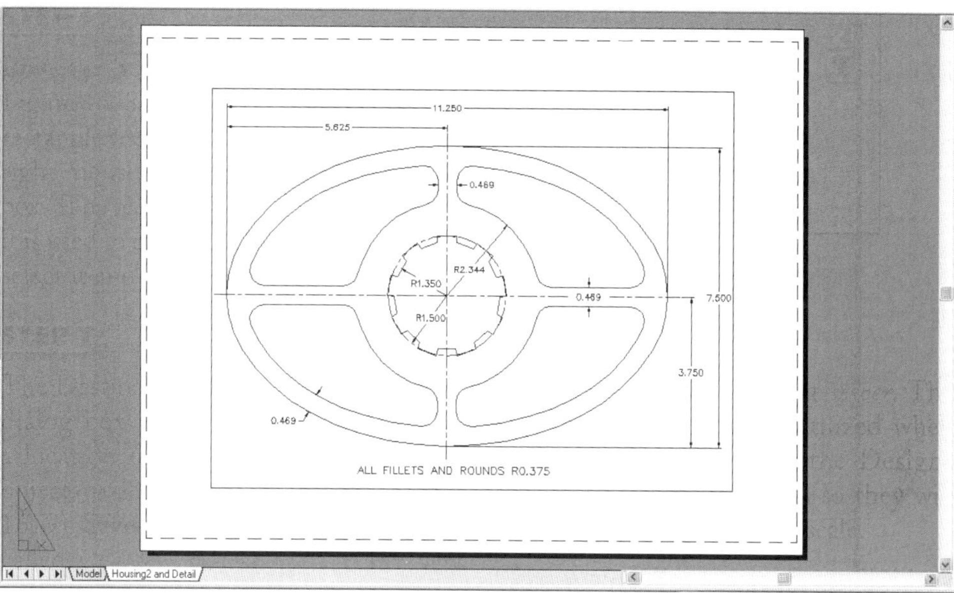

Figure 19–5

STEP 3

Insert an ANSI-C title block on the Border layer, using the INSERT command or the AutoCAD DesignCenter. You can find a title block in AutoCAD's Template folder. Your display should be similar to Figure 19–6.

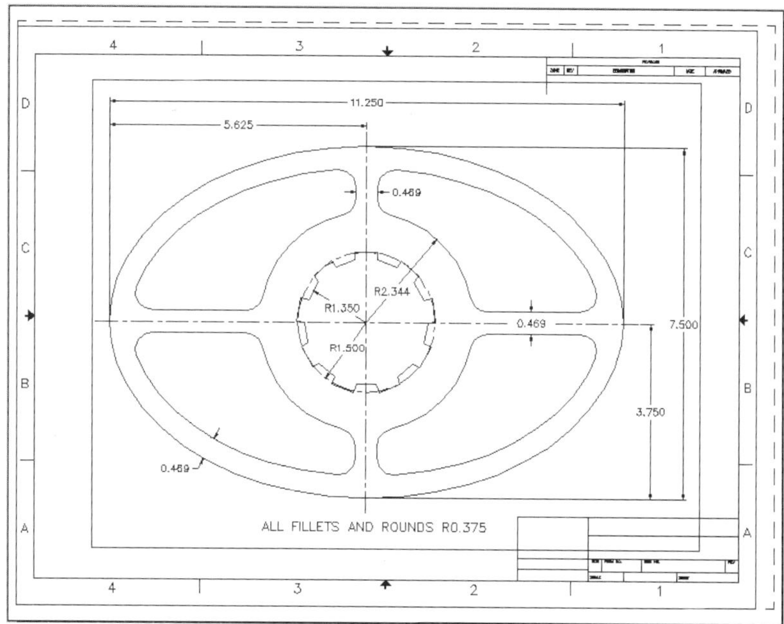

Figure 19–6

STEP 4

To make room for a second viewport that will contain a detail, resize the viewport using grips. Your drawing should appear similar to Figure 19–7. Next, create an additional viewport inside of Paper Space using the -VPORTS or MVIEW command. This can be entered at the command prompt, or by choosing Viewports> from the View pull-down menu and then 1 Viewport, or by selecting Single Viewport from the Viewports toolbar, as in Figure 19–8. The size of each viewport is not critical at this stage. Use the illustration in Figure 19–9 as a guide for how large to construct the new viewport.

Command: **-VPORTS**
[view>viewports>1 viewport]
Specify corner of viewport or [ON/OFF/Fit/Hideplot/Lock/Object/Polygonal/Restore/2/3/4] <Fit>: *(Click on "A," as shown in Figure 19–9)*
Specify opposite corner: *(Click on "B," as shown in Figure 19–9)*
Regenerating model.

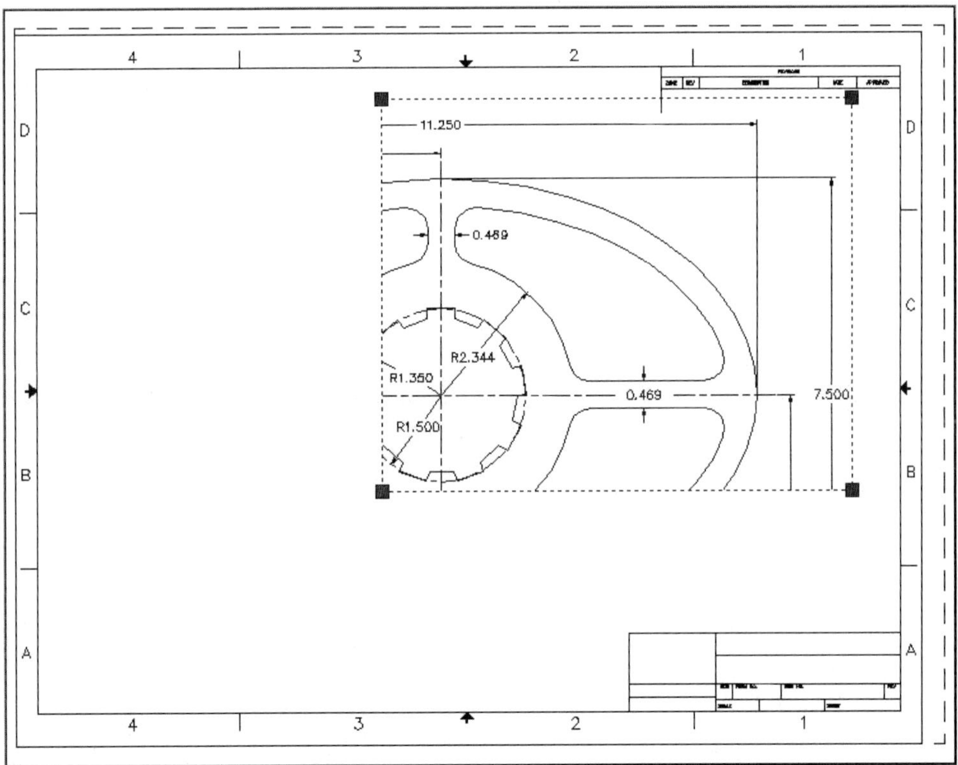

Figure 19–7

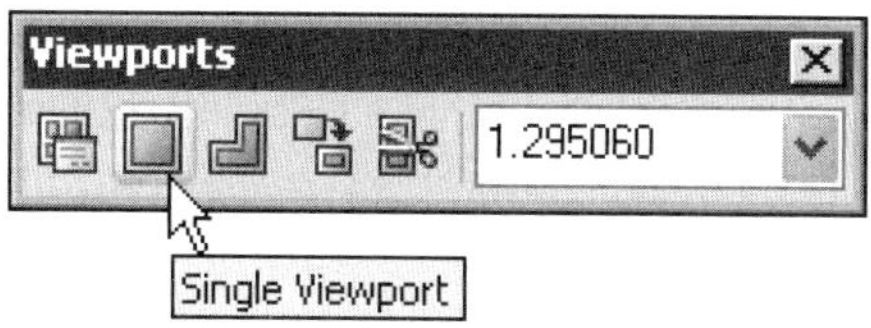

Figure 19-8

Do not be concerned if the viewports overlap each other. This is acceptable inside the Paper Space environment. After creating these viewports, notice that the original image of the drawing of 19_Housing2 reappears inside both viewports. The small viewport will contain the entire image of 19_Housing2; the larger viewport will contain an enlargement of one circular gear tooth, which will be dimensioned inside the detail viewport.

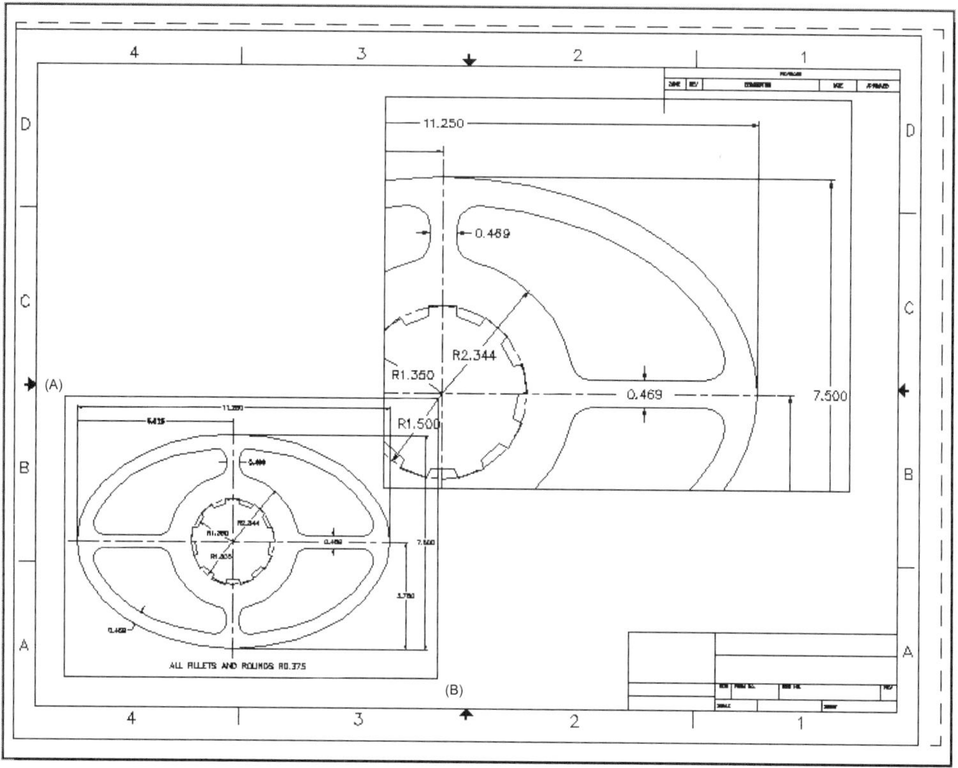

Figure 19-9

Before continuing to the next step, change both viewports to the Vports layer; their color should now be gray (assigned to the number 8).

STEP 5

Switch to floating Model Space by entering MS at the command prompt, or by clicking on the PAPER/MODEL toggle button located in the status bar, as shown in Figure 19–10, or by double-clicking a viewport.

Command: **MS** *(For MSPACE)* [choose model on status bar]
(Switches to Floating model space)

Figure 19–10

Notice that the User Coordinate System icon is present in both viewports. However, only one viewport is now active. Click inside the large viewport, shown in Figure 19–11, to make this the active viewport. Notice how the viewport outline appears thicker to signify that it is the current viewport. The image inside of this viewport will be scaled at full size in relation to Paper Space units. While in the large viewport, use the Viewports toolbar or issue the ZOOM command to scale the image inside the viewport to Paper Space units. Since the image is designed to be scaled at full size, enter a value of 1:1 in the toolbar or 1XP for the ZOOM command scale factor.

Command: **Z** *(For ZOOM)* [view>zoom]
Specify corner of window, enter a scale factor (nX or nXP), or [All/Center/Dynamic/Extents/Previous/Scale/Window] <real time>: **1XP**

The image inside the large viewport is now scaled full size, as in Figure 19–11. Use the PAN command to move the image in the viewport as required. It might also be necessary to switch to Paper Space and use grips to size the viewport to the image.

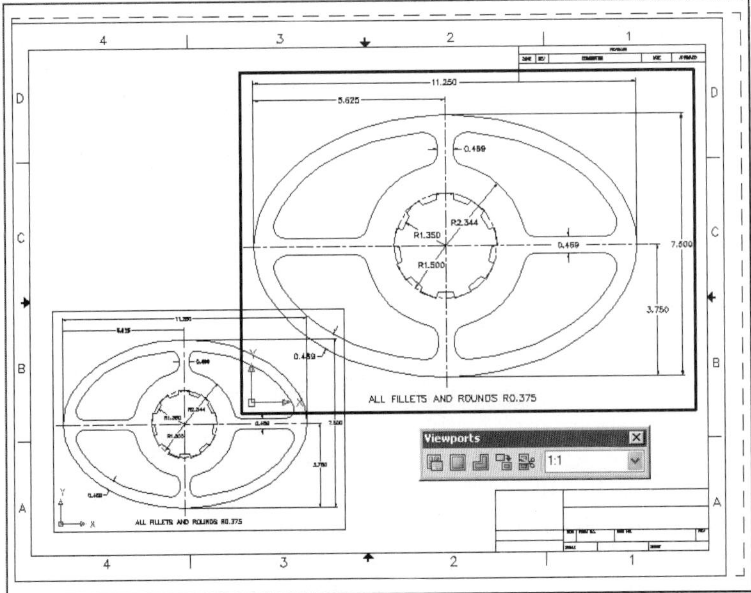

Figure 19–11

STEP 6

Be sure you are in floating Model Space; now move to the smaller viewport and make it active by clicking inside it, as shown in Figure 19–12. Again, this viewport outline will appear thicker, signifying that it is the current viewport. This image needs to be enlarged to four times its original size for viewing the gear tooth andto be dimensioned. Before increasing the size, use the PAN command to center the tooth area that is to be enlarged in the viewport. With this small viewport active, use the Viewports toolbar or issue the ZOOM command to scale the image inside the small viewport to four times the Paper Space units. Since this image is designed to be viewed at four times its normal size, enter a value of 4:1 in the toolbar or 4XP for the ZOOM command scale factor.

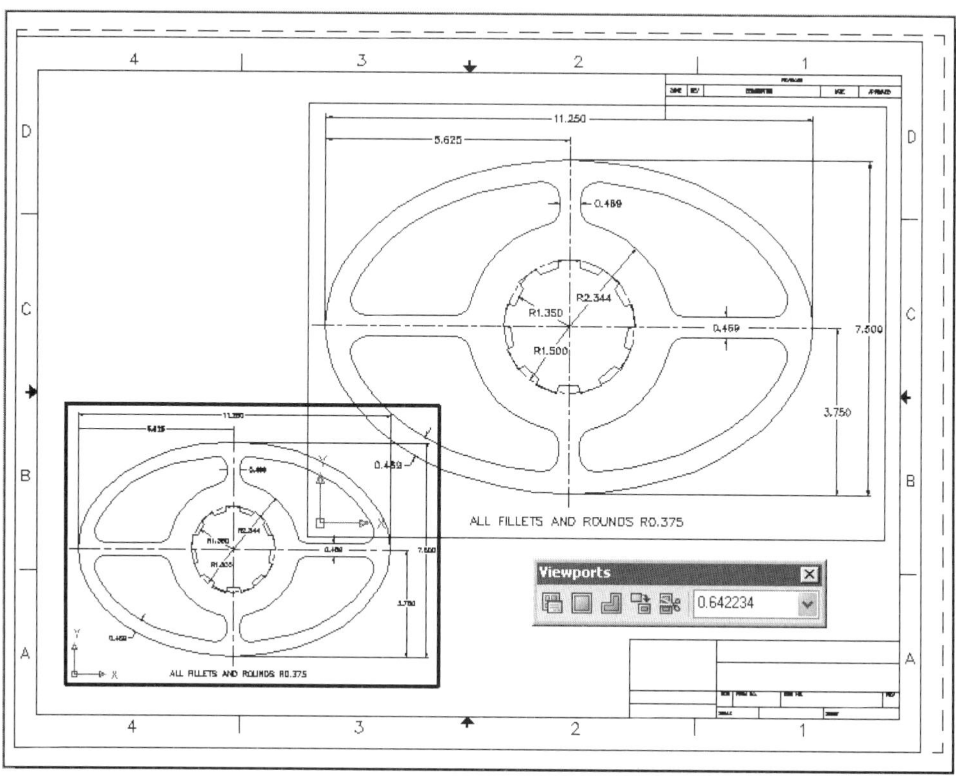

Figure 19–12

Command: **Z** *(For ZOOM)* [view>zoom]
Specify corner of window, enter a scale factor (nX or nXP), or [All/Center/Dynamic/Extents/Previous/Scale/Window] <real time>: **4XP**

Notice that the image increases in size. Use the PAN command to locate the gear tooth located at the 90° position (see Figure 19–13). If necessary, switch to Paper Space and use grips to size the viewport to the image.

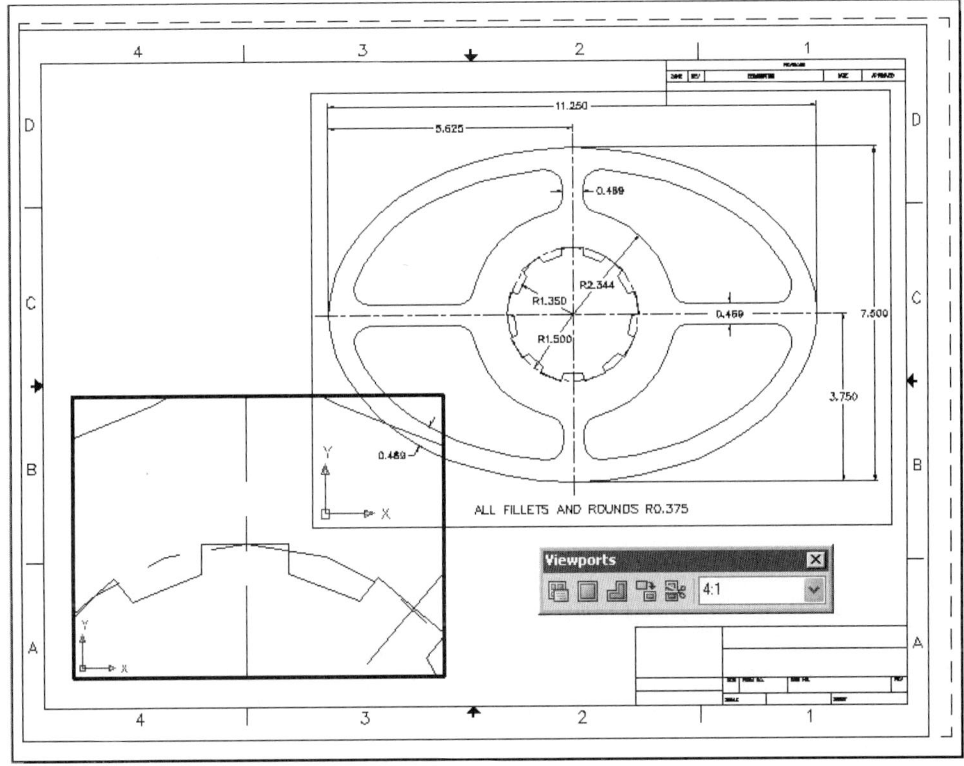

Figure 19–13

STEP 7

Before continuing with the next operation, be sure floating Model Space is active and the small viewport containing the detail is the current viewport. Activate the Layer Properties Manager dialog box and use Figure 19–14 as a guide to freeze the CEN layer in the current viewport. This will freeze only the centerlines located in the detail viewport; the centerlines located in the large viewport will remain visible (see Figure 19–15).

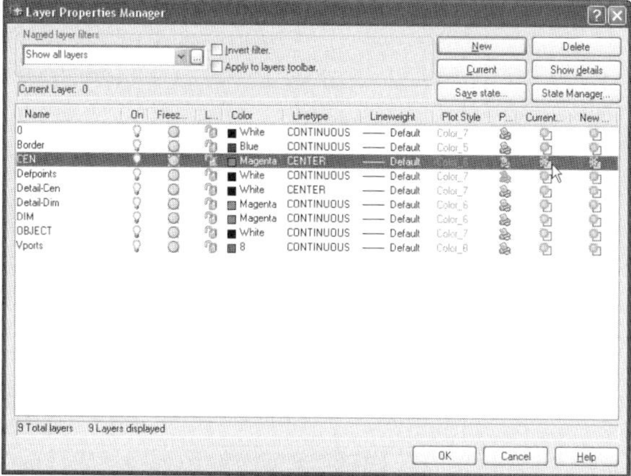

Figure 19–14

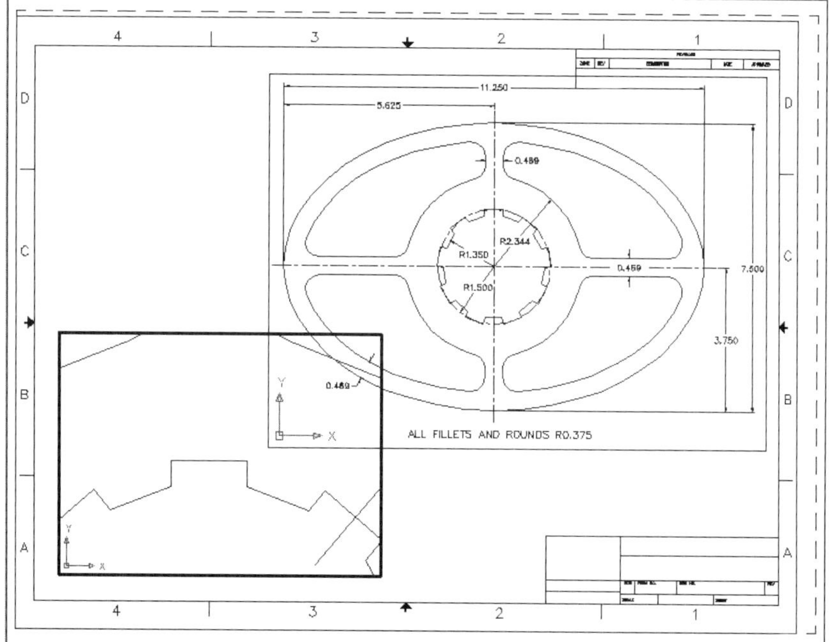

Figure 19–15

STEP 8

Make Detail-Cen the new current layer using the Layer Properties Manager dialog box shown in Figure 19–16.

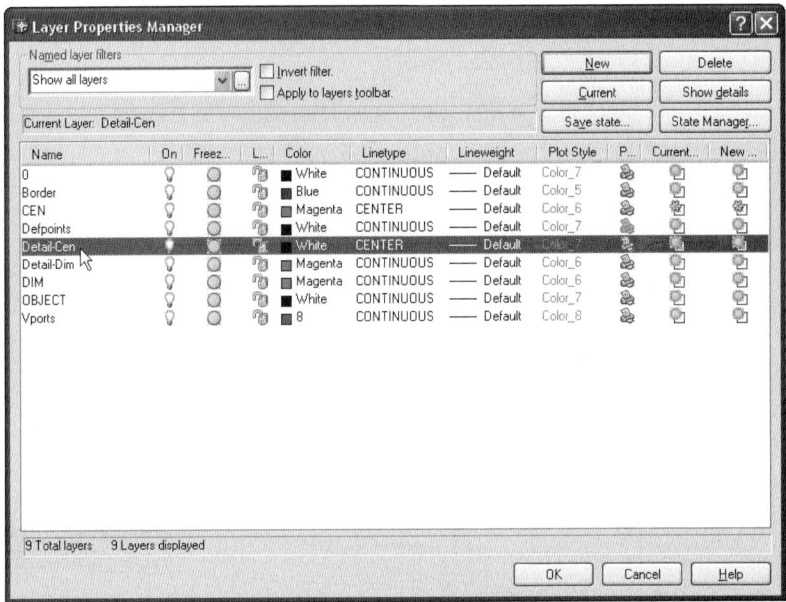

Figure 19–16

Next, construct a centerline in the detail view from the midpoint of the gear tooth at "A" a distance of 0.75 units in the 270 direction using a polar coordinate or the Polar Tracking mode, as shown in Figure 19–17.

Command: **L** *(For LINE)*[draw>line]
Specify first point: **Mid**
of *(Click the midpoint of the line at "A")*
Specify next point or [Undo]: **@0.75<270**
Specify next point or [Undo]: *(Press* ENTER *to exit this command)*

By default, the PSLTSCALE system variable is turned on and allows Paper Space to control the linetype scale, depending on the scale of the image inside the viewport. If PSLTSCALE were turned off, the line would resemble the continuous linetype. Because the scale of the viewport increased by a factor of four, the linetype would be too large to display its dashes.

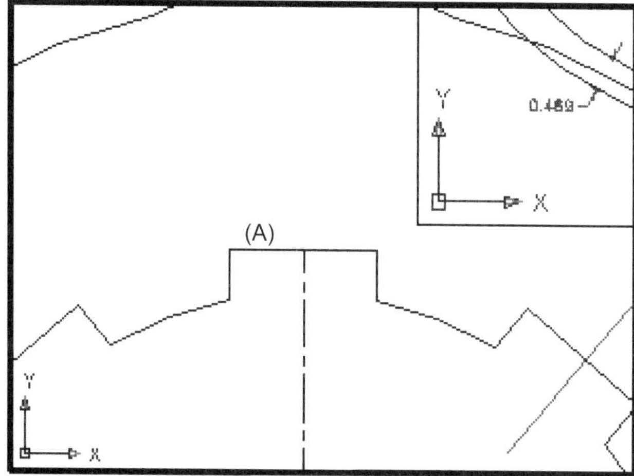

Figure 19–17

STEP 9

Before you begin to dimension inside the viewport containing the detail of the gear tooth, a few items need to be prepared. First, make Detail-Dim the new current layer, as shown in Figure 19–18.

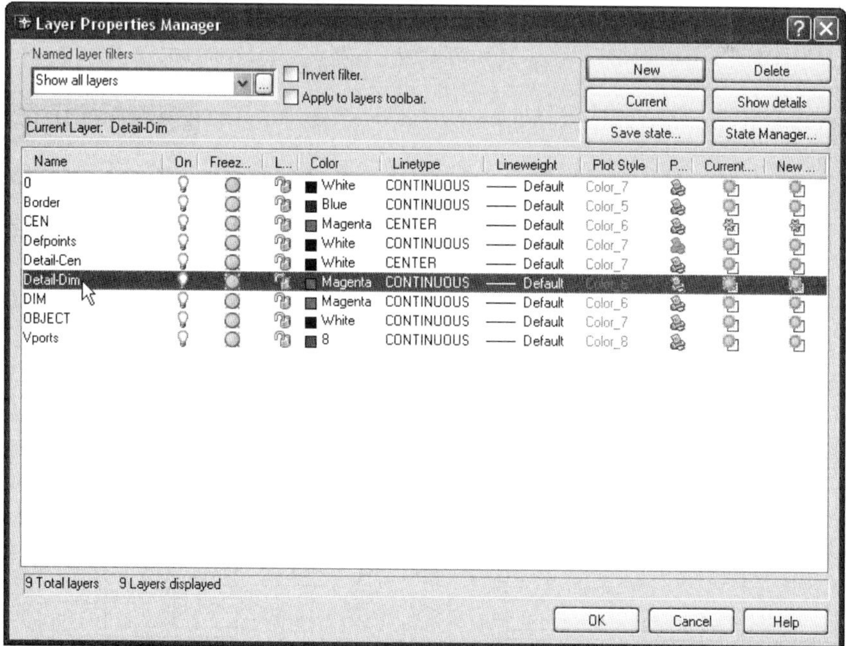

Figure 19–18

Next, create a new dimension style: display the Dimension Style Manager dialog box using the DDIM command (D is the command's alias). Click the New button to display the Create New... Dimension Style dialog box, as shown in Figure 19–19. In the New Style Name edit box, change the name to **DETAIL** and click on the Continue button. This will create a new dimension style called DETAIL from the information contained in the STANDARD dimension style.

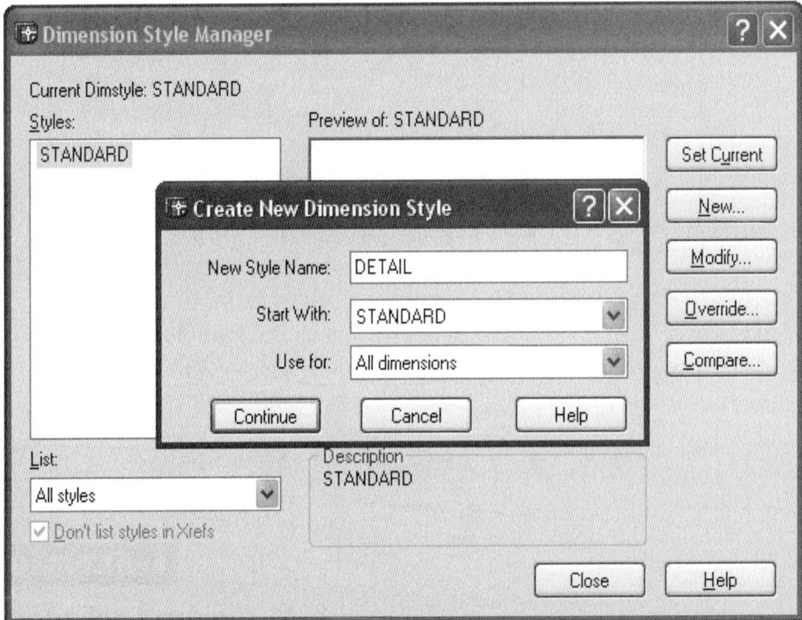

Figure 19–19

Click on the Fit tab in the New Dimension Style dialog box, shown in Figure 19–20. In the Scale for Dimension Features area, select the radio button for Scale dimensions to layout (paperspace). This will automatically make the dimension settings relative to the scale of the Paper Space viewport and ensure all dimension text, arrowheads, and other dimension settings will be identical in appearance to the dimensions located in the main viewport.

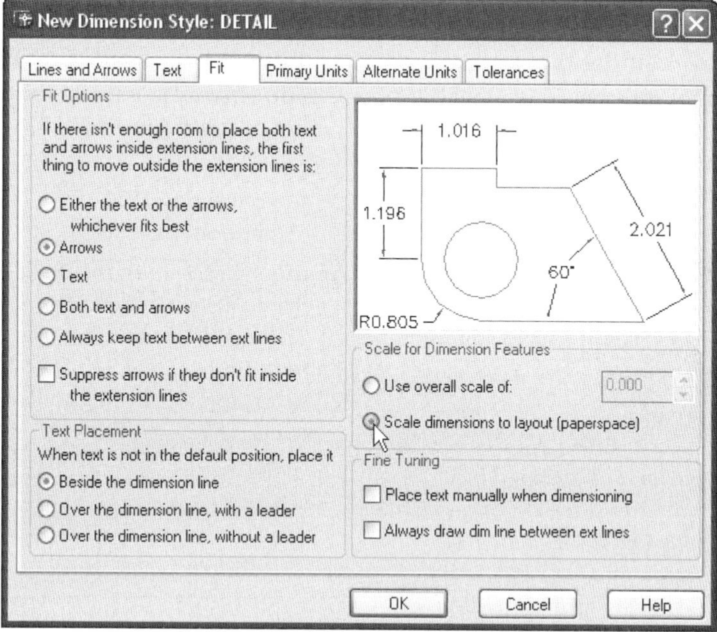

Figure 19–20

Save this change to the DETAIL dimension style by clicking OK, as shown in Figure 19–20. In the Dimension Style Manager dialog box, click the Set Current button to make DETAIL the current dimension style (see Figure 19–21), and then click the Close button.

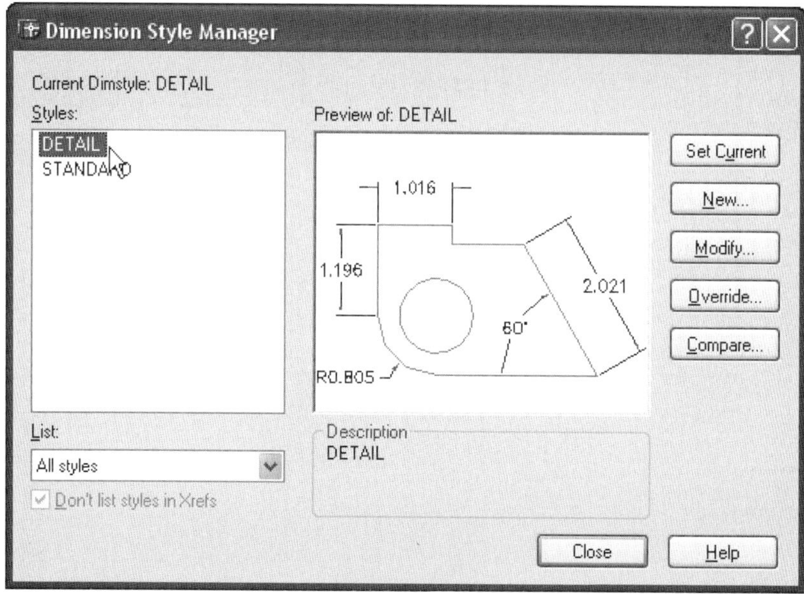

Figure 19–21

STEP 10

With Detail-Dim as the current layer and with the enlargement of the gear tooth as the current viewport in floating Model Space, place a horizontal dimension from the corner of the gear tooth to the endpoint of the centerline, shown in Figure 19–22.

Command: **DLI** *(For DIMLINEAR)* [dimension>linear]

Specify first extension line origin or <select object>: **End**
of *(Pick the endpoint of the line at "A")*
Specify second extension line origin: **End**
of *(Pick the endpoint of the line at "B")*
Specify dimension line location or [Mtext/Text/Angle/Horizontal/Vertical/Rotated]: *(Pick a point at "C")*
Dimension text = 0.250

With the linear dimension placed, add a baseline dimension using the DIMBASELINE command in Figure 19–22.

Command: **DBA** *(For DIMBASELINE)* [dimension>baseline]

Specify a second extension line origin or [Undo/Select] <Select>): **End**
of *(Pick the endpoint of the line at "D")*
Dimension text = 0.500 Specify a second extension line origin or [Undo/Select] <Select>): *(Press ENTER to exit this mode)*

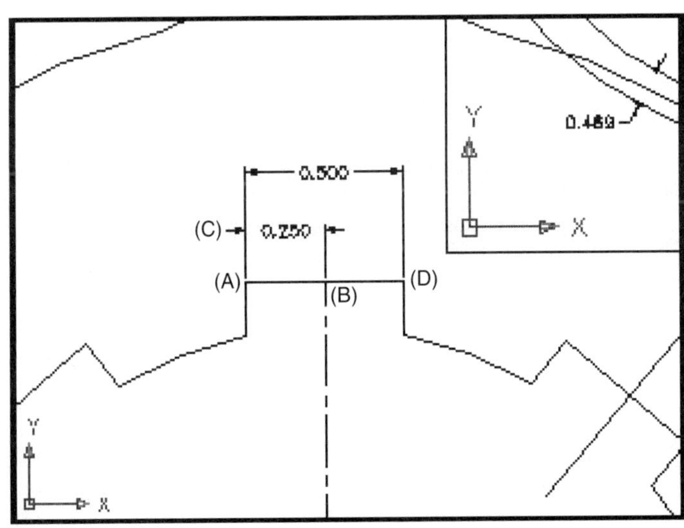

Figure 19–22

STEP 11

Notice how the original dimension of R2.344 still appears inside the small viewport. Just as the CEN layer was frozen in this viewport back in step 7, take this opportunity to freeze the DIM layer only in the small viewport through the Layer Properties Manager dialog box shown in Figure 19–23. Remember that this operation freezes the layer only in the small viewport while keeping the DIM layer visible in the large viewport.

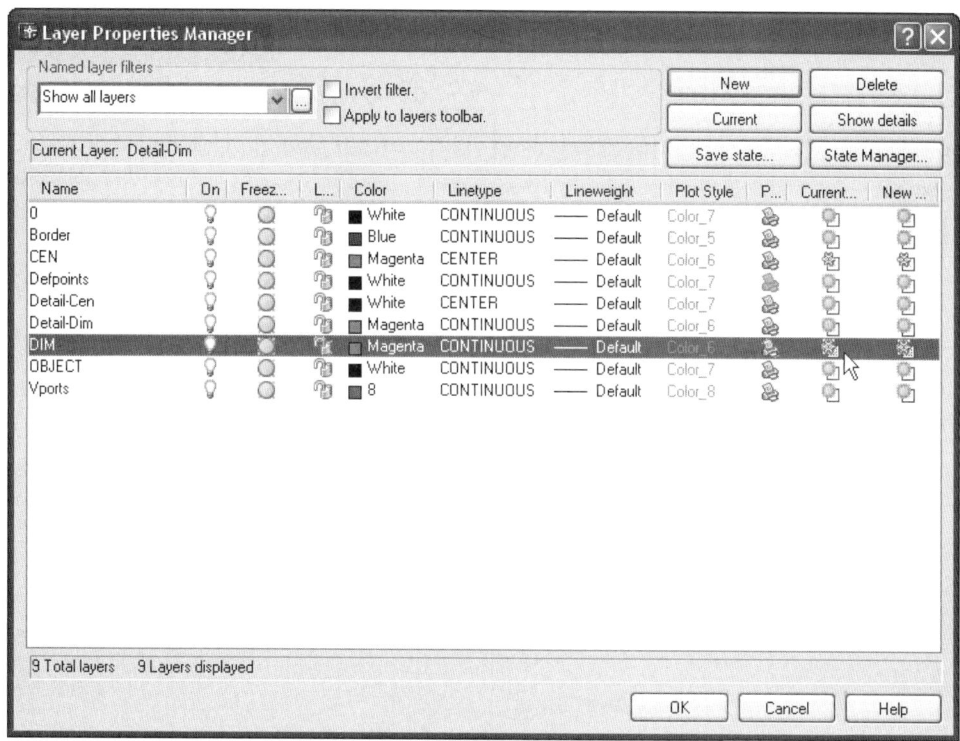

Figure 19–23

STEP 12

Before you complete the layout of this drawing, a few layers need to be frozen in the main viewport holding 19_Housing2. Because a centerline was added to the detail on the Detail-Cen layer and two dimensions were added to the detail in the Detail-Dim layer, the images of all objects drawn on these layers are present in the main viewport (see Figure 19–24).

To fix this, first be sure the main viewport is active by clicking inside it anywhere. Then activate the Layer Properties Manager dialog box as illustrated in Figure 19–25 and freeze the Detail-Cen and Detail-Dim layers in the current viewport. Notice that both dimensions and centerline disappear from the main viewport.

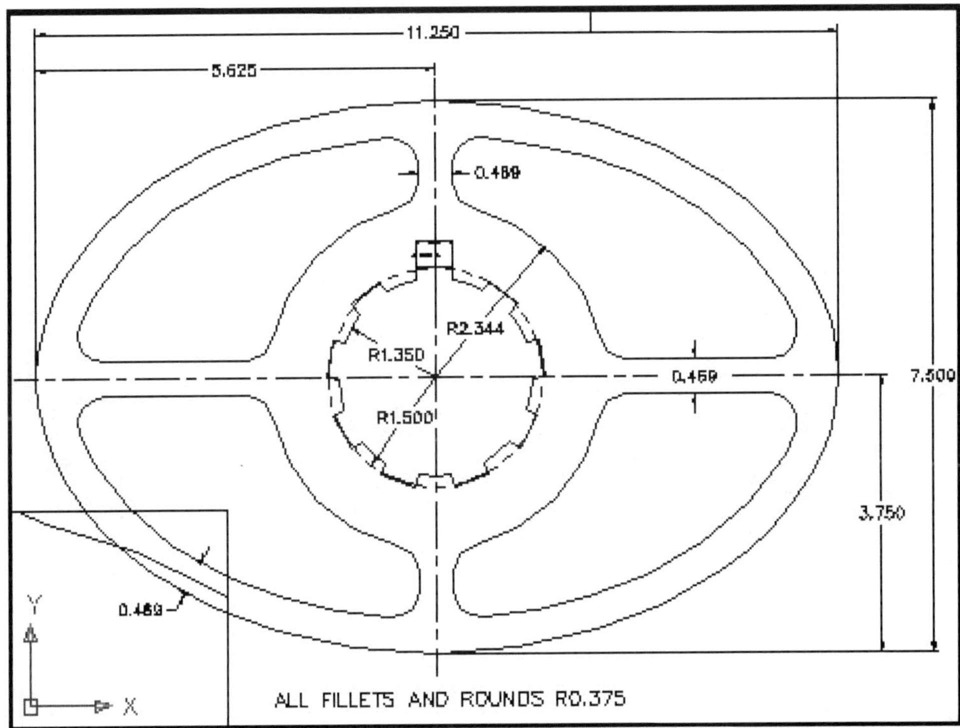

Figure 19–24

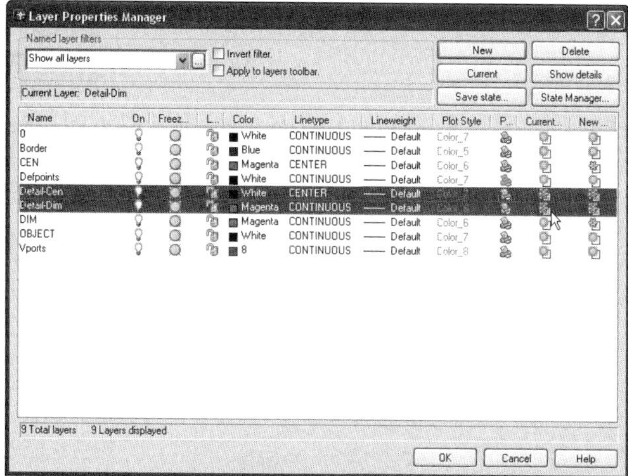

Figure 19–25

STEP 13

Complete the layout of the drawing by switching to Paper Space and adjusting all viewports using grips until the desired effect is achieved, as in Figure 19–26.

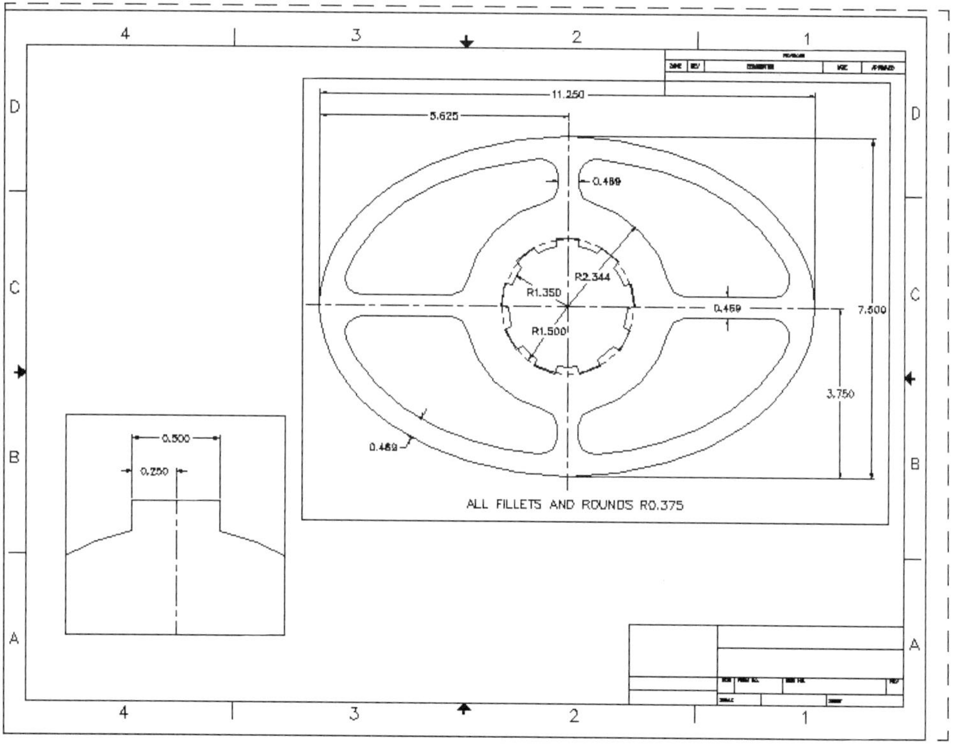

Figure 19–26

STEP 14

Turn off the Vports layer. Scale notes need to be added to both views. This should be accomplished in Paper Space with the current layer set to DIM. Use the MTEXT or DTEXT commands to place the note "SCALE: 4:1" under the detail view and the note "SCALE: 1:1" under the main view of 19_Housing2, shown in Figure 19–27. Use a text height of 0.18 units. The completed drawing is illustrated in Figure 19–28 and can be plotted out at a scale of 1=1.

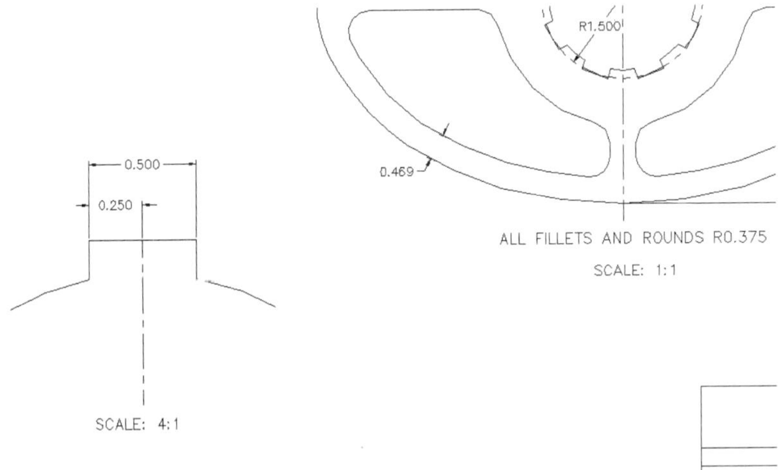

Figure 19–27

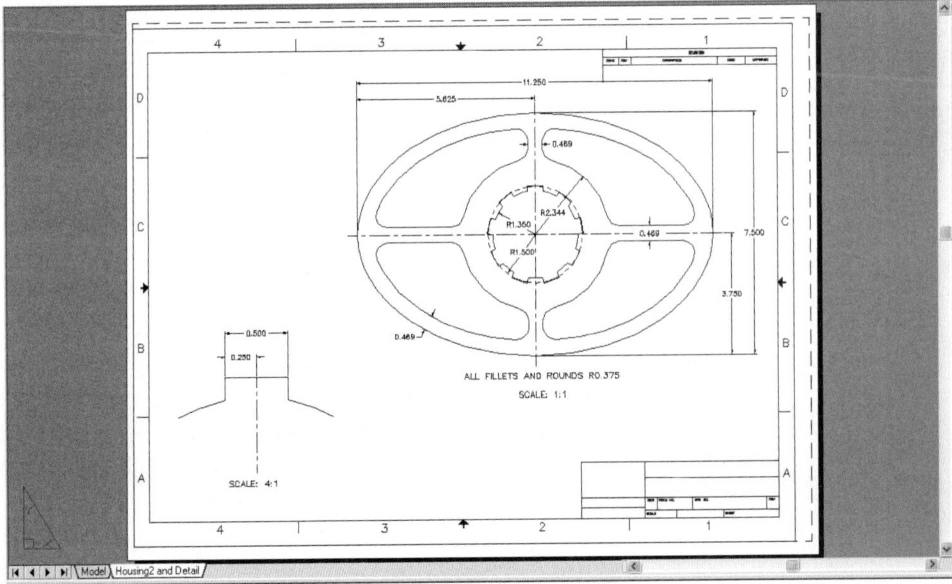

Figure 19–28

Beginning Level Drawings

PROBLEM 19–1

Directions for problem 19-1

Using the drawing *19_Fence detail.dwg* found on the CD, create the layout as shown in Figure 19-29. The scale of each viewport is 1"=1'; all layers and drawing objects are supplied. (special note: before plotting, change the plotting device using the right mouse click on layout 1 and select page setup.)

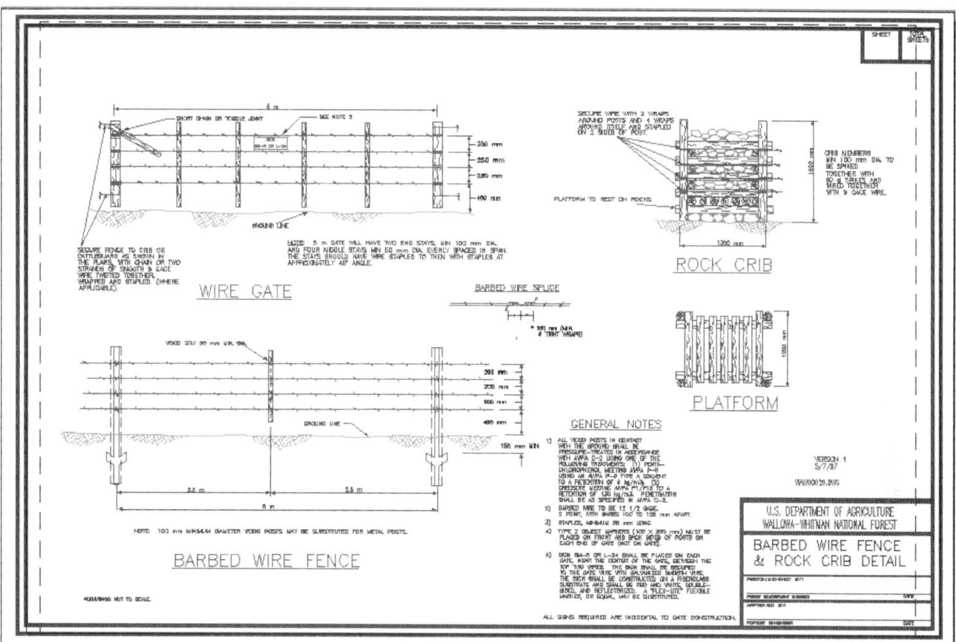

Figure 19–29

PROBLEM 19–2

Arrange Problems 13–1 and 13–2 from Chapter 13, "Analyzing 2D Drawings" on the same sheet in Paper Space. Create layers called Vports and Xref. Use *ANSI_C.dwg* as the title block inside Paper Space. Scale each image inside its floating Model Space viewport at a scale of 1XP or full size.

PROBLEM 19–3

Arrange Problems 13–5 and 13–6 from Chapter 13, "Analyzing 2D Drawings" on the same sheet in Paper Space. Create layers called Vports and Xref. Use *ANSI_C.dwg* as the title block inside Paper Space. Scale each image inside its floating Model Space viewport at a scale of1XP or full size.

PROBLEM 19–4

Arrange Problems 13–7 and 13–8 from Chapter 13, "Analyzing 2D Drawings" on the same sheet in Paper Space. Create layers calledVports and Xref. Use *ANSI_D.dwg* as the title block inside Paper Space. Scale problem 13–7 inside its floating Model Space viewport at a scale of 1/25.4XP Scale 13-8 inside its floating Model Space viewport at a scale of1XP or full size.

Intermediate Level drawings

PROBLEM 19–5

Arrange Problems 13–11 and 13–12 from Chapter 13, "Analyzing 2D Drawings" on the same sheet in Paper Space. Create layers calledVports and Xref. Use *ANSI_C.dwg* as the title block inside Paper Space. Scale Problem 13–3 inside its floating Model Space viewport at a scale of 1/4XP; scale metric Problem 13–4 inside its floating Model Space viewport at a scale of 1/25.4XP.

PROBLEM 19–6

Arrange Problems 13–13 and 13–14 from Chapter 13, "Analyzing 2D Drawings" on the same sheet in Paper Space. Create layers called Vports and Xref. Use *ANSI_D.dwg* as the title block inside Paper Space. Scale each image inside its floating Model Space viewport at a scale of1XP or full size.

PROBLEM 19–7

Arrange Problems 13–15 and 13–16 from Chapter 13, "Analyzing 2D Drawings" on the same sheet in Paper Space. Create layers calledVports and Xref. Use *ANSI_D.dwg* as the title block inside Paper Space. Scale metric Problem 13–15 inside its floating Model Space viewport at a scale of 1/25.4XP; scale Problem 13–16 inside its floating Model Space viewport at a scale of 1XP.

PROBLEM 19–8

Arrange Problems 13–17 and 13–18 from Chapter 13, "Analyzing 2D Drawings" on the same sheet in Paper Space. Create layers called Vports and Xref. Use *ANSI_D.dwg* as the title block inside Paper Space. Scale each image inside its floating Model Space viewport at a scale of1XP or full size.

PROBLEM 19–9

Arrange Problems 13–19 and 13–20 from Chapter 13, "Analyzing 2D Drawings" on the same sheet in Paper Space. Create layers calledVports and Xref. Use *ANSI_D.dwg* as the title block inside Paper Space. Scale Problem 13–19 inside its floating Model Space viewport at a scale of 1XP; scale metric Problem 13–20 inside its floating Model Space viewport at a scale of 1/25.4XP.

PROBLEM 19–10

Arrange Problems 13–21 and 13–22 from Chapter 13, "Analyzing 2D Drawings" on the same sheet in Paper Space. Create layers called Vports and Xref. Use *ANSI_D.dwg* as the title block inside Paper Space. Scale each image inside its floating Model Space viewport at a scale of1XP or full size.

PROBLEM 19–11

Arrange Problems 13–23 and 13–24 from Chapter 13, "Analyzing 2D Drawings" on the same sheet in Paper Space. Create layers called Vports and Xref. Use *ANSI_D.dwg* as the title block inside Paper Space. Scale each image inside its floating Model Space viewport at a scale of1XP or full size.

PROBLEM 19–12

Arrange Problems 13–25 and 13–26 from Chapter 13, "Analyzing 2D Drawings" on the same sheet in Paper Space. Create layers called Vports and Xref. Use *ANSI_D.dwg* as the title block inside Paper Space. Scale each image inside its floating Model Space viewport at a scale of1XP or full size.

PROBLEM 19–13

Arrange Problems 13–27 and 13–28 from Chapter 13, "Analyzing 2D Drawings" on the same sheet in Paper Space. Create layers called Vports and Xref. Use *ANSI_D.dwg* as the title block inside Paper Space. Scale each image inside its floating Model Space viewport at a scale of1XP or full size.

PROBLEM 19–14

Arrange Problems 13–9 and 13–29 from Chapter 13, "Analyzing 2D Drawings" on the same sheet in Paper Space. Create layers calledVports and Xref. Use *ANSI_D.dwg* as the title block inside Paper Space. Scale Problem 13–9 inside its floating Model Space viewport at a scale of ¼"=1'; scale Problem 13–29 inside its floating Model Space viewport at a scale of ¼"=1'.

PROBLEM 19–15

Arrange Problem 13–31 from Chapter 13, "Analyzing 2D Drawings" in Paper Space. Use *ArchEng.dwg* as the title block inside Paper Space. Scale Problem 13–31 inside its floating Model Space viewport at a scale of 1:200.

PROBLEM 19–16

Arrange Problem 13–32 from Chapter 13, "Analyzing 2D Drawings" in Paper Space. Use *ArchEng.dwg* as the title block inside Paper Space. Scale Problem 13–32 inside its floating Model Space viewport at a scale of 1/2" = 1'-0".

PROBLEM 19–17

Arrange Problem 13–33 from Chapter 13, "Analyzing 2D Drawings" in Paper Space. Use *ArchEng.dwg* as the title block inside Paper Space. Scale Problem 13–33 inside its floating Model Space viewport at a scale of 1:4.

PROBLEM 19–18

Arrange Problem 13–36 from Chapter 13, "Analyzing 2D Drawings" in Paper Space. Use *ArchEng.dwg* as the title block inside Paper Space. Scale Problem 13–36 to best fit the titleblock.

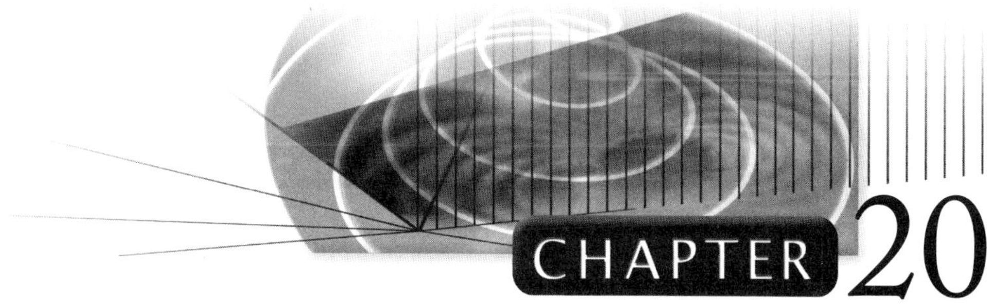

Solid Modeling Fundamentals

The exercises in this chapter introduce:

- The concepts of working in 3D
- Developing a 3D solid model
- Viewing objects in 3D
- Working with the User Coordinate System
- Using Modify commands on 3D solid models
- Analysing 3D solids models

TUTORIAL EXERCISE: BPLATE.DWG

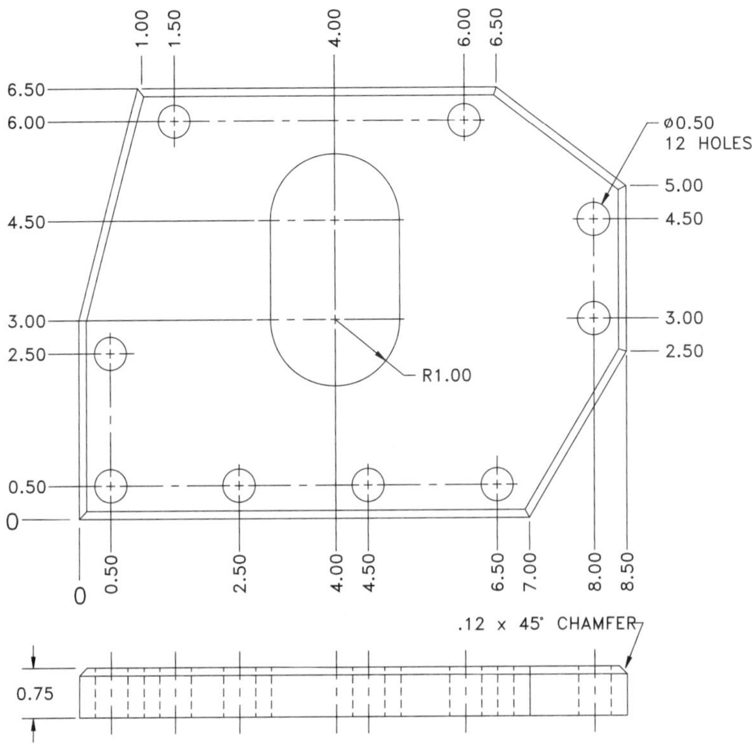

Figure 20–1

Purpose

The purpose of this tutorial is to produce a solid model of the Bplate shown in Figure 20–1.

System Settings

Keep the current limits settings of (0,0) for the lower-left corner and (12,9) for the upper-right corner. Change precision from four to two using the Drawing Units dialog box. Keep all remaining system settings.

Layers

Special layers do not have to be created for this tutorial exercise.

Suggested Commands

Begin this tutorial by constructing the profile of the Bplate using polylines. Add all circles; use the EXTRUDE command to extrude all objects at a thickness of 0.75 units. Chamfer the top edge of the plate using the CHAMFER command. Use the SUBTRACT command to subtract all cylinders from the Bplate forming the holes in the plate. Perform hidden line removals using the HIDE command and create shaded models using the SHADEMODE command.

Whenever possible, substitute the appropriate command alias in place of the full AutoCAD command in each tutorial step. For example, use "CP" for the COPY command, "L" for the LINE command, and so on. The complete listing of all command aliases is located in Table 1–2.

STEP 1

Begin the Bplate by establishing a new coordinate system using the UCS command. Define the origin at 2.00,1.50. Use the UCSICON command, if necessary, to display the User Coordinate System icon at the new origin. Use the PLINE command to draw the profile of the Bplate. See Figure 20–2.

 Command: **UCS** [tools>new UCS]

Current ucs name: *WORLD* Enter an option [New/Move/ orthoGraphic/ Prev/Restore/Save/Del/ Apply/?/World] <World>: **M** *(For Move)*

Specify new origin point or [Zdepth] <0,0,0>: **2.00,1.50**

Command: **UCSICON** [view>display>ucsicon]

Enter an option [ON/OFF/All/Noorigin/ ORigin/Properties] <ON>: **OR** *(For Origin)*

 Command: **PL** *(For PLINE)* *[draw>polyline]*

Specify start point: **0,0**

Current line-width is 0.00 Specify next point or [Arc/Halfwidth/Length/Undo/ Width]: **@7.00<0**

Specify next point or [Arc/Close/ Halfwidth/Length/Undo/Width]: **@1.50,2.50**

Specify next point or [Arc/Close/ Halfwidth/Length/Undo/Width]: **@2.50<90**

Specify next point or [Arc/Close/ Halfwidth/Length/Undo/Width]: **@-2.00,1.50**

Specify next point or [Arc/Close/ Halfwidth/Length/Undo/Width]: **@5.50<180**

Specify next point or [Arc/Close/ Halfwidth/Length/Undo/Width]: **@-1.00,-3.50**

Specify next point or [Arc/Close/ Halfwidth/Length/Undo/Width]: **C** *(To close the polyline and return to the command prompt)*

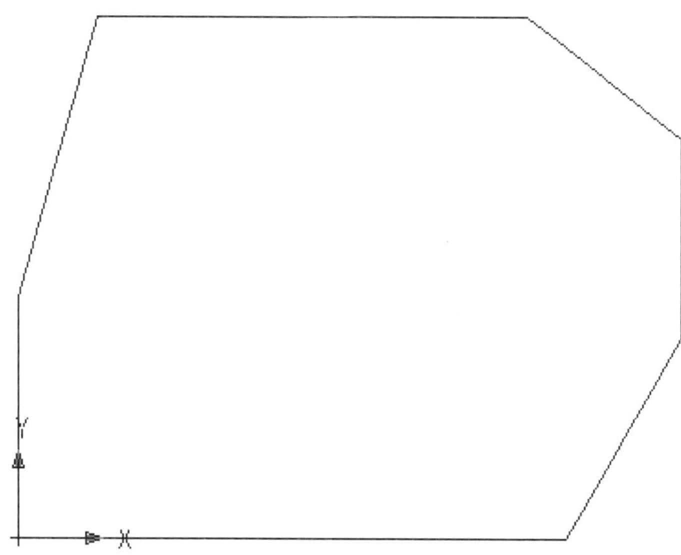

Figure 20–2

STEP 2

Draw the nine circles of 0.50 diameter by placing one circle at "A" and copying the remaining circles to their desired locations (see Figure 20–3). Use the Multiple option of the COPY command to accomplish this.

Command: **C** *(For CIRCLE)* [draw>circle]

Specify center point for circle or [3P/2P/Ttr (tan tan radius)]: **0.50,0.50**

Specify radius of circle or [Diameter]: **D** *(For Diameter)*

Specify diameter of circle: **0.50**

Command: **CP** *(For COPY)* [modify>copy]

Select objects: **L** *(for the last circle)*

Select objects: *(Press ENTER to continue with this command)*

Specify base point or displacement, or [Multiple]: **M** *(For Multiple)*

Specify base point: @ *(References the center of the 0.50 circle)*

Specify second point of displacement or <use first point as displacement>: **2.50,0.50**

Specify second point of displacement or <use first point as displacement>: **4.50,0.50**

Specify second point of displacement or <use first point as displacement>: **6.50,0.50**

Specify second point of displacement or <use first point as displacement>: **8.00,3.00**

Specify second point of displacement or <use first point as displacement>: **8.00,4.50**

Specify second point of displacement or <use first point as displacement>: **0.50,2.50**

Specify second point of displacement or <use first point as displacement>: **1.50,6.00**

Specify second point of displacement or <use first point as displacement>: **6.00,6.00**

Specify second point of displacement or <use first point as displacement>: (Press ENTER to exit this command)

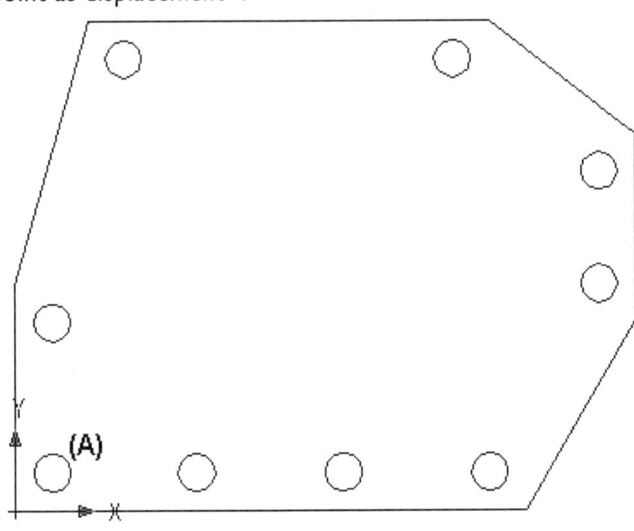

Figure 20–3

STEP 3

Form the slot by placing two circles using the CIRCLE command followed by two lines drawn from the quadrants of the circles using the OSNAP-Quadrant mode. See Figure 20–4.

Command: **C** *(For CIRCLE)* [draw>circle]

Specify center point for circle or [3P/2P/Ttr (tan tan radius)]: **4.00,3.00**

Specify radius of circle or [Diameter] <0.25>: **1.00**

Command: **C** *(For CIRCLE)* [draw>circle]

Specify center point for circle or [3P/2P/Ttr (tan tan radius)]: **4.00,4.50**

Specify radius of circle or [Diameter] <1.00> *(Press* ENTER *to accept the default value)*

Command: **L** *(For LINE)* [draw>line]

Specify first point: **Qua**

of *(Select the quadrant of the circle at "A")*

Specify next point or [Undo]: **Qua**

of *(Select the quadrant of the circle at "B")*

Specify next point or [Undo]: *(Press* ENTER *to exit this command)*

Command: **L** *(For LINE)* [draw>line]

Specify first point: **Qua**

of *(Select the quadrant of the circle at "C")*

Specify next point or [Undo]: **Qua**

of *(Select the quadrant of the circle at "D")*

Specify next point or [Undo]: *(Press* ENTER *to exit this command)*

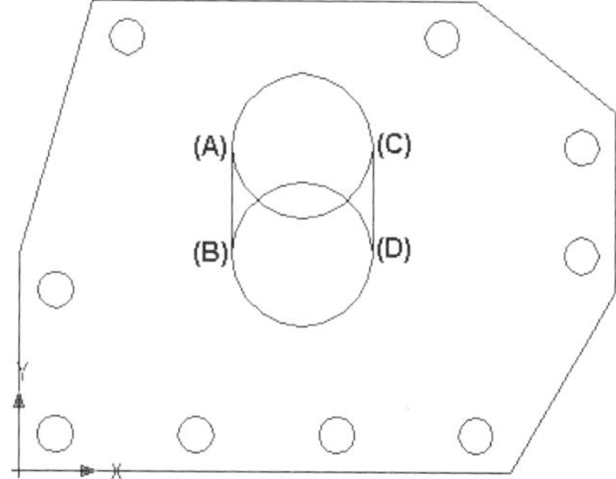

Figure 20–4

STEP 4

Use the TRIM command to trim away any unnecessary arcs to form the slot. See Figure 20–5.

Command: **TR** (For TRIM) [modify>trim]

Current settings: Projection=UCS Edge=None Select cutting edges ... Select objects: *(Select both vertical lines at "A" and "B")*

Select objects: *(Press* ENTER *to continue with this command)*

Select object to trim or shift-select to extend or [Project/Edge/Undo]: *(Select the circle at "C")*

Select object to trim or shift-select to extend or [Project/Edge/Undo]: *(Select the circle at "D")*

Select object to trim or shift-select to extend or [Project/Edge/Undo]: *(Press* ENTER *to exit this command).*

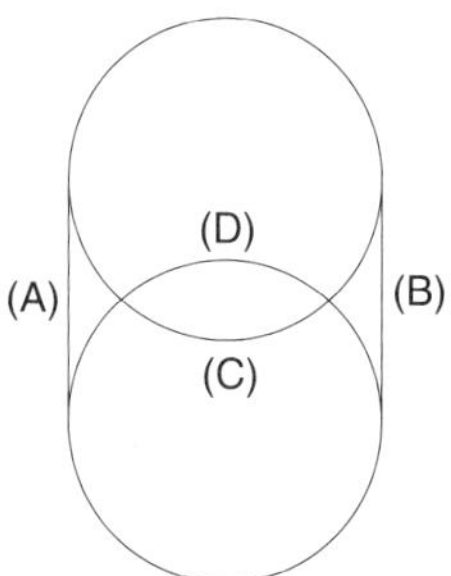

Figure 20–5

STEP 5

The next step illustrates the use of the EXTRUDE command, to give the plate a thickness of 0.75 units. However, this command operates on closed single entity objects. Currently, all objects can be extruded except for the two arcs and lines representing the slot (see Figure 20–6). Use the PEDIT command to convert these objects to a single polyline. This command can be selected by choosing Object from the Modify menu and then Polyline.

Command: **PE** *(For PEDIT)* [modify>polyline]

Select polyline or [Multiple]: *(Select the bottom arc at "A")*

Object selected is not a polyline Do you want to turn it into one? <Y>

(Press ENTER *to accept this default value)*

Enter an option [Close/Join/Width/Edit vertex/Fit/Spline/Decurve/Ltype gen/ Undo]: **J** *(For Join)*

Select objects: (Select the objects labeled "B," "C," and "D")

Select objects: *(Press* ENTER *to perform the joining* operation)

3 segments added to polyline Enter an option [Open/Join/Width/Edit vertex/ Fit/Spline/Decurve/Ltype gen/ Undo]: *(Press* ENTER *to exit this command)*

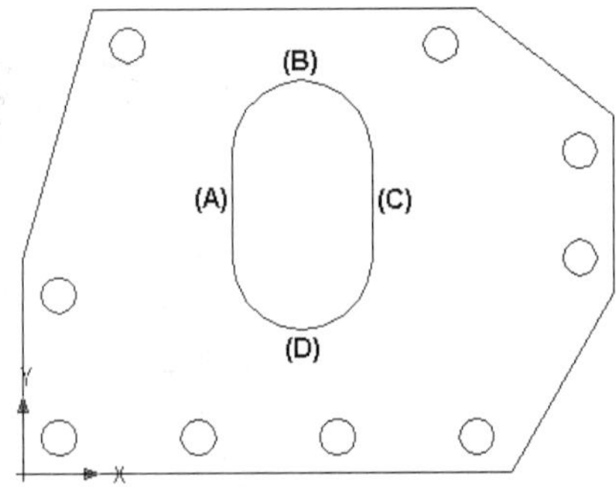

Figure 20–6

STEP 6

Use the EXTRUDE command to give Bplate thickness. Type this command from the keyboard or choose Solids from the Draw menu and then Extrude. Use the All option to select all circles and polyline outlines. Type a value of **0.75** as the height of the extrusion. Keep the default value for the extrusion taper angle. When finished with this operation, turn off the UCS icon with the UCSICON command. See Figure 20–7.

Command: **EXT** (For EXTRUDE) [draw>solids>extrude]

Current wire frame density: ISOLINES=4 Select objects: **All**

Select objects: (Press ENTER to continue with this command)

Specify height of extrusion or [Path]: **0.75**

Specify angle of taper for extrusion <0>:

(Press ENTER *to execute the extrusion operation)*

Command: **UCSICON** [view>display>UCS icon]

Enter an option [ON/OFF/All/Noorigin/ ORigin/Properties] <ON>: **Off**

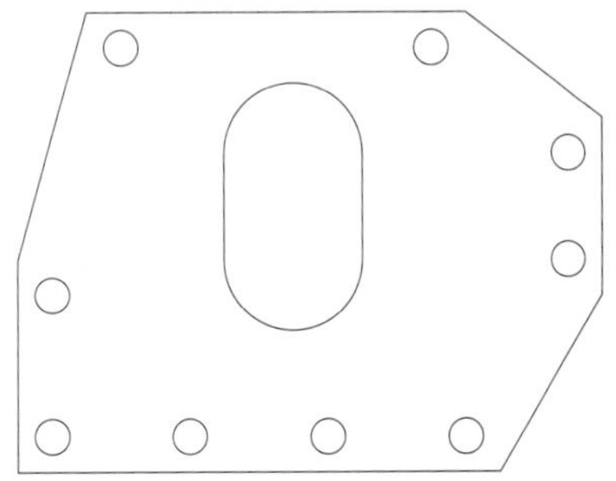

Figure 20–7

STEP 7

Use the VPOINT command to view the plate in three dimensions using the following prompt sequence, see Figure 20–8. Or, choose 3D Views from the View menu and then SE Isometric to achieve the same results. Next, use the CHAMFER command to place a chamfer of 0.12 units along the top edge of the plate. If the top surface is not highlighted when you are prompted for the base surface, use the Next option until the top surface is highlighted. After entering the chamfer distances, use the Loop option to create the chamfer along the entire edge without having to pick each individual segment of the top edge.

Command: **VPOINT** [view>3D views>viewpoint]
Switching to the WCS
Current view direction: VIEWDIR=0.00,0.00,1.00 Specify a view point or [Rotate] <display compass and tripod>: **1,-1,1**

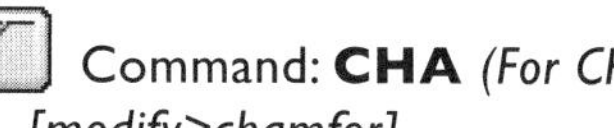

Command: **CHA** *(For CHAMFER)* *[modify>chamfer]*

(TRIM mode) Current chamfer Dist1 = 0.50, Dist2 = 0.50 Select first line or [Polyline/Distance/ Angle/Trim/ Method]: *(Select the line at "A")*

Base surface selection... Enter surface selection option [Next/OK (current)] <OK>: **N** *(For the next surface, which should be the top surface)*

Enter surface selection option [Next/ OK (current)] <OK>: *(Press ENTER to accept the top surface as the base)*

Specify base surface chamfer distance <0.5000>: **0.12**

Specify other surface chamfer distance <0.5000>: **0.12**

Select an edge or [Loop]: **L** *(For loop)*

Select an edge loop or [Edge]: *(Select the edge along the top at "B")*

Select an edge loop or [Edge]: *(Press ENTER to perform the chamfer operation)*

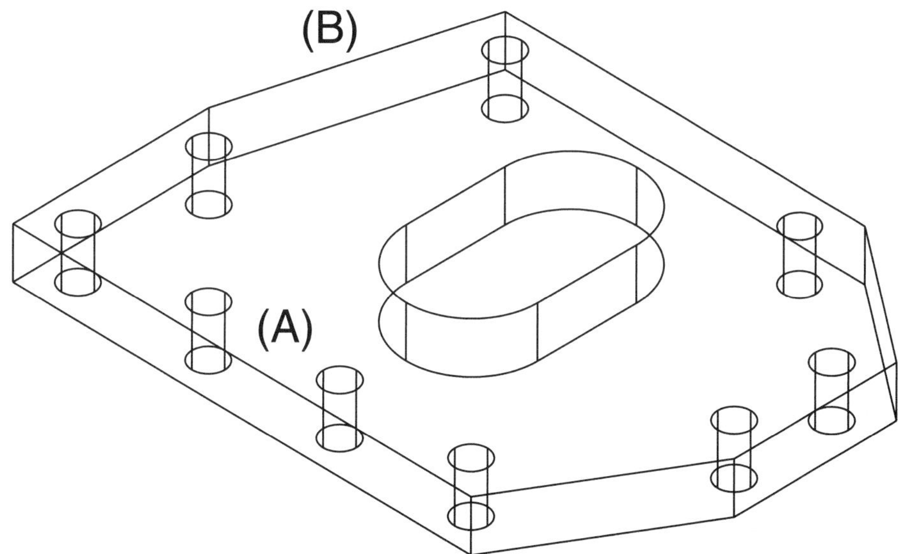

Figure 20–8

STEP 8

Because all of the objects have been extruded a distance of 0.75 units, the cylinders and rounded end bar that we created are considered individual solid objects that do not yet belong to the base. Use the SUBTRACT command to subtract these objects and create the holes and slot in the base plate (see Figure 20–9). This operation will resemble the drilling of holes and the milling for the slot. The SUBTRACT command can be selected from the Solids Editing cascading menu on the Modify menu.

Command: **SU** *(For SUBTRACT)*
[modify>solids editing>subtract]

Select solids and regions to subtract from... Select objects: *(Select the base of the Bplate along any edge)*

Select objects: *(Press ENTER to continue with this command)*

Select solids and regions to subtract... Select objects: (Select the 9 holes and the slot)

Select objects: (Press ENTER to perform the subtraction operation)

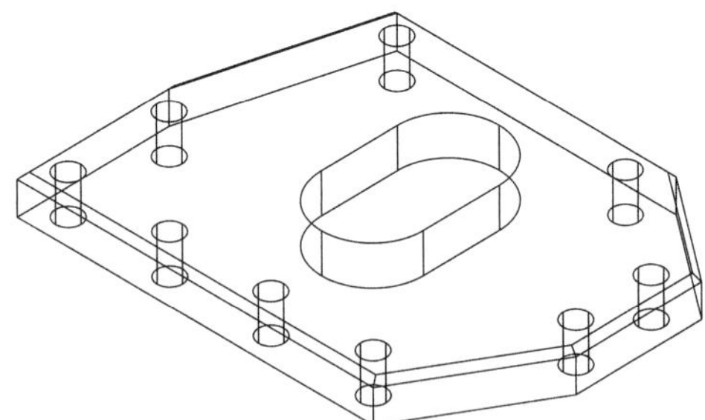

Figure 20–9

STEP 9

To see the object with all hidden edges removed, use the HIDE command. See Figure 20–10. This command can be selected from the View menu.

Command: **HI** (For HIDE)
[view>shade>hide]

Regenerating model.

Regenerating the display will return the model to its wireframe mode.

Command: **RE** (For REGEN)
[view>regen]

Regenerating model.

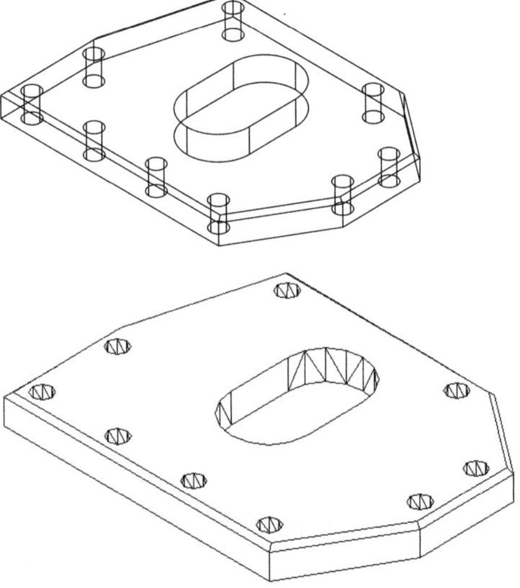

Figure 20–10

STEP 10

As illustrated in Figure 20–10, in the previous step, the circles and slot look somewhat irregular after you perform the HIDE operation. To smooth out the circles and arcs representing the slot, use the FACETRES system variable. This stands for facet resolution and it controls the density at which circles and arcs are displayed. Set the facet resolution to a new value of 2, perform a HIDE operation, and observe the results. See Figure 20–11.

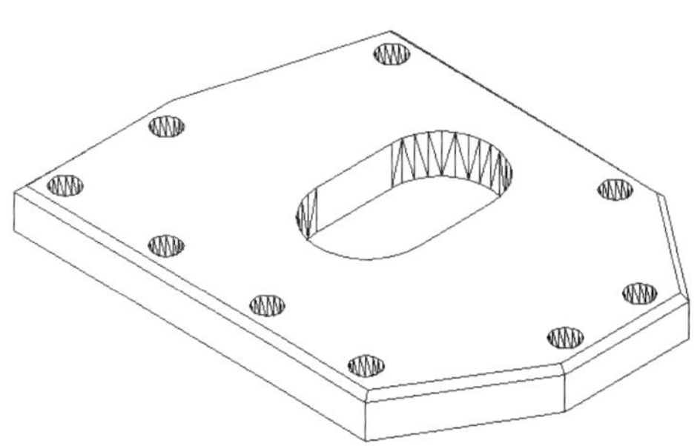

Figure 20–11

Command: **FACETRES**
Enter new value for FACETRES <0.5000>: **2**
Command: **HI** *(For HIDE)*
Regenerating model.
Notice the circles are better defined due to the increased density. Regenerating the display will return the model to its wireframe mode.
Command: **RE** *(For REGEN)* [view>regen]
Regenerating model.

STEP 11

FACETRES controls the density of circles and arcs when you perform hidden line removals. The ISOLINES system variable controls the density of lines that represent circles and arcs in the wireframe mode. These lines are called tessellation lines. Change the value of ISOLINES from 4 to a new value of 15, regenerate the screen, and observe the results. See Figure 20–12.

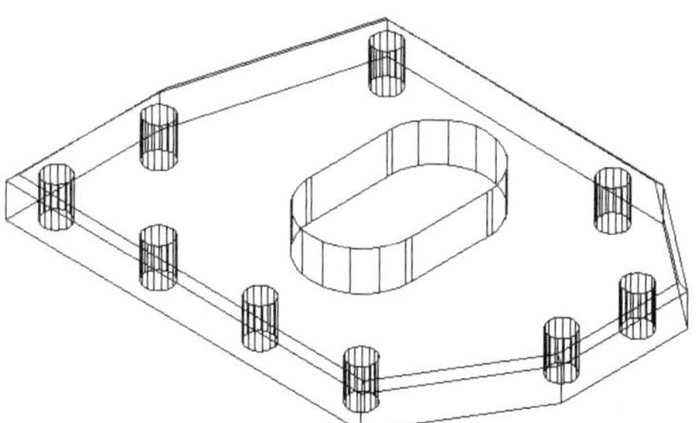

Figure 20–12

Command: **ISOLINES**
Enter new value for ISOLINES <4>: **15**
Command: **RE** *(For REGEN)* [view>regen]
Regenerating model.

STEP 12

Use the SHADEMODE command with the Flat + Edges option to produce a shaded image consisting of hidden line removal in addition to the shading being performed in the original color of the model. (**Note:** A better shaded image often results with the use of a lighter color. Try changing your layer color to Cyan and shade the image again.) Notice that the regen command does not return the image to a wireframe. To return to wireframe mode, use the SHADEMODE command with the 2D wireframe option. See Figure 20–13.

 Command: **SHADEMODE** [view>shade]
Current mode: 2D wireframe Enter option [2D wireframe/3D wireframe/ Hidden/Flat/Gouraud/ fLat+edges/ gOuraud+edges] <Flat+Edges>: **L** *(For fLat+edges)*

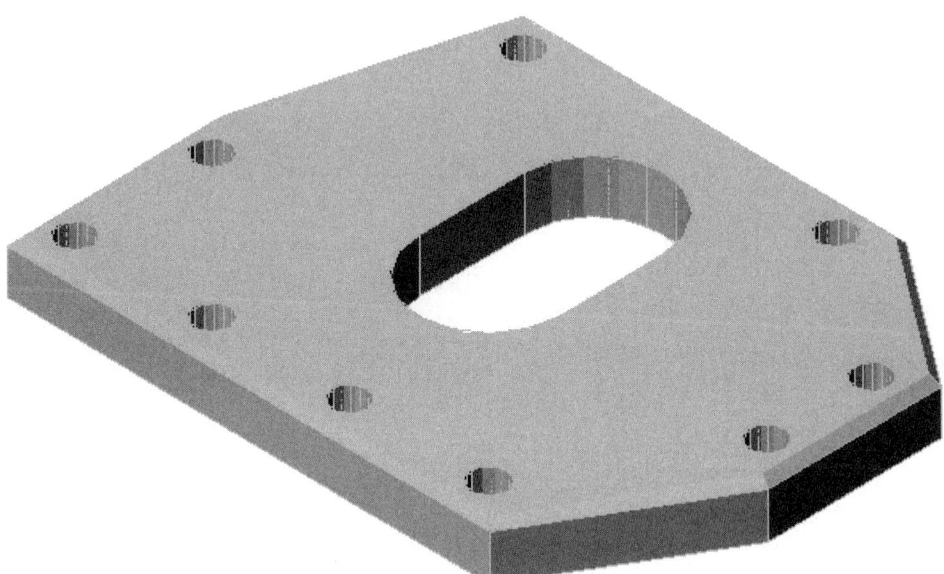

Figure 20–13

TUTORIAL EXERCISE: LEVER.DWG

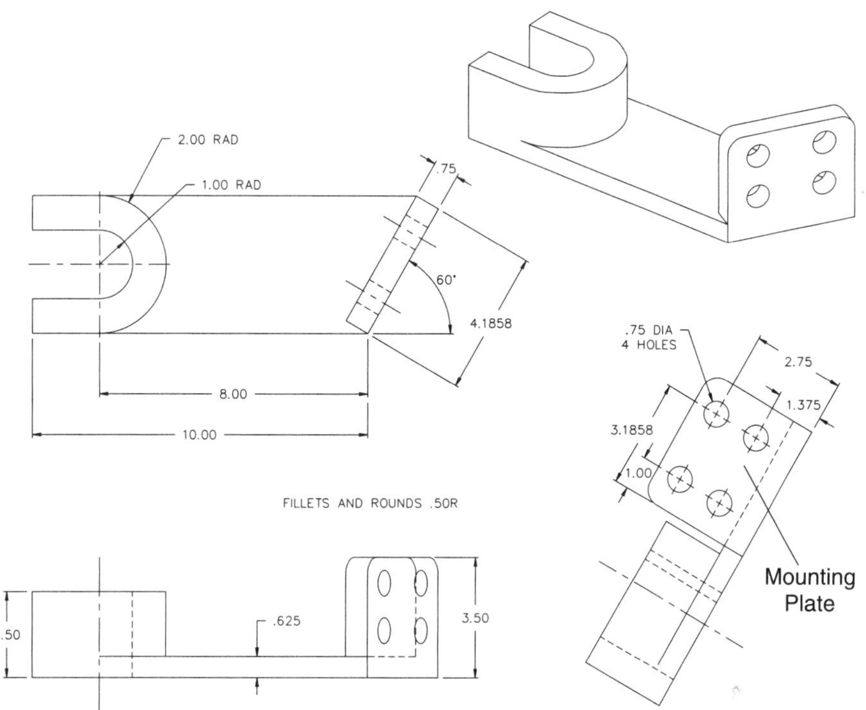

Figure 20–14

Purpose

This tutorial is designed to use the UCS command to construct a 3D solid model of the Lever in Figure 20–14.

System Settings

Begin a new drawing called "Lever." Use the Drawing Units dialog box to change the precision from four to two. Keep all default values for the UNITS command. Using the LIMITS command, keep (0,0) for the lower-left corner and change the upper-right corner from (12,9) to (15.50,9.50). Perform a ZOOM ALL.

Layers

Create the following layer with the format:

Name	Color	Linetype
Model	Cyan	Continuous

Suggested Commands

Begin the layout of this problem by constructing the plan view of the Lever. Use the EXTRUDE command to create the height of the individual components that make up the lever.

Position the UCS to construct the mounting plate.

Whenever possible, substitute the appropriate command alias in place of the full AutoCAD command in each tutorial step. For example, use "CP" for the COPY command, "L" for the LINE command, and so on. The complete listing of all command aliases is located in Table 1–2.

STEP 1

Begin this drawing by making the Model layer current. Then, construct two circles of radius values 1.00 and 2.00 using the CIRCLE command with 4.00,5.00 as the center of both circles (see Figure 20–15). Use the @ symbol to identify the last known point (4.00,5.00) as the center of the circle.

Command: **C** (For CIRCLE) [draw>circle]

Specify center point for circle or [3P/2P/Ttr (tan tan radius)]: **4.00,5.00**

Specify radius of circle or [Diameter]: **1.00**

Command: **C** (For CIRCLE) [draw>circle]

Specify center point for circle or [3P/2P/Ttr (tan tan radius)]: @ (To identify the last point)

Specify radius of circle or [Diameter] <1.00>: **2.00**

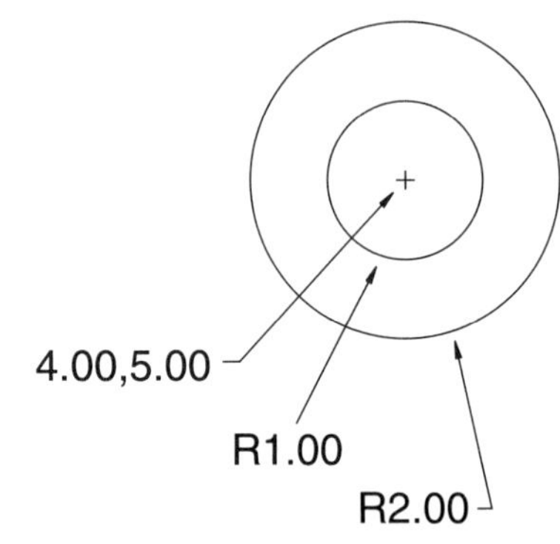

Figure 20–15

STEP 2

Draw a vertical line 2 units to the left of the center of the circles and 4 units long in the 270 degree direction. In the command sequence below, use OSNAP-Tracking and Direct Distance mode to accomplish this. Turn on OSNAP, OTRACK, and ORTHO at the status bar. Set Quadrant as a running object snap. See Figure 20-16.

 Command: **L** *(For LINE)* [draw>line]

Specify first point: (Move the cursor over point "A" to acquire the quadrant for tracking, then move the cursor over point "B" to acquire the quadrant for tracking. As you move the cursor to point "C," two dotted lines will appear; pick a point at this intersection for the starting point of your line. Move the cursor toward point "D" to use Direct Distance mode for the next prompt.)

Specify next point or [Undo]: **4**

Specify next point or [Undo]: *(Press* ENTER *to exit this command)*

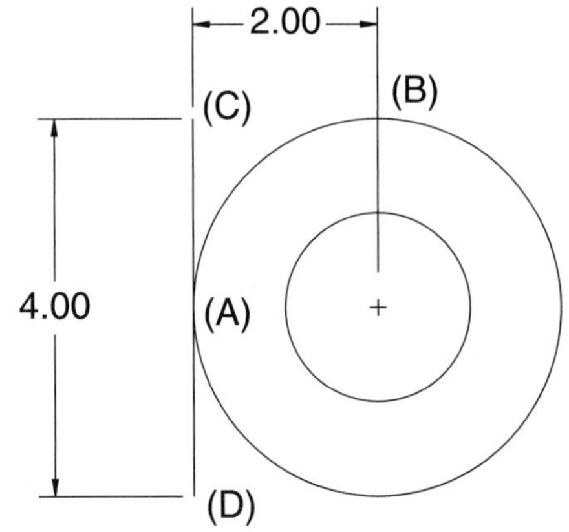

Figure 20–16

STEP 3

Draw four horizontal lines from quadrant points on the two circles a distance of 2 units. Polar coordinates or the Direct Distance mode can be used to accomplish this. These lines should intersect with the vertical line drawn in the previous step. See Figure 20–17.

 Command: **L** *(For LINE)* [draw>linee]

Specify first point: **Qua**

of *(Select the quadrant of the large circle at "A")*

Specify next point or [Undo]: **@2<180**

Specify next point or [Undo]: *(Press* ENTER *to exit this command)*

Repeat the previous procedure and draw three more lines from points "B," "C," and "D" as shown in Figure 20–17. Using the same polar coordinate value for the length, namely, @2<180. An alternate method would be to use the Multiple option of the COPY command to duplicate the three remaining lines.

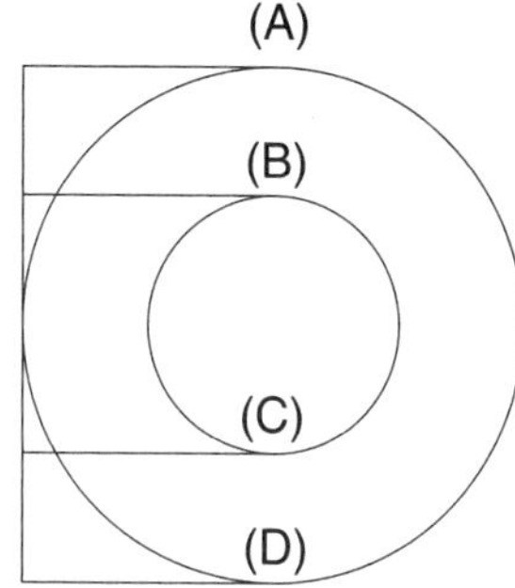

Figure 20–17

STEP 4

Use the TRIM command, select the two dashed lines shown in Figure 20–18 as cutting edges, and trim the left side of the large circle.

 Command: **TR** *(For TRIM)* *[modify>trim]*

Current settings: Projection=UCS Edge=None Select cutting edges ... Select objects: *(Select the two dashed lines shown in Figure 20–18)*

Select objects: *(Press* ENTER *to continue with this command)*

Select object to trim or shift-select to extend or [Project/Edge/Undo]: *(Pick the circle at "A")*

Select object to trim or shift-select to extend or [Project/Edge/Undo]: *(Press* ENTER to exit this command)

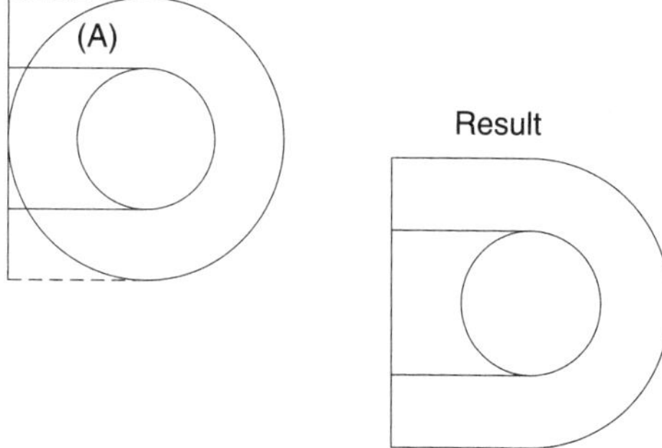

Figure 20–18

STEP 5

Use the TRIM command again and select the two dashed lines in Figure 20–19 as cutting edges. Select the left side of the small circle and the middle of the vertical line as the objects to trim.

Command: **TR** *(For TRIM)* *[modify>trim]*

Current settings: Projection=UCS Edge=None Select cutting edges ... Select objects: *(Select the two dashed lines in Figure 20–19)*

Select objects: *(Press* ENTER *to continue with this command)*

Select object to trim or shift-select to extend or [Project/Edge/Undo]: *(Select the vertical line at "A")*

Select object to trim or shift-select to extend or [Project/Edge/Undo]: *(Select the small circle at "B")*

Select object to trim or shift-select to extend or [Project/Edge/Undo]: *(Press* ENTER *to exit this command)*

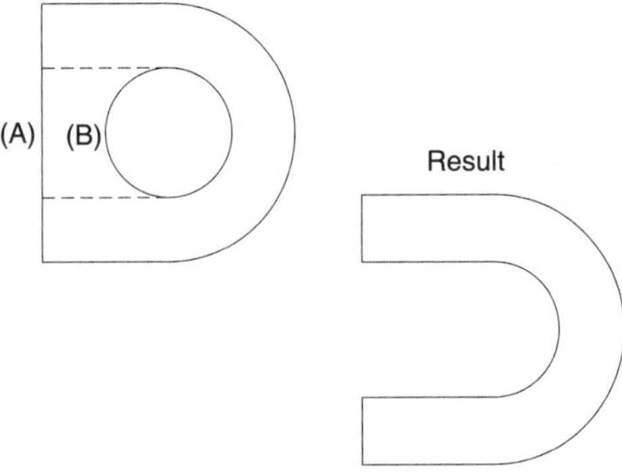

Figure 20–19

STEP 6

Complete a partial plan view by using the LINE command and Figure 20–20 to draw the four lines. Use OSNAP-Endpoint in combination with polar coordinates.

 Command: **L** *(For LINE)* [draw>line]

Specify first point: **End**

of *(Select the endpoint of the line or arc labeled "Start")*

Specify next point or [Undo]: **@8.00<0**

Specify next point or [Undo]: **@4.1858<60**

Specify next point or [Undo]: **@0.75<150**

Specify next point or [Undo]: **End**

of *(Select the endpoint of the line or arc labeled "End")*

Specify next point or [Undo]: *(Press* ENTER *to exit this command)*

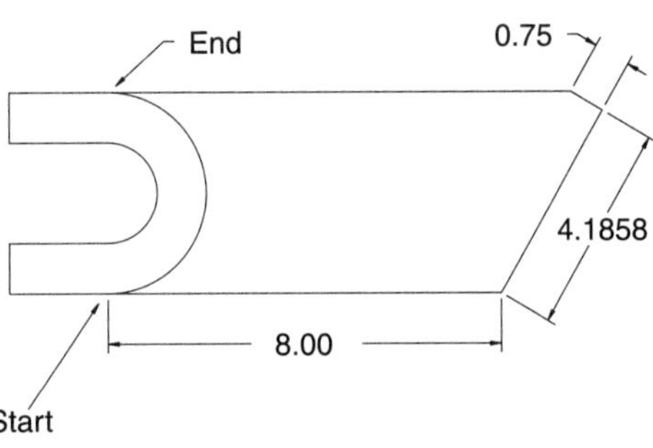

Figure 20–20

STEP 7

View the model in 3D using the VPOINT command. Use a viewing position of 1,-1,1 or choose 3D Views from the View menu and then SE Isometric. See Figure 20–21.

Command: **VPOINT** [view>3D views> view point]
Current view direction: VIEWDIR=0.00,0.00,1.00 Specify a view point or [Rotate] <display compass and tripod>: **1,-1,1**

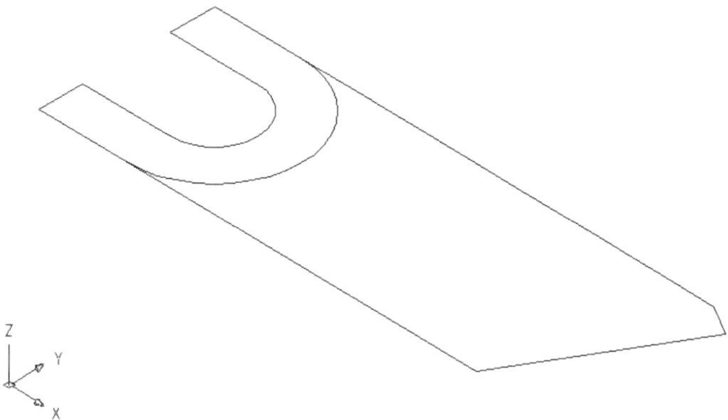

Figure 20–21

STEP 8

Temporarily move the four lines that represent the base away from the U-shaped figure as shown in Figure 20–22. Since both figures are at different heights, they will be treated as different objects when you extrude. The move operation will help in creating the profiles that will be extruded.

Command: **M** *(For MOVE) [modify>move]*
Select objects: *(Select all four lines labeled "A," "B," "C," and "D")*
Select objects: *(Press* ENTER *to continue with this command)*
Specify base point or displacement: **End**
of *(Pick the endpoint near "A")*
Specify second point of displacement or <use first point as displacement>:
@2.00<0

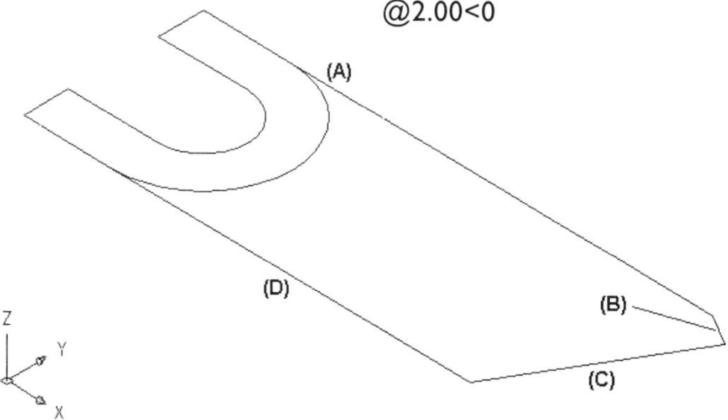

Figure 20–22

STEP 9

Copy arc "A" to the endpoint of the line at "C." This will allow the shape represented by the lines to be closed through the placement of the arc. See Figure 20–23.

 Command: **CP** *(For COPY)* [modify>copy]

Select objects: *(Select arc "A")*

Select objects: *(Press* ENTER *to continue with this command)*

Specify base point or displacement, or [Multiple]: **End**

of *(Pick the endpoint of the arc at "B")*

Specify second point of displacement or <use first point as displacement>: **End**

of *(Pick the endpoint of the arc at "C")*

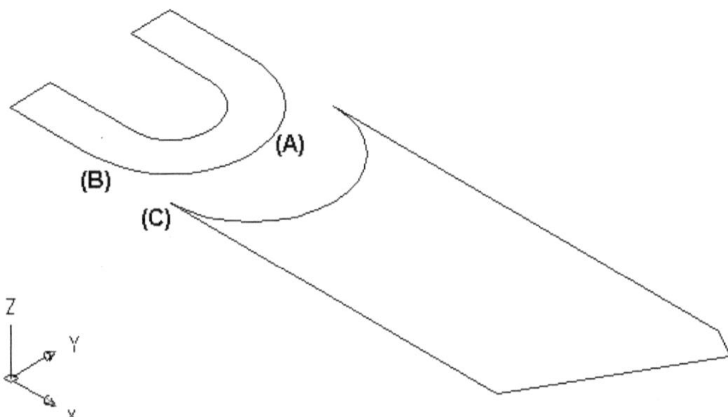

Figure 20–23

STEP 10

Convert both shapes to polylines using the PEDIT command. Refer also to Figure 20–24.

 Command: **PE** *(For PEDIT)* [modify>object>poly line]

Select polyline or [Multiple]: *(Select the arc at "A")*

Object selected is not a polyline Do you want to turn it into one? <Y>

(Press ENTER *to accept the default)*

Enter an option [Close/Join/Width/Edit vertex/Fit/Spline/Decurve/Ltype gen/ Undo]: **J** *(For Join)*

Select objects: *(Select all arc and line segments that make up the outline of the U-shaped item)*

Select objects: *(Press* ENTER *to perform the join operation)*

7 segments added to polyline Enter an option [Open/Join/Width/Edit vertex/ Fit/Spline/Decurve/Ltype gen/ Undo]: *(Press* ENTER *to exit this command)*

 Command: **PE** *(For PEDIT)* [modify>object>poly line]

Select polyline or [Multiple]: *(Select the line at "B")*

Object selected is not a polyline Do you want to turn it into one? <Y>

(Press ENTER *to accept the default)*

Enter an option [Close/Join/Width/Edit vertex/Fit/Spline/Decurve/Ltype gen/ Undo]: **J** *(For Join)*

Select objects: *(Select the other line and arc segments that make up the outline of the other shape)*

Select objects: *(Press* ENTER *to perform the join operation)*

4 segments added to polyline Enter an option [Open/Join/Width/Edit vertex/ Fit/Spline/Decurve/Ltype gen/ Undo]: *(Press* ENTER *to exit this command)*

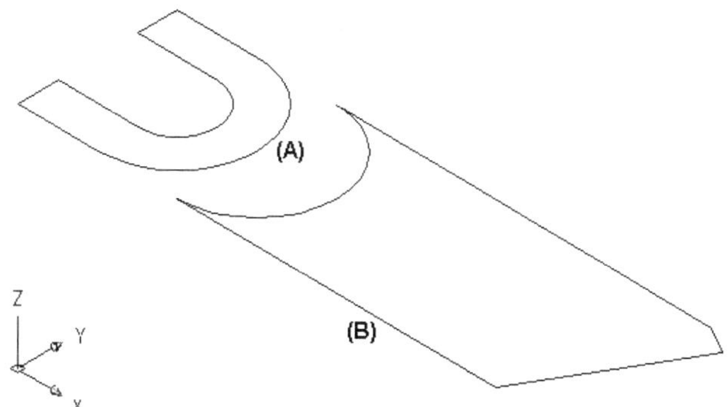

Figure 20–24

STEP 11

Extrude the U-shaped figure to a height of 2.50 units. Extrude the other shape to a height of 0.625 units. Perform both operations using the EXTRUDE command and Figure 20–25.

 Command: **EXT** *(For EXTRUDE)* [draw>solids>extrude]

Current wire frame density: ISOLINES=4
Select objects: *(Select the U-shaped figure)*
Select objects: *(Press* ENTER *to continue with this command)*
Specify height of extrusion or [Path]: **2.50**
Specify angle of taper for extrusion <0>: *(Press* ENTER *to perform the extrusion operation)*

Command: **EXT** *(For EXTRUDE)* [draw>solids>extrude]

Current wire frame density: ISOLINES=4
Select objects: *(Select the other shape at "A")*
Select objects: *(Press* ENTER *to continue with this command)*
Specify height of extrusion or [Path]: **0.625**
Specify angle of taper for extrusion <0>: *(Press* ENTER *to perform the extrusion operation)*

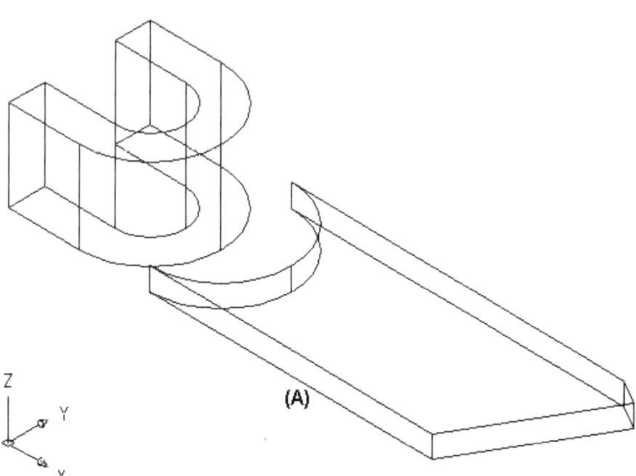

Figure 20–25

STEP 12

With both shapes extruded, connect both shapes back together from the quadrant of one shape to the quadrant of the other shape using the MOVE command. See Figure 20–26.

Command: **M** *(For MOVE)*
[modify>move]

Select objects: *(Select shape "A")*
Select objects: *(Press* ENTER *to continue with this command)*
Specify base point or displacement: **Qua**
of *(Pick the bottom quadrant of the arc at "B")*
Specify second point of displacement or <use first point as displacement>: **Qua**
of *(Pick the bottom quadrant of the arc at "C")*

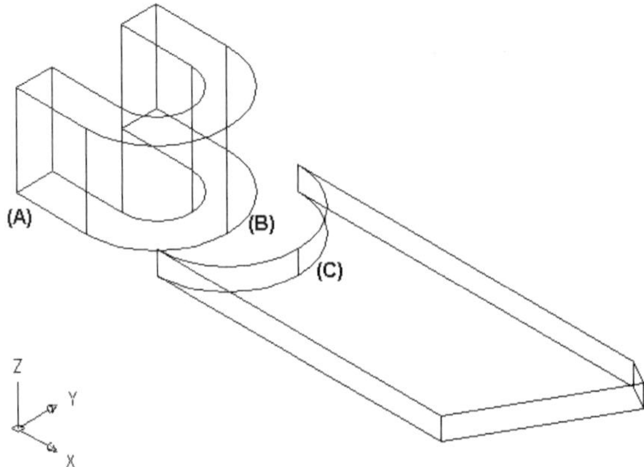

Figure 20–26

STEP 13

Join both shapes into one using the UNION command, as shown in Figure 20–27.

Command: **UNI** *(For UNION)*
[modify>solids editing>union]

Select objects: *(Select shape "A" and "B")*
Select objects: *(Press* ENTER *to perform the union operation)*

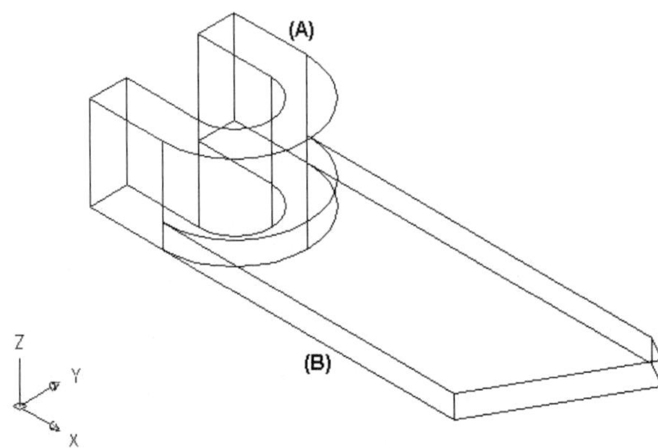

Figure 20–27

STEP 14

Define a new origin for the User Coordinate System by placing it at the endpoint at "A" in Figure 20–28. If the icon does not move to the new location, use the UCSICON command and the Origin option to update the position of the User Coordinate System icon. If the icon still does not move, zoom out so that it can be displayed at the origin.

 Command: **UCS** [tools>move ucs]

Current ucs name: *WORLD* Enter an option [New/Move/ orthoGraphic/ Prev/Restore/Save/Del/ Apply/?/World] <World>: **M** *(For Move)*
Specify new origin point or [Zdepth] <0,0,0>: **End**
of *(Pick the line at "A" to find the endpoint)*
Command: **UCSICON** [view>display>ucsicon]
Enter an option [ON/OFF/All/Noorigin/ ORigin] <ON>: **OR** *(For Origin)*

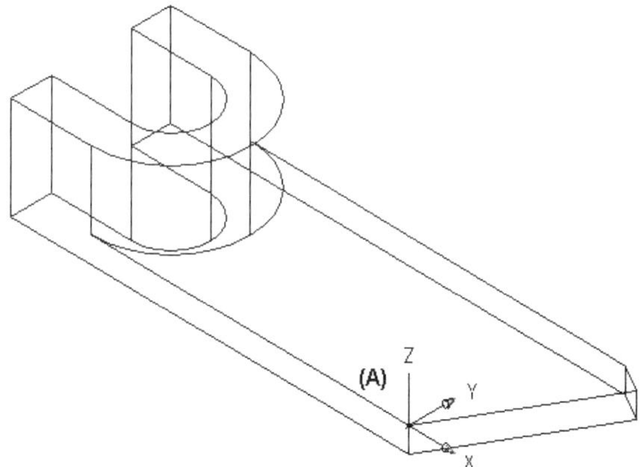

Figure 20–28

STEP 15

Before beginning the construction of the mounting plate consisting of the four holes, rotate the UCS icon 60° about the Z axis and then 90° about the X axis. See Figure 20–29.

 Command: **UCS** [tools>new ucs]

Current ucs name: *NO NAME* Enter an option [New/Move/ orthoGraphic/ Prev/Restore/Save/Del/ Apply/?/World] <World>: **N** *(For New)*
Specify origin of new UCS or [ZAxis/ 3point/OBject/Face/View/X/Y/Z] <0,0,0>: **Z**
Specify rotation angle about Z axis <90>: **60**

 Command: **UCS** [tools>new ucs]

Current ucs name: *NO NAME* Enter an option [New/Move/ orthoGraphic/ Prev/Restore/Save/Del/ Apply/?/World] <World>: **N** *(For New)*
Specify origin of new UCS or [ZAxis/ 3point/OBject/Face/View/X/Y/Z] <0,0,0>: **X**
Specify rotation angle about X axis <90>: *(Press* ENTER *to accept default)*

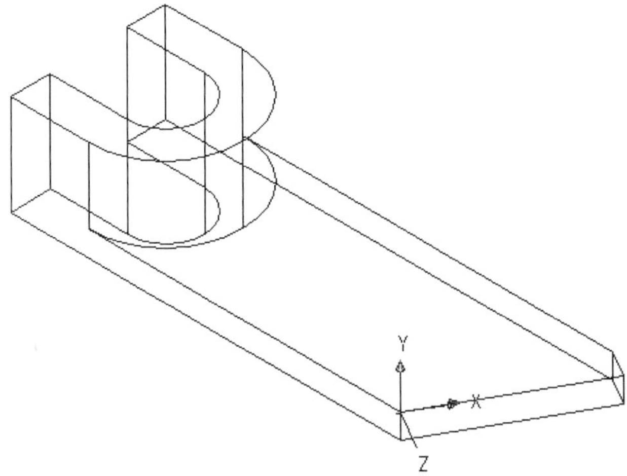

Figure 20–29

STEP 16

Construct a polyline representing the front face of the mounting plate (see Figure 20–30). Polar coordinates or the Direct Distance mode can be used to accomplish this step. Round off the two top corners of the mounting plate using the FILLET command.

Command: **PL** *(For PLINE)*
[draw>polyline]

Specify start point: **0,0,0**
Current line-width is 0.00 Specify next point or [Arc/Halfwidth/ Length/Undo/ Width]: **@3.50<90**
Specify next point or [Arc/Close/ Halfwidth/Length/Undo/Width]:
@4.1858<0
Specify next point or [Arc/Close/ Halfwidth/Length/Undo/Width]:
@3.50<270
Specify next point or [Arc/Close/ Halfwidth/Length/Undo/Width]: **C** *(To close the polyline and exit the command)*

Command: **F** *(For FILLET)*
[modify>fillet]

Current settings: Mode = TRIM Radius = 0.50 Select first object or [Polyline/ Radius/ Trim]: *(Pick the polyline at "A")*
Select second object: *(Pick the polyline at "B")*

Repeat the FILLET command to round off the other corner.

Command: **F** *(For FILLET)*
[modify>fillet]

Current settings: Mode = TRIM Radius = 0.50 Select first object or [Polyline/ Radius/ Trim]: *(Pick the polyline at "B")*
Select second object: *(Pick the polyline at "C")*

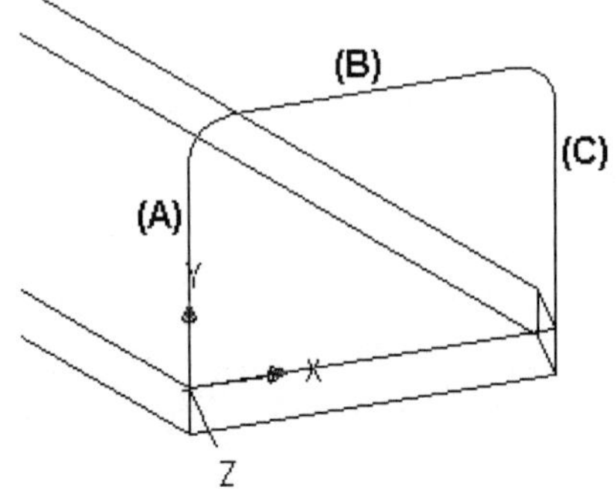

Figure 20–30

STEP 17

Extrude the face of the mounting plate back a distance of -0.75 units to give it thickness using the extrude command. Once the mounting plate has been extruded, join the plate to the Lever using the union command. See Figure 20–31.

 Command: **EXT** *(For EXTRUDE)* [draw>solids>extrude]

Current wire frame density: ISOLINES=4
Select objects: *(Select the face of the mounting plate)*
Select objects: *(Press ENTER to continue with this command)*
Specify height of extrusion or [Path]: **-0.75**
Specify angle of taper for extrusion <0>: *(Press ENTER to perform the extrusion operation)*

 Command: **UNI** *(For UNION)* [modify>solids editing>union]

Select objects: *(Select the mounting plate and the Lever)*
Select objects: *(Press ENTER to perform the union operation)*

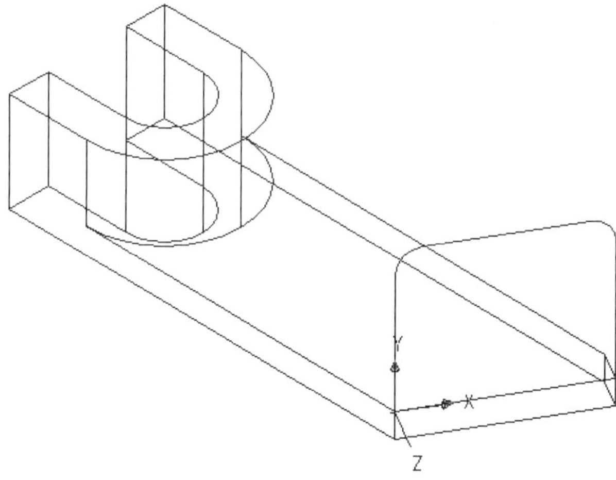

Figure 20–31

STEP 18

Create one of the holes in the mounting plate using the cylinder command. Use the OSNAP-From option to establish a reference point that will allow you to construct the cylinder -1.00 units in the X axis and 1.375 units along the Y axis. See Figure 20–32.

 Command: **CYLINDER** [draw>solids>cylinder]

Current wire frame density: ISOLINES=4
Specify center point for base of cylinder or [Elliptical] <0,0,0>: **From**
Base point: **End**
of *(Pick the endpoint at "A")*
<Offset>: **@-1.00,1.375**
Specify radius for base of cylinder or [Diameter]: **D** *(For Diameter)*
Specify diameter for base of cylinder: **0.75**
Specify height of cylinder or [Center of other end]: **-0.75**

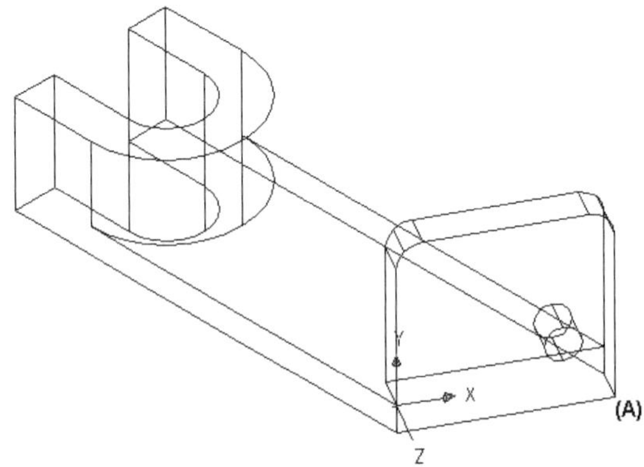

Figure 20–32

STEP 19

Use the ARRAY command (-AR) or Array dialog box to create a rectangular pattern consisting of two rows and two columns. The distance between the rows will be 1.375 units and the distance between columns will be -2.1858 units. Then subtract the four cylinders from the Lever to complete the object using the SUBTRACT command. See Figure 20–33.

Command: **-AR** *(For ARRAY)*
[modify>array]
Select objects: *(Select the cylinder just created)*
Select objects: *(Press* ENTER *to continue with this command)*
Enter the type of array [Rectangular/Polar] <R>: **R** *(For rectangular)*
Enter the number of rows (—) <1>: **2**
Enter the number of columns (|||) <1>: **2**
Enter the distance between rows or specify unit cell (—): **1.375**
Specify the distance between columns (|||): **-2.1858**

Command: **SU** *(For SUBTRACT)*
[modify>solids editing>subtract]
Select solids and regions to subtract from... Select objects: *(Select the object at "A")*
Select objects: *(Press* ENTER *to continue with command)*
Select solids and regions to subtract... Select objects: *(Select the four individual cylinders just created through the ARRAY command)*
Select objects: *(Press* ENTER *to perform the subtraction operation)*

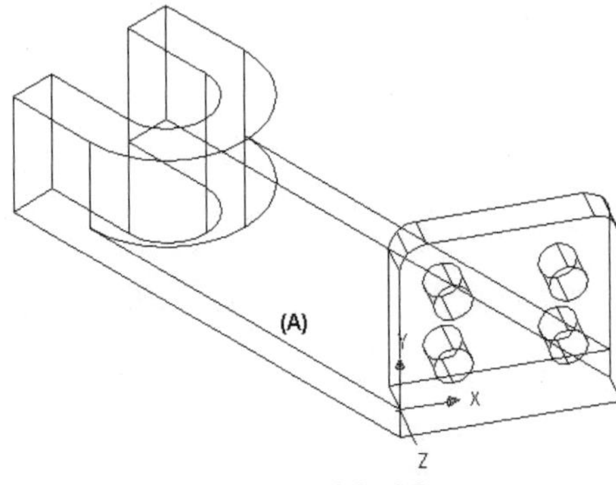

Figure 20–33

STEP 20

Set FACETRES to a value of 2 and perform a hidden line removal to get a better look at the model. See Figure 20–34.

Command: **FACETRES**

Enter new value for FACETRES <0.5000>: **2**
Command: **HI** *(For HIDE)*

Perform a REGEN to convert the model back to its wireframe state.

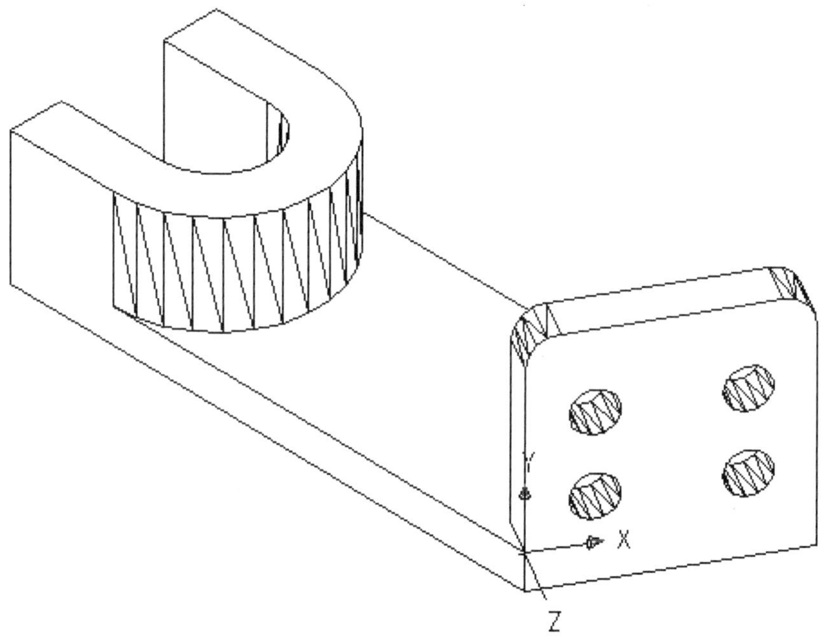

Figure 20–34

STEP 21

Return the User Coordinate System icon to its home position by using the World option of the UCS command. Then use the MASSPROP command to calculate the mass properties of the lever. See Figure 20–35. Variances may occur in a few of the data fields when you use this command.

Command: **UCS** [tools>new ucs>world]
Current ucs name: *NO NAME* Enter an option [New/Move/ orthoGraphic/ Prev/Restore/Save/Del/ Apply/?/World] <World>: *(Press ENTER to accept the default as WORLD)*
Command: **MASSPROP** [tools>inquiry>mass properties]
Select objects: *(Select anywhere along the model)*
Select objects: *(Press ENTER to perform the mass property calculation)*

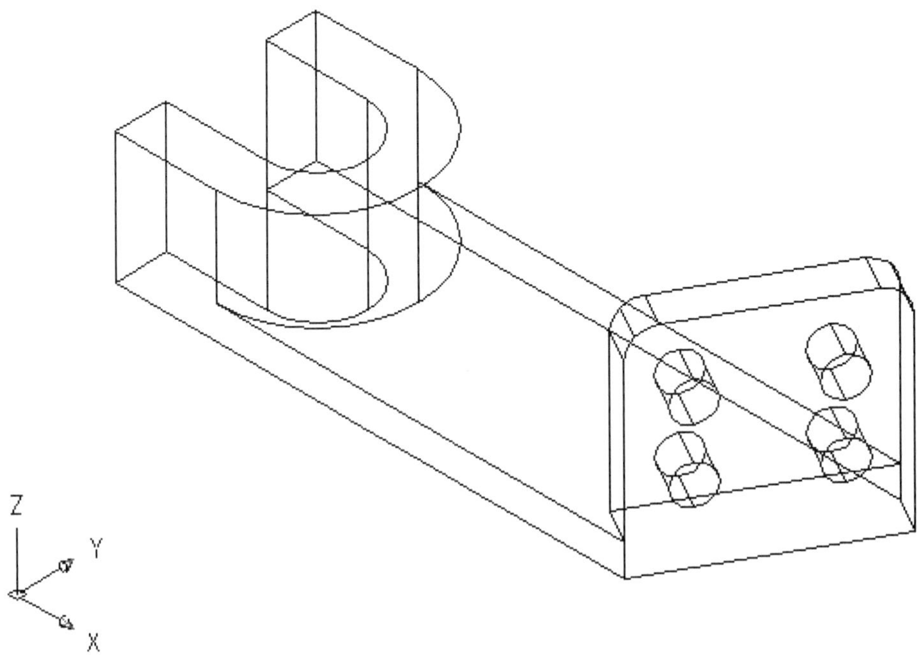

Figure 20–35

```
---------------- SOLIDS ----------------
Mass:                    50.22
Volume:                  50.22
Bounding box:            X: 4.00 -- 16.09
                         Y: 3.00 -- 7.00
                         Z: 0.00 -- 4.13
Centroid:                X: 9.71
                         Y: 5.03
                         Z: 1.12
Moments of inertia:      X: 1461.27
                         Y: 5552.27
                         Z: 6786.70
Products of inertia:     XY: 2469.73
                         YZ: 281.34
                         ZX: 571.53
Radii of gyration:       X: 5.39
                         Y: 10.51
                         Z: 11.62
Principal moments and X-Y-Z directions about centroid:
                         I: 124.12 along [1.00 0.02 0.04]
                         J: 756.75 along [-0.02 1.00 -0.08]
                         K: 782.06 along [-0.04 0.08 1.00]
```

Beginning Level Projects

Directions for Problems 20–1 through 20–11

Create a 3D solid model of each object on layer "Model."

When completed, calculate the volume of the solid model using the MASSPROP command.

PROBLEM 20–1

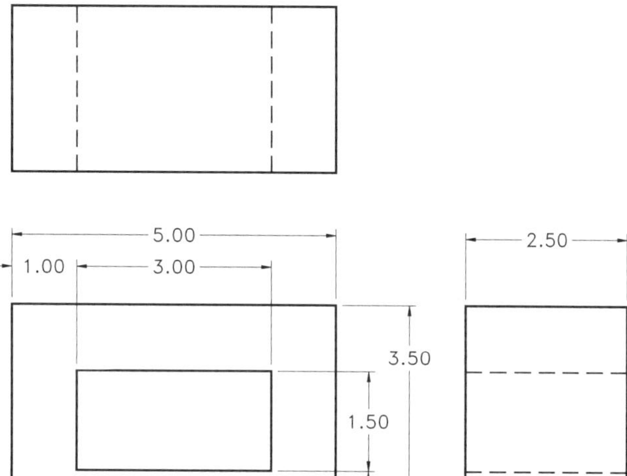

PROBLEM 20–2

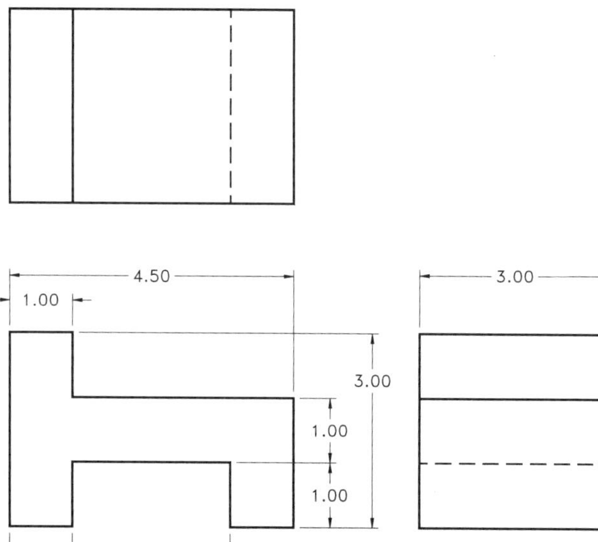

PROBLEM 20–3

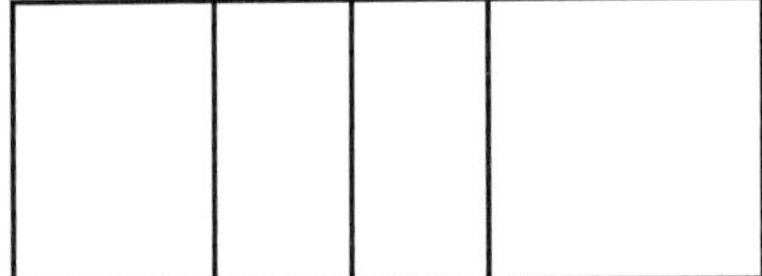

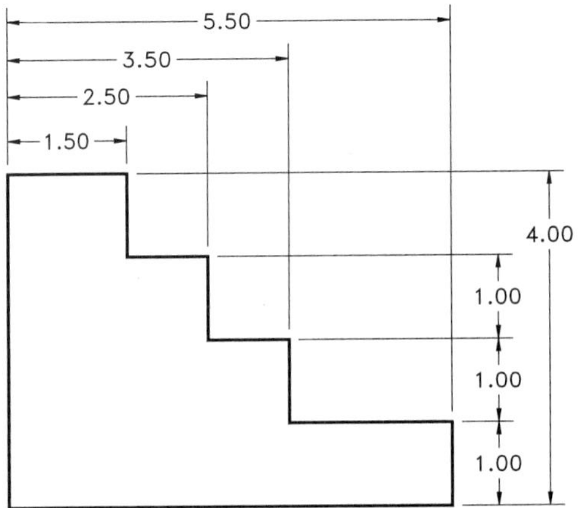

PROBLEM 20–4

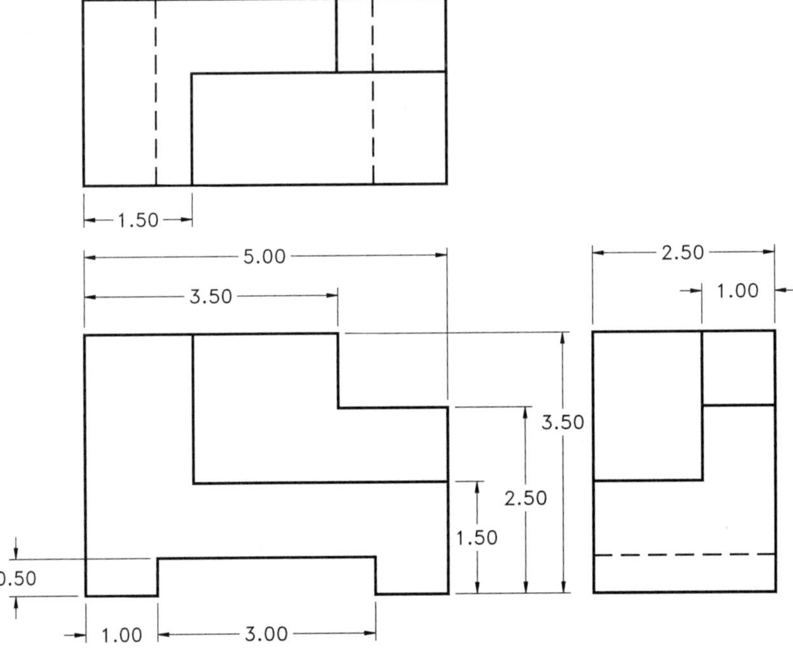

PROBLEM 20–5

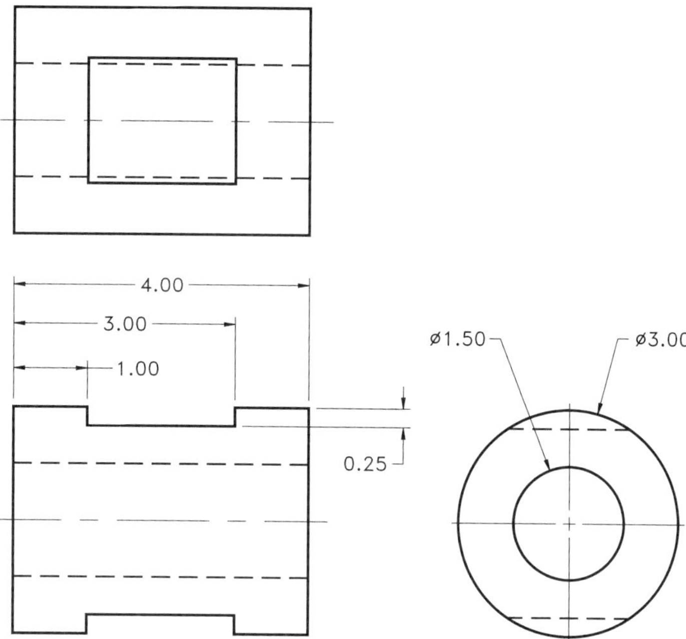

PROBLEM 20–6

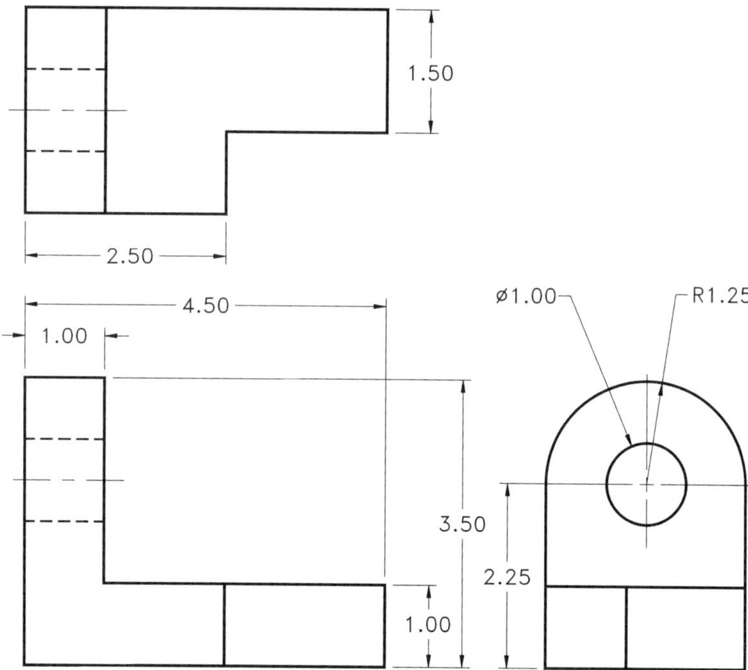

PROBLEM 20–7

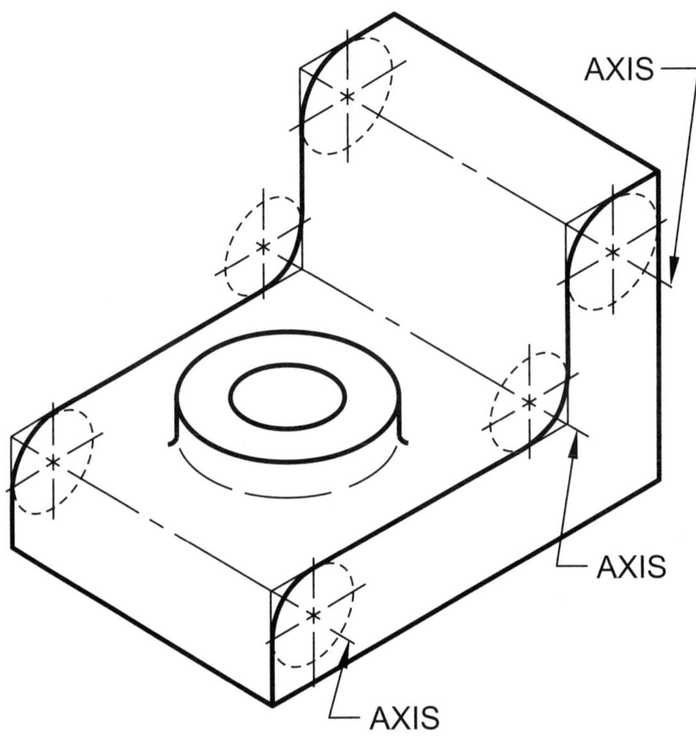

PROBLEM 20–8

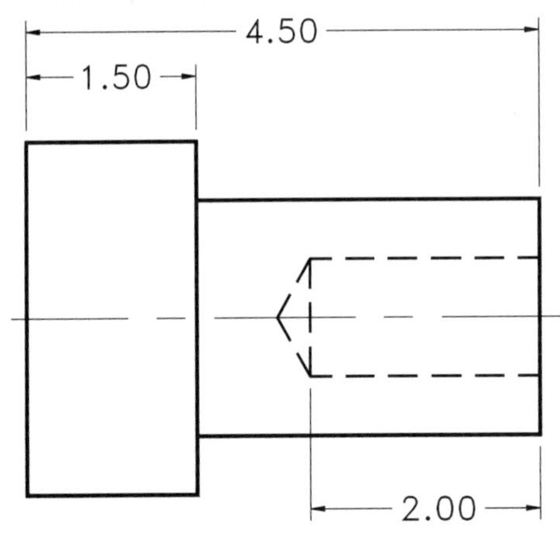

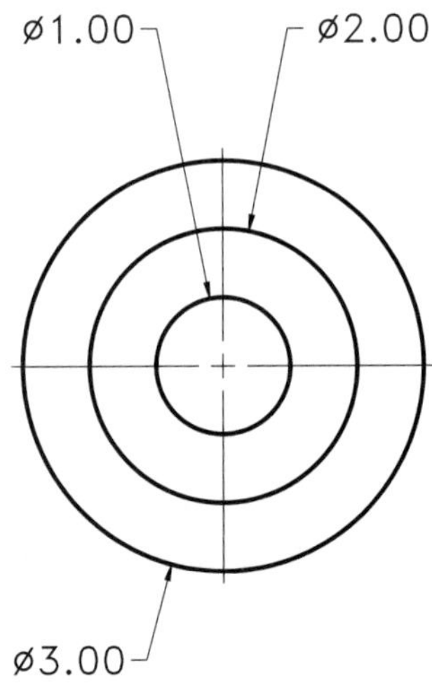

PROBLEM 20–9

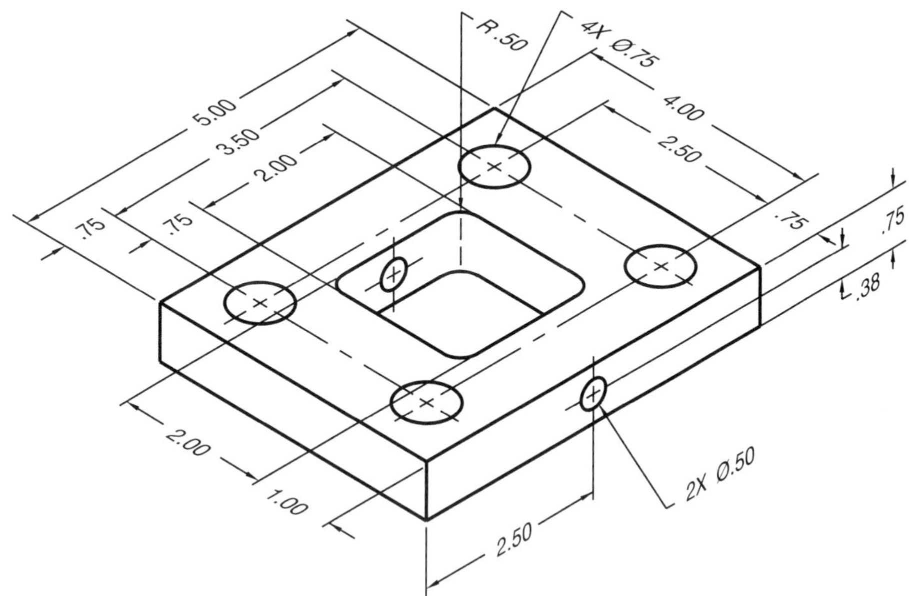

PROBLEM 20–10

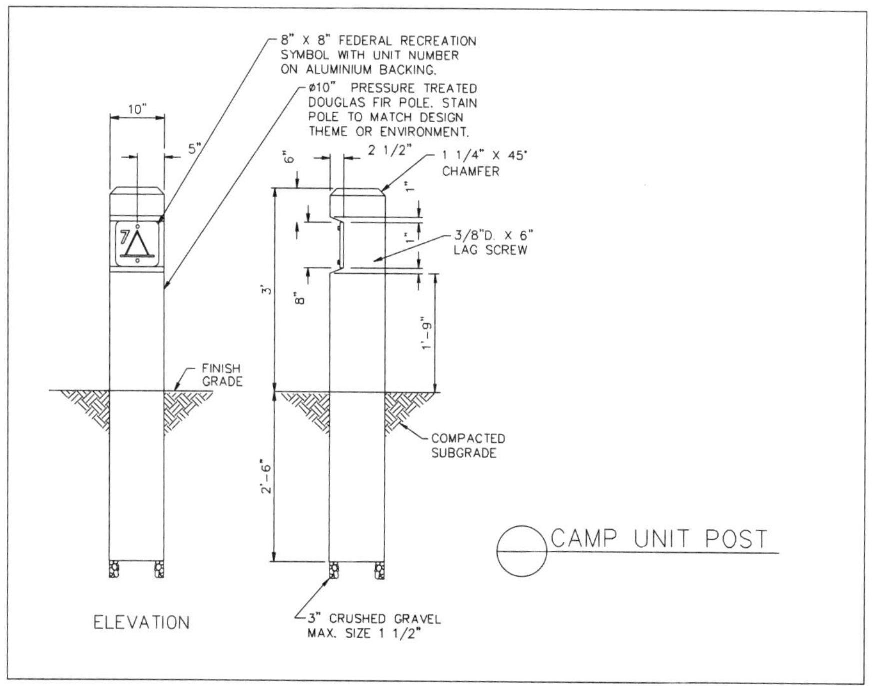

Directions for Problems 20–11 through 20–20

Create a 3D solid model of each object on Layer "Model."

When completed, calculate the volume of the solid model using the MASSPROP command.

PROBLEM 20–11

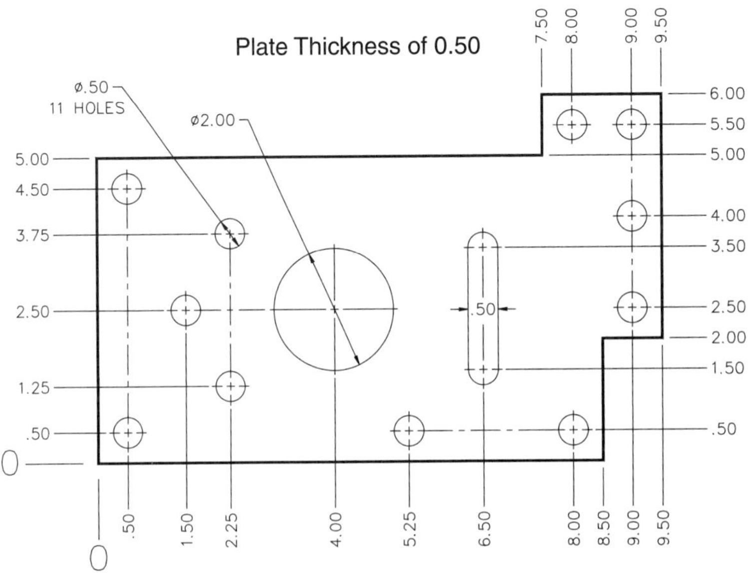

PROBLEM 20–12

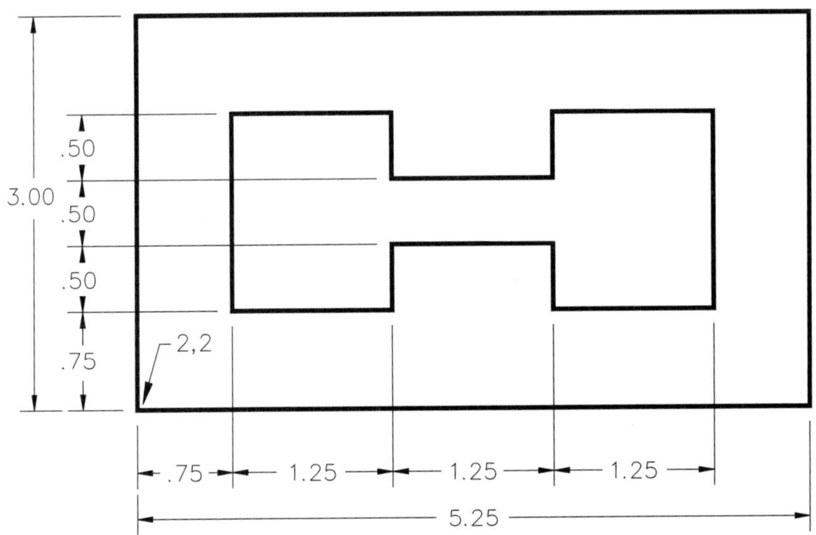

PROBLEM 20–13

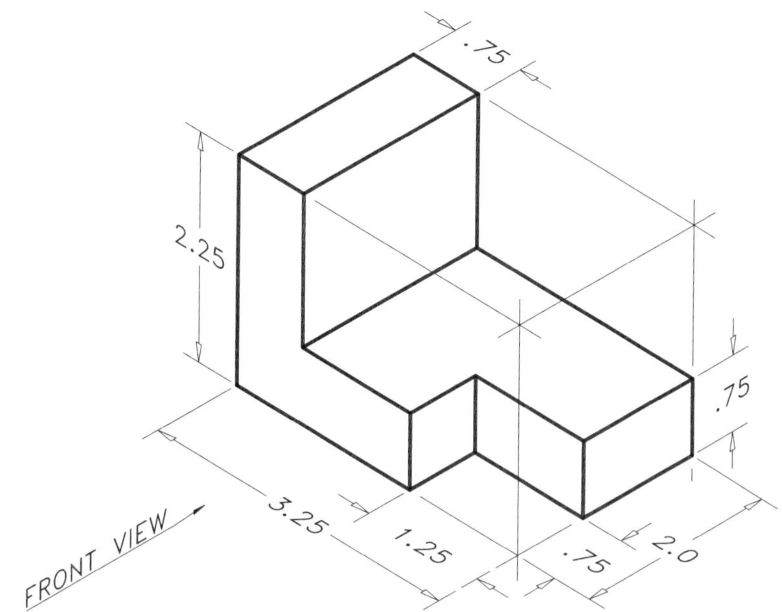

PROBLEM 20–14

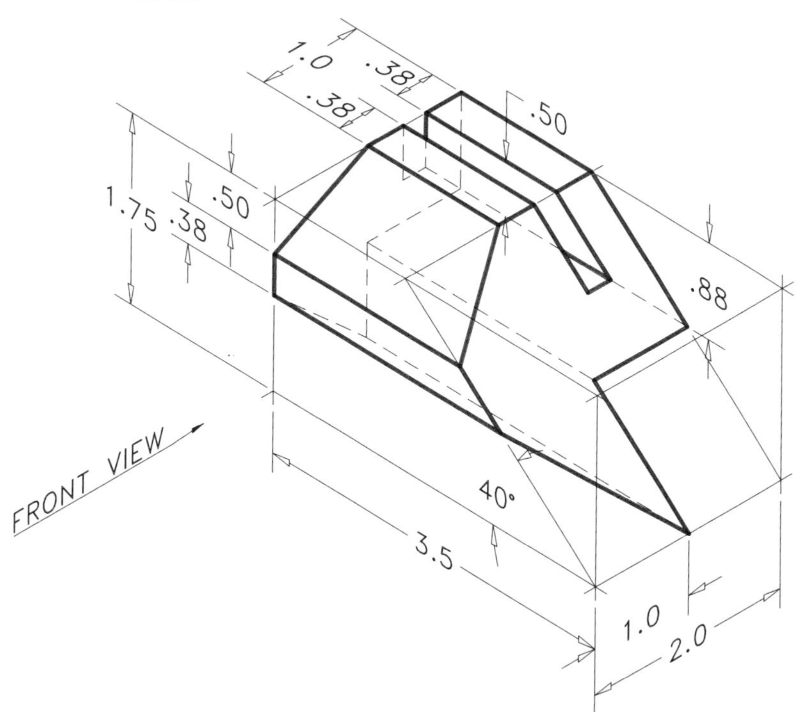

PROBLEM 20–15

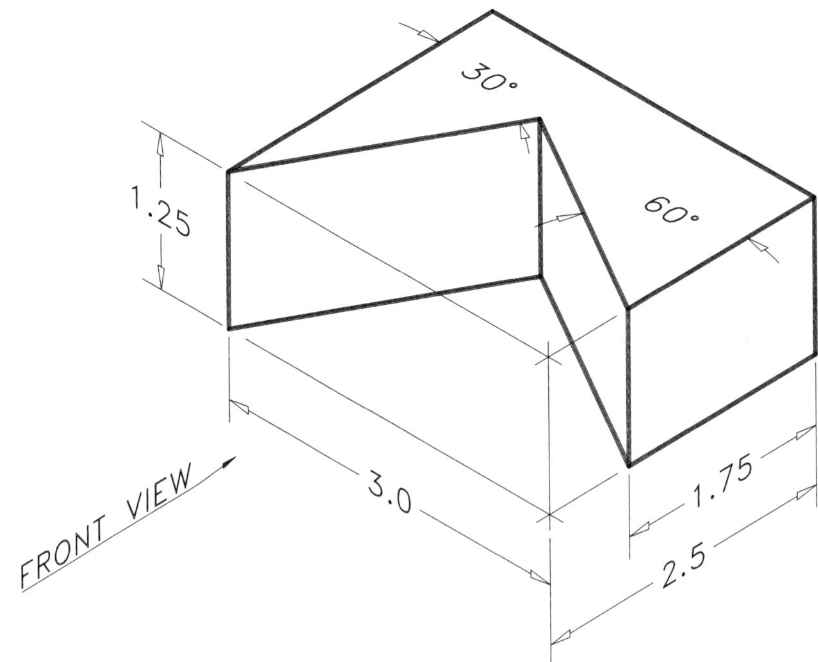

PROBLEM 20–16

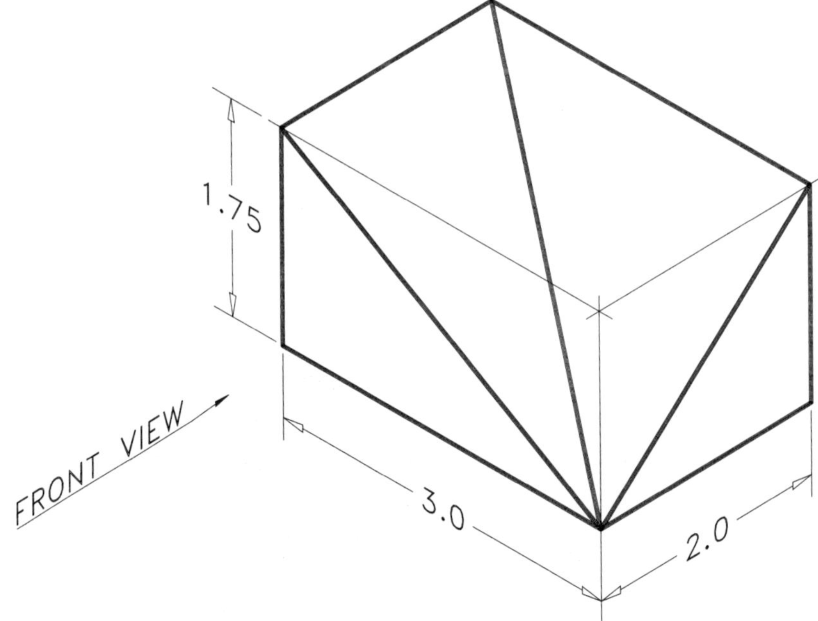

PROBLEM 20–17

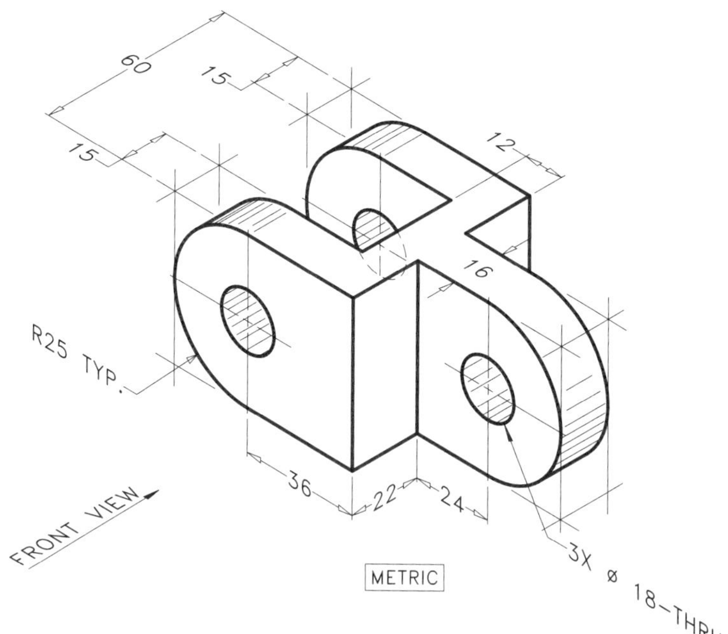

PROBLEM 20–18

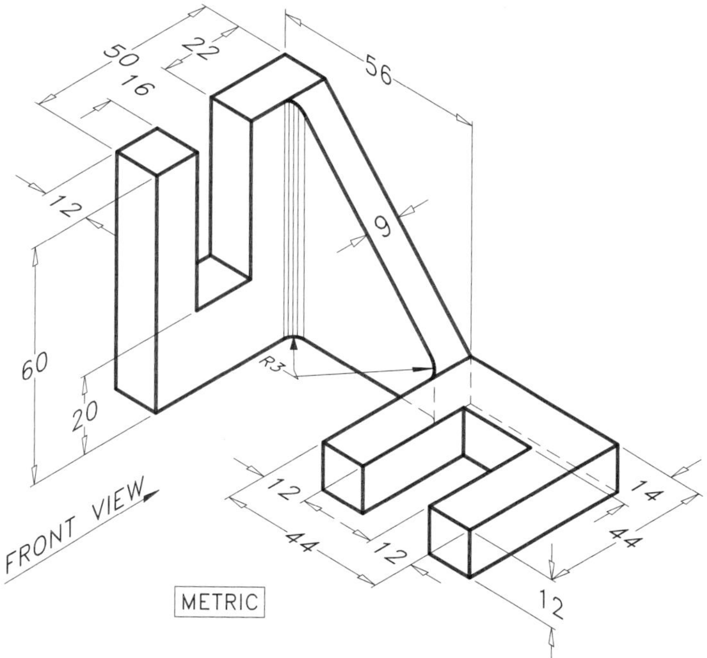

PROBLEM 20–19

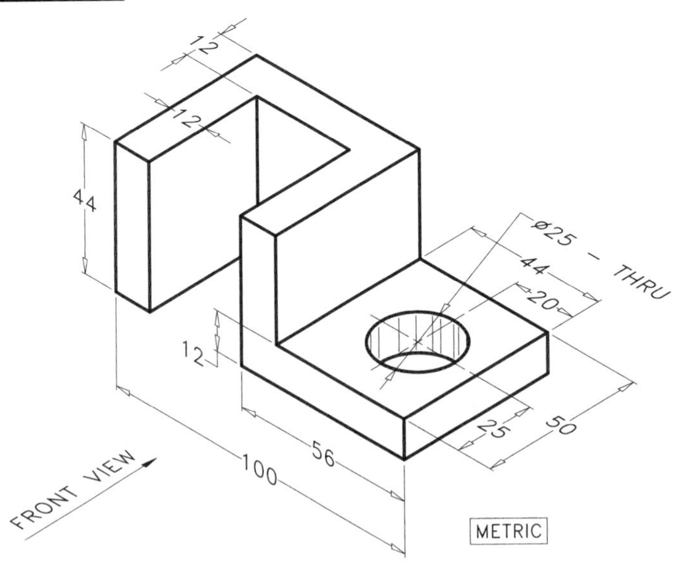

PROBLEM 20–20

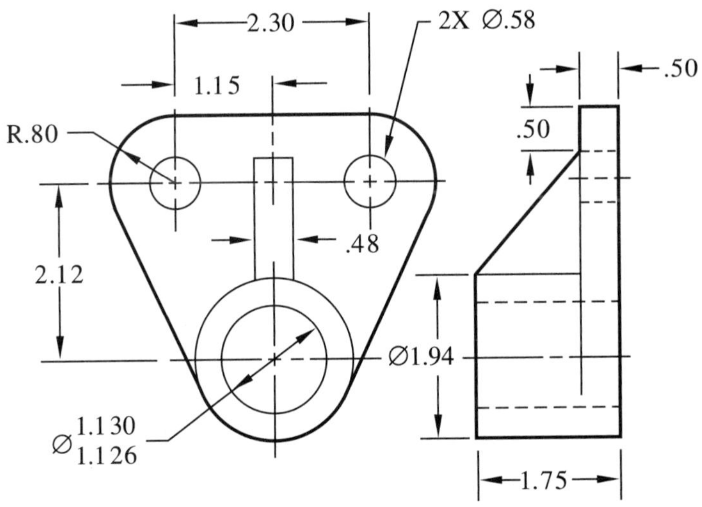

Intermediate Level Drawings
Directions for Problems 20–20 through 20–33

Create a 3D solid model of each object. Use the necessary layers for project completion.

PROBLEM 20–21

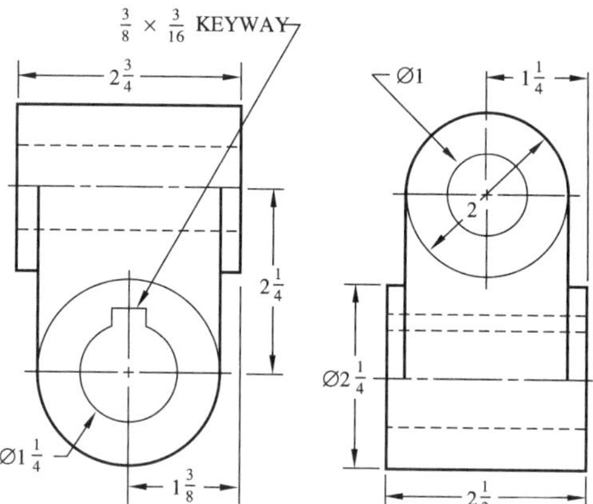

PROBLEM 20–22

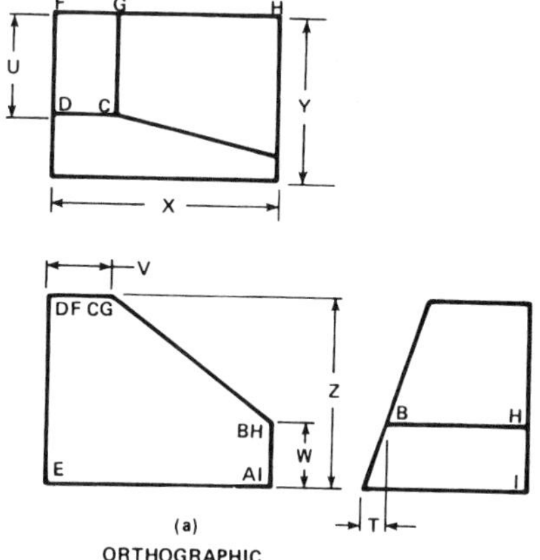

(a)
ORTHOGRAPHIC

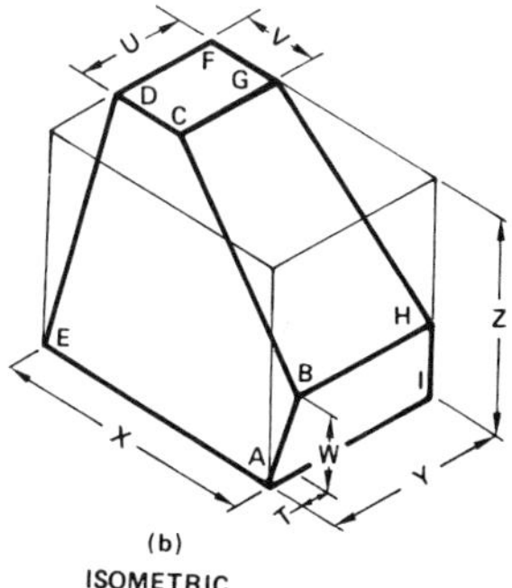

(b)
ISOMETRIC

PROBLEM 20–23

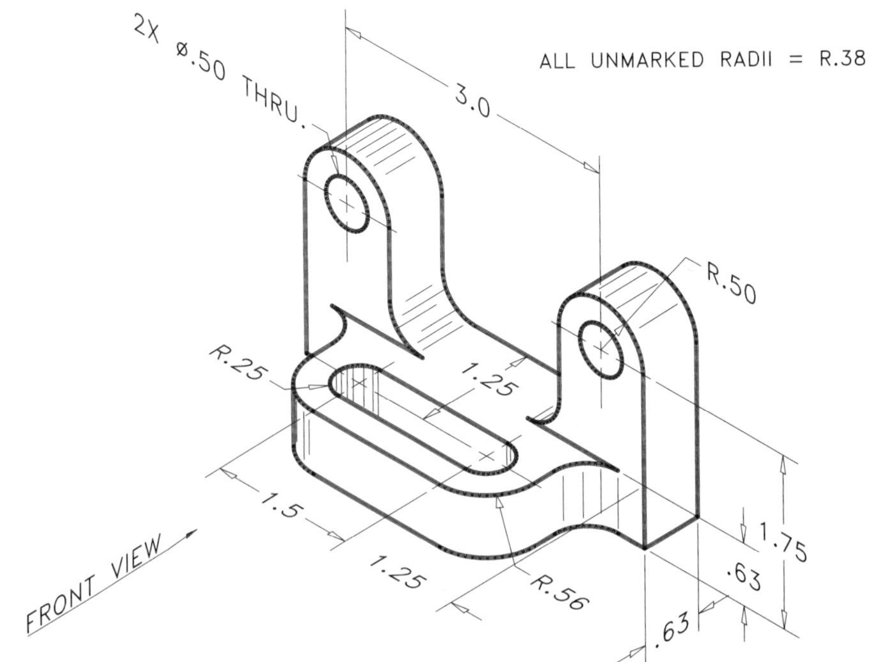

PROBLEM 20–24

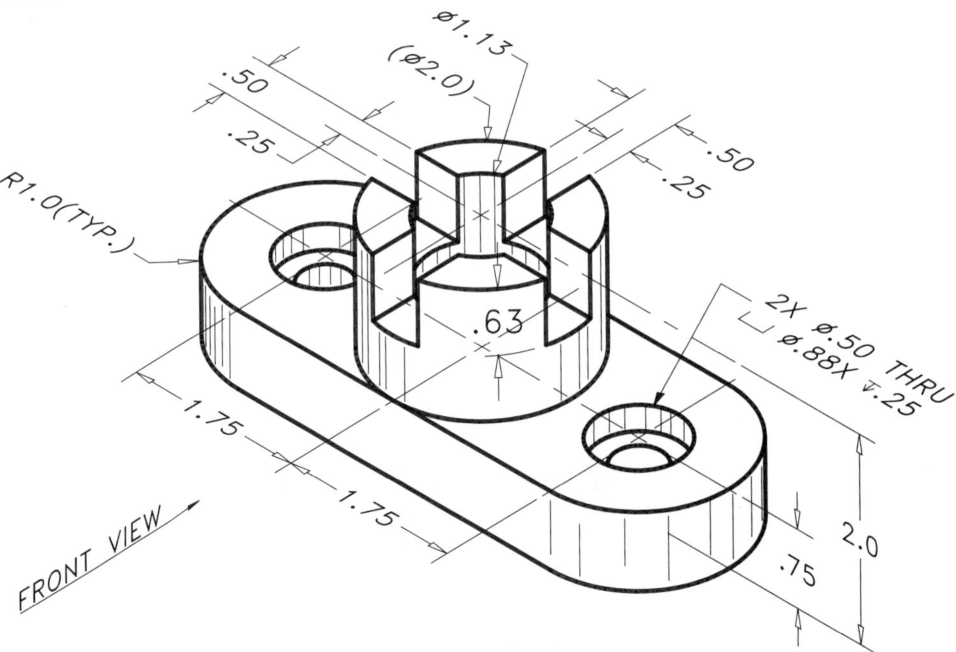

PROBLEM 20–25

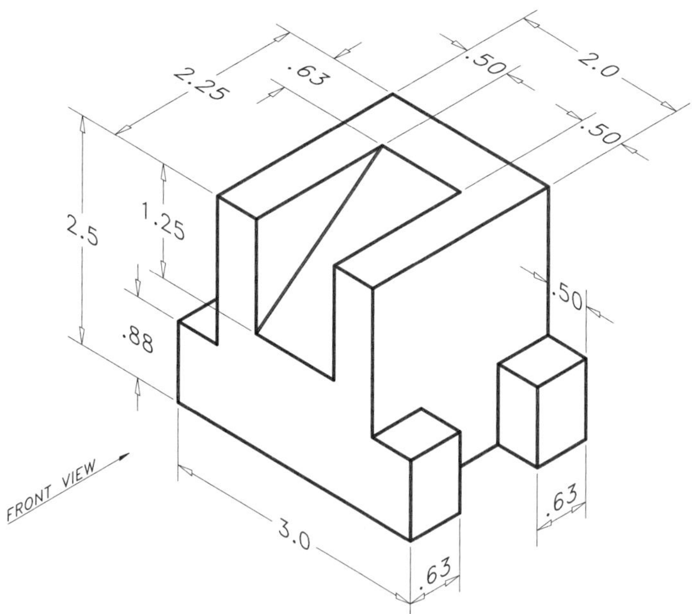

PROBLEM 20–26

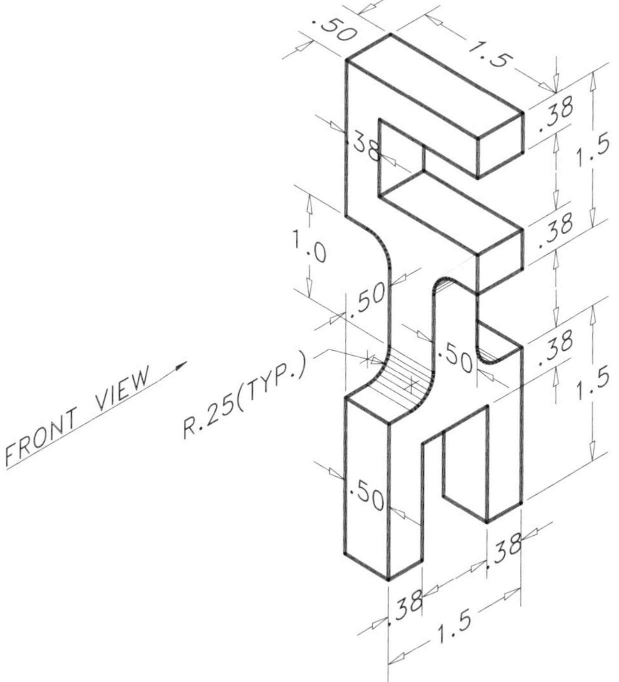

PROBLEM 20–27

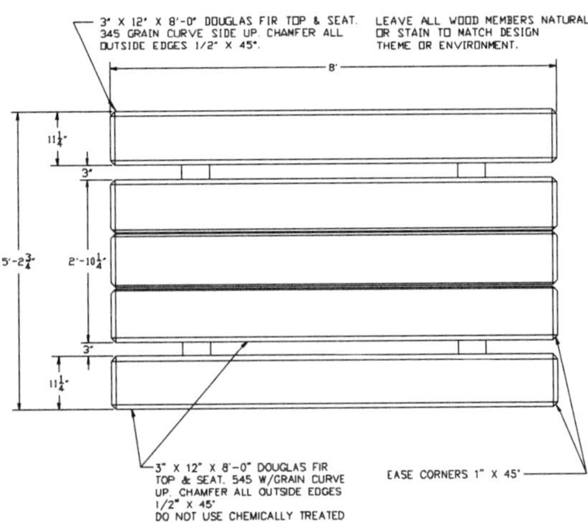

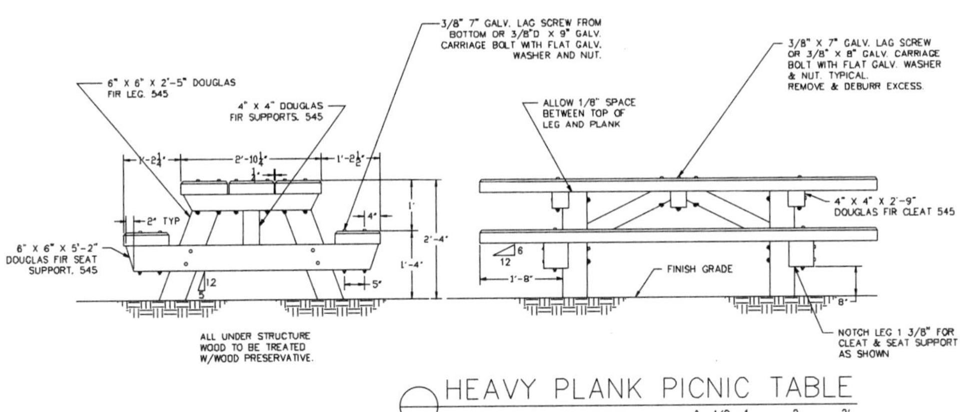

PROBLEM 20–28

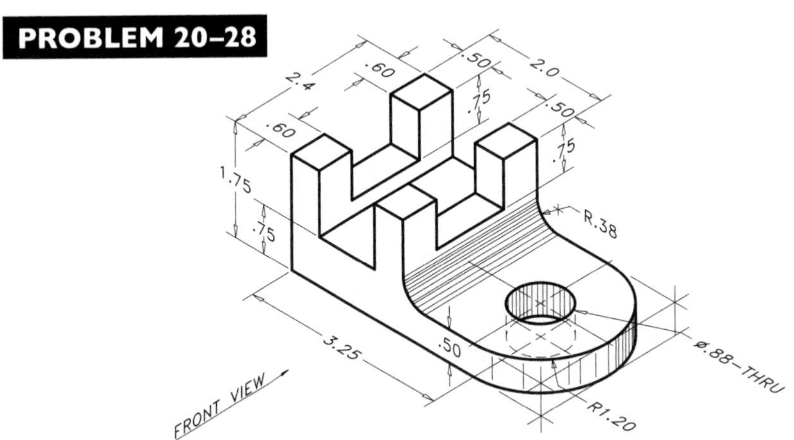

PROBLEM 20–29

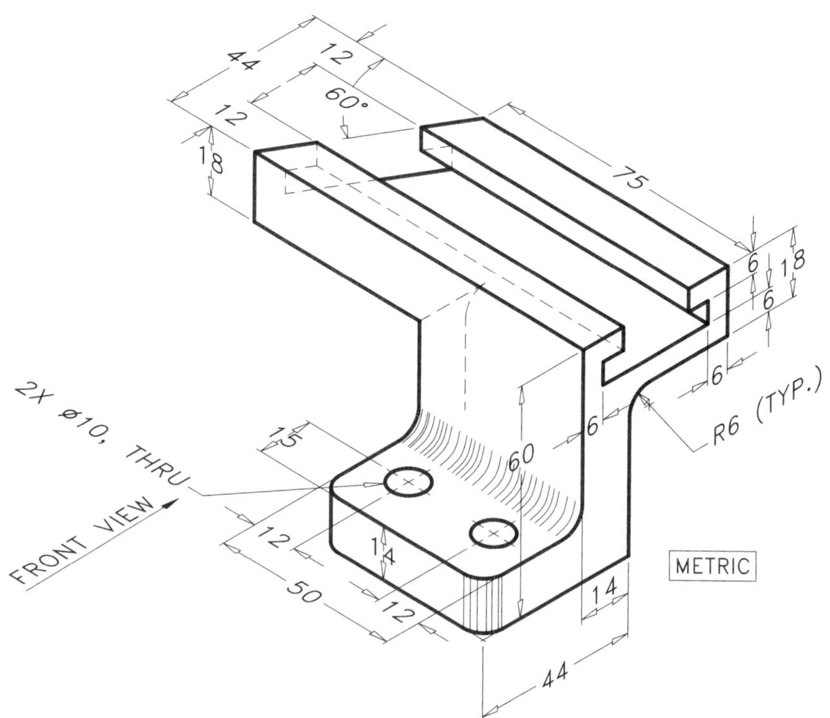

PROBLEM 20–30

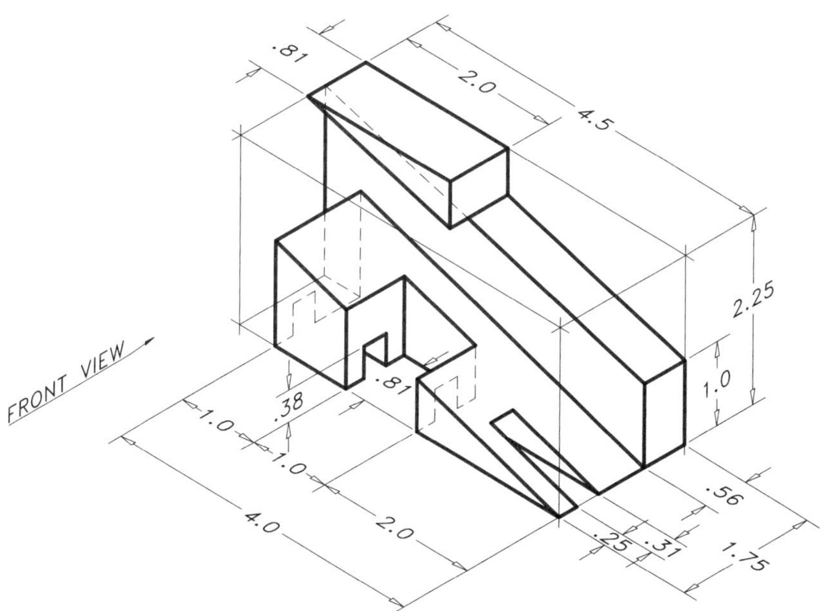

PROBLEM 20–31

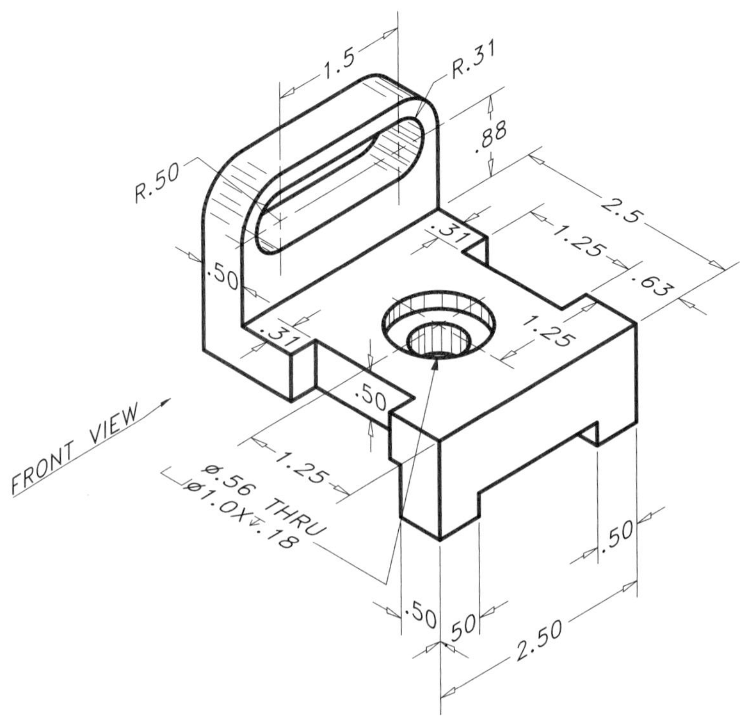

PROBLEM 20–32

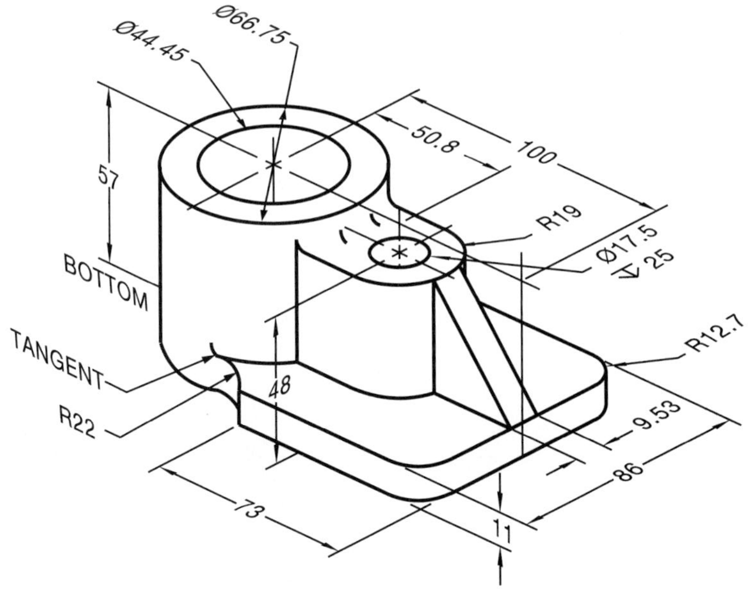

PROBLEM 20–33

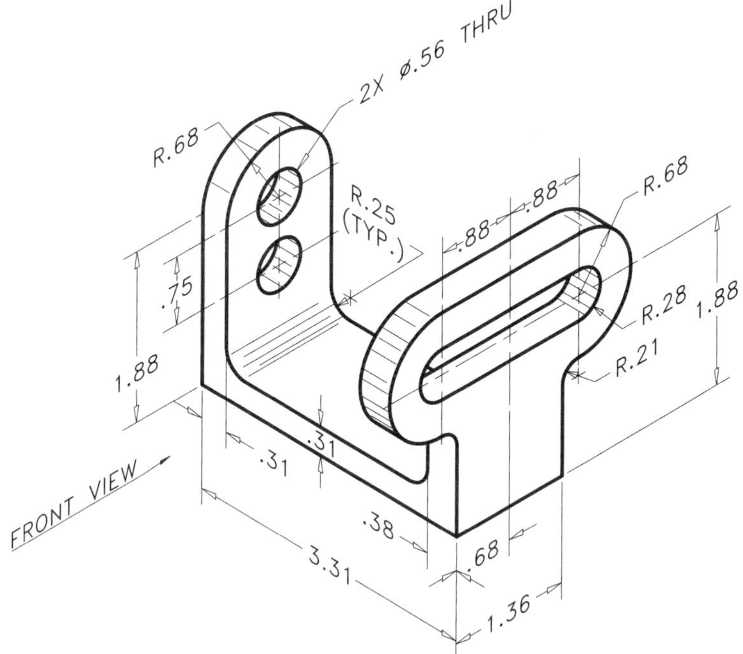

PROBLEM 20–34

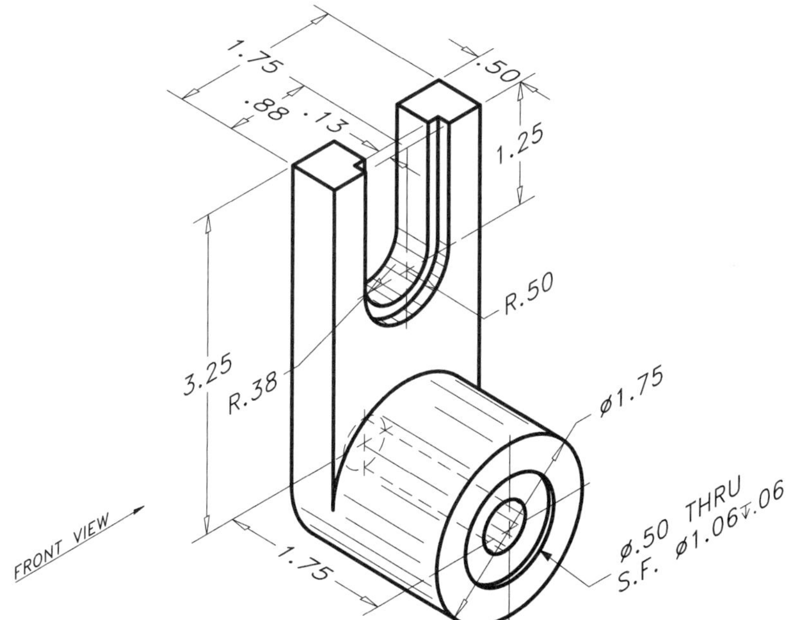

Advanced Level Drawings
Directions for Problems 20–35 through 20–46

Create a 3D solid model of each object. Use the necessary layers for project completion.

PROBLEM 20–35

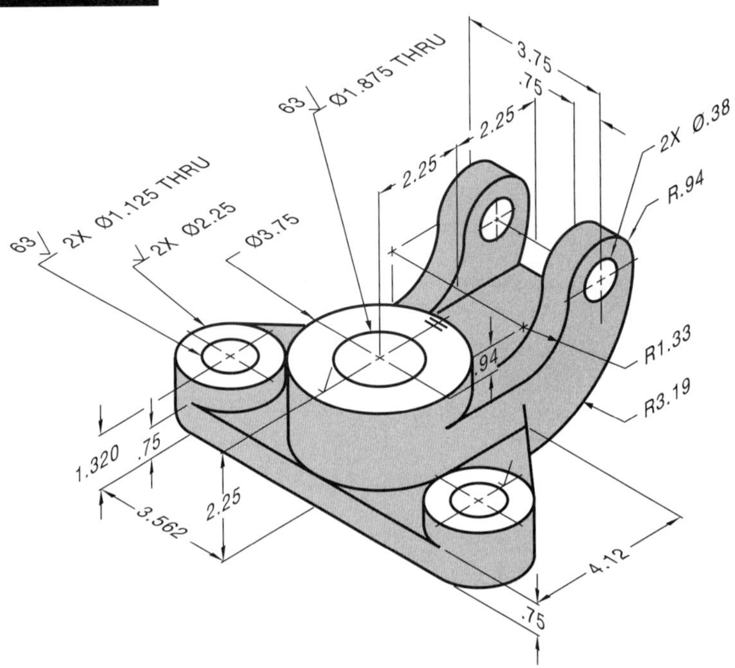

PROBLEM 20–36

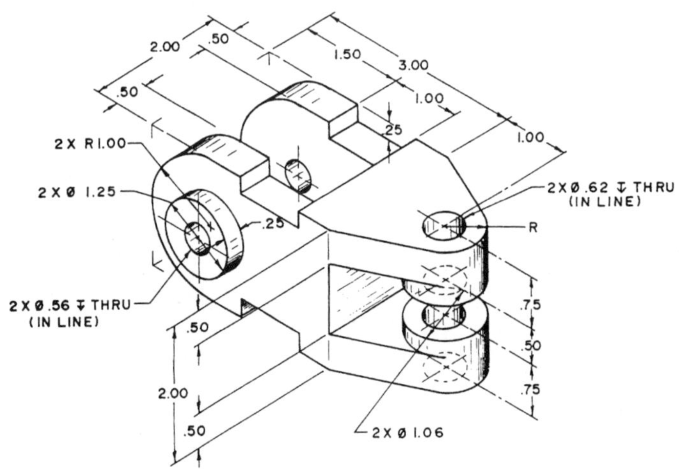

PROBLEM 20–37

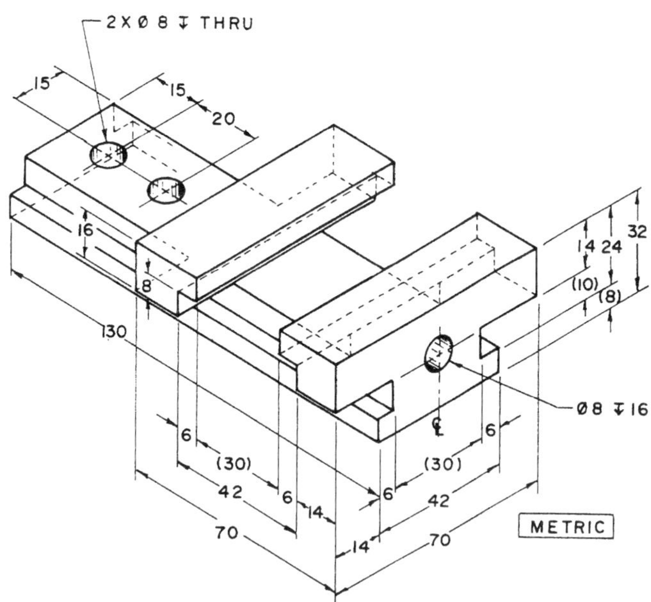

PROBLEM 20–38

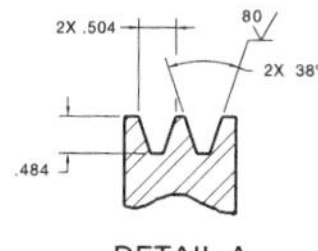

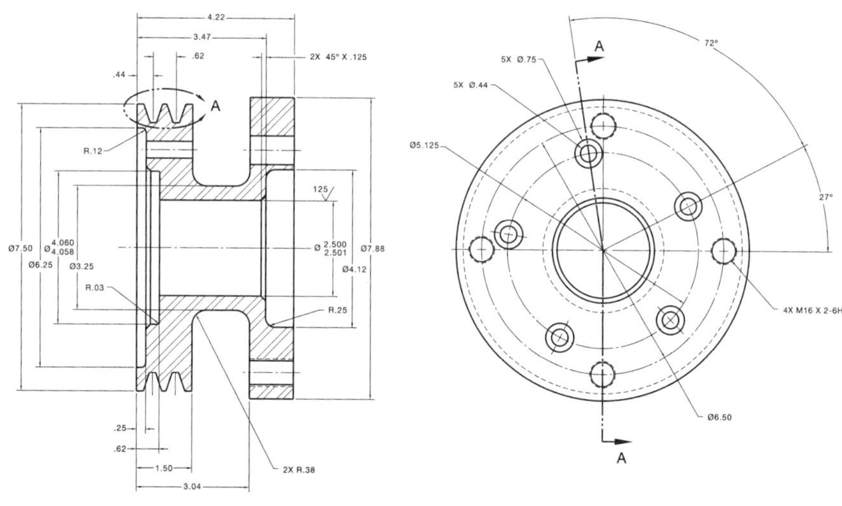

PROBLEM 20–39

FRONT VIEW

EVENLY SPACED

6 X ⌀.44 THRU

ALL UNMARKED RADII = R.13

PROBLEM 20–40

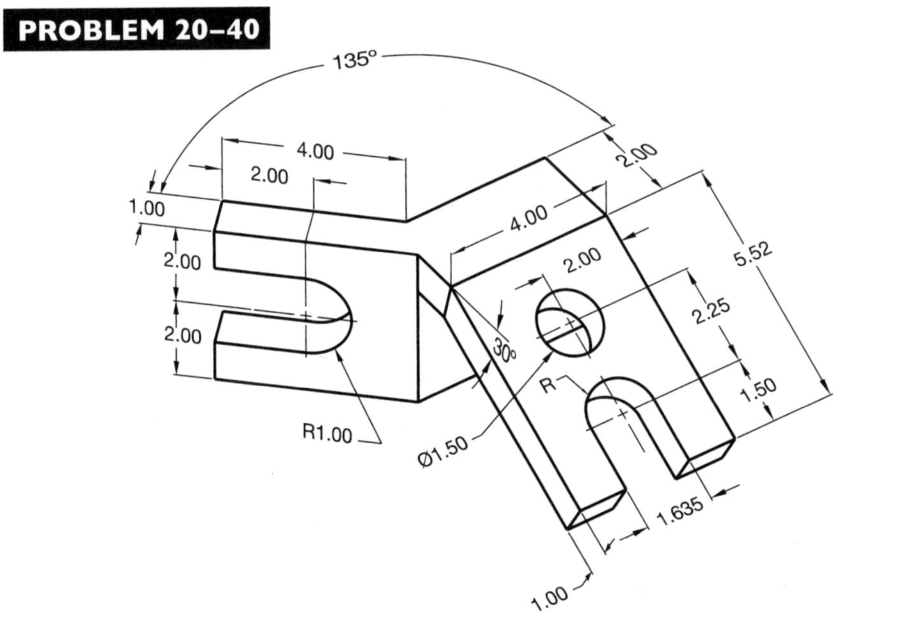

PROBLEM 20–41

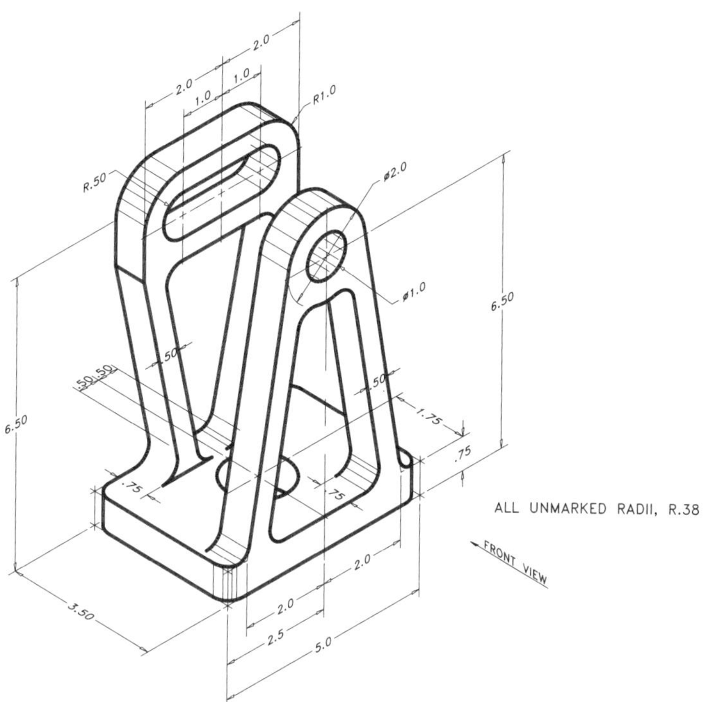

PROBLEM 20–42

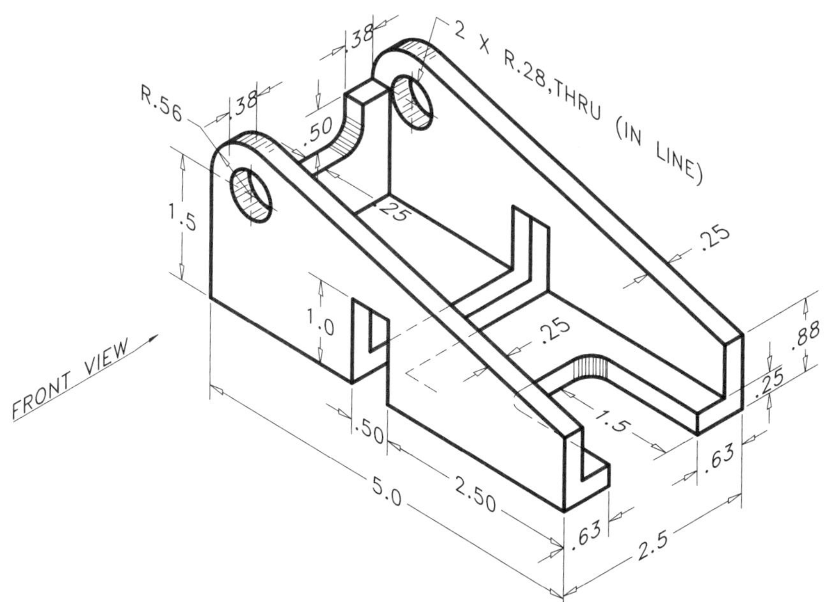

PROBLEM 20–43

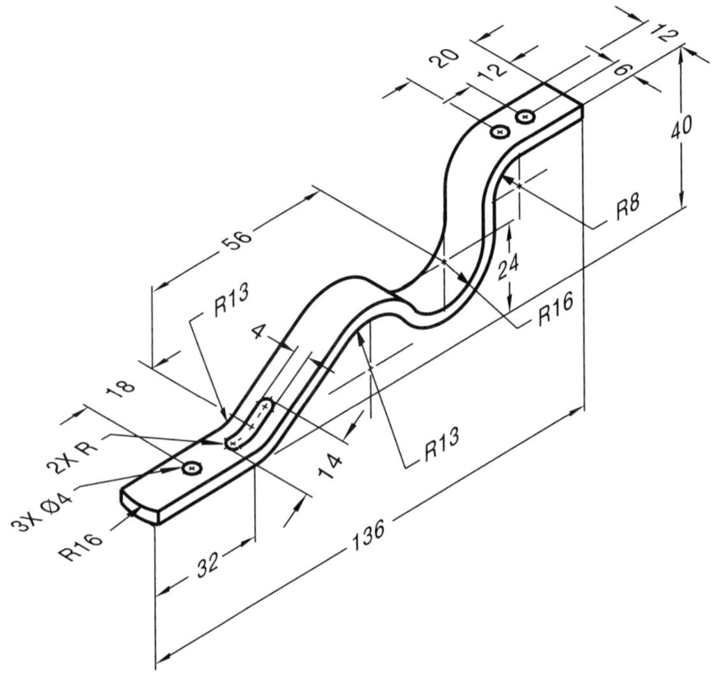

PROBLEM 20–44

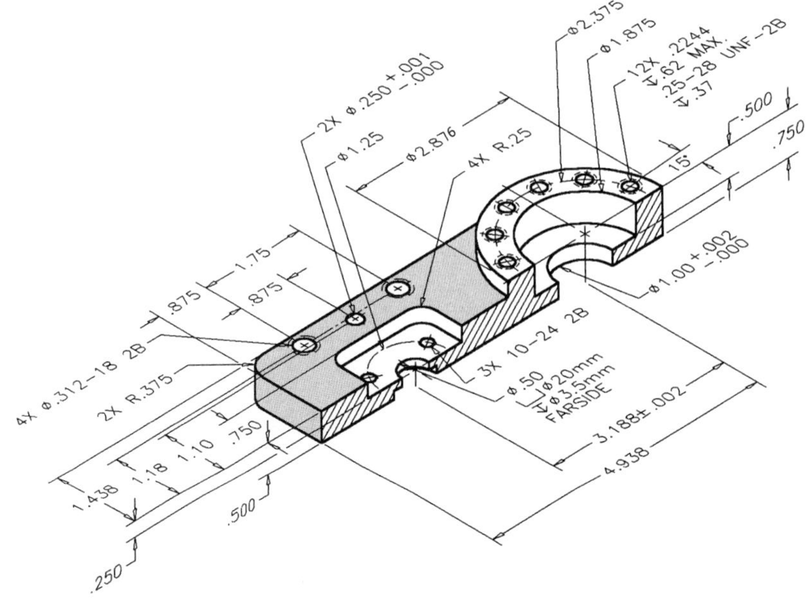

PROBLEM 20–45

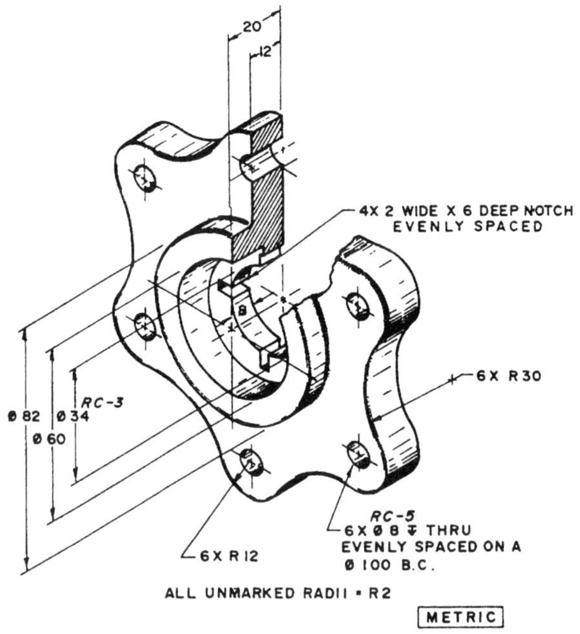

PROBLEM 20–46

Directions for Problems 20–46

Create the manhole cover and the circular watertight manhole frame.

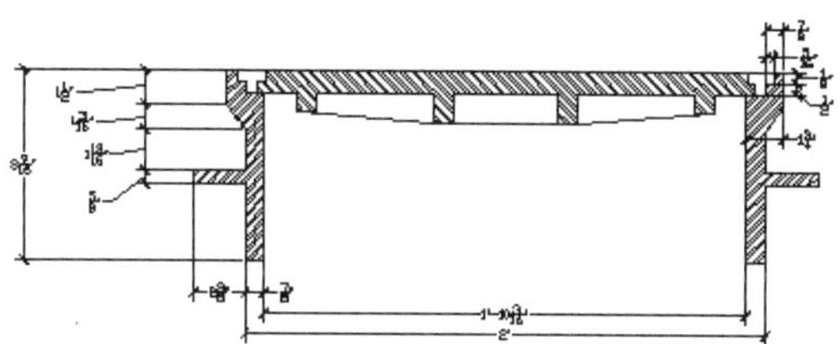

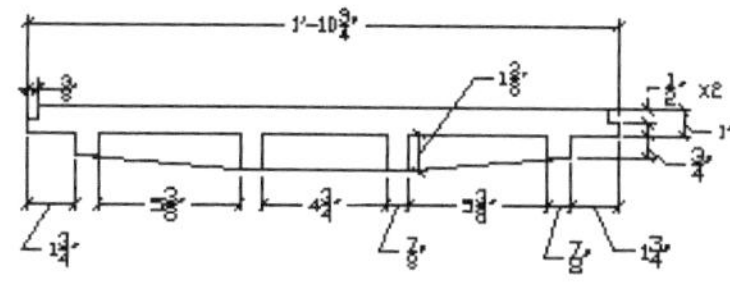

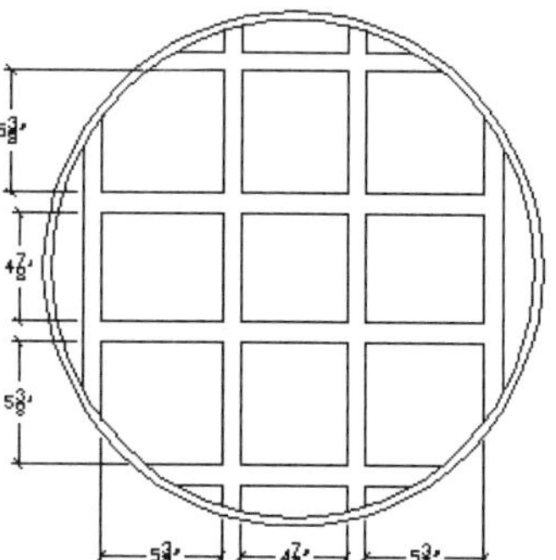

Editing Solid Models

The exercises in this chapter introduce:

- Solids editing techniques, such as:
  - Face editing
  - Edge editing
  - The ROATE3D, MIRROR and ARRAY commands
- Shelling parts to create thin wall designs

TUTORIAL EXERCISE: 21_ROTATE3D.DWG

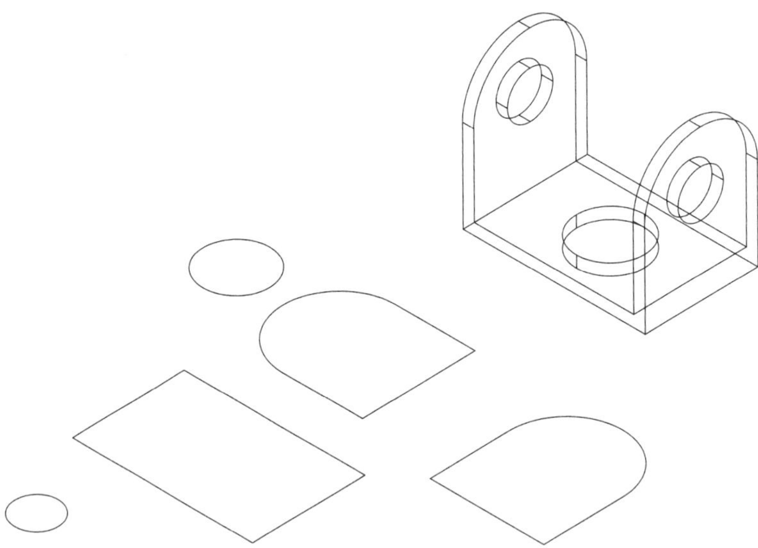

Figure 21–1

Purpose

This tutorial exercise is designed to produce 3D objects and then use the ROTATE3D command to rotate the objects into position before creating unions or performing subtractions. See Figure 21–1.

System Settings

The drawing units, limits, grid, and snap values are already set for this drawing. Check to see that the following Object Snap modes are already set: Endpoint, Extension, Intersection, Center. Also check to see that the POLAR, OSNAP, and OTRACK modes are turned on. These are located in the bottom status bar.

Layers

Make sure the current layer is Model.

Suggested Commands

Open the drawing file *21_Rotate3D.dwg.*

All 2D objects have been converted to polyline objects. The EXTRUDE command is used to create a thickness of 0.50 units for most objects. Then the ROTATE3D command is used to rotate the sides to the proper angle before they are moved into place and joined together. The cylinders are also rotated and moved into place before being subtracted from the base and sides to create holes.

Whenever possible, substitute the appropriate command alias in place of the full AutoCAD command in each tutorial step. For example, use "CP" for the COPY command, "L" for the LINE command, and so on. The complete listing of all command aliases is located in Chapter 1, Table 1–2.

STEP 1

Extrude the base, two side panels, and large circle to a height of 0.50 units, as shown in Figure 21–2. Then extrude the small circle to a height of 8 units. This represents the total length of the base and will be used to cut a hole through the two side panels. Use the EXTRUDE command for both operations.

Command: **EXT** *(For EXTRUDE)*
[draw>solids>extrude]

Current wire frame density: ISOLINES=4
Select objects: *(Select the base at "A", the two side panels at "B" and "C", and the large circle at "D")*
Select objects: *(Press ENTER to continue)*
Specify height of extrusion or [Path]: **0.50**
Specify angle of taper for extrusion <0>: *(Press ENTER to perform the extrude operation)*

Command: **EXT** *(For EXTRUDE)*
[draw>solids>extrude]

Current wire frame density: ISOLINES=4
Select objects: *(Select the small circle at "E")*
Select objects: *(Press ENTER to continue)*
Specify height of extrusion or [Path]: **8.00**
Specify angle of taper for extrusion <0>: *(Press ENTER to perform the extrude operation)*

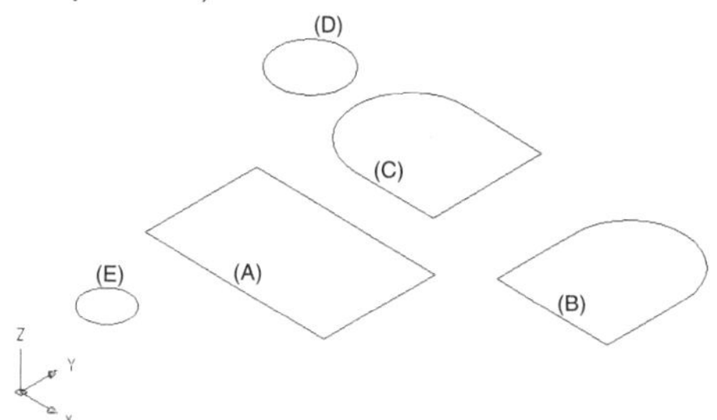

Figure 21–2

STEP 2

Use the MOVE command to position the large cylinder in the center of the base. You may also use the geometry calculator and the MEE function to center the cylinder in the base. See Figure 21–3.

Command: **M** *(For MOVE)*
[modify>move]

Select objects: *(Select the large cylinder at "A")*
Select objects: *(Press ENTER to continue)*
Specify base point or displacement: *(Pick the top face of the large cylinder at "A" to identify its center)*
Specify second point of displacement or <use first point as displacement>: **'CAL**
Initializing...>> Expression: **MEE** *(For Midpoint and two Endpoints)*
>> Select one endpoint for MEE: *(Pick a point along the top corner of the base at "B")*
>> Select another endpoint for MEE: *(Pick a point along the opposite corner of the base at "C")*
(10.6466 0.320423 0.5)

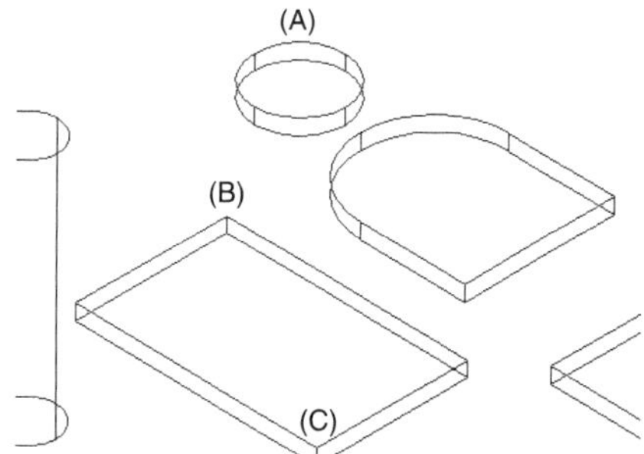

Figure 21–3

STEP 3

Begin rotating the remaining shapes into place using the ROTATE3D command (see Figure 21–4). Being careful to observe the current position of the User Coordinate System, rotate one of the side panels at an angle of 90° using the Y axis as the axis of rotation.

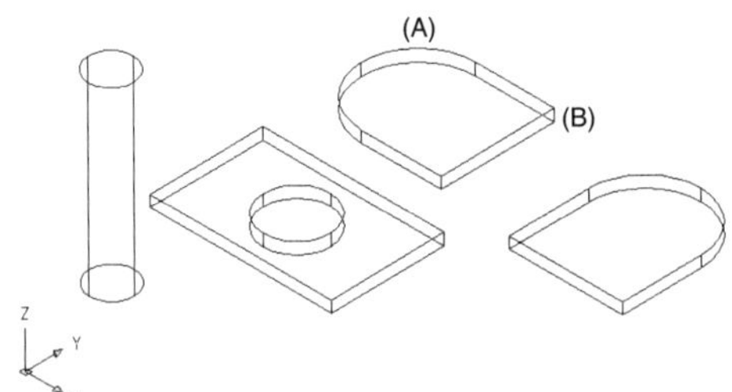

Figure 21–4

Command: **ROTATE3D** [modify>3D operations>rotate 3D]
Current positive angle:
ANGDIR=counterclockwise
ANGBASE=0
Select objects: *(Select the side panel labeled "A")*
Select objects: *(Press* ENTER *to continue)*
Specify first point on axis or define axis by [Object/Last/View/Xaxis/Yaxis/Zaxis/2points]: **Y** *(For Y axis)*
Specify a point on the Y axis <0,0,0>: *(Pick the endpoint of the side panel at "B")*
Specify rotation angle or [Reference]: **90**

STEP 4

Position the side panel you just rotated on the top of the base, using the MOVE command. See Figure 21–5.

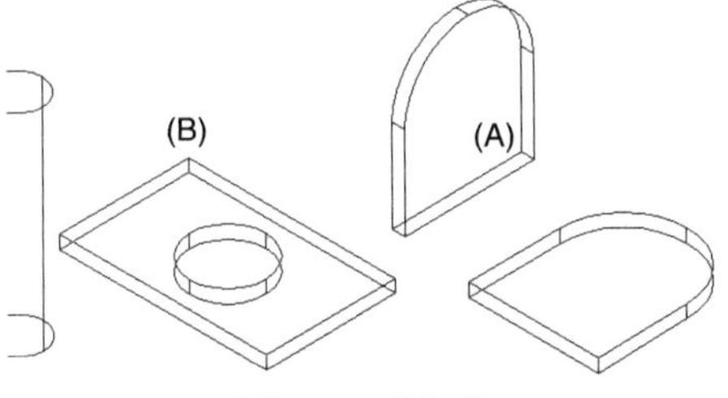

Figure 21–5

Command: **M** *(For MOVE)* [modify>move]
Select objects: *(Select the side panel at "A")*
Select objects: *(Press* ENTER *to continue)*
Specify base point or displacement: *(Pick the endpoint of the side panel at "A")*
Specify second point of displacement or <use first point as displacement>: *(Pick the endpoint of the base at "B")*

STEP 5

Now, rotate the second side panel into position. This will take two steps to accomplish.

First, rotate the panel 90° along the X axis. See Figure 21–6.

Command: **ROTATE3D** [modify>3D operations>rotate 3D]
Current positive angle: ANGDIR=counterclockwise ANGBASE=0
Select objects: *(Select the second side panel)*
Select objects: *(Press ENTER to continue)*
Specify first point on axis or define axis by [Object/Last/View/Xaxis/Yaxis/Zaxis/2points]: **X** *(For X axis)*
Specify a point on the X axis <0,0,0>:
(Pick the endpoint of the side panel at "A")
Specify rotation angle or [Reference]: **90**

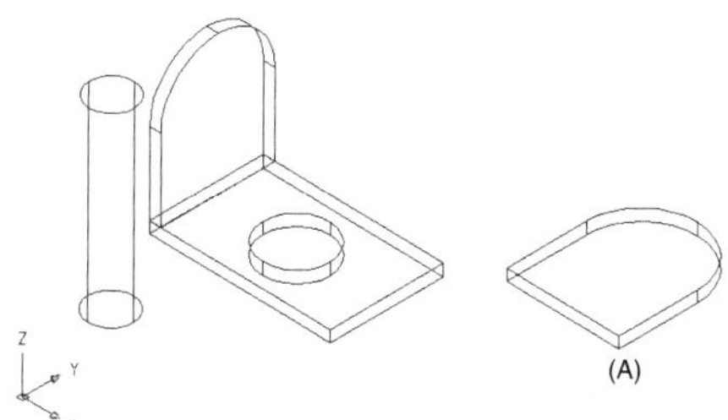

Figure 21–6

STEP 6

Now, rotate the same panel 90° along the Z Axis. See Figure 21–7.

Command: **ROTATE3D** [modify>3D operations>rotate 3D]
Current positive angle: ANGDIR=counterclockwise ANGBASE=0
Select objects: *(Select the same side panel)*
Select objects: *(Press ENTER to continue)*
Specify first point on axis or define axis by [Object/Last/View/Xaxis/Yaxis/Zaxis/2points]: **Z** *(For Z axis)*
Specify a point on the Z axis <0,0,0>:
(Pick the endpoint of the side panel at "A")
Specify rotation angle or [Reference]: **90**

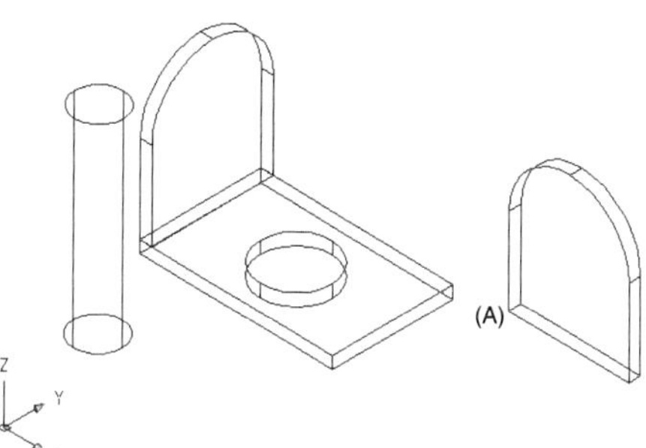

Figure 21–7

STEP 7

Finally, position the panel along the top of the base using the MOVE command. See Figure 21–8.

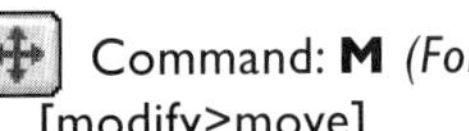
Command: **M** *(For MOVE)* [modify>move]

Select objects: *(Select the side panel at "A")*
Select objects: *(Press ENTER to continue)*
Specify base point or displacement: *(Pick the endpoint of the side panel at "A")*
Specify second point of displacement or <use first point as displacement>: *(Pick the top endpoint of the base at "B")*

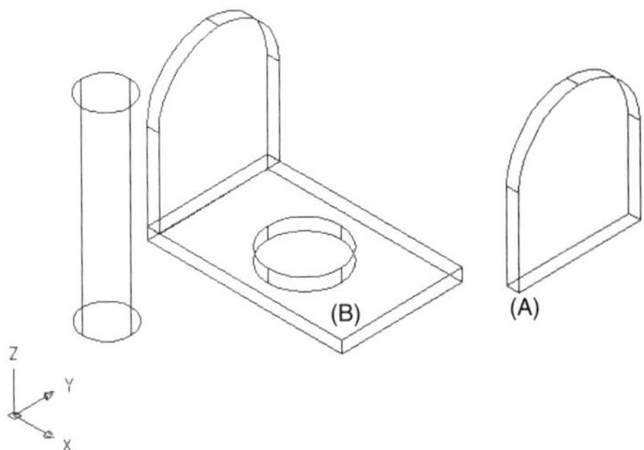

Figure 21–8

STEP 8

Rotate the long cylinder 90° using the Y axis as the axis of rotation, as shown in Figure 21–9.

Command: **ROTATE3D** [modify>3D operations>rotate 3D]
Current positive angle:
ANGDIR=counterclockwise
ANGBASE=0
Select objects: *(Select the long cylinder)*
Select objects: *(Press ENTER to continue)*
Specify first point on axis or define axis by [Object/Last/View/Xaxis/Yaxis/Zaxis/2points]: **Y** *(For Y axis)*
Specify a point on the Y axis <0,0,0>: *(Pick the edge of the cylinder at "A" to identify its center)*
Specify rotation angle or [Reference]: **90**

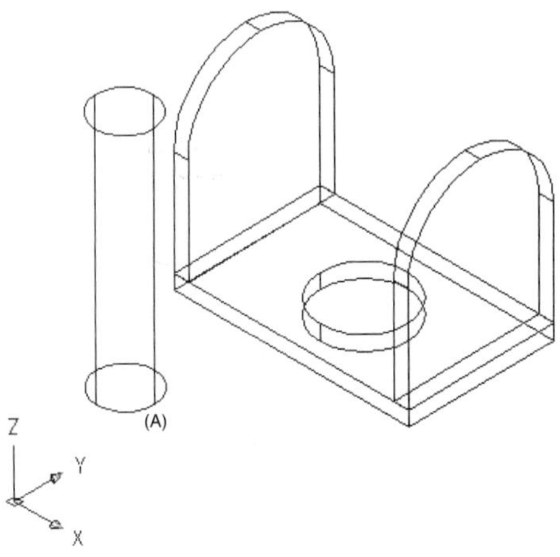

Figure 21–9

STEP 9

Move the long cylinder into position; it should span the complete length of the object and should be lined up with the center of the side panels, as shown in Figure 21–10.

Command: **M** *(For MOVE)*
[modify>move]

Select objects: *(Select the long cylinder)*
Select objects: *(Press* ENTER *to continue)*
Specify base point or displacement: *(Pick the edge of the cylinder at "A" to identify its center)*
Specify second point of displacement or <use first point as displacement>: *(Pick the outer edge of the side panel at "B" to identify its center)*

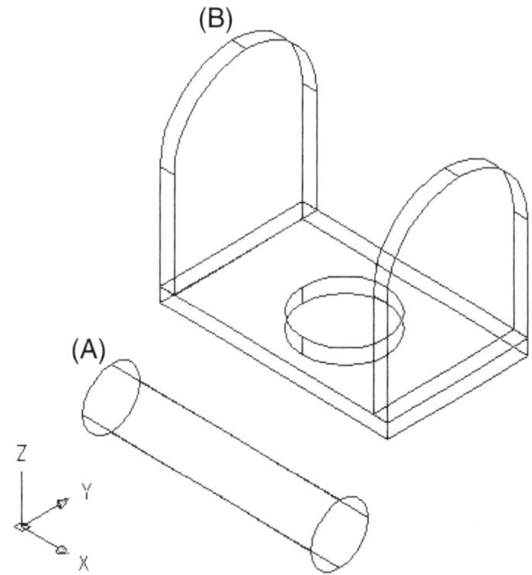

Figure 21–10

STEP 10

Finally, use the SUBTRACT command to subtract the two cylinders from the object. The side panels will automatically be unioned to the base when you complete this command. See Figure 21–11.

Command: **SU** *(For SUBTRACT)*
[modify>solids editing>subtract]

Select solids and regions to subtract from
Select objects: *(Select the two side panels and the base)*
Select objects: *(Press* ENTER *to continue)*
Select solids and regions to subtract ..
Select objects: *(Select the two cylinders to subtract)*
Select objects: *(Press* ENTER *to perform the union/subtraction operations)*

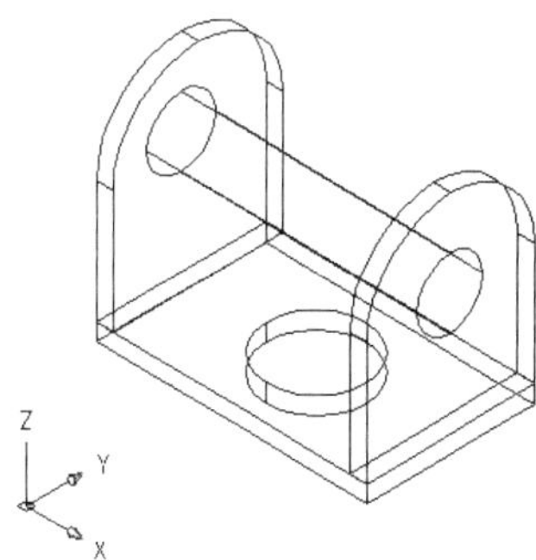

Figure 21–11

The completed object is displayed in Figure 21–12. The ALIGN command could also have been used to rotate and move the extruded shapes into their proper positions.

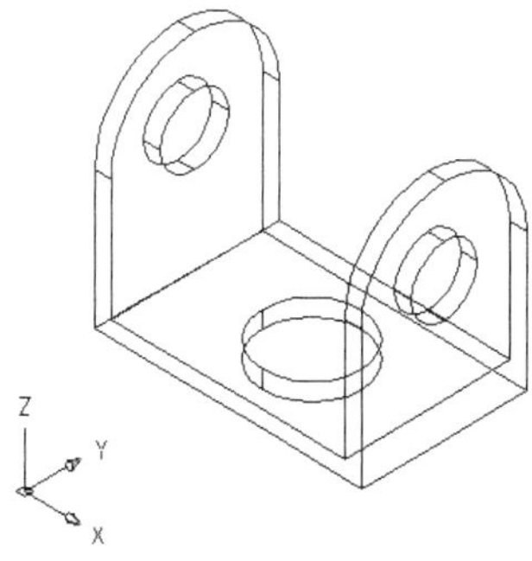

Figure 21–12

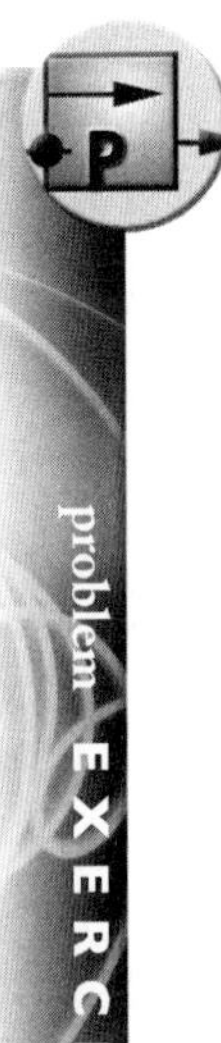

PROBLEM 21–1

Utilize the skills that are laid out in this chapter on problems that are found in Chapters 14, 15, 20, and Appendix A Working Drawing Problems (on the accompanying CD). Many of the intermediate and advanced projects in these chapters will apply the editing skills found in this chapter.

One option is to take a drawing that was completed as a 2 dimensional drawing, and re-create it as a 3D solid.

Creating 2D Multiview Drawings from a Solid Model

The exercises in this chapter introduce:

- The process of creating 2D drawings in a orthographic layout from a 3D solids part
- The creation of an isometric view angle viewport
- Managing viewports:
    - Linetypes, lineweight, colors
    - The control of objects viewed in each of the viewports
    - Scale of the orthographic and isometric viewports
    - Placement of the viewports

TUTORIAL EXERCISE: 22_COLUMN.DWG

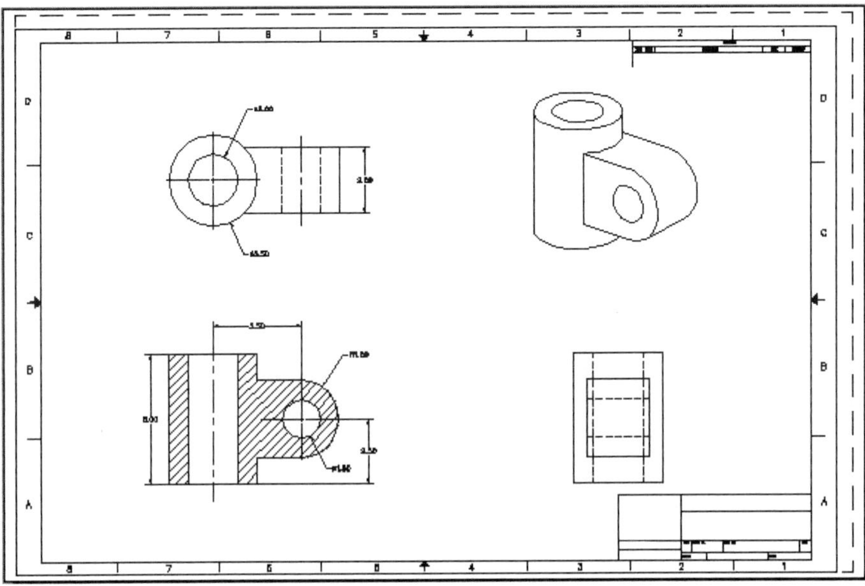

Figure 22–1

Purpose

This tutorial exercise is designed to generate 2D multiview projections of the solid model "22_Column" using the SOLVIEW and SOLDRAW commands. See Figure 22–1.

System Settings

Open the drawing file *22_Column.Dwg*. Keep all remaining default settings, including units and limits.

Layers

The following layers are already created:

Name	Color	Linetype
Model	White	Continuous
Title Block	White	Continuous

Suggested Commands

Use the SOLVIEW command to lay out the Top, Front Section, Right Side, and Isometric views of the column. Next, draw the views in two dimensions using the SOLDRAW command. Dimensioning techniques will be explained. This exercise switches you from Model Space to Layout Mode to accomplish the necessary tasks in building the solid model and extracting the views.

Whenever possible, substitute the appropriate command alias in place of the full AutoCAD command in each tutorial step. For example, use "CP" for the COPY command, "L" for the LINE command, and so on. The complete listing of all command aliases is located in Chapter 1, Table 1–2.

Phase I—Drawing Preparation

STEP 1

Check to make sure that the Hidden linetype is loaded before you extract any views (see Figure 22–2). Also, before using SOLVIEW and SOLDRAW, turn off running OSNAP (OSNAP should be deselected on the Status Bar) and verify that SHADEMODE is set to 2D Wireframe (the solid model is not shaded – the UCS icon is not tri-color).

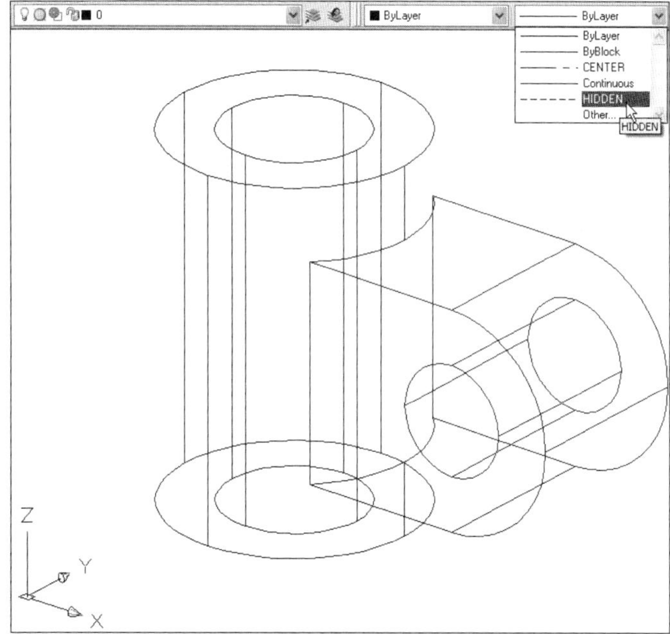

Figure 22–2

STEP 2

Activate the Options dialog box and click on the Display tab. In the Layout elements area, remove the check from the box for Create viewport in new layouts, as shown in Figure 22–3. This will turn off the automatic creation of a viewport when you first enter Layout Mode. The viewports for this drawing will all be created by the SOLVIEW command.

Note: When you have finished completing this tutorial exercise, turn the Create viewport in new layouts feature back on.

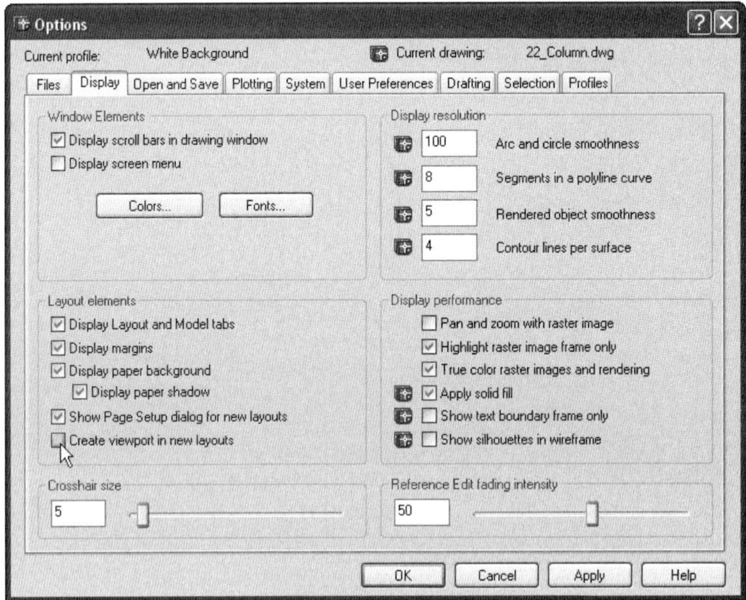

Figure 22–3

Phase II—Extracting the Orthographic Views

With the completed model of the Column loaded, begin the view extraction process by first noting the position of the User Coordinate System icon, shown in Figure 22–4. This position will be shown to extract the first view, the Top view (looking down the Z axis). Once the Top view is created, the Front Section and Right Side views follow.

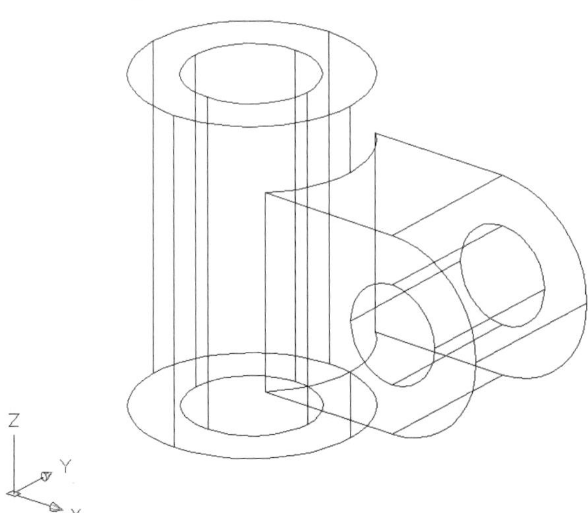

Figure 22–4

STEP 3

Activate the SOLVIEW command, which automatically switches you to the Paper Space environment, shown in Figure 22–5. You will build an initial view (Top view) based on the current position of the User Coordinate System. Once you locate the view on the drawing, you can change the scale of the view, give the view a name, and construct a viewport around the view. When constructing the viewport, make it large enough to accommodate the dimensions that need to be added later. See Figure 22–5 and the following prompts for performing this task.

 Command: **SOLVIEW**
[draw>solids>setup>view]

Enter an option [Ucs/Ortho/Auxiliary/Section]: **U** *(For Ucs)*

Enter an option [Named/World/?/Current] <Current>: *(Press* ENTER*)*

Enter view scale <1.0000>: *(Press* ENTER*)*

Specify view center: *(Locate the view approximately in the upper left corner of the drawing sheet)*

Specify view center <specify viewport>: *(Press* ENTER *to place the view)*

Specify first corner of viewport: *(Pick a point approximately at "A")*

Specify opposite corner of viewport: *(Pick a point approximately at "B")*

Enter view name: **TOP**

UCSVIEW = 1 UCS will be saved with view

Enter an option [Ucs/Ortho/Auxiliary/Section]: *(Press* ENTER *to exit this command)*

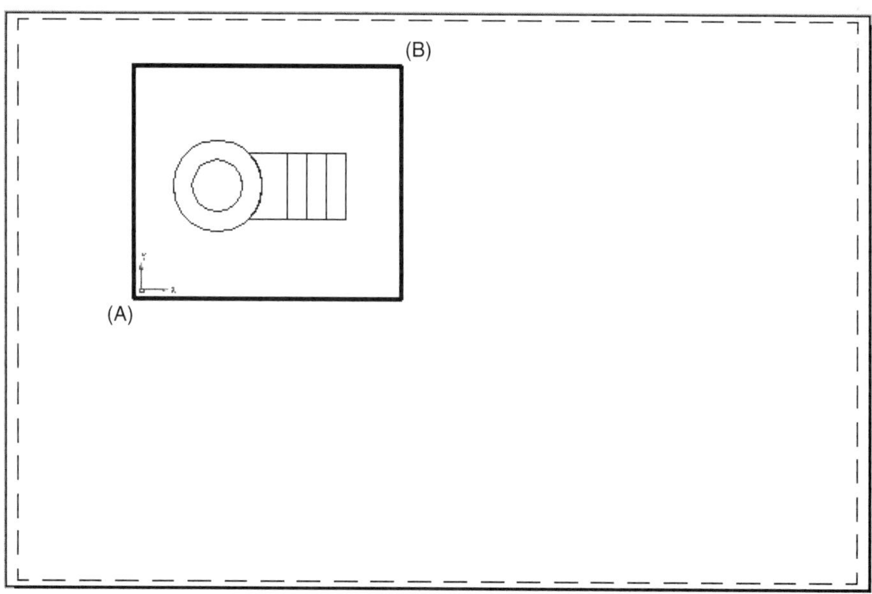

Figure 22–5

STEP 4

Before continuing with the SOLVIEW command to extract the other views, examine the layers at this point through the Layer Properties Manager dialog box, shown in Figure 22–6. Four layers were automatically created as a result of laying out the first view through SOLVIEW: TOP-DIM, TOP-HID, TOP-VIS, and VPORTS. TOP-DIM will hold all dimension information located in the Top view. TOP-HID will hold all hidden line information for the Top view, and TOP-VIS will hold all visible line information in the Top view. The VPORTS layer will hold all viewports created. The "TOP" layers are automatically visible in the current viewport and will be automatically frozen in any other new viewports; the presence of the snowflake symbol in the New VP Freeze column indicates this. Now return to the SOLVIEW command and lay out the remaining orthographic views.

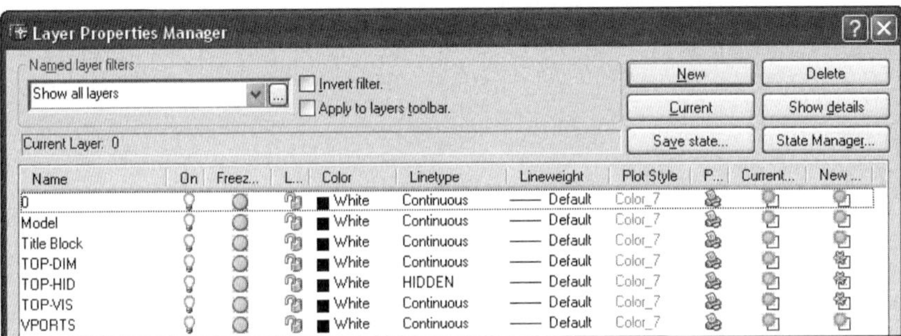

Figure 22–6

STEP 5

Use SOLVIEW and the Section option to create the Front view in full section, based on the Top view already placed. Establish the cutting plane of the section from the Quadrant of the cylinder at "A" to the midpoint of the line at "B" in Figure 22–7. Pick a point approximately at "C" to identify the side from which to view the section. As the new view is being located, Ortho mode is automatically turned on to keep both views lined up with each other in the same orientation. Create a viewport by following the prompts to clip a first corner and other corner. Again, you will add dimensions to this view later on, so make the viewport a good size. Name the view "FRONT". Continue by laying out the R_SIDE view, using the Ortho option of the SOLVIEW command. Create a small viewport because no dimensions will be placed on this view. When finished, your drawing should look similar to Figure 22–8.

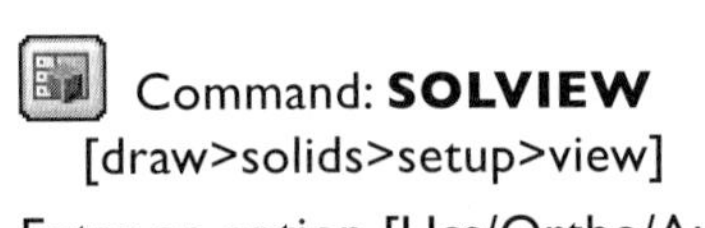

Command: **SOLVIEW**
[draw>solids>setup>view]

Enter an option [Ucs/Ortho/Auxiliary/Section]: **S** *(For Section)*

Specify first point of cutting plane: **Qua** of *(Pick the quadrant at "A")*

Specify second point of cutting plane: **Mid**
of *(Pick the midpoint at "B")*
Specify side to view from: *(Pick a point at "C")*
Enter view scale <1.0000>: *(Press* ENTER*)*
Specify view center: *(Pick a location below the TOP view)*
Specify view center <specify viewport>: *(Press* ENTER *to place the view)*
Specify first corner of viewport: *(Pick a point at "D")*
Specify opposite corner of viewport: *(Pick a point at "E")*
Enter view name: **FRONT**
UCSVIEW = 1 UCS will be saved with view
Enter an option [Ucs/Ortho/Auxiliary/Section]: **O** *(For Ortho)*
Specify side of viewport to project: *(Pick the viewport at "F")*
Specify view center: *(Pick a point to the right of the FRONT view)*
Specify view center <specify viewport>: *(Press* ENTER *to place the view)*
Specify first corner of viewport: *(Pick a point at "G")*
Specify opposite corner of viewport: *(Pick a point at "H")*
Enter view name: **R_SIDE**
UCSVIEW = 1 UCS will be saved with view
Enter an option [Ucs/Ortho/Auxiliary/Section]: *(Press* ENTER *to exit this command)*

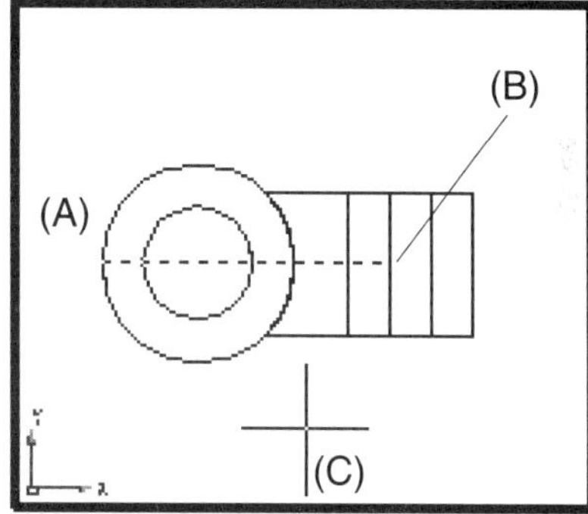

Figure 22–7

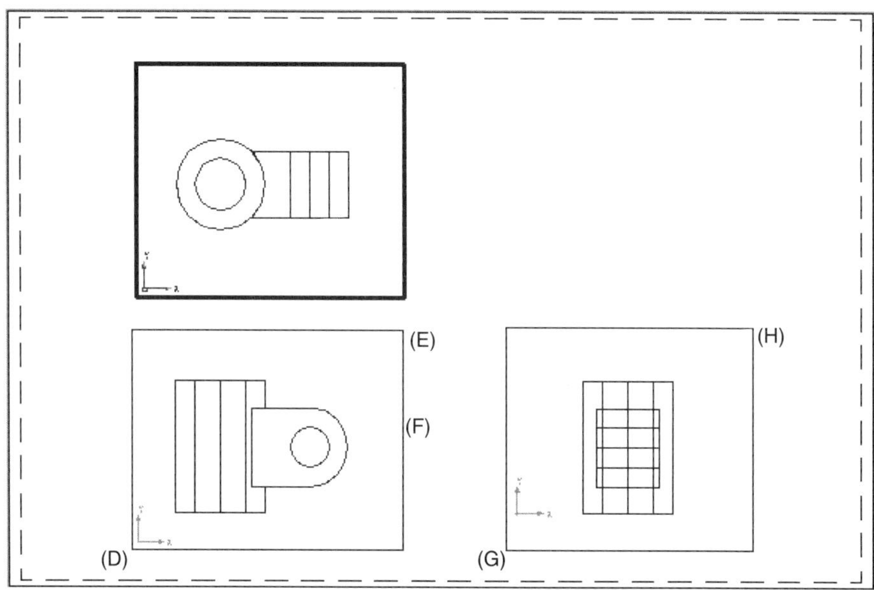

Figure 22–8

STEP 6

To create a projection of an isometric view, return to Model Space and the solid model by clicking on the Model tab. Then use the UCS command to create a new User Coordinate System based on the current view. See Figure 22–9. Also save this User Coordinate System as ISO using the Save option of the UCS command. ISO is the current User Coordinate System.

Command: **UCS** [tools>new ucs>world]

Current ucs name: *WORLD*

Enter an option [New/Move/ orthoGraphic/Prev/Restore/Save/Del/ Apply/?/World] <World>: **N** *(For New)*

Specify origin of new UCS or [ZAxis/ 3point/OBject/Face/View/X/Y/Z] <0,0,0>: **V** *(For View)*

Command: **UCS**

Current ucs name: *NO NAME*

Enter an option [New/Move/ orthoGraphic/Prev/Restore/Save/Del/ Apply/?/World] <World>: **S** *(For Save)*

Enter name to save current UCS or [?]: **ISO**

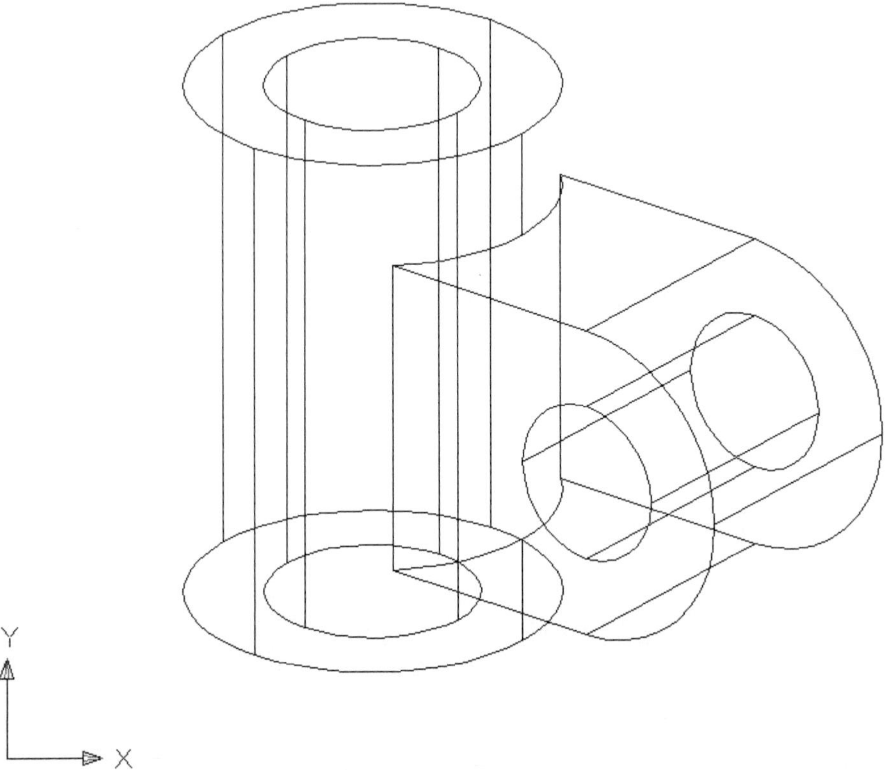

Figure 22–9

STEP 7

Return to Layout Mode by clicking on the Layout1 tab. Issue the SOLVIEW command and use the UCS option to lay out the Isometric view of the column based on the current User Coordinate System. See Figure 22–10.

(Click on the Layout1 tab)

 Command: **SOLVIEW**
[draw>solids>setup>view]

Enter an option [Ucs/Ortho/Auxiliary/Section]: **U** *(For Ucs)*
Enter an option [Named/World/?/Current] <Current>: **N** *(For Named)*
Enter name of UCS to restore: **ISO**
Enter view scale <1.0000>: *(Press* ENTER*)*
Specify view center: *(Pick a point to locate the view)*
Specify view center <specify viewport>: *(Press* ENTER *to place the view)*
Specify first corner of viewport: *(Pick a point at "A")*
Specify opposite corner of viewport: *(Pick a point at "B")*
Enter view name: **ISO**
UCSVIEW = 1 UCS will be saved with view
Enter an option [Ucs/Ortho/Auxiliary/Section]: *(Press* ENTER *to exit this command)*

All views have been successfully laid out. However, none of the orthographic views show hidden features defined by hidden lines. Also, the Front view in full section does not show section lines to illustrate which surfaces were cut by the imaginary cutting plane line. The SOLVIEW command is used only as a layout tool. To actually draw the view complete with visible, object, and section lines, use the SOLDRAW command. This will be covered in Phase III of this project.

Tip: Be careful when using the UCS option in the SOLVIEW command to create an Isometric view. The "current" UCS that will be used to create a view, might be selected from one of viewports previously created. Using a Named UCS will always create predictable results.

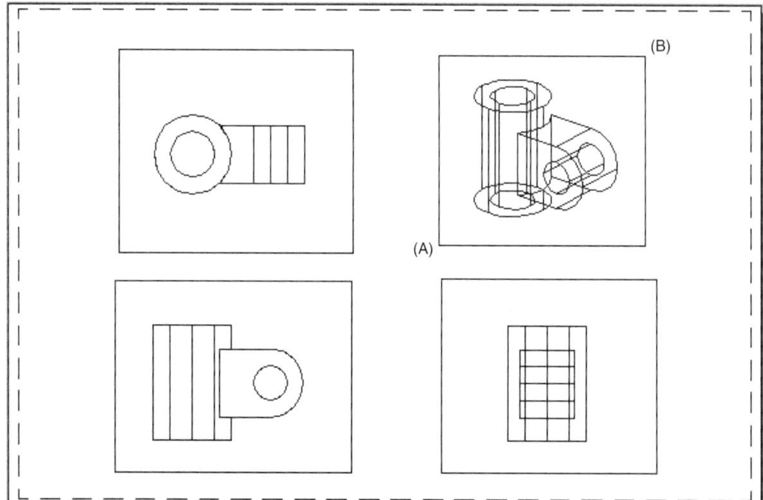

Figure 22–10

Phase III—Using SOLDRAW to Construct the 2D Views

STEP 8

The SOLDRAW command is designed to follow the SOLVIEW command. SOLVIEW is used as a layout tool; it allows you to position the Top, Front Section, R_Side, and Isometric views in the Paper Space environment. SOLDRAW is used to draw object, hidden, and section lines (see Figure 22–11). Dimensions will be placed in the drawing during Phase IV of our project.

 Command: **SOLDRAW** [draw>solids>setup>drawing]

Select viewports to draw..

Select objects: **All** *(To select all viewports)*

Select objects: (Press ENTER to draw the views)

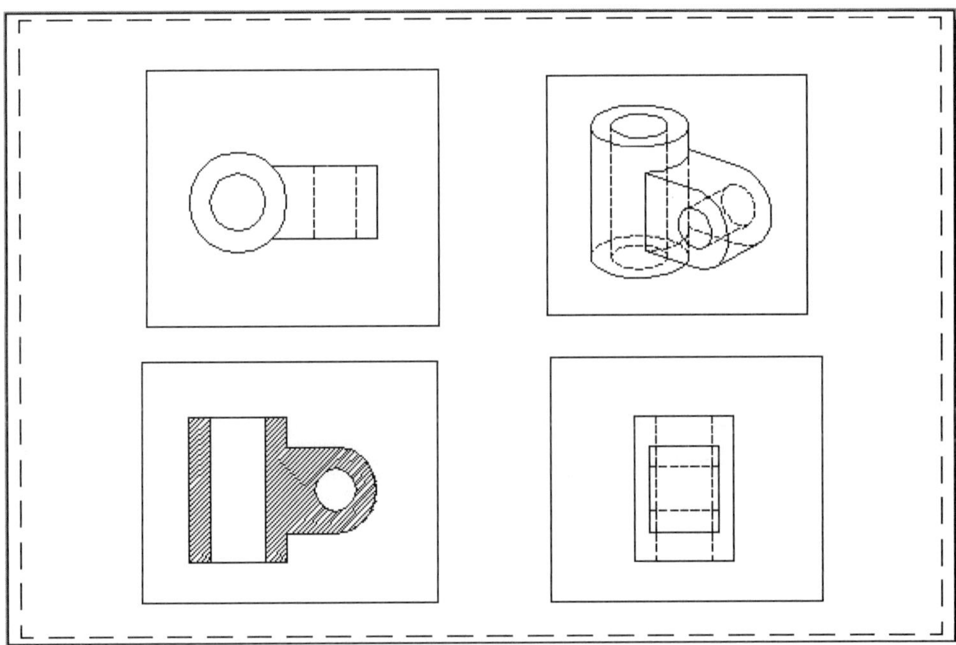

Figure 22–11

STEP 9

Notice that the Isometric view, shown in Figure 22–11, has been drawn complete with hidden lines; these lines need to be made invisible in the isometric view. To accomplish this, first click inside the viewport holding the isometric view, to make it active. Next, activate the Layer Properties Manager dialog box, shown in Figure 22–12, and freeze the layer containing the hidden lines, ISO-HID. This will freeze the hidden lines in the Isometric view, only (see figure 22–13).

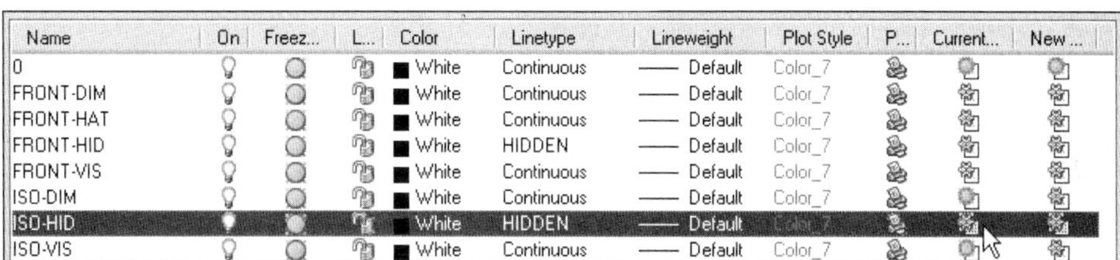

Name	On	Freez...	L...	Color	Linetype	Lineweight	Plot Style	P...	Current...	New ...
0				White	Continuous	—— Default	Color_7			
FRONT-DIM				White	Continuous	—— Default	Color_7			
FRONT-HAT				White	Continuous	—— Default	Color_7			
FRONT-HID				White	HIDDEN	—— Default	Color_7			
FRONT-VIS				White	Continuous	—— Default	Color_7			
ISO-DIM				White	Continuous	—— Default	Color_7			
ISO-HID				White	HIDDEN	—— Default	Color_7			
ISO-VIS				White	Continuous	—— Default	Color_7			

Figure 22–12

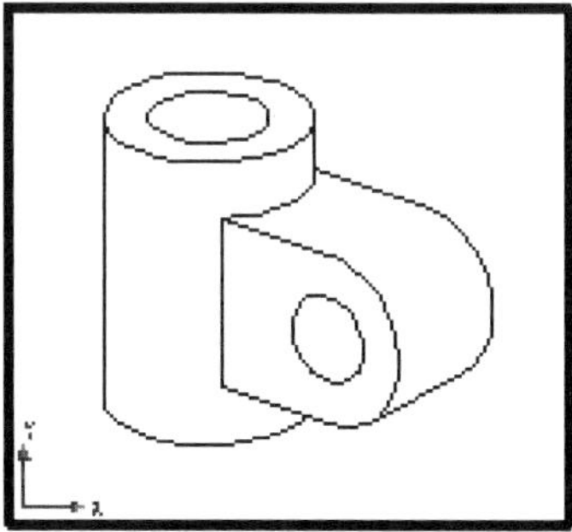

Figure 22–13

STEP 10

Activate the Front viewport, holding the section view information, by clicking inside that viewport. The hatching looks too dense and needs to be scaled up in size using the Hatch Edit dialog box in Figure 22–14. In the Scale edit box of the Hatchedit dialog box, change the hatch pattern Scale to a new value of 2 units. Click the OK button to change the scale of the hatch pattern. The Front view should appear similar to Figure 22–15.

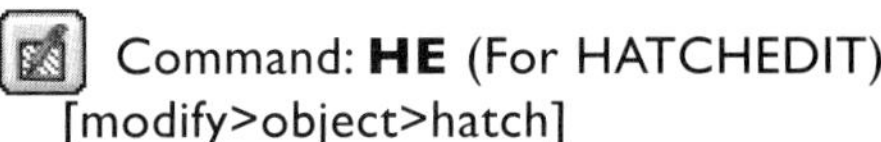

Command: **HE** (For HATCHEDIT) [modify>object>hatch]

Select associative hatch object: (Select the hatch pattern located in the viewport in Figure 22–15)

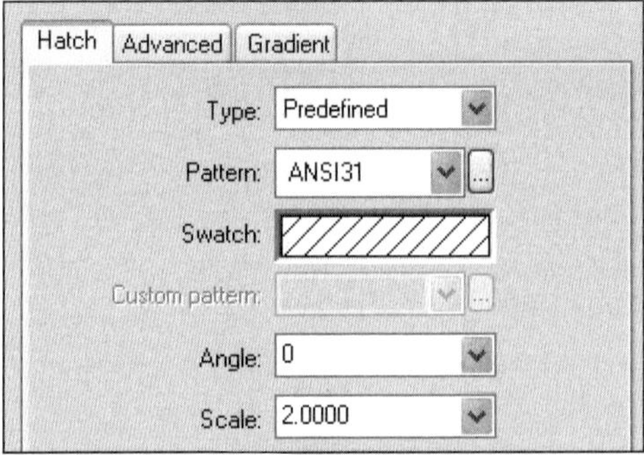

Figure 22–14

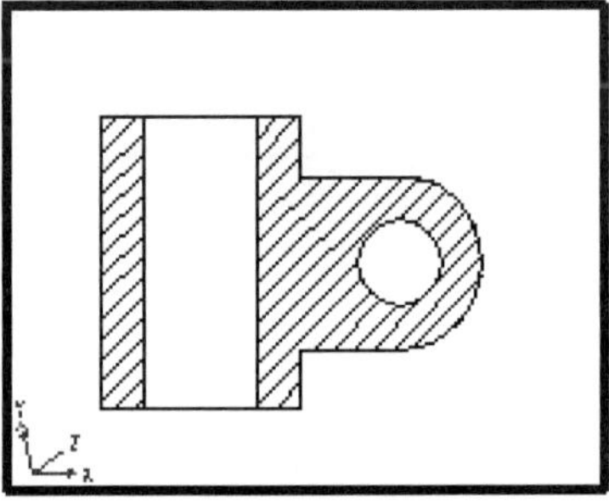

Figure 22–15

PHASE IV—Adding Dimensions to the Views

STEP 11

Begin adding dimensions to the Front Section view; do this while still in floating Model Space. First, click in the viewport containing the Front Section view information. Then, issue the UCS command along with the View option. This will line up the User Coordinate System icon parallel to the display screen, enabling you to place the dimensions as in a normal 2D drawing. Once the icon has been changed, make the FRONTDIM layer current and turn off the FRONT-HAT layer, as shown in Figure 22–16. As you place the dimensions in the Front Section view, they should not appear in other viewports. The SOLVIEW command automatically creates dimension layers and freezes the layers in the viewports in which they do not apply.

(Click in the Front section viewport at "A" in Figure 22–16 to make it active)

Command: **UCS** [tools>new ucs]

Current ucs name: *NO NAME*

Enter an option [New/Move/orthoGraphic/Prev/Restore/Save/Del/Apply/?/World] <World>: **N** *(For New)*

Specify origin of new UCS or [ZAxis/3point/OBject/Face/View/X/Y/Z] <0,0,0>: **V** *(For View)*

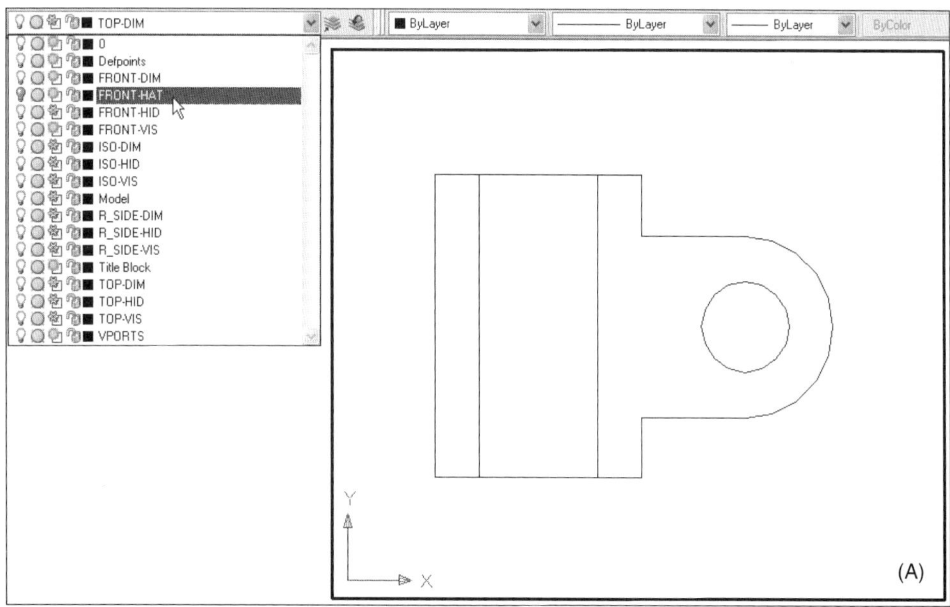

Figure 22–16

STEP 12

Draw a series of centerlines on the FRONT-DIM layer. To accomplish this, use the DIMCENTER command and pick the edge of the arc at "A." Then turn ORTHO mode on and construct a vertical centerline from the midpoint at "B" to the approximate point at "C," as shown in Figure 22–17.

Command: **DCE** *(For DIMCENTER)* [dimension>center mark]

Select arc or circle: (Select the edge of the arc at "A")

Command: **L** *(For LINE)* [draw>line]

Specify first point: **Mid**

of *(Select the line at "B")*

Specify next point or [Undo]: *(Pick a point at "C")*

Specify next point or [Undo]: *(Press* ENTER *to exit this command)*

Use grips to extend the top end of the line as shown in Figure 22–17. Then change the highlighted lines in the figure to the Center linetype (this linetype was previously loaded).

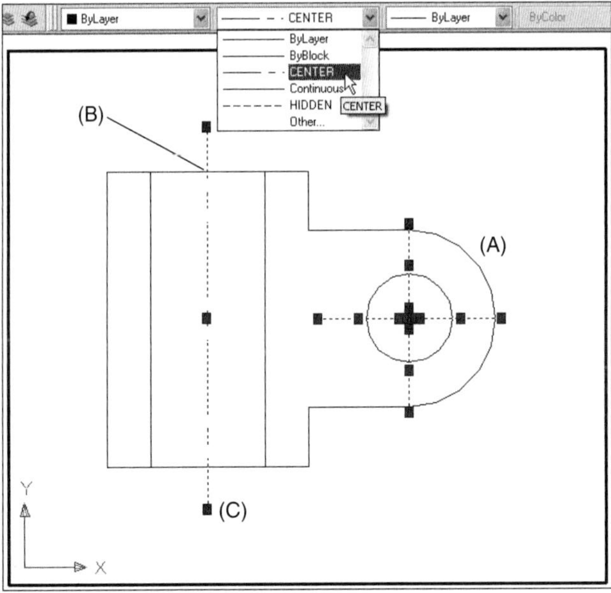

Figure 22–17

STEP 13

Begin adding dimensions to the Front view. Before performing this operation, turn on OSNAP. Set the dimension center mark control (DIMCEN) to 0. This will prevent the placement of duplicate center marks when you place diameter and radius dimensions. Check that the Endpoint and Intersection modes are set, because numerous intersections will be required in dimensioning. With the Front viewport active and the User Coordinate System set to the current view, begin placing the linear, radius, and diameter dimensions, using Figure 22–18 as a guide.

Command: **DIMCEN**
[dimension>center mark]
Enter new value for DIMCEN <-0.0900>: **0**

Command: **DLI** *(For DIMLINEAR)*
[dimension>linear]
Specify first extension line origin or <select object>: *(Pick the endpoint at "A")*
Specify second extension line origin: *(Pick the endpoint at "B")*
Specify dimension line location or [Mtext/Text/Angle/Horizontal/Vertical/Rotated]: *(Pick the endpoint at "C")*
Dimension text = 5.00

Command: **DLI** *(For DIMLINEAR)*
[dimension>linear]
Specify first extension line origin or <select object>: *(Pick the endpoint at "D")*
Specify second extension line origin: *(Pick the endpoint at "E")*
Specify dimension line location or [Mtext/Text/Angle/Horizontal/Vertical/Rotated]: *(Pick the endpoint at "F")*
Dimension text = 2.50

Command: **DLI** *(For DIMLINEAR)*
[dimension>linear]
Specify first extension line origin or <select object>: *(Pick the endpoint at "G")*
Specify second extension line origin: *(Pick the endpoint at "H")*
Specify dimension line location or [Mtext/Text/Angle/Horizontal/Vertical/Rotated]: *(Pick the endpoint at "I")*
Dimension text = 3.50

Command: **DRA** (For DIMRADIUS)
[dimension>radius]
Select arc or circle: (Select the edge of the arc at "J")
Dimension text = 1.50 Specify dimension line location or [Mtext/Text/Angle]: *(Pick a point at "K")*

Command: **DDI** *(For DIMDIAMETER)*
[dimension>diameter]
Select arc or circle: *(Select the edge of the circle at "L")*
Dimension text = 1.50 Specify dimension line location or [Mtext/Text/Angle]: *(Pick a point at "M")*

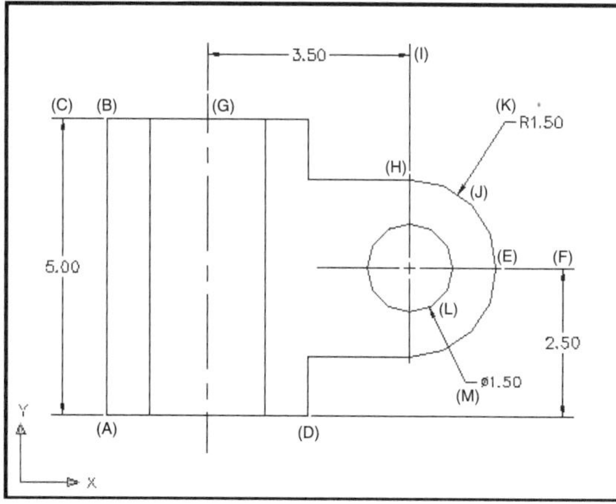

Figure 22–18

Before continuing to the next step, activate the Layer Control box, shown in Figure 22–19, and turn the FRONT-HAT layer back on.

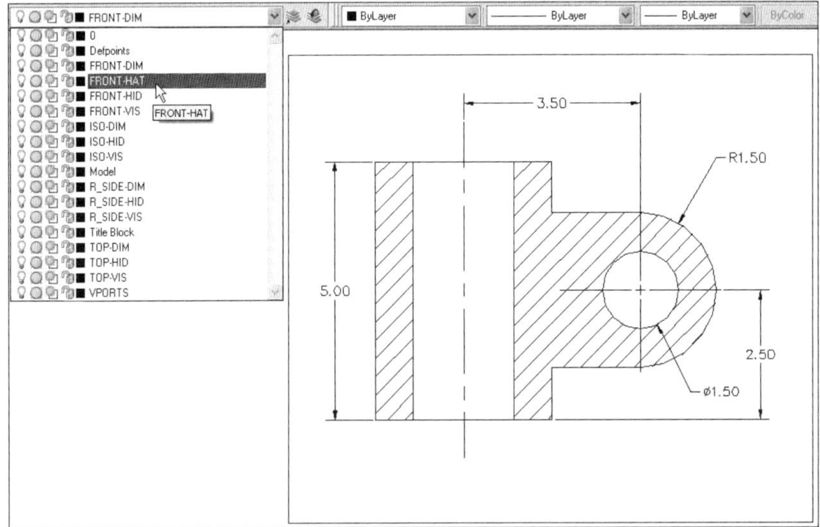

Figure 22–19

STEP 14

Switch back to Paper Space by double-clicking in the area at "A" in Figure 22–20, which can be anywhere outside the viewport. Then pick the viewport; it will highlight and grips will appear. Right-click to display the shortcut menu. Choose Display Locked and then Yes to lock the viewport. This will prevent any accidental zooming of the image inside the Paper Space viewport.

When finished, press ESC to remove the object highlight and grips from the screen.

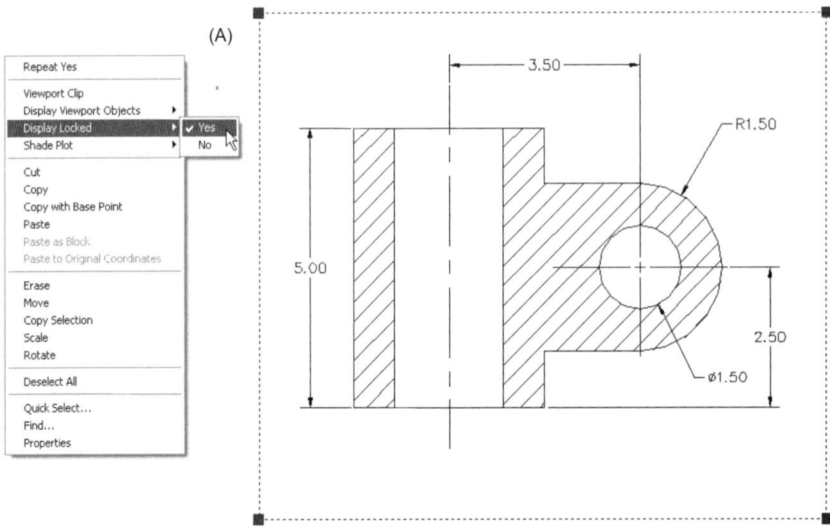

Figure 22–20

STEP 15

Activate the Top view by double-clicking inside the top viewport. Issue the UCS command along with the View option. This will ensure that the position of the User Coordinate System icon is parallel to the display screen. This will also enable the dimensions to be placed as in a normal 2D drawing. Once the icon has been aligned to the view, make the TOP-DIM layer current as in Figure 22–21. As the dimensions are being placed, they will not appear in other viewports. The SOLVIEW command automatically creates dimension layers and freezes the layers in the viewports in which they do not apply. Begin adding dimensions to the Top view; do this while still in floating Model Space.

(Double-click in the Top View viewport at "A" in Figure 22–21, which switches you to floating Model Space and sets the viewport active)

Command: **UCS** [tools>new ucs]
Current ucs name: *WORLD*
Enter an option [New/Move/orthoGraphic/Prev/Restore/Save/Del/Apply/?/World] <World>: **N** *(For New)*
Specify origin of new UCS or [ZAxis/3point/OBject/Face/View/X/Y/Z] <0,0,0>: **V** *(For View)*

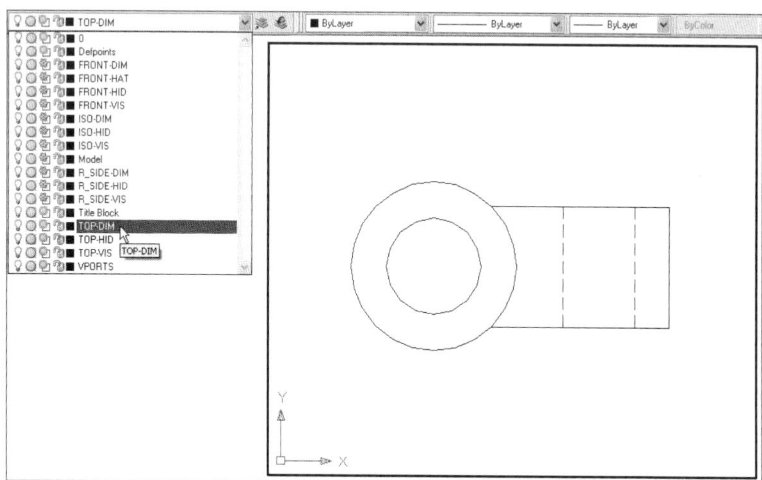

Figure 22–21

STEP 16

Draw a series of centerlines on the TOP-DIM layer. First use DIMCEN to reset the center mark setting to a value of -0.09 units. Use the DIMCENTER command and pick the edge of the circle at "A." Then turn ORTHO mode on and construct a vertical centerline from "D" to "E," as in Figure 22–22, with assistance from the Geometry Calculator and the MEE (Midpoint-Endpoint-Endpoint) option.

Command: **DIMCEN**
Enter new value for DIMCEN <0.0000>: **-0.09**

Command: **DCE** *(For DIMCENTER)* [dimension>center mark]

Select arc or circle: *(Select the edge of the arc at "A")*

Command: **L** *(For LINE)*

Specify first point: **'CAL**

Initializing...>> Expression: **MEE**

>> Select one endpoint for MEE: *(Select the endpoint of the hidden line at "B")*

>> Select another endpoint for MEE: *(Select the endpoint of the hidden line at "C")*

(11.3822 5.55593 0.0)

Specify next point or [Undo]: *(Pick a point approximately at "D")*

Specify next point or [Undo]: *(Press* ENTER *to exit this command)*

(Use grips to stretch the line to "E" in Figure 22–22. Then change the highlighted lines in the figure to the Center linetype.)

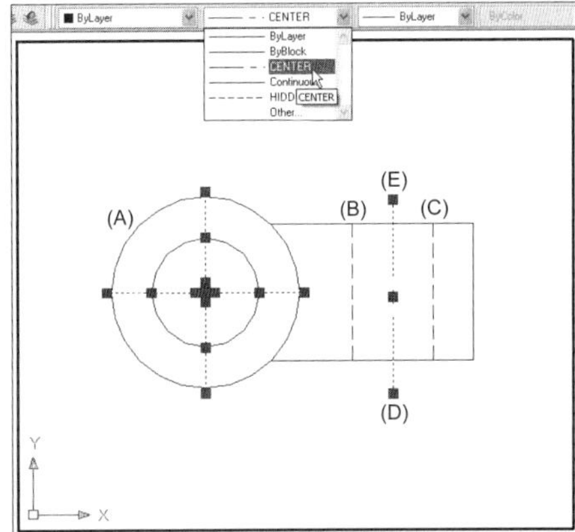

Figure 22–22

STEP 17

Change the DIMCEN command back to 0, which will prevent a center mark from being drawn when placing diameter and radius dimensions. Begin adding linear, radius, and diameter dimensions to the Top view. The OSNAP Endpoint and Intersection modes should still be active and running. See Figure 22–23.

 Command: **DIMCEN**

Enter new value for DIMCEN <-0.0900>: **0**

 Command: **DLI** *(For DIMLINEAR)* [dimension>linear]

Specify first extension line origin or <select object>: *(Select the endpoint at "A")*

Specify second extension line origin: *(Select the endpoint at "B")*

Specify dimension line location or [Mtext/Text/Angle/Horizontal/Vertical/Rotated]: *(Pick a point at "C")*

Dimension text = 2.50

Command: **DDI** *(For DIMDIAMETER)* [dimension>diameter]

Select arc or circle: *(Select the edge of the circle at "D")*

Dimension text = 2.00 Specify dimension line location or [Mtext/Text/Angle]: *(Pick a point at "E")*

 Command: **DDI** *(For DIMDIAMETER)* [dimension>diameter]

Select arc or circle: *(Select the edge of the circle at "F")*

Dimension text = 3.50 Specify dimension line location or [Mtext/Text/Angle]: *(Pick a point at "G")*

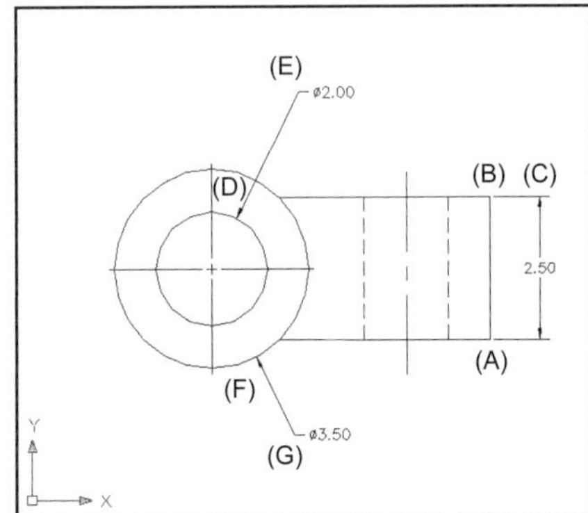

Figure 22–23

STEP 18

Switch back to Paper Space by double-clicking in the area at "A, shown" in Figure 22–24. Then pick the viewport; it will highlight and grips will appear. Right-click to display the shortcut menu. Choose Display Locked and then Yes to lock the viewport. This will prevent any accidental zooming of the image inside the Paper Space viewport. When finished, press ESC to remove the object highlight and grips from the screen.

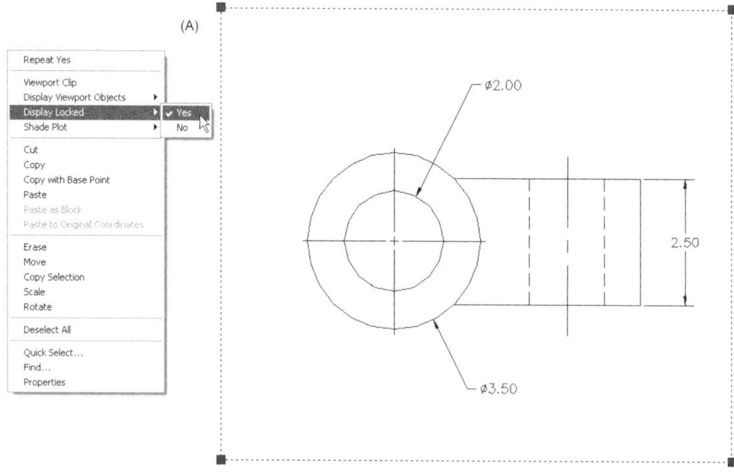

Figure 22–24

STEP 19

Activate the Layer Control box and turn off the VPORTS layer, shown in Figure 22–25. This will display only the views without the rectangular viewports present.

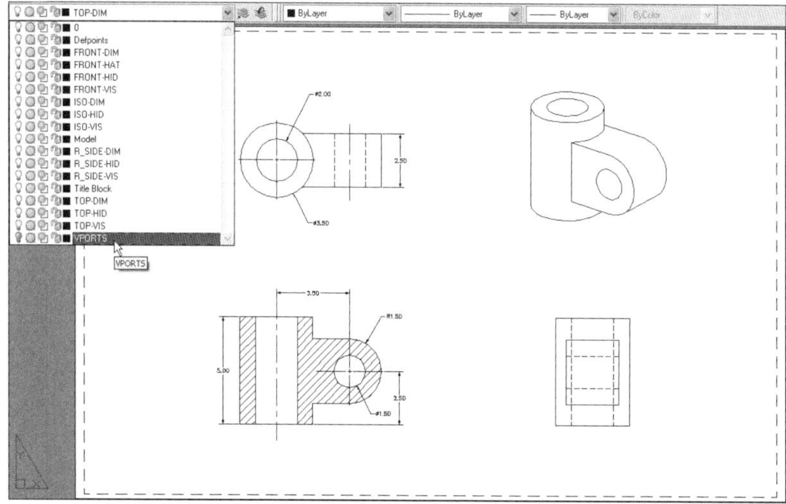

Figure 22–25

STEP 20

An ANSI-D title block can be inserted in the completed drawing, as shown in Figure 22–26. It is considered good practice to also change the layout name from "Layout1" to a different name such as "Orthographic Views". This gives more meaning to the layout.

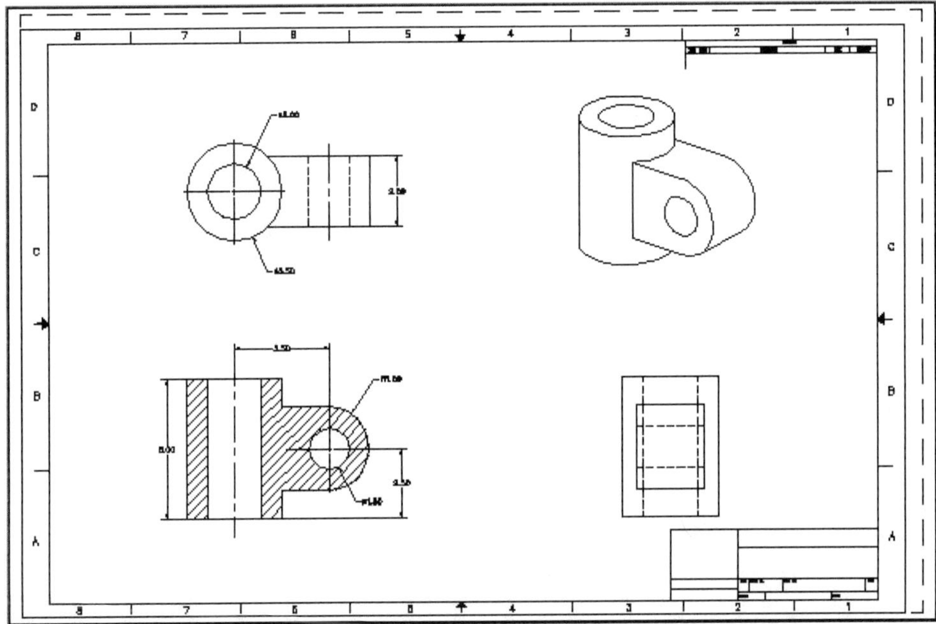

Figure 22–26

PROBLEM 22–1

Beginning and Intermediate Levels

1. Using the Tutorial Exercise of 22_Column.Dwg as a guide, create an engineering drawing of Problems 20–1 through 20–45 consisting of Front, Top, Right Side, and Isometric views.
2. Properly dimension the engineering drawing.

Advanced Level

Use the projects in Appendix A Working Drawing Problems (on the accompanying CD) to create drawings that utilize the same process as the tutorials in Chapter 22. Select from any of the levels and complete the project.

AutoCAD Rendering

The exercises in this chapter introduce:

- Using the various styles of rendering
- Applying lighting characteristics to a 3D drawing

TUTORIAL EXCERSICE: 23_INTERIOR LIGHTING.DWG

Figure 23–1

Purpose

This tutorial exercise is designed to add lights to the 3D model illustrated in Figure 23–1.

System Settings

Special system settings do not have to be made for this drawing.

Layers

A number of layers to cover the furniture, lamps, and walls, etc have already been created.

Suggested Commands

A series of four lights will be installed in this drawing in order to produce special lighting effects. A point light will be situated in the ceiling of the room. This light will simulate a typical overhead light in any room. A distant light will be added for the purpose of illuminating the entire room without casting any shadows. Two spotlights will be installed; the first to simulate light cast out of the top of the lamp shade and the second to simulate light shining down from the lamp onto the table top. The Photo Realistic rendering type will be used throughout this tutorial to obtain the desired results when rendering the view.

Whenever possible, substitute the appropriate command alias in place of the full AutoCAD command in each tutorial step. For example, use "CP" for the COPY command, "L" for the line command, and so on. The complete listing of all command aliases is located in Chapter 1, Table 1–2.

STEP 1

Open the file *23_Interior Lighting.dwg* from the CD-ROM that comes with this book. The model in this file, which is shown in Figure 23–1, is actually a 3D surface model room constructed with a ceiling, door, lamp, dresser, chair, and wall mirror. We will install a point light, a direct light, and two spotlights in the room and observe the results when rendering the view.

STEP 2

Use AutoCAD's VIEW command to restore the view named PERSPECTIVE_01 as shown in Figure 23–2. This is a perspective view that has a front clipping plane that allows us to see inside the room.

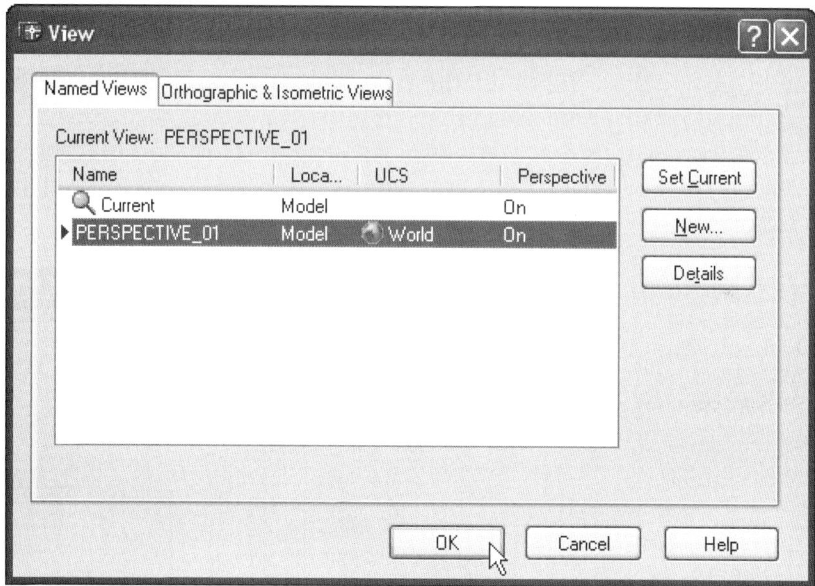

Figure 23–2

STEP 3

Four lights will be installed in this room—one direct light, one point light, and two spotlights. Rather than go through the steps to install these lights, we will simply list the parameters you should use for them. Notice that feet and inch dimensions are given for point locations because this drawing file uses architectural units. Leave Ambient Light Intensity at its default setting of 0.30, and leave the color of each light at white, as shown in Figure 23–3.

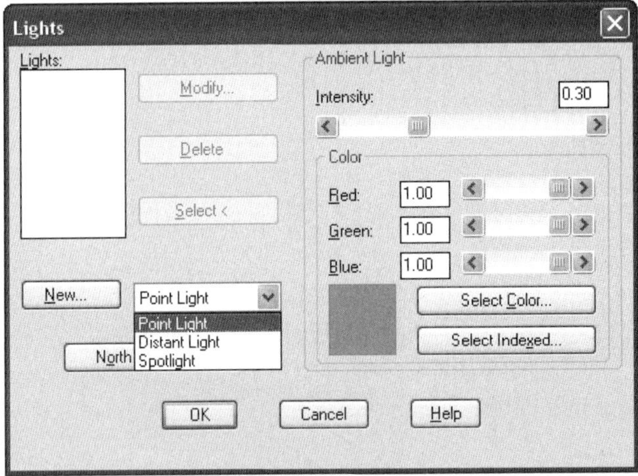

Figure 23–3

You should install these lights in the listed order; and make a rendering after each light is installed so that you can see the results of the light. Also, using multiple viewports will help you compare results. Use the Photo Real rendering type and make certain that shadows are turned on.

STEP 4

The first light to create is a point light that will simulate a light installed in the ceiling of the room. Study the information below and in the New Lights dialog box in Figure 23–4 for creating this light.

First Light	
Type:	Point Light
Suggested Name:	PT_01
Intensity:	120.0
Attenuation:	Inverse Linear
Shadow:	On
Shadow type:	Volumetric
Light Location:	X = 4' (48")
	Y = 5' (60")
	Z = 12' (144")

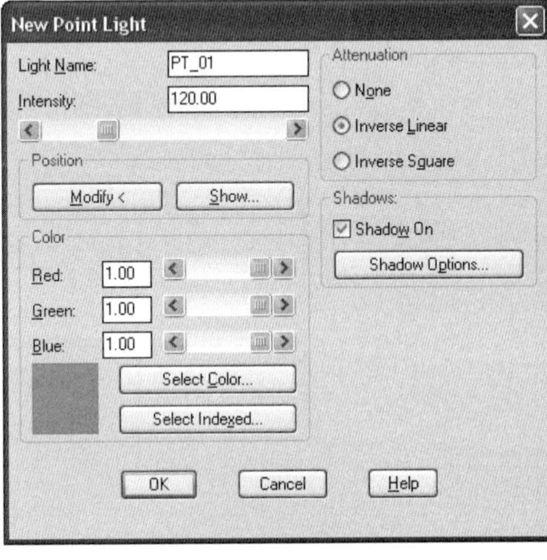

Figure 23–4

After you create this light and exit the Lights dialog box, activate the Render dialog box, check to see that you are rendering using the Photo Realistic type, and perform the render operation on the room with this single light. Your display should appear similar to Figure 23–5.

Figure 23–5

STEP 5

The second light created will be a Distant Light. This light will provide general room illumination. Shadows will be turned off. Study the settings to the right below and in the New Point Lights dialog box, shown in Figure 23–6, for creating this light.

Second Light	
Type:	Distant Light
Suggested Name:	DIST_01
Intensity:	0.50
Azimuth:	150.0
Altitude:	45.0
Shadow:	Off

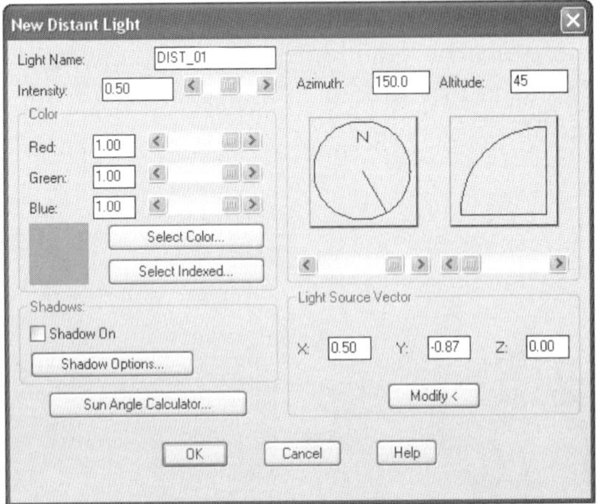

Figure 23–6

Again, after creating this light, activate the Render dialog box and perform the render operation on the room with both lights. Your display should appear similar to Figure 23–7.

Figure 23–7

STEP 6

The third light will be in the form of a spotlight. This light will simulate light radiating from the top of the lampshade. It will point directly up and will be somewhat dim compared to the other lights. Because of the nature of installing a spotlight, you will be prompted for a target location (where the spotlight will focus on) and an actual light location (where the spotlight will originate from. Study the information below and in the New Spotlight dialog box, shown in Figure 23–8, for creating this light.

Third Light	
Type:	Spotlight
Suggested Name:	SPOT_01
Intensity:	5.00
Attenuation:	Inverse Linear
Hotspot:	90.00
Falloff:	160.00
Shadow:	Off
Target Location:	X = 4'4 (52") Y = 8'4 (100") Z = 8' (96")
Light Location:	X = 4'4 (52") Y = 8'4 (100") Z = 5"6 (66")

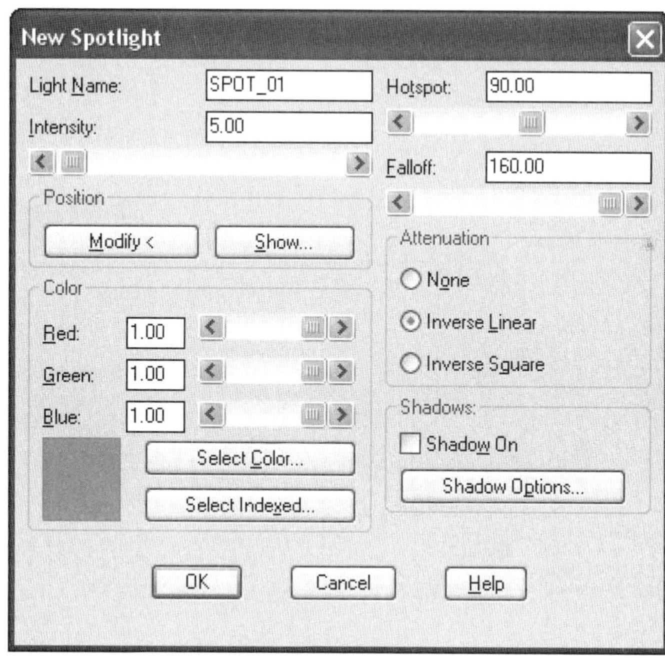

Figure 23–8

STEP 7

The final light to create and install will be another spotlight. This light will be point directly down, simulating light radiating from the bottom of the lampshade. As with all of the other lights that you created, study the settings to the right and in the New Spotlight dialog box in Figure 23–9 for creating this light.

Fourth Light	
Type:	Spotlight
Suggested Name:	SPOT_02
Intensity:	100.0
Attenuation:	Inverse Linear
Hotspot:	0.00
Falloff:	160.00
Shadow:	On
Shadow type:	Volumetric
Target Location:	X = 4'4 (52")
	Y = 8'4 (100")
	Z = 0' (0")
Light Location:	X = 4'4 (52")
	Y = 8'4 (100")
	Z = 4' (48")

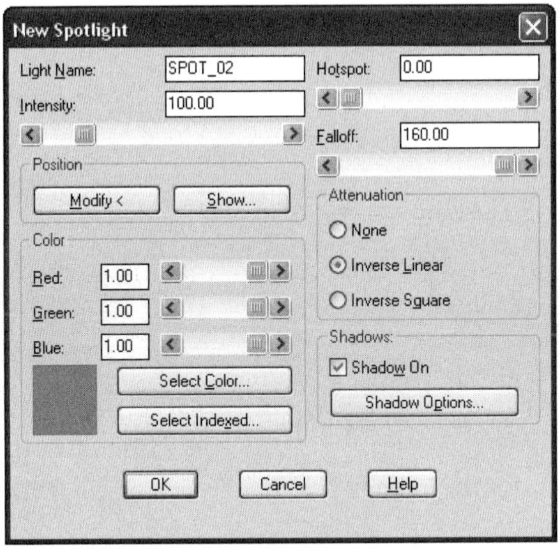

Figure 23–9

STEP 8

With all lights created and installed in the room, activate the Render dialog box one final time and perform the rendering operation on the room. Your display should appear similar to Figure 23–10.

Note: Choosing and installing lights is a subjective process. There are many other light installations for this room that would be equally good, and you may want to use this surface model room for experiments of your own. Also, the colors used for objects in this file are extremely basic, and you may want to change them to textures, which will give the room a more realistic appearance.

Figure 23–10 shows, in grayscale, the rendered room when all four lights are in use.

Figure 23–10

PROBLEM 23–1

Beginning Level Drawings

Apply rendering techniques to create a rendered image of a solid model component from Chapter 20.The rendered image should be in a perspective or isometric style view, showing three sides of the object.

Save the rendered image as a JPG or TIFF file and insert the raster image on your layout in place of the 3D image viewport found on the layout-plotting sheet.

Intermediate Level Drawings

Select one of the assembly drawings from Appendix A Working Drawing Problems (on the accompanying CD). Create 3D solids models of all the objects. Develop the detailed layout sheets of all the components using a rendered image in place o the 3D isometric image on the sheet. Additionally, render a separate assembly drawing of all the components in the assembly.

PROBLEM 23–2

Advanced Level Drawing

Using Problem 9 from Appendix A Working Drawing Problems (on the accompanying CD) as a guide, create a rendered "picture book" of the house. Select several of the rooms to render. To increase the difficulty add 3D model furniture in a room.

SHEET SETS AND FIELDS

SHEET SET SETUP

Sheet sets are a drawing management tool. They allow users the ability to save several sets of drawing iterations of a project as it progresses through the development and building stages. On the Manufacturing floor, using sheet sets can provide a long term trend analysis of shop floor layouts and product movement.

The following tutorial creates an example sheet set, applies the sheet set the drawings and adds field text to the AutoCAD 2004 sample drawings of the 8th floor of a building set.

1. Open the sheets set manager from the tools pull down and select "new sheet set" to create a new sheet set.

2. The sheet set we are setting up is an example sheet set for our project. We will then apply the sheet set to the existing drawings and select the example sheet set option and choose next

3. Next, choose what template / title block should be used in our sheet set. These sheet set example files are more than just a drawing template; they provide the title block and have built in Field text. Choose Architectural Imperial sheet set and choose next.

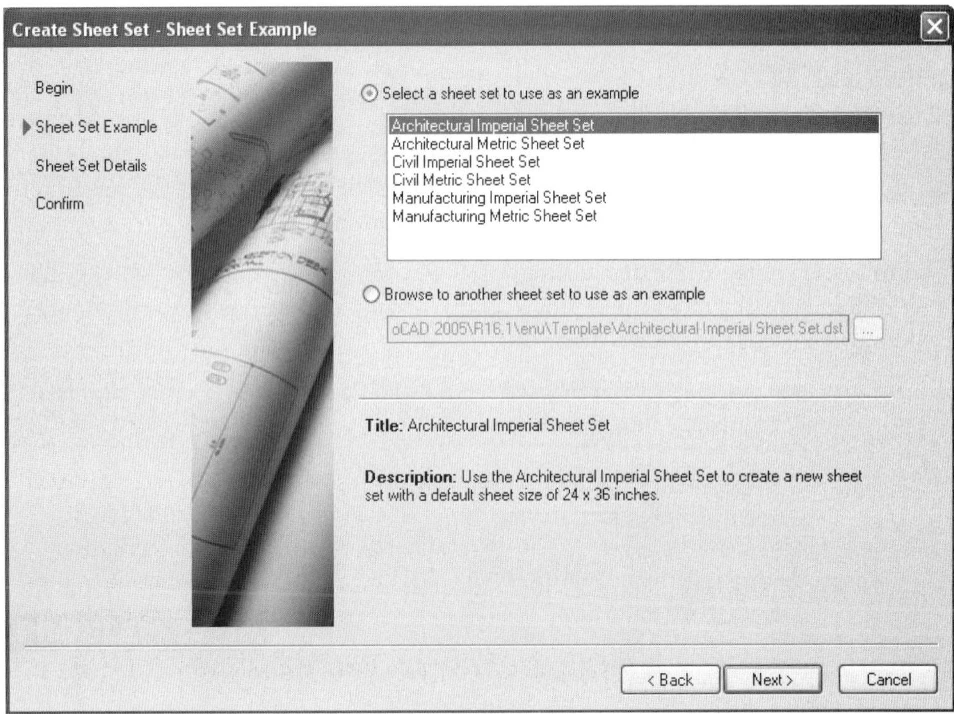

4. Providing a sheet set name, description, and location of the sheet set coordination file (.dst) is done in the next dialogue window. Name the project "8th floor project" and verify the location of the file placement. Click on Next.

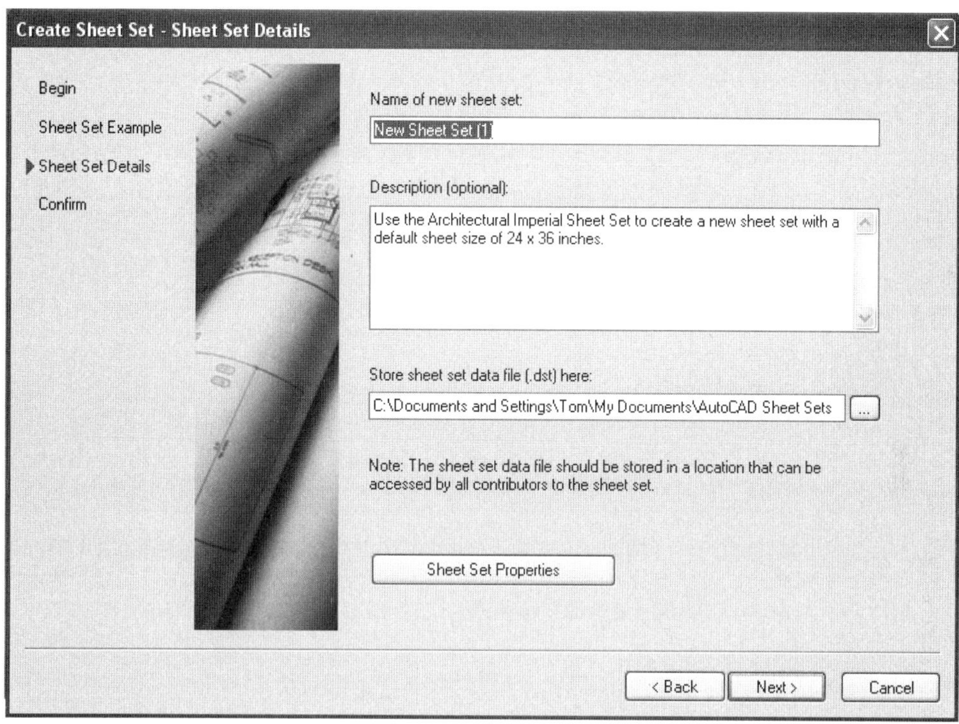

5. In the sheet set properties, the user can apply values to the pre-set fields that are set up in the current architectural template title block. Additional custom fields can be setup within the sheet set properties by using the edit custom properties and the Add button. For this project, fill in the client name, address and project information as listed below in the sheet set properties dialogue box.

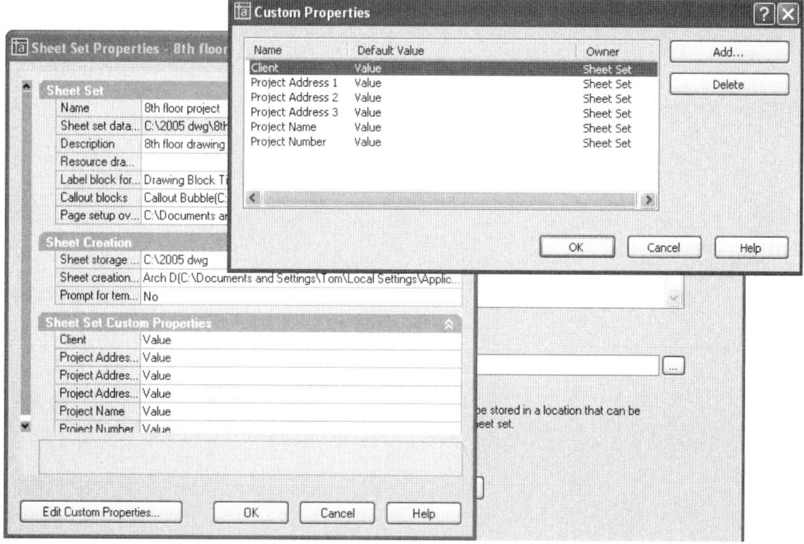

Project Name: Tutor Tower
Project Number: TT98765
Client: Fred Gwynn
Address: 1313 Mockingbird Lane

Sheet Set Custom Properties	
Client	Fred Gwynn
Project Addres...	1313 Mockingbird Lane
Project Addres...	Value
Project Addres...	Value
Project Name	Tutor tower
Project Number	TT98765

Choose OK to close the Sheet set Properties and Next to move onto the Sheet set preview.

6. Choosing Finish will create the sheet set with the listed subsets. We will create the sheets within this set by importing drawing layouts. The only tab that will not transfer from a drawing is the model tab.
7. Before we can import the layouts into the sheet set we need to make sure we have placed the Architectural sheet set template into the drawing. We need to add a new layout that is template based to the drawing file 8th floor plan.dwg. Right mouse click on any of he layout tabs and choose From A Template

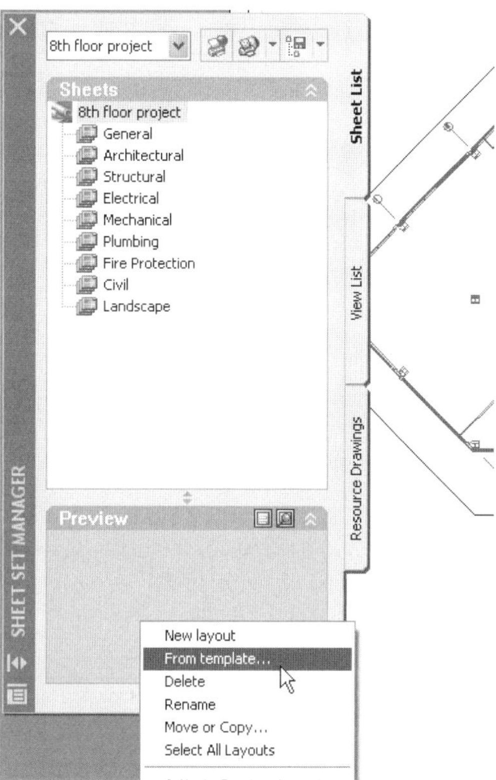

Navigate under the sheet sets folder and chose Architectural imperial. Save the drawing after the layout is in place.

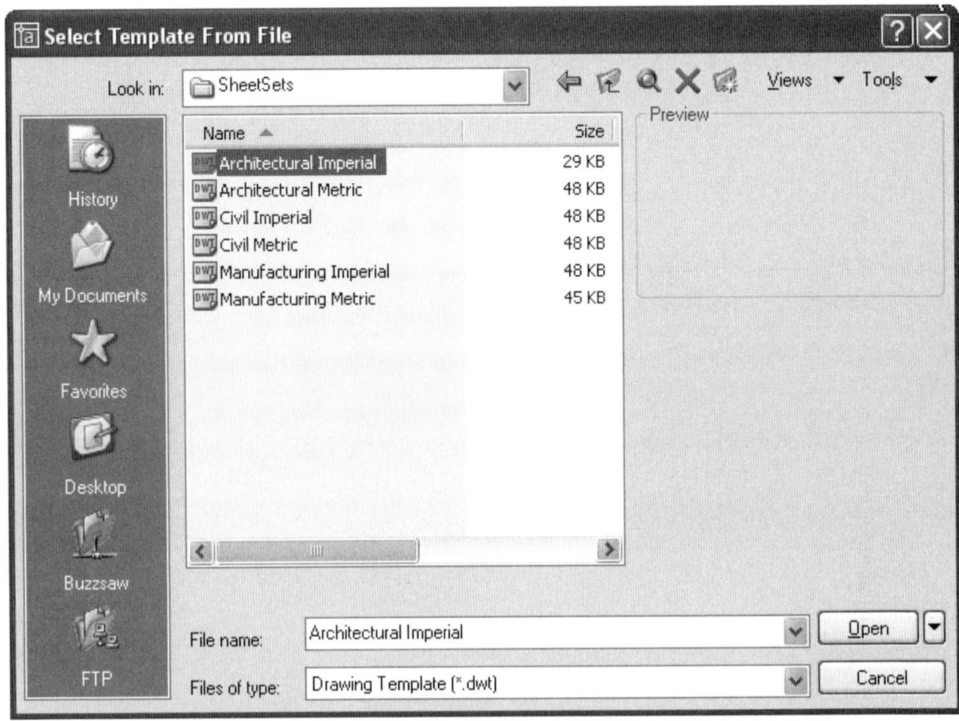

8. Once Arch D layout is created, select the tab to make it current. Place a Single viewport into the layout. Use the view pull down and move to view>viewports>1 viewport.

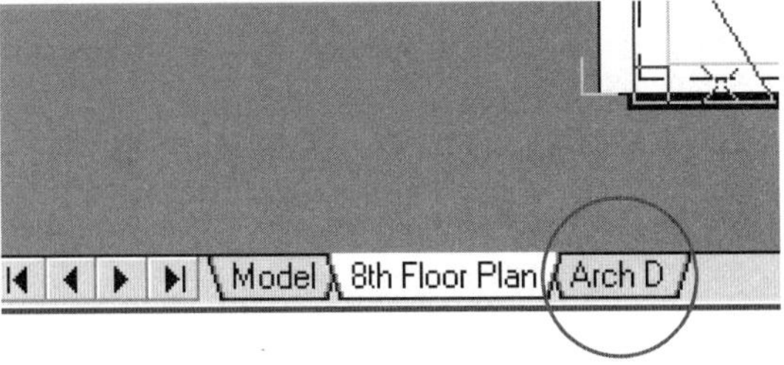

9. Create a viewport on the Arch D layout tab. Use the "View pull down>viewports >1 viewport" command to create it. It is best to place the viewport on a separate layer so it can be turned off. Turn off the layer once the scale and view location look appropriate.

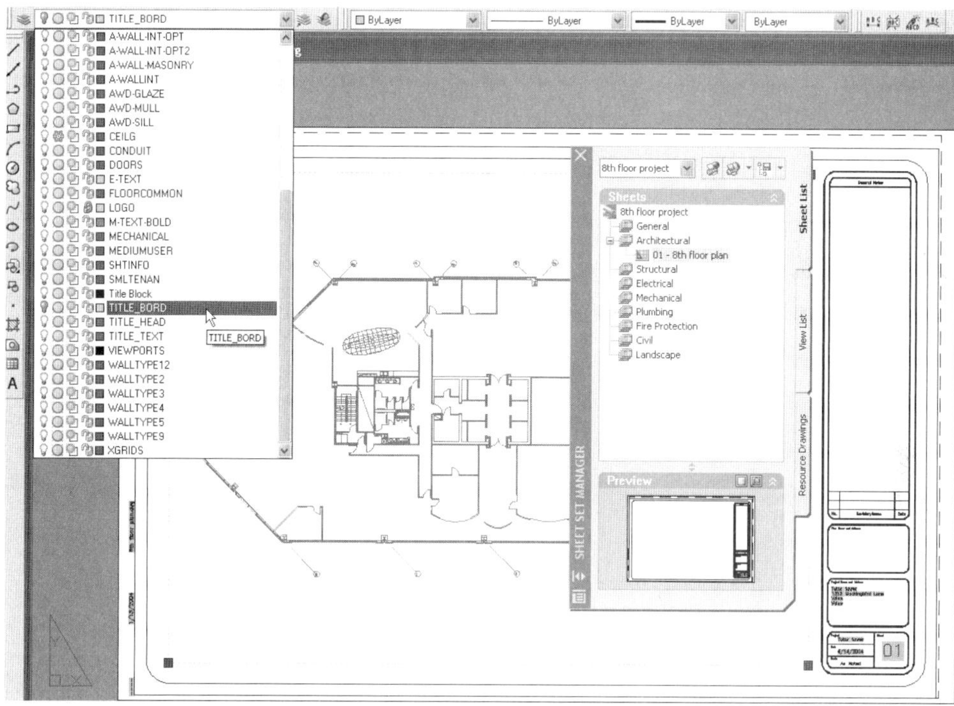

10. Save the Drawing.

11. Move back to the sheet set manager to import the layout. When importing a layout, make sure the appropriate subset is selected to place the drawing image under. In our sample the first sheet will be the floor plan and it will be placed under the Architectural subset.

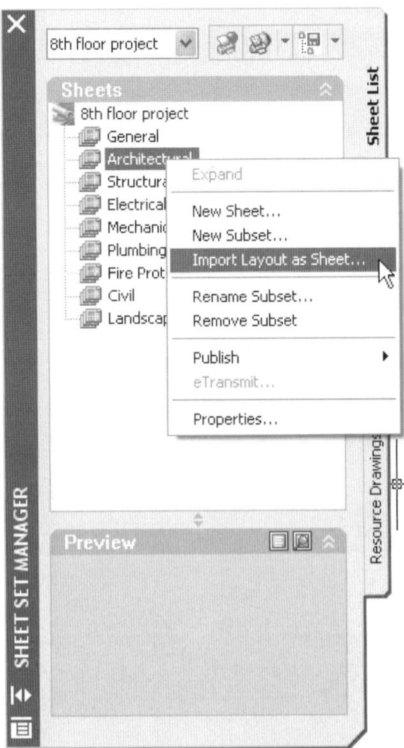

12. When importing the layouts, the ones available are listed in the Import Layouts as Sheets. Select the Arch D Layout to import. Then select OK.

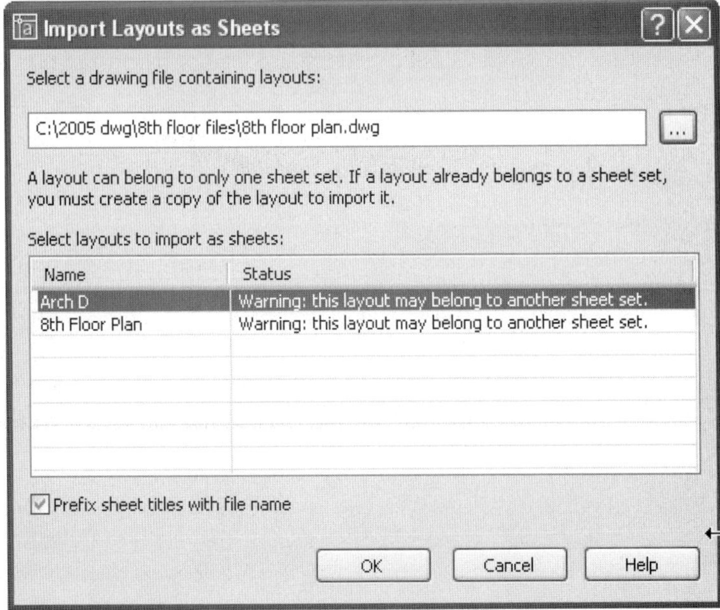

13. The sheet needs to have its name and sheet number reviewed for consistency. Right mouse click on the imported sheet and choose rename and renumber.

14. Add the number 01 to the number field and remove the "-Arch D" from the sheet title. This controls fields that are automatically updated on the drawing layout in the sheet set.

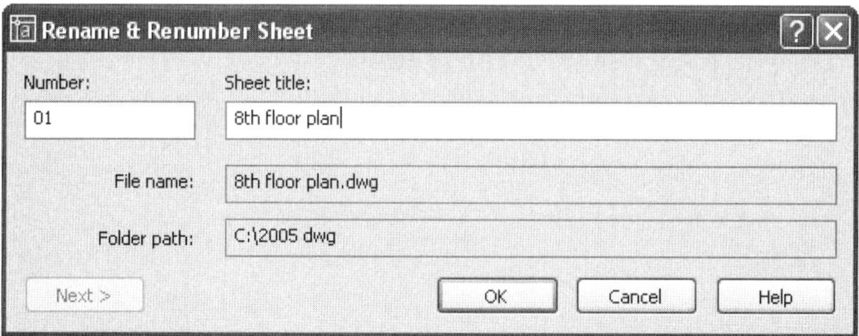

15. Fields allow the user to place text information that is generated as part of the AutoCAD drawing as a text field. A field is a supercharged attribute. This field can be edited once placed with an Mtext editor. Place the plot date on out current layout sheet. Choose the insert pull down and choose Fields.

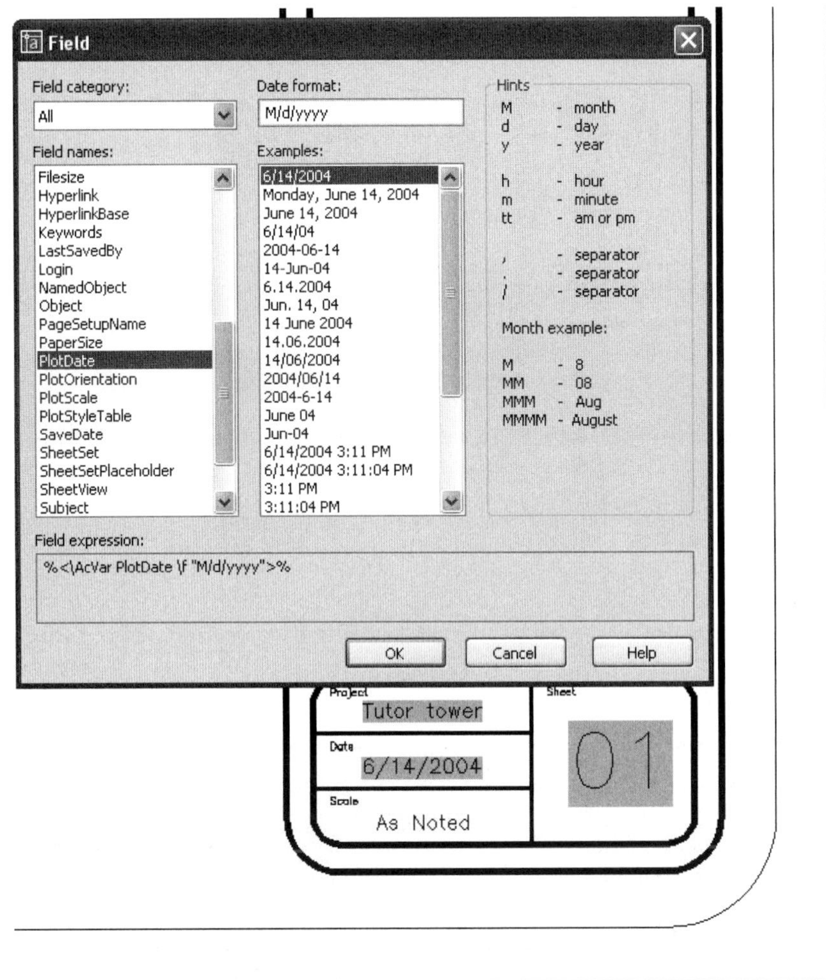

16. In the Fields Dialogue choose plotdate and the date format that is desired. Once selected. Place this field under the title block

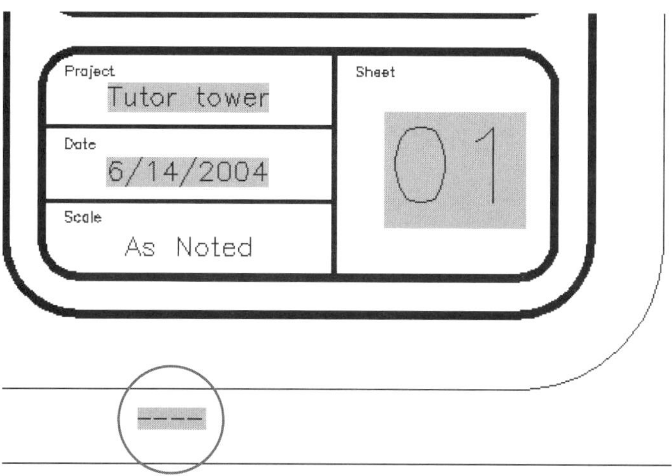

The plot date field is still a series of ---- until it is actually plotted, then a date will be placed within the file.

17. Sheet Sets can be published as a plot or a DWF (Drawing Web Format). The plots are handled without the drawing actually being opened. The DWF publish creates a web optimized drawing file output with each sheet as a separate page.

18. To practice the application of fields and sheet sets use table below to create sheets within the current sheet set for the additional support drawings. Some of the drawings have the Arch D layout, others need to have the layout added. This will complete a full sheet set for this building floor. Drawing files are available on the CD.

Table 24–1

Drawing File Name	Sheet Set / Subset Location
8th floor HVAC	Mechanical
8th floor Plumbing	Plumbing
8th floor Furniture	General
8th floor Lighting	Electrical
8th floor Power	Electrical

APPENDIX a

Working Drawing Problems

There are 15 problems in Appendix A that can be used for further exploration of the many AutoCAD features introduced in this Project Manual. The complete Appendix A can be found on the accompanying CD, in PDF format.

LICENSE AGREEMENT FOR AUTODESK PRESS

A Thomson Learning Company

Educational Software/Data

You the customer, and Autodesk Press incur certain benefits, rights, and obligations to each other when you open this package and use the software/data it contains. BE SURE YOU READ THE LICENSE AGREEMENT CAREFULLY, SINCE BY USING THE SOFTWARE/DATA YOU INDICATE YOU HAVE READ, UNDERSTOOD, AND ACCEPTED THE TERMS OF THIS AGREEMENT.

Your rights:

1. You enjoy a non-exclusive license to use the enclosed software/data on a single microcomputer that is not part of a network or multi-machine system in consideration for payment of the required license fee, (which may be included in the purchase price of an accompanying print component), or receipt of this software/data, and your acceptance of the terms and conditions of this agreement.
2. You own the media on which the software/data is recorded, but you acknowledge that you do not own the software/data recorded on them. You also acknowledge that the software/data is furnished "as is," and contains copyrighted and/or proprietary and confidential information of Autodesk Press or its licensors.
3. If you do not accept the terms of this license agreement you may return the media within 30 days. However, you may not use the software during this period.

There are limitations on your rights:

1. You may not copy or print the software/data for any reason whatsoever, except to install it on a hard drive on a single microcomputer and to make one archival copy, unless copying or printing is expressly permitted in writing or statements recorded on the diskette(s).
2. You may not revise, translate, convert, disassemble or otherwise reverse engineer the software/data except that you may add to or rearrange any data recorded on the media as part of the normal use of the software/data.
3. You may not sell, license, lease, rent, loan, or otherwise distribute or network the software/data except that you may give the software/data to a student or and instructor for use at school or, temporarily at home.

Should you fail to abide by the Copyright Law of the United States as it applies to this software/data your license to use it will become invalid. You agree to erase or otherwise destroy the software/data immediately after receiving note of Autodesk Press' termination of this agreement for violation of its provisions.

Autodesk Press gives you a LIMITED WARRANTY covering the enclosed software/data. The LIMITED WARRANTY can be found in this product and/or the instructor's manual that accompanies it.

This license is the entire agreement between you and Autodesk Press interpreted and enforced under New York law.

Limited Warranty

Autodesk Press warrants to the original licensee/ purchaser of this copy of microcomputer software/ data and the media on which it is recorded that the media will be free from defects in material and workmanship for ninety (90) days from the date of original purchase. All implied warranties are limited in duration to this ninety (90) day period. THEREAFTER, ANY IMPLIED WARRANTIES, INCLUDING IMPLIED WARRANTIES OF MERCHANTABILITY AND FITNESS FOR A PARTICULAR PURPOSE ARE EXCLUDED. THIS WARRANTY IS IN LIEU OF ALL OTHER WARRANTIES, WHETHER ORAL OR WRITTEN, EXPRESSED OR IMPLIED.

If you believe the media is defective, please return it during the ninety day period to the address shown below. A defective diskette will be replaced without charge provided that it has not been subjected to misuse or damage.

This warranty does not extend to the software or information recorded on the media. The software and information are provided "AS IS." Any statements made about the utility of the software or information are not to be considered as express or implied warranties. Delmar will not be liable for incidental or consequential damages of any kind incurred by you, the consumer, or any other user.

Some states do not allow the exclusion or limitation of incidental or consequential damages, or limitations on the duration of implied warranties, so the above limitation or exclusion may not apply to you. This warranty gives you specific legal rights, and you may also have other rights which vary from state to state. Address all correspondence to:

AutodeskPressExecutive Woods5 Maxwell DriveClifton Park, NY 12065Albany, NY 12212-5015